T0187896

MECHATRONICS IN ENGINEERING DESIGN AND PRODUCT DEVELOPMENT

MECHATRONICS IN ENGINEERING DESIGN AND PRODUCT DEVELOPMENT

Dobrivoje Popovic
University of Bremen
Bremen, Germany

Ljubo Vlacic
Griffith University
Brisbane, Australia

CRC Press
Taylor & Francis Group
Boca Raton London New York

CRC Press is an imprint of the
Taylor & Francis Group, an **informa** business

CRC Press
Taylor & Francis Group
6000 Broken Sound Parkway NW, Suite 300
Boca Raton, FL 33487-2742

First issued in paperback 2019

© 1999 by Taylor & Francis Group, LLC
CRC Press is an imprint of Taylor & Francis Group, an Informa business

No claim to original U.S. Government works

ISBN-13: 978-0-8247-0226-7 (hbk)
ISBN-13: 978-0-367-40029-3 (pbk)

Library of Congress Cataloging-in-Publication Data

Mechatronics in engineering design and product development / [edited by]
    Dobrivojie Popovic, Ljubo Vlacic.
        p.   cm.
    Includes index.
    ISBN 0-8247-0226-3 (alk. paper)
    1. Mechatronics. 2. Engineering design. 3. New products. I. Popovic,
Dobrivojie. II. Vlacic, Ljubo.
    TJ163.12.M435   1999
    621.3—dc21   98-38127
                                                                        CIP

Visit the Taylor & Francis Web site at
http://www.taylorandfrancis.com

and the CRC Press Web site at
http://www.crcpress.com

Preface

In order to compete in the worldwide marketplace, manufacturers must ensure that their products will fulfill consumers' desired performance and quality requirements. Being at the synergetic intersection of a number of technologies, mechatronics has become the most instrumental tool in facilitating such company goals. Hence, many companies are gathering data on the use of mechatronic processes in engineering design and product development. Our long experience in teaching engineering at the university level and as R&D engineers with industry has led us to the belief that a practical and accessible presentation of mechatronics know-how is both possible and necessary. This book, presenting both theory and practice, is the result of this belief.

A universally accepted definition of the term mechatronics is: the integration of a number of disciplines such as mechanics, electronics, electrical, computer, control, and software engineering using microelectronics to control mechanical devices. In addition to product design, mechatronics as a design philosophy penetrates and is applied to production design, monitoring, and control with the objective of achieving high-quality products at optimal running conditions. To achieve this, mechatronics integrates advanced semiconductor technology, computer and communications technology, robotics, computer vision, and intelligent neuro-fuzzy technology.

The process of mechatronics and its interdisciplinary synergy is best explained by professionals from all the disciplines involved. This book provides systematic and comprehensive information to practicing engineers in industry and to advanced students. It also helps them to adapt this expert knowledge to their own unique situations and thus become more productive. We therefore expect that this text will serve as a reference book for professionals from the automotive, process, production and aviation industries, robotics, consumer electronics, medicine, manufacturing, and CAD centers. We also expect the text to be used in various pedagogical settings. All who are dedicated to improving engineering design and product development can make use of this book.

The design of a mechatronic product is a challenge for design engineers. It is an exciting and satisfying journey with a specific destination — that of making a product to meet the needs of the customer. While the destination is well defined, the approaches to it are different because each mechatronic product is specific in its design, with a unique composition of the disciplines involved. Having this in mind, this book aims to provide the reader with specific knowledge on how to integrate all pertinent disciplines into a mechatronic product.

Recent progress across all disciplines pertinent to mechatronics has significantly affected future mechatronic product design trends and has fast-forwarded technological enhancement of existing products. It has also led to the design of totally new classes of products such as mechatronic machines. This has been necessary because the conventional approach to machine design has limited further machine improvement due to restrictions imposed upon mechanical components and mechanical subsystems of machines. If a machine is considered to be a mechatronic system then the machine can be designed based on: (a) a minimum number of mechanical and electronic components — those hardware components that are considered absolutely necessary; (b) an intelligent unit to process all information and machine-related functions; and (c) sensors to receive and actuators to respond to this information.

The intelligent SIMOVERT Master Drives Motion Control Servo Converter of Siemens, Germany, is being designed in line with mechatronic machine design principles. As Siemens describes it: "To start with, each mechatronic machine will have its own drive. Mechanical coupling such as cams and kingpins will be eliminated. The electronics will precisely coordinate the various movements. Motion sequences will also be coordinated. Each drive will know what the other drives are doing based on interlinks. Typical applications such as start-up, positioning, synchronous operations and cam can be called upon as standard functional blocks. All these functions, now performed by software, are to be configured to suit a specific machine application. The response so far from the end users of pioneering mechatronic machines is that implementation of the mechatronic approach to machine design has reduced the operating and maintenance costs of the machines and increased their application flexibility."

Mechatronic process is therefore a cross-disciplinary design process which can be properly applied if, and only if, the specialists from all pertinent disciplines work together from very early in the design process. Marketing specialists and production engineers should also participate. This interdisciplinary team-based work is usually called concurrent engineering due to the simultaneous interrelationships of the disciplines and their influence on the final solution. A mechatronic product is therefore not only a simple collection but a synergistic composition of all pertinent disciplinary knowledge.

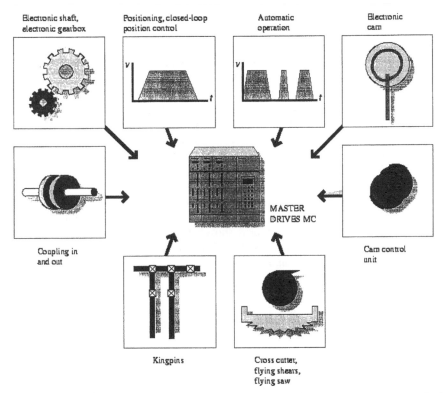

Electronic shaft, electronic gearbox

Positioning, closed-loop position control

Automatic operation

Electronic cam

Coupling in and out

MASTER DRIVES MC

Cam control unit

Kingpins

Cross cutter, flying shears, flying saw

Frequently used applications are integrated into the drive converter as standard mechatronic functions. (Courtesy of Siemens, Germany.)

It is up to the design team to find and define the composition, harmony and balance of the disciplines involved.

With the above in mind, this book is presented in four parts. Part One of the book addresses the core technologies necessary for the design and development of the mechatronic product. Chapter 1 discusses the transducers (sensors and actuators) most commonly used in mechatronics with the main emphasis on sensors, especially position/displacement sensors. Chapter 2 is dedicated to the advanced solutions recently developed in the area of microsensors and microactuators.

Chapter 3 describes the design philosophy of the microcontroller, a form of microcomputer, which is the intelligent core of a mechatronic system and is responsible for processing information received by the mechatronic product via its sensors. This chapter focuses particularly on designing microcontroller and associated circuits for a target system and discusses prototype implementation techniques. Three prototype design case studies are explained in detail.

The scope of real-time information processing, which is required to be performed by the mechatronic product, may comprise simple measurement and/or control functions but also may consist of complex supervisory, optimization, knowledge-based, and intelligent control functions. The theory behind the machine's ability to be intelligent is explained in Chapter 4.

Part One ends with Chapter 5, an introduction to communications technology. This knowledge is considered essential when integrating the mechatronic product. Some specific bus systems, local area networks, and related inter-networking elements such as network bridges and network gateways are presented and their applications discussed.

Part Two deals with some design approaches, including conceptual design, and relies on the distributed structure of production systems, summarized in Chapter 6. Chapter 7 shows how an automotive engine controller can be developed using computer-aided design tools. A step-by-step explanation of the whole computer-aided design process shows how the design can be coded into the target processor and tested. Chapter 7 also describes the state of the art for automatic code generation, analysis, and synthesis of mechatronic systems that are controlled by computers.

If the controller for a mechatronic product is to be designed, developed, and prototyped in hardware as an Application Specific Integrated Circuit (ASIC), the know-how needed can be found in Chapter 8. The chapter presents an overview of methods, tools, and the latest technology used in the rapid prototyping of mechatronic systems. The main emphasis is on design tools for the rapid prototyping of the mechanical and electronic components of a mechatronic product. Both of these components need to be rapidly prototyped early in the design stage of mechatronic product development in order to evaluate the performance expected to be fulfilled by the target system, to evaluate likely customer response, desired reliability, and so forth.

Related mechatronic product design aspects are grouped in Chapters 9, 10, and 11, which belong to Part Three of the book. In Chapter 9, the design aspect of system integration, optimality, and compatibility of the system elements is presented. Guidelines to the selection and interface of the system elements and the measurement of resulting reliability and robustness of the integrated system are also provided. In Chapter 10, system performance aspects are discussed, with particular attention to production and product quality monitoring, quality assurance, and control. An issue presented within Chapter 11, system software, is the crucial design issue in relation to the real-time application of mechatronic products for which effective interaction between the system and its immediate environment is essential to the performance required of the product.

Part Four, consisting of three chapters, addresses some mechatronic products application-related issues. Chapter 12 describes the versatility of mechatronic system applications. Among the case studies, explained are mechatronic development in gear measuring technology, automatic calibration system for an angular encoder, a construction robot for marking of the ceiling, and, finally, musician robots playing a trio (recorder, violin, and cello) of chamber music. Chapter 13 discusses control and optimization of mechatronic processes describing an operator's model which is applicable to many complex industrial processes that require human intervention. Application examples derived from a pH neutralization process and gear ratio control problems are discussed. These demonstrate applicability of the model to a variety of industrial, manufacturing, and other dynamic large-scale processes where the operator is called upon to exercise corrective actions based upon experience. Chapter 14 is dedicated to the ethics of product design and introduces the reader to the realm of ethical problem solving, emphasizing the similarities between it and the design process with which most engineers are familiar. It shows that ethical considerations can both drive and constrain engineering design and concludes with an examination of a case directly related to the field of mechatronics.

Our profound gratitude goes to all chapter authors. Without their enthusiasm and strong dedication, the manuscript for the book would not have been completed nor its high quality achieved.

Having done our best to review the manuscript carefully, we share the blame for any shortcomings. We give full credit to the authors for the value of their contributions that enabled us to bring this volume to the community of engineers and scientists involved in mechatronic product design and development.

When all is said and done, the book could not have been produced without the able assistance of the production editors, at Marcel Dekker, Inc., Matthew MacIsaac and Brian Black. Their patience and expert guidance were instrumental in converting the manuscript into a book. For this, we are sincerely appreciative.

Dobrivoje Popovic
Ljubo Vlacic

Contents

Contributors

Stuart Bennett   The University of Sheffield, Sheffield, United Kingdom

Periklis Christodoulou   CSIRO Manufacturing Science and Technology, Brisbane, Australia

Michael Goldfarb   Vanderbilt University, Nashville, Tennessee

Richard J. Gran   Mathematical Analysis Company, Northborough, Massachusetts

Makoto Kajitani   University of Electro-Communications, Tokyo, Japan

Vojislav Kecman   The University of Auckland, Auckland, New Zealand

Sungshin Kim*   Georgia Institute of Technology, Atlanta, Georgia

Brendon Lilly   Griffith University, Brisbane, Queensland, Australia

Dobrivoje Popovic   University of Bremen, Bremen, Germany

Adam Postula   University of Queensland, Brisbane, Queensland, Australia

Michael J. Rabins   Texas A & M University, College Station, Texas

Donald R. Searing   Texas A & M University, College Station, Texas

George Vachtsevanos   Georgia Institute of Technology, Atlanta, Georgia

Ljubo Vlacic   Griffith University, Brisbane, Queensland, Australia

Wanjun Wang   Louisiana State University, Baton Rouge, Louisiana

---

Current affifiation: Pusan National University, Pusan, Korea.

1
Sensors and Actuators in Mechatronics

Wanjun Wang
Louisiana State University, Baton Rouge, Louisiana

In the past several decades, partly because of the rapid development of the microelectronics industry and the ever-increasing applications of microcomputers and the automation of various industries, demands for transducers (sensors/actuators) have increased exponentially. This trend is expected to continue as global competition for higher productivity and better quality forces companies in every industry to be constantly looking for ways to reduce cost and improve the quality of their production. In one respect, the revolution in microelectronics and computers has dramatically reduced the cost of automation and control, and has therefore led to broad applications of these technologies in areas where these technologies were not deemed to be economically feasible. On the other hand, the developments in the microelectronics industry and, more recently, the fast development of microelectromechanical systems (MEMS), have generated a wide variety of sensors and actuators at ever-lowering costs, therefore opening up opportunities for applications in some areas not feasible in the past. The fast development of microelectronics has also dramatically improved signal processing and computation capabilities. A signal processing job that might have required a large box of electronic components decades ago can now be carried out by a single IC chip. The availability, the simplicity, and the performance of circuit modules such as active filters, analog dividers, sample/hold devices, function generators, lock-in amplifiers, etc., have made it very convenient to integrate sensors and actuators in a mechanical system and has made the lives of application engineers much easier. As a consequence, integration of sensors and actuators into a system and dealing with signal conditioning, work that was deemed to be the province of an electrical engineer many years ago, can now be done by mechanical, civil, or chemical engineers. This helps expand further the application of advanced sensors in industry.

Nowadays, the application of sensors is so pervasive that it is difficult to find any machines or appliances that do not have integrated sensors. A typical car now has more than 70 sensors and the number grows continuously as efforts towards better performance are made. As the trend towards intelligent vehicles continues, more advanced or smarter sensors will be implemented, for example, ultrasonic sensors for collision prevention, bar-coded intelligent highways and vehicles with optical scanning sensors and computers as well as vehicles with built-in global satellite positioning systems and other sensors for location and guidance. Other common examples of mechatronics products include home appliances such as washing machines and dryers. They are no longer the very simple appliances as they were years ago and have become intelligent machines with many sensors and functions. Go to any toy store, and you will be amazed to find how many of the toys have integrated sensors and are really intelligent toys.

Transducers (sensors and actuators) are to mechatronics systems as the sensing organs and hands and feet are to human beings. As a matter of fact, in most cases the overall performance of a system is set by the performance of the sensors and actuators used. In an instrument, the sensor transforms the physical parameter to be measured into a signal as shown in Figure 1. In most modern instruments, the physical parameter being measured is transduced into an electrical signal. This signal is then processed by a signal conditioning circuit and displayed on a panel or stored for future use or processing. Obviously, the overall sensitivity of the instrument will not be higher than that of the sensor used regardless of the quality of the remaining parts of the system.

The same argument holds for a closed-loop control system. Take a position control system as shown in Figure 2 for example. A target position is set by the operator. Based on the control strategy adopted, a command is generated and sent to the actuators which then drive the plant to the target position. The output position is then measured with a sensing system and compared with the target position. If there is a difference between the measured and the target positions, a correction command is generated and sent out to compensate for the error. This check-correction action continues until the target position is reached. The revolution in microelectronics and microcomputers has made the implementation of advanced control systems and high quality electronic components readily available at ever-lowering costs, and they have therefore become a less vital part of an automatic system. Consequently, researchers and engineers have found that the overall performance of the system is nearly always limited by the performance of the sensors and actuators, especially the sensors. If the sensing system cannot

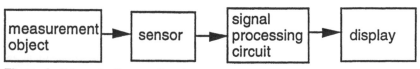

Figure 1   Schematic diagram of an A instrument.

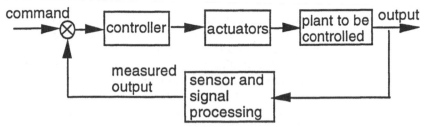

Figure 2 Schematic diagram of an A closed-loop control system.

accurately measure the actual position output, the wrong control command will be generated no matter how fancy the controller is. Similarly, if the actuators are not accurate enough, the correct control commands can still lead to an erroneous output. Also, the dynamic characteristics of the whole system are affected by the characteristics of the transducers used. Therefore it is very important to study the dynamic behavior of the sensors and actuators.

In this chapter, we are going to discuss the transducers (sensors/actuators) most commonly used in mechatronics, with the main emphasis on sensors, especially position/displacement sensors. In the following sections, classification of sensors and actuators will be introduced, followed by the basic concepts and common terms used in sensors and actuators. Then the fundamental principles of several major categories of transducers will be presented. In particular, two of the most commonly used position/displacement sensors, LVDTs and PSDs, will be discussed in detail.

I. CLASSIFICATION OF SENSORS AND ACTUATORS

A. Two Categories of Transducers: Sensors and Actuators

The term *transducer* is widely used in science and engineering. However, there are several different definitions, each having its own advantages and disadvantages, and they are all used in the field. Middlehoek [1] defines a transducer as a device that transforms non-electrical energy into electrical energy or vice versa. Karnopp et al. [2] defines a transducer as a device that transforms energy from one domain to another. The last, and the broadest definition was proposed by Busch-Vishniac [3]: a transducer is a device that transforms energy from one form to another, and it does not matter whether the energy belongs to different domains or the same domains. For example, a load cell, which is commonly used in mechanical measurement, can be classified as a transducer even though both the input energy and output energy are in the mechanical domain. This definition is more general than the one suggested by Middlehoek. If we insist a transducer must have an electrical output, a regular thermostat will not be classified as a transducer because it transforms thermal energy into mechanical energy and no electrical energy is involved. A load cell will not be classified as a transducer using this criterion. Strictly speaking, even a strain gauge, one of the widely used mechanical sensors, cannot be called a transducer (sensor) because it only transforms the mechanical energy (strain) into a change in resistance, not an electrical signal.

Transducers are normally classified using two categories: sensors and actuators. In most cases, they are reciprocal. For example, a capacitive transducer may function as a position sensor that transforms mechanical energy into capacitance change, and therefore electrical signal output (electrical charge or voltage change). It may also function as an actuator when an electrical signal (voltage or charge) is supplied and an electrostatic force or displacement is delivered as output. Both cases have practical applications: for example, the rangefinder of the auto-focusing system for a popular type of Polaroid camera [4]. This camera uses an electrostatic type transducer as both an actuator to generate the ultrasonic pulses and a sensor to receive the acoustic signal echoed back. Similarly, a piezo-electric transducer can either be used as an actuator, transforming an input voltage signal into a controlled displacement (or to deliver a driving force) as commonly used for precision positioning applications such as driving atomic force microscope (AFM) tips. When a force (or pressure) is applied on a piezo-electric disk, a voltage signal is generated across the piezo-electric disk. By measuring the voltage signal, the force or the pressure can be measured. This principle is used for pressure measurement and in accelerometers. Of course, not every type of transducer can be used with equal effectiveness in sensing and actuation. As a matter of fact, optimization of the performances of sensors and actuators often calls for totally different, or most possibly, conflicting requirements in choosing design parameters. In general, the optimization of a sensor requires: (1) a minimum change in the object parameter (called *measurand*) can cause a maximum change in the state of the sensor output; (2) the influence of the sensor on the object being measured is minimal, that is, the status of the object being measured should not be affected by the presentation of the sensor itself; and (3) the output of the sensor is only affected by the desired input, not by any other parameters or environmental conditions. For example, when a thermostat is used to measure the temperature of an object, it is definitely not desired that the temperature of the object is significantly affected by the presence of the thermostat, or that the thermostat readout is influenced by the surrounding magnetic field. Strictly speaking, a thermostat will always change the temperature of the object being measured from the basic principle of the measurement. A thermostat absorbs heat from the measurement object when they come into contact; its temperature goes up as a result of this heat transfer. Therefore a large sized thermostat is always avoided because large physical size almost always means large thermal capacitance, and consequently a significant amount of heat needs to be transferred from the measurement object to the thermostat to bring the thermostat to the same temperature as the object being measured. This significant heat transfer means that the temperature of the object will inevitably be changed by the presentation of the thermostat. The smaller the thermal capacitance of the thermostat, the less significantly the temperature of the object will be changed. This also explains why sensors with smaller physical sizes and weight are always preferred in engineering practice. For an actuator, optimization calls for them being able to impose the desired state on an object regardless of the load applied to them. Taking a DC motor as an example, the ideal performance for the motor means a motor should be able to drive anything connected to it in a specified speed independent of the load it has to overcome. This means an unlimited supply of power is required, a case that can never be achieved. Because actuators are always required to deliver certain power, their physical sizes will be power-limited in most practical cases. In most cases, actuators of extremely small size will not be very useful because of their very limited power capability.

From the foregoing discussions, it is obvious that smaller physical size and lower power requirement is always preferred for sensors while for actuators this is not always true. Partly because of this, progress made in the field of microelectromechanical systems (MEMS) and solid state technologies has greatly advanced the state-of-the-art for sensors while very limited success has been achieved in developing actuators.

B. Classifications of Sensors and Actuators

Because of the great variety of sensors and actuators used in the field of mechatronics, it is very difficult to discuss all of them in detail in the limited space available here. In this section, a very general classification of the most commonly used sensors and actuators in mechatronics will be provided, with emphasis on sensors. The classification of transducers can be done in several different ways. The first approach may be to classify sensors according to their applications, or the physical quantities the sensors can be used to measure. This is probably the most popular method of classification. Using this method, the sensors can be classified as, for example: position or displacement sensing; pressure sensing; temperature sensing; magnetic field sensing; flow measurement; torque sensing; stress or strain sensing; gas sensing; humidity sensing; chemical sensing; biological sensing; velocity or acceleration sensing; acoustic sensing (e.g., sound intensity sensing); radiation sensing. Some representative categories of sensors are shown in Table 1. One major advantage of this classification is that all the sensors for the same application can be introduced as one category and compared by performance and limitations, which can be very convenient for application engineers. There are several major disadvantages in using this classification. First, this method of presentation may become simply a review of what is commercially available today, which is not very useful to engineers who may want to study the fundamentals in sensors and actuators and learn how to develop similar new sensors. Secondly, sensors for same applications may be based on totally different principles, therefore in-depth discussions may lead to unnecessary repetitions and cannot be organized easily into the limited space here.

Another commonly used approach is to classify the sensors according their basic operation principles: optical sensing; capacitive sensing; inductive sensing; acoustic sensing (e.g., ultrasound sensing devices); fiber-optic sensing; Hall-effect

Table 1 Classification of Sensors According to Their Application

Mechanical sensors	Position (linear and rotational), displacement (linear and rotational), velocity (linear and rotational), acceleration (linear and rotational), vibration (linear and rotational), stress and strain, force, torque, pressure, surface topography or roughness and flatness, roundness, etc.
Electrical and magnetic sensors	Voltage, current, resistance, capacitance, inductance, magnetic, radiation, etc.
Acoustic and flow sensors	Sound intensity (pressure), viscosity, flow rate, frequency, ultrasound nondestructive detection, etc.
Chemical and biological sensors	pH, enzymes, ions, gases, concentration, humidity, biological, frequency shift or Doppler, etc.
Optical sensors	Intensity, wavelength, phase, vision and image (e.g., CCD camera), interference, polarization, reflectance, transmittance, scattering, refractive index, spectrum, etc.
Thermal sensors	Temperature, infrared radiation image, etc.

sensing; eddy current; and etc. as shown in Table 2. It should be noted that the list in Table 2 is only representative, not inclusive. There are some sensors not listed because of the limited space here. For example, resonant sensors, magnetoresistive sensors, etc. This method of classification is convenient for in-depth discussions of each category of sensors and is very suitable for efficient presentation of the wide variety of sensors within the confines of this chapter.

We will concentrate our discussions on sensors. Actuators will be discussed only when they are reciprocals of a particular type of sensor. Because of the limited space here, very brief discussion will be provided for each category of sensors. In each category of sensors, only one or two examples will be provided. In addition, because it is estimated that almost 80% of the sensors used in industry are for position measurement, see Luo [5], we will spend most of our efforts on studying position/displacement sensors. In particular, we will discuss in detail two of the best position sensors: linear variable differential transformers (LVDTs) and lateral effect position sensitive detectors (PSDs).

II. PERFORMANCE PARAMETERS OF A SENSOR

In this section, we will briefly discuss the terms commonly used in science and engineering practice to describe the performance of sensors. The databooks of commercial sensor products tend to use different terms to describe the same parameters of products and this makes it quite difficult to compare their performance. To avoid possible confusion, we will first define the fundamental terms used in this chapter. Efforts have been made to define these terms consistently with common usage.

Table 2 Classification of Sensors to Their Fundamental Principles

Capacitive	Position (linear and rotational), displacement (linear and rotational), electrostatic driving and deflection sensing for MEMS devices, chemical sensing, etc.
Inductive	Electric sensors, position/displacement, proximity, magnetic field detection, electromagnetic relay, etc.
Ultrasonic	Range-finders, nondestructive testing, thickness measurement, image scanning, flow measurement (Doppler), etc.
Photoelectric, PSD, CCD	Displacement/position, temperature, vision, and image (e.g., CCD camera), light intensity
Optical and fiber-optic	Optical encoders, gyrator (fiber-optic), temperature, magnetic, fiber-optic interferometer for phase shift measurement and any physical parameters that can modulate the phase shift, proximity sensors, etc.
Eddy current	
Hall effect	Magnetic field detection, proximity sensor
Piezo-electric	Actuators, pressure, force, and torque sensing, etc.

Generally speaking, in choosing sensors we must decide what the sensor is to do and what results we expect. We will discuss some of the criteria that must be considered in selecting and using different kinds of sensors for different kinds of applications in an automation system. The most important parameters are defined and discussed in the following subsections.

A. Sensitivity

Sensitivity is defined as the ratio of change of output to change in input. Suppose the output of a transducer is y for a given input x, that is, $y = f(x)$. This ratio is:

For example, if a 0.01 mm displacement in input gives rise to a 0.5 volt

$$S = \frac{\Delta y}{\Delta x} \tag{1}$$

For example, if an 0.01 mm displacement in input gives rise to an 0.5 volt change in output, then the sensitivity is 50 volt/mm. Some people prefer to use sensitivity to indicate the smallest input that can be detected by the sensor, but in most cases, another term, *resolution,* has been used for this purpose. Normally, the maximum sensitivity is always desired if other parameters such as *linearity* and *accuracy* would not be sacrificed.

B. Linearity

The term *linearity* is used to indicate the constancy of the ratio of output to input. If the output and input of a sensor system have a perfectly linear relationship, it would mean that in the following equation:

$$y = cx \tag{2}$$

where y is output, x is input, and c would be constant. If there exists some nonlinearity, c would be a function of x. That is, instead of the previous equation, the actual relationship between the output and input of the sensors would be described by the following nonlinear equation:

$$y = f(x) \tag{3}$$

There are several ways to describe linearity. One way to measure linearity is to use the absolute maximum error between the output predicted by using the linear equation and the actual output over the whole working space:

$$distortion = Max|f(x) - cx| \tag{4}$$

Here *distortion* is used to indicate the linearity. The smaller the distortion, the higher the linearity is. Another way to describe the linearity is to use a relative value for *distortion:*

$$distortion = \frac{Max|f(x) - cx|}{cx} \tag{5}$$

This definition is fairly popular and has been adopted in the discussion throughout this chapter.

C. Range

Range is a measure of the difference between the maximum and minimum values measured. For example, a thermometer might be able to measure values over the range of -40 degrees centigrade to 100 degrees centigrade. A large working range is always preferred. Normally there is always some kind of trade-off between the range and accuracy. It is generally quite expensive to achieve both high accuracy and large range in a sensor.

D. Accuracy

Accuracy is the term used to indicate the difference between the measured and the actual values. It can either be defined as an absolute value which is the maximum error within the working range, or a relative value which is the ratio of the maximum error to the range of measurement and is a dimensionless value. An accuracy of ±0.01 mm means that, under any circumstances, the value measured by the sensor would be within 0.01 mm of the actual value. Accuracy is a specification very hard to check. For example, if a robot end-effector is claimed to have an accuracy of ±0.01 mm, to verify this accuracy would require very careful measurement of the end-effector with respect to the base coordinates under specified operating conditions with regard to speed, temperature, force, torque, and load. The overall error must be within the specified range. Accuracy is one of the most rigorous technical specifications.

E. Repeatability

Repeatability is used to indicate the difference in value between two successive measurements under the same environmental and operational conditions. Repeatability is specified in most robot systems. Compared with accuracy, it is a far less stringent criterion. In most cases, no matter how poor the other parameters are, a relatively better repeatability can be expected even if the operational and environmental conditions are maintained.

F. Resolution

Resolution is defined as the minimum change of input that can be detected at the output of the sensing system, or the minimum change of the output parameter that can be achieved for an actuator. For a position sensor, the resolution would be the minimum displacement that can be detected. For a linear motor (a commonly used actuator), the resolution is defined as the minimum position displacement it can be controlled to move. Some people also prefer to define resolution as the number of measurements within the range from minimum to maximum. Resolution is one of the most important parameters of sensors and actuators.

G. Output

The type of *output* is also very important for sensors. It can be electrical current or voltage. It can also be a mechanical movement, a pressure change, liquid level variation, or resistance change. In most of today's applications, a voltage output signal is preferred because computers are being increasingly used to control the systems. If the output signal is not a voltage signal, it is often converted to a voltage signal.

H. Dynamic Characteristics

All the performance parameters discussed above are steady-state characteristics. However, in most mechatronics applications, in addition to the requirement for satisfactory steady-state characteristics, a transducer also must have satisfactory dynamic characteristics.

Bandwidth

Bandwidth is one of the most important dynamic parameters of a transducer. Ideally, we would prefer a transducer to have the same amplitude of output for any input signals with the same amplitude, independent of their frequencies. However, this ideal case can never be achieved. In reality, the output of a transducer is dependent on both the amplitude and frequency of an input signal because of the limited bandwidth of the transducer. Both the amplitude and the phase of the output signal can be a function of the amplitude and frequency of the input signal. This dynamic characteristic can be better discussed in the frequency domain, in term of the *transfer function*. Assuming a sinusoidal input signal is supplied to a transducer, we should expect the output of the transducer to be a sinusoidal function. The ratio of the output and input signals is therefore defined as the *transfer function* of the transducer. Therefore the transfer function of the transducer is a function of the frequency of the input signal. If the amplitude of the transfer function is plotted out as the function of the input frequency it should, in most cases, look like the response curve shown in Figure 3. Below a certain frequency limit, this amplification factor of the transducer, $|H(\omega)|$, should be nearly independent of the input frequency α. As the frequency of the input signal increases, the amplitude of the output signal will decay. If the transducer can be modeled as a second (or higher) order system, there may exist a resonant frequency, or multiple resonant frequencies.

The *bandwidth* of a transducer is defined as the frequency range in which the amplification factor or the amplitude of the transfer function will not decay significantly. In Figure 3, the bandwidth is α_B.

Response Time

Another term frequently used to describe the dynamic characteristics of a transducer is its *response time*. In control theory, the response time is called settling time. The response time can be defined as the time required for a change in input to become observable as a stable change in output. In most cases, the output of the sensor would oscillate for a short time before it finally reaches and stays within a specified percentage of the final (steady-state) value. In these cases, we measure response time from the start of an input change to the time when the output has settled to the specified range. The response time of the sensor system is determined by its time constant.

III. CAPACITIVE SENSORS AND ACTUATORS

In this and following sections, a unified energy approach will be adopted in the derivation of all the fundamental equations. This energy approach has been

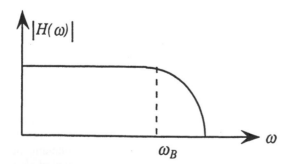

Figure 3 The bandwidth of a transducer.

used extensively in the theory of physical system dynamics by Karnopp et al. [2] and has been applied in systematic modeling and analysis of electromechanical sensors and actuators by Busch-Vishniac [3]. Because of the limited space here, the discussion will follow Busch-Vishniac's approach to this type of problems and will be brief. Interested readers are referred to the book by Busch-Vishniac [3].

The electrical charge q, the capacitance C, and the electrical voltage across the two plates of a capacitor e, are related by the following equation:

$$q = C \cdot e \tag{6}$$

For a parallel plate capacitor, the capacitance can be derived as the function of the distance d between the two plates, the cross-sectional area A of the plates, and the dielectric constant M as:

$$C = \frac{\varepsilon A}{d}. \tag{7}$$

Suppose one plate of the capacitor is stationary and another is moving as shown in Figure 4A and the gap between the two plates is x, the effective cross-sectional area is $A = w \cdot y$, where w is the width of the capacitor plate. The electric energy stored in the capacitor can be obtained using

$$E = \frac{1}{2C}q2 = \frac{xq2}{2\varepsilon A} = \frac{xq2}{2\varepsilon wy}. \tag{8}$$

This expression for energy can then be differentiated with respect to time to obtain an expression for the power involved as

$$P = \frac{dE}{dt} = \frac{\partial E}{\partial q}\frac{dq}{dt} + \frac{\partial E}{\partial x}\frac{dx}{dt} + \frac{\partial E}{\partial y}\frac{dy}{dt}. \tag{9}$$

Because the left side of the equation is power, and dq/dt is electrical current, then the first term $\partial E / \partial q$ must be the voltage. Similarly, the fact that dx/dt and dy/dt are velocities of the moving plate in the horizontal and vertical directions respectively means that the other two terms, $\partial E / \partial x$ and $\partial E / \partial y$ must be the mechanical forces in the x and y directions respectively in order to produce power. Therefore, we can obtain expressions for the voltage e across the capacitor and the forces F_x and F_y required to drive the moving plate of the capacitor in the x and y directions respectively, using:

$$e = \frac{\partial E}{\partial q} = \frac{qx}{\varepsilon wy}, \tag{10}$$

and

$$F_x = \frac{\partial E}{dx} = \frac{q2}{2\varepsilon wy} = \frac{\varepsilon wy}{2x2}e2, \tag{11}$$

and

$$F_y = \frac{\partial E}{\partial y} = \frac{q^2}{2\varepsilon wy^2} = -\frac{\varepsilon w}{2x}e^2. \tag{12}$$

Equation (10) shows that this parallel plate capacitor can be used for sensing purposes. There are several possible sensing mechanisms: (a) with one plated fixed and another moving vertically (with y = constant), the voltage e across them is then the function of the gap x between them as shown in Figure 4A; (b) with the gap between the two plates fixed (with x = constant), the bottom plate fixed and the top one moving laterally, then the effective area $A = w \cdot y$ changes as a function of the lateral position y as shown in Figure 4B; and (c) with both plates fixed (no vertical and lateral movement permitted) the effective dielectric constant of the medium can be changed by either moving a solid dielectric material between two capacitor plates as shown in Figure 4B. An example of sensors based on the principle shown in Figure 4B is a rain gauge or fluid level sensor. As the fluid (for example, water) level rises, the equivalent dielectric constant changes, and therefore the voltage across the two plates changes. This last scheme may also be used for chemical sensing because different materials may have different dielectric constants.

Equations (11) and (12) show that the parallel plate capacitor may also be used as an actuator. When a charge is supplied (and a voltage generated), either a vertical force F_x (perpendicular to the two plates) or lateral force F_y can be generated. Of course, the scheme shown in Figure 4B can also be used. In this scheme, a solid block of dielectric material is sliding in and out between the two capacitor plates; the capacitor can be treated as two capacitors with different dielectric materials in parallel. Its total energy needs to be differentiated with respect to y to obtain the force.

The capacitive sensing and driving principles have been widely used in the field of microelectromechanical systems (MEMS). Various types of microdevices and systems have been developed based on this principle. A typical example is the microfabricated pressure sensor that uses a capacitive sensor, see ref. [6]. Another example is a microvalve with electrostatic actuation described by Bosch et al. [7].

The foregoing discussion is only illustrative, and not exclusive. Other types of capacitive sensors are used also widely in engineering practice. Obviously a cylindrical capacitor with a hollow tube as one plate and a solid rod or hollow tube

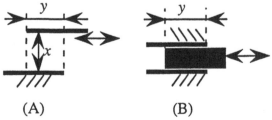

(A)   (B)

Figure 4   Several ways to use a parallel plate capacitor as a sensor.

with smaller diameter sliding inside as the other plate may also be used. The basic function is still $q = C \cdot e$ as in the case of the parallel plate capacitor, only the expression for the capacitance C is changed. The basic equations governing it can also be derived with the same energy approach as the one just shown for the parallel plate capacitor.

Another example of capacitive position sensors is a proximity sensor. In a proximity sensor, the two capacitor plates are in the same plane instead of in parallel. As the distance between the two plates and the object changes, the effective dielectric constant between the two capacitor plates changes; then the capacitance measured will be modulated by the position of the object. Proximity sensors based on this principle are commercially available.

The major advantage of capacitive position sensors is high precision. The main disadvantage of them is their small working range. They are widely used in vibration and displacement measurement. They are also one of the most commonly used transduction principles used in the fast developing MEMS field.

IV. INDUCTIVE SENSORS/ACTUATORS AND LVDT

A. General Principle

The analogy between the electrical field and the magnetic field makes the derivation of the fundamental relations in magnetic field very easy. Again the energy approach derivation will be adopted and a much more detailed discussion of this type of transducer can be found in the book by Busch-Vishniac [3].

An expression for the energy stored in a given volume of isotropic material in a magnetic field can be represented in an equation very similar to Equation (8).

$$E = \frac{\phi 2}{2p} \tag{13}$$

where ϕ is the magnetic flux and p is the permeance, or magnetic capacitance (a term similar to electric capacitance). The operation of typical inductive transducers (sensors and actuators) can be best explained by referring to the schematic diagrams shown in Figure 5. In Figure 5, two pieces of ferromagnetic rods are facing each other with the magnetic flux passing from one to the other through the air gap between them.

The air gap between the two facing electromagnetic cores shown in Figure 5 has a magnetic capacitance of almost exactly the same form as that of a parallel capacitor. If we assume the left-side element is fixed and the right-side element can move either vertically or horizontally, with the effective cross-sectional area $A = w \cdot y$, then its magnetic capacitance or permeance can be obtained as:

$$p = C = \frac{\mu A}{x} = \frac{\mu w y}{x}, \tag{14}$$

where T is the magnetic permeability of the isotropic medium.

Plug Eq. (14) into Eq. (13) and the magnetic energy stored in the air gap can be written as:

$$E = \frac{\phi 2x}{2\mu A} = \frac{\phi 2x}{2\mu wy}. \tag{15}$$

For a dynamic system, the power then can be derived by differentiating Eq. (15) with respect to time:

$$P = \frac{sE}{dt} = \frac{\partial E}{\partial \phi}\frac{d\phi}{dt} + \frac{\partial E}{\partial x}\frac{dx}{dt} + \frac{\partial E}{\partial y}\frac{dy}{dt}, \tag{16}$$

where $d\phi/dt$ is the flux rate, dx/dt and dy/dt are the velocity components of the moving magnetic element in the horizontal and vertical directions respectively. Power in the mechanical domain is equal to the product of force and velocity, and power in the magnetic domain is equal to the product of magnetic flux rate $d\phi/dt$ and *magnetomotance*, or *magnetomotive force M*. Therefore, the magnetomotive force M and force F_x in the x direction and F_y in the y direction can be derived as:

$$M = \frac{\partial E}{\partial \phi} = \frac{\phi x}{\mu A} = \frac{\phi x}{\mu wy} \tag{17}$$

and

$$F_x = \frac{\partial E}{\partial x} = \frac{\phi 2}{2\mu A} = \frac{\phi 2}{2\mu wy} = \frac{\mu wy}{2x2}M2 \tag{18}$$

and

$$F_y = \frac{\partial E}{\partial y} = -\frac{\phi 2x}{2\mu wy2} = -\frac{\mu w}{2x}M2 \tag{19}$$

The magnetomotive force in a magnetic element is a function of the magnetic field H and the length of the magnetic path, or as a function of electric current I and the number of turns in the coil according to Ampere's law:

$$M = Ni = Hl. \tag{20}$$

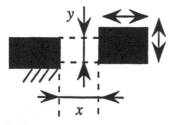

Figure 5 Basic principle of an inductive transducer.

Equation (17) represents the basic operation principle for inductive position sensing. A displacement x in the horizontal direction (with y fixed) or a displacement y in the vertical direction (with x fixed) can be transduced into a change in magnetomotance M, which can then be measured indirectly by measuring induced current i using a pick-up coil as shown in Eq. (20). Of course, a change in magnetomotance can also be achieved by changing permeability μ. This approach can also be taken in proximity sensing in similar design to the capacitive proximity sensors. Another example of using the variable permeability principle in position sensing is the widely used linear variable differential transformers (LVDTs). Detailed discussion of the LVDT and its transduction principle will be provided later in this section.

Equations (18) and (19) represent the fundamental principle for inductive actuators. They show that an electromagnetic driving force can be generated by a changing magnetomotance, that is, by supplying current to a driving coil.

In an inductive transducer (sensor/actuator), a magnetic bias must be applied. There are two ways to do this. The first way is to use a permanent magnet. The second method is to apply an electric bias, that is, to use an electromagnetic coil and supply a sinusoidal current to it. In most cases, this second approach is preferred because of its potential in securing a linear transducer relationship and the benefit to signal processing. A typical example is the linear variable differential transformer (LVDT) widely used in mechatronics applications. The advantages and wide applications of LVDT definitely justify a detailed discussion of it.

B. Operation Principle of LVDT

The schematic design of an LVDT is shown in Figure 6. One primary coil and two secondary coils are arranged as shown. A magnetic core with high permeability is inserted into the coils. At the beginning, the magnetic core is located at the middle point.

Again, we use the energy approach suggested by Busch-Vishniac [3]. Because the LVDT is symmetric, we can start the analysis by studying the left half first. When the magnetic core is in the neutral position as shown in Figure 6A, the section of the coil with air core has a magnetic capacitance:

$$C_l = \frac{\mu_1 A}{x_0} \tag{21}$$

and the magnetic capacitance of the section with the magnetic core inserted is:

$$C_l = \frac{\mu_2 A}{L - x_0} \tag{22}$$

where A is the cross-sectional area of the magnetic core, μ_1 is the permeability of the air core, and μ_2 is the permeability of the magnetic core. The equivalent capacitance of the combined air and magnetic cores can be calculated using the parallel rule and the total energy stored in the left-side coil can be obtained as:

$$E = \frac{1}{2C} = \phi_2 = \frac{\mu_1(L - x_0) + \mu_2 x_0}{2\mu_1\mu_2 A}\phi_2 \qquad (23)$$

The magnetomotance M can be derived by differentiating energy with respect to ϕ:

$$M = \frac{[\mu_1 L + (\mu_2 - \mu_1)x_0]\phi}{\mu_1\mu_2 A}. \qquad (24)$$

Now suppose the magnetic core moves from the neutral position to the right-side as shown in Figure 6B, then in the left-side, $x_0 \to x_0 + x$, and in the right-side, $x_0 \to x_0 - x$. Then the magnetomotance in the left-side of the LVDT can be obtained by substituting $(x_0 + x)$ in Eq. (24) for x_0 as:

$$M_l \frac{\phi[\mu_1 L + (\mu_2 - \mu_1)(x_0 + x)]}{\mu_1\mu_2 A} \qquad (25)$$

and the magnetomotance in the right-side of the LVDT can also be obtained by substituting $(x - x)$ in Eq. (24) for x_0 as:

$$M_r = \frac{\phi[\mu_1 L + (\mu_2 - \mu_1)(x_0 - x)]}{\mu_1\mu_2 A}. \qquad (26)$$

Then the magnetomotance difference of the left-side and the right-side can be obtained as:

$$\Delta M = M_l - M_r = \frac{2x(\mu_2 - \mu_1)}{\mu_1\mu_2 A}\phi. \qquad (27)$$

If a sinusoidal signal is supplied in the primary coil, a sinusoidal magnetic field will be generated. This means that ϕ in Eq. (29) is also going to be a sinusoidal function of time:

$$\phi = \phi_0\sin\omega t \qquad (28)$$

where ϕ_0 is the amplitude of the magnetic flux.

Then the magnetomotance difference between the two secondary coils is also going to be a sinusoidal function of time:

$$\Delta M = M_l - M_r = \frac{2x(\mu_2 - \mu_1)}{\mu_1\mu_2 A}\phi_0\sin\omega t \qquad (29)$$

From Eq. (29), it can be seen that the magnetomotance difference in the secondary coils is of the same frequency as the current supplied to the primary coil, and its amplitude is a function of the permeability of the air and the magnetic core, the cross-sectional area of the magnetic core, the flux amplitude of the primary coil and, most importantly, the displacement x. Since all other parameters are fixed

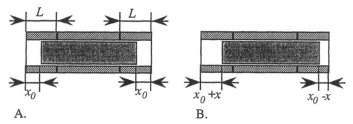

Figure 6   A schematic design of an LVDT: (A) magnetic core in neutral position; (B) magnetic core moves to the right side.

during the operation of LVDT, the amplitude of this magnetomotance difference is only modulated by the displacement to be measured. It should be noted that this position dependence is perfectly linear. The position-dependent sinusoidal signal can be demodulated easily and the position signal restored.

LVDTs have many advantages and are widely used in mechatronics products. They provide almost perfect linear responses over unlimited working range, high precision, and long mechanical life. There are many manufacturers of LVDTs and many different types of LVDTs are commercially available. There are also rotational variable differential transformers (RVDTs) available for measuring the rotational displacement.

V. ULTRASONIC SENSORS

Ultrasonic transducers are based on the measurement of sound as it travels between a transducer and an object being measured. The term ultrasonic is used to distinguish between general acoustic waves and sound in the audible spectrum (20Hz–20 kHz); most acoustic measurement is done in the ultrasonic (non-audible range) using frequencies higher than 20 kHz. Most of the ultrasonic position/displacement sensors work on the same principle as the classical sonar system. The basic design always includes an ultrasonic source (a wave generator) and a receiver, such as a microphone. There are two basic types of measurement modes: pitch-catch mode (also known as through transmission) in which a wave travels between a separate source and sensor, and pulse-echo mode in which the ultrasonic source is also used as the receiver. In the pitch-catch mode an ultrasonic wave is sent by the source and sensed by the receiver located on the target whose position is being measured. The range or distance d to the target is calculated based on the time of flight of the signal t and the speed of the ultrasonic wave propagation ($c = 110,000$ inches per second in air):

$$d = ct \tag{30}$$

In the pulse-echo mode the ultrasonic transducer listens for the reflected portion of the signal as it bounces off an object in its path. This echo is caused by an acoustic impedance mismatch between the medium through which the wave is propagating, and an object located in the wave's path. Every material has a characteristic acoustic impedance z which depends upon its density and the speed at which sound propagates:

$$Z = c\Psi \tag{31}$$

Depending upon the ratio of the impedance of two materials, a portion of the sound impinging on the boundary will be reflected and a portion transmitted into the second material. More sound is reflected in cases where the impedance mismatch is larger: an impedance ratio of one means that none of the sound will be reflected back, hence no echo will be returned to the transducer. Perhaps the best known ultrasonic position sensor which operates in the pulse-echo mode is the range-finder for the auto-focusing system of the Polaroid camera [4]. It uses a single electrostatic transducer (capacitive) as both the actuator to generate the ultrasonic wave and the sensor for receiving the signal. As the exposure button is pushed, the sensor on the camera emits an ultrasonic pulse, and then listens for the return echo (caused by the impedance mismatch between the air and the object being photographed). The distance to the object is calculated as:

$$d = \frac{ct}{2} \tag{32}$$

and the camera is then properly focused before exposing the film. The range-finder in the Polaroid camera uses a group of wave pulses at different frequencies to make sure the distance measurement is accurate.

In addition to performing range-finding measurement in air, ultrasonic sensors can be sued to evaluate position in solid materials. As in the air, both pitch-catch and pulse-echo modes can be used to measure positions. Pulse-echo, however, is more popular for the measurement of position in solids. In many testing applications, such as the detection of cracks in a reactor wall, there is access to only one side of the solid material. In such a case, the ultrasonic wave will bounce off any cracks or voids in the material, as well as the far boundary of the solid. Each reflector will be detected separately by the transducer.

The accuracy with which position can be measured ultrasonically depends upon the frequency of the signal. In general, measurement resolution is determined by the signal wavelength Σ which is in turn determined by the signal frequency:

$$c = f\Sigma \tag{33}$$

where f is the frequency.

This would imply that high frequencies should be used for range-finding in order to maximize the accuracy of the measurement. Unfortunately frequency cannot be increased without bound, due to acoustic attenuation which increases as a function of frequency. For instance, a 1 MHz signal will only travel a few inches in air before it becomes indistinguishable from the background noise. A good treatment of issues pertaining to ultrasonic measurement is provided by Krautkramer and Krautkramer [8].

VI. FIBER-OPTIC SENSORS AND OPTICAL INCODERS

A. Fiber-Optic Sensors

Fiber-optic sensors have some significant advantages: for example, freedom from electromagnetic interference, flexibility, low signal attenuation, rugged structure and smaller physical sizes compared with conventional optical system, and potentially high accuracy. Probably the most important fiber-optic position sensors is the fiber-optic interferometer. Compared with a conventional interferometer, a fiber-optic interferometer has the advantages of, for example, much smaller physical size, stable performance (less sensitive to environmental conditions such as vibrations), and ease of assembly.

There are two main categories of fiber-optic sensors. The first category is intrinsic, and the second category is extrinsic. In an intrinsic fiber-optic sensor the transmission properties of the optic-fiber are modulated by the physical parameter to be measured. For example, if a piece of optic-fiber coated with magnetostrictive material can be used for magnetic field detection because a change in magnetic field can cause an extension in the optic-fiber, this extension of fiber can then cause a change in the total length of the light-path and can be measured by measuring the phase difference of the light signal using the interferometric principle. Similarly, a fiber-optic temperature sensor can also be designed by introducing another transduction mechanism to transform the temperature change into a change in the length of an optic-fiber.

In extrinsic types of fiber-optic sensors, the optic-fiber is only used to transmit the light signal. A typical example is a proximity sensor that can be built with a photoelectric sensing cell and an LED pig-tailed with optic-fibers. In this case, the optic fibers are used as the sensing head. Compared with a regular proximity sensor with photoelectric sensor and LED, this type of sensor has unique advantages in applications that require position sensing in limited space.

B. Optical Encoders

Optical encoders are high-precision rotational position sensing devices that are widely used in mechatronics products, especially in industrial robots. They are mounted on a rotary shaft to generate a digital signal to sense its rotational position. There are two types of optical encoders: absolute and incremental, see Kuo [9]. In an absolute encoder, multiple concentric rings of binary code are etched on a code-wheel. The code along any radial line on this code-wheel represents an absolute value of the rotational position. One light source and one light detector are used to read the code. Using absolute encoders, it is necessary to know only one radial line of code to determine the position of the shaft. Absolute encoders can be very accurate. Angular position can be measured to an accuracy of one part in 2^{10}. The incremental optical encoders are simpler and cheaper compared with the absolute ones. Two pairs of photoelectric sensor and light sources are used to sense both the direction and size of the motion. There are two optional designs. One uses a one-track code-disk and another uses a two-track (one inner and one outer track) code-disk. When a two-track disk is used, each track on the code-disk requires one pair of photoelectric sensor and light source. When a one-track disk is used, the two photoelectric sensors are arranged in such a way that one is displaced from the other by one and one-half slot widths. The pulse trains from the photoelectric sensors are then counted and interpreted by an electronic circuit to obtain the rotational position and direction.

VII. PHOTOELECTRIC SENSORS AND LATERAL-EFFECT PSD

A. Photoelectric Sensors

Photoelectric sensors are another major category of sensor that is used in proximity and position sensing. The commonly used photosensors include photoelectric tubes, photodiodes, phototransistors, photoconductive transducers, photovoltaic cells, CCDs, and lateral-effect position sensitive detectors (PSDs). In this section, we will briefly introduce the general principles of position and proximity sensing with photoelectric sensing cells, and then spend most of the time discussing the lateral-effect position detectors (PSDs) because of their advantages in providing highly precise, highly linear position measurements.

Most of the photoelectric sensors can only provide an output electric signal dependent on the total power of light falling on the sensing cells. In general, there are two ways to use them for position sensing as shown in Figure 7. The first method is to arrange the light source (for example, light emitting diodes, LEDs), and the photoelectric sensor in opposite positions as shown in Method A of Figure 7. By moving into or out of the light beam, the position, presence, or absence of an object can be sensed. Another common method of using photoelectric cells in position sensing is shown in Method B of Figure 7. In this arrangement, the light source and the photo-sensor are located in the same plane and normally fabricated in one package. Many manufacturers supply this type of proximity sensor. They are supplied by many manufacturers at very low prices. The moving object whose position is being sensed reflects a light beam back onto the surface of the photoelectric sensing cell. Typically, this type of proximity sensor has a larger working range and lower precision compared with the capacitive sensors. The precision may be increased by reducing the measurement range.

Silicon image sensors such as charge-couple devices (CCDs) belong to another important category of photoelectric sensors that have very wide mechatronics applications. The CCD can be used to transfer an optical image or frame from light-sensitive array to a digital output. It has become popular in the manufacturing industry because of the wide applications of microcomputers. The major disadvantage of the CCD cameras is that they are quite expensive.

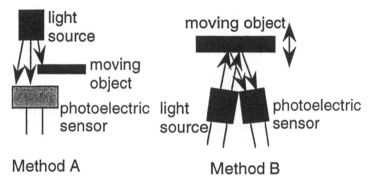

Figure 7 Two methods to use the photoelectric cells for position measurement.

B. Lateral Effect Position Sensitive Detector (PSD)

The lateral-effect position sensitive detector (PSD) is widely used for high precision position measurement. This type of position detector has some prominent advantages compared with other optical and photoelectric sensors. It can measure displacement in a spatially continuous manner, unlike other types of large sensitive area detectors such as charge-coupled devices (CCDs). It provides either one-dimensional or two-dimensional, highly linear, spatially continuous, fast, real time position/displacement measurement. Submicron resolution over a large working range can be obtained. These advantages make this type of position sensor very suitable for mechatronics applications that require very high precision position/displacement measurement over a large working space.

Operation Principle and Different Types of PSDs

It is commonly known that a light beam projected onto the surface of a p-n junction produces a photopotential on each plane of the junction. The photopotential will induce photocurrent if there are electrodes on the boundary of a junction plane. The induced photocurrent will flow laterally toward the electrodes on the boundary because of the photopotential gradient in the lateral direction. This phenomenon is called the lateral photoeffect and was first discovered by Schottky [10] in 1930. The discovery did not draw enough attention until 'rediscovered' by Wallmark [11] in 1959. Wallmark designed a device composed of a Ge-In p-n junction in which a light beam projected onto the surface causes a position dependent photopotential difference between the point contacts on the sensor surface.

Locovsky [12] derived the fundamental equations describing the diffusion and recombination processes for the steady-state and the small-signal transient cases for a modern PSD. Connors [13] investigated the one-dimensional model of a reverse-biased PSD. Woltring [14] extended Connors' analysis to the two-dimensional rectangular PSD operated in both the unbiased (small signal) and the fully reverse-biased modes. The fundamental equation for the potential distribution in a PSD that is in steady-state and fully reverse-biased provided by Woltring [14] has a very simple form as follows:

$$\frac{\partial^2 U(x,y)}{\partial x^2} + \frac{\partial^2 U(x,y)}{\partial^2 y} = -\frac{\rho}{w} I_s(x,y) \qquad (34)$$

where $U(x,y)$ is the electric potential at point (x,y) generated by the light beam projected on the sensor surface, Ψ the surface resistivity, w the thickness of the resistive layer of the sensor surface, and $I(x,y)$ the generated photocurrent density. Using this equation and boundary conditions set by the design of electrodes, the potential distribution can be solved either analytically or numerically. Then the photocurrents flowing to electrodes can also be obtained. Woltring [14] provided a very detailed analysis for two types of the most commonly used PSDs: the duolateral and tetralateral, whose geometries are shown in Figures 8A and 8B. Wang and Busch-Vishniac [15] studied the linearity for all three types of PSDs (duolateral, tetralateral, and pin-cushion) commercially available today, and proposed an alternative, clover geometry design.

In the schematic diagrams shown in Figure 8, the shaded area is the effective (useful) device area, and the dark areas the electrodes. A duolateral, duoaxis PSD has two electrodes on each side of the p-n junction. A tetralateral type PSD has all four electrodes on one side of the p-n junction. A pin-cushion type PSD shown in Figure 8c has curved boundaries and the effective transducer area is formed by a rectangle whose sides are tangent to the innermost points on the edges. The electrodes are point contacts located at the four corners.

To have the best performance and prevent any possible recombination across the p-n junction upon illumination, it is always preferable to operate PSDs in the fully reverse-biased mode. In the fully reverse-biased mode, the photocurrent generated by a falling light beam distributes to electrodes on the boundaries of the sensor surface. These photocurrents are then measured for position sensing of the lightbeam. Full reverse-biasing of lateral-effect PSDs can increase the thickness of the depletion layer of the p-n junction, and reduce the current flowing across the p-n junction to almost zero, therefore preventing signal loss due to surface recombination. It also reduces the effective capacitance of the p-n junction, therefore increasing the response speed of the PSD.

In the fully reverse-biased mode, the duoaxis duolateral PSD is inherently linear because the x and y contacts are on opposite sides of the junction. Compared to the duolateral PSD, the tetralateral PSD with four extended ohmic contacts (electrodes) on one side of the p-n junction has disadvantages, namely, the electrode structure makes cross-talk and nonlinearity inevitable. In a tetralateral type PSD, the generated photocurrent is divided into four parts instead of two parts as in a duolateral type PSD, so that the resolution is about half that of the duolateral one for the same noise level. However, the tetralateral PSD also has advantages when compared to the duolateral PSD: a faster response, a much lower dark current, an easier reverse-bias application, and a lower fabrication cost. The pin-

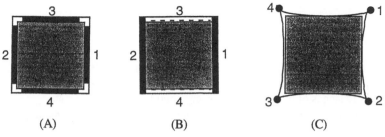

Figure 8   (A) Tetralateral position sensitive device. All four circuits are on a single surface. Shaded area is photosensitive, and the heavy solid lines indicate electrodes. The effecctive device area is shown hatched; (B) duolateral duoaxis position sensitive device. Two of the electrodes are on the p-surface and two on the n-surface; (C) Pin-cushion type position sensitive device. Curved boundary with four electrodes located in corners.

cushion type of PSD by Hamamatsu [16] has a performance somewhere between those of the duolateral and tetralateral ones.

All three types of PSDs can provide the cartesian two-dimensional location of a light spot. Geometries suitable for other coordinate systems have been studied, including work by Xing and Boeder [17] on a PSD for measurement of angular position.

Signal Processing for PSD

When a light beam scans across the surface of a two-dimensional PSD, the photocurrents flowing to the four electrodes will change as a function of the light spot position. A power fluctuation of the light beam or variations of environmental illumination may also cause a change in the total photocurrent generated, therefore resulting in changes of photocurrents flowing to electrodes. A normalization technique is always adopted to eliminate the influences of beam intensity or environmental illumination on the position signals. Suppose I_1, I_2, I_3, and I_4 are the photocurrents from PSD electrodes as numbered in Figure 8; then the normalized x and y position signals for a duolateral or a tetralateral PSD are, see Woltring [14]:

$$x = K_x \frac{I_1 - I_2}{I_1 + I_2}, \tag{35}$$

$$y = K_y \frac{I_3 - I_4}{I_3 + I_4}, \tag{36}$$

where K_x and K_y are the amplification factors of the x and y channels in the signal processing circuit.

Similarly, the normalized x and y position signals for a pin-cushion type PSD are, see Hamamatsu [16]:

$$x = K_x \frac{(I_1 - I_2) - (I_3 - I_4)}{I_1 + I_2 + I_3 + I_4}. \tag{37}$$

$$y = K_y \frac{(I_1 + I_4) - (I_2 + I_3)}{I_1 + I_2 + I_3 + I_4}. \tag{38}$$

Generally, the photocurrents from the electrodes of a PSD are amplified with preamplifiers first, then additions, subtractions and divisions are carried out in the following stages of the signal processing circuit.

The resolution of a position sensing system using a PSD as a sensing cell is ultimately limited by the signal to noise ratio (S/N) of the system. The signal can be boosted by increasing the power of the light beam. However, the intensity of the light beam is limited by the saturation level. The other option is to reduce the total noise of the system. To reduce the noise level would not be an easy task because it is very hard to separate the signal from the noise without using modulation techniques. This means that, if further improvement of the signal to noise ratio is desired, a modulation technique has to be used.

Different Ways to Use PSDs

Because PSDs can only detect the relative movement of a light spot on its surface, to use it for noncontact position sensing it is necessary to convert movement of the object into the relative movement of a light spot on the sensor surface. There are three ways to accomplish this as shown in Figure 9. The first method is to fix the PSD on the moving object to be sensed while keeping the light source stationary. The second method is to fix a light source such as an LED or a semiconductor laser on the moving object to be measured while holding the PSD stationary. Both approaches require tethering power wires to the moving object, either supplying to PSDs or to the light sources. In some applications, these wires may not be accepted. The third method is to fix a rigid, opaque plate with a pin-hole onto the object to be sensed while both a highly collimated light source and a PSD are kept stationary. The light emitted from the light source is blocked by the opaque plate except at the pin-hole. When the object being sensed moves around, a different part of the light beam passes through the pin-hole, causing a light spot to move on the PSD surface.

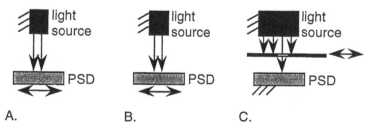

Figure 9   Three ways to use PSDs for noncontact position sensing: (A) light source fixed; (B) PSD fixed; and (C) both PSD and light source fixed.

One major advantage with the third method is that there is no cross-talk between the motion within the $x - y$ plane and the out-of-plane rotation. This is possible because a PSD detects the location of the centroid of the light spot falling on its surface. When the object has an out-of-plane rotation, the projection of the pin-hole on the PSD surface becomes an ellipse instead of a circle, but both have the same centroid if the light beam is of uniform intensity. Consequently only the total amount of light falling on the surface of the PSD is decreased. This does not affect the output position signals because they are normalized by the signal processing circuit. The disadvantage of the third approach is that the measurement accuracy depends heavily on the uniformity of the light source. Ideally, we would like to use a highly collimated light source, like a laser.

Modulation Technique for PSD

The essence of modulation technology is to narrow the frequency bandwidth of the transmitted signal, so that all other background noises can be eliminated with appropriate filtering. This approach, therefore, can dramatically improve the signal to noise ratio of a sensing system. There are many kinds of modulation methods. The two most popular methods used in radio communication are frequency modulation (FM) and amplitude modulation (AM). Both the frequency modulation and amplitude modulation of a continuous light beam are not suitable because it would be very difficult to have a light beam whose intensity is a sinusoidal function. Other alternative options include pulse modulation approaches. In pulse modulation, the carrier signal is a pulse train instead of a sinusoidal wave. There are three kinds of pulse modulation approaches commonly in use. One is called pulse position modulation (PPM), where the pulse position is modulated by the signal to be transmitted. This method is not suitable because it is very difficult to design a measurement scheme in which the pulse position can be modulated by the displacement of a measurement object. The second method is called pulse duration modulation (PDM) where the pulse duration must be modulated by the moving object, which is very difficult to implement. The third method is the pulse amplitude modulation (PAM), that is, the pulse amplitude is modulated by the signal to be transmitted. This method is recommended by the manufacturer of the PSD, Woltring [14], and has been used by Wang and Busch-Vishniac [18], and a resolution down to 0.1 micrometers has been achieved over a large work space. Interested readers are referred to the reference list at the end of this chapter.

The principle of PAM is explained in Figure 10. Shown in Figure 10A is a pulse train signal supplied to a light source, such as an LED, to generate a light beam which switches on and off continuously at the modulation frequency. If a light beam such as an LED is fixed on a moving object whose motion is being monitored while the PSD is stationary, then the photocurrents flowing to the PSD electrodes are also pulse trains at the same frequency as the input light beam. Therefore the amplitudes of the photocurrents arriving at the electrodes are modulated functions of the object movement. Two synchronized delayed pulse trains with the same frequency shown in Figure 10B and Figure 10C are used to trigger the sample/hold amplifiers to restore the DC mode. One has its rising edge corresponding to the light beam on period. The other has its rising edge located at the light beam 'off' period. The output signals are then subtracted from each other to obtain the amplitude of the position signal. Detailed design of the circuit is not able to be provided here. Interested readers may find more information about the circuit design in the paper by Wang and Busch-Vishniac [18].

Measuring Multiple Beams with One PSD

Normally, one PSD is needed to measure the position/displacement of one light beam. However, if an optical measurement system uses multiple beams and multiple detectors, the system hardware can become large and redundant. Because of the cost of the PSDs, the system cost may become prohibitively high. Further, alignment difficulties increase dramatically as more PSDs are used. Finally, if several PSDs are used in a sensing system and each one of them needs to be calibrated separately, calibration of the sensing system will demand significant efforts. Therefore it is highly desirable to use a single PSD to measure the positions/displacements of multiple light beams instead of one SD for each light source. The resulting advantages include: a more compact system, lower cost, faster calibration, and preservation of almost identical physical and environmental conditions for every sensing element.

A suitable modulation technique which permits the measurement of multiple light spots using a single optical sensor has been provided by Tian et al. [19]. Borrowing from technology widely used in the telecommunication industry, a modulation method is used to permit one sensor to monitor multiple light sources. Each light source is modulated at a different frequency. When there are multiple beams irradiating a single PSD the total photo-voltage generated will equal the sum of the photo-voltages generated by all beams if the PSD is working in the linear region, i.e., if the reverse-biased voltage is large enough to prevent recombination. The

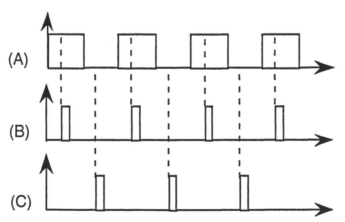

Figure 10 Pulse amplitude modulation technique for position sensing with PSD.

position-sensitive photocurrents from the PSD sensor are then superpositions of the photocurrents generated by all light beams. The position/displacement information on a specific light beam can be obtained through demodulation. An alternative mean of accomplishing separation of signals from multiple light sources would be to switch light sources on and off sequentially. While this method may be acceptable for some applications, it is generally less desirable than a frequency separation approach because the time separation method introduces transients into the system. In order to prevent the system transients from causing saturation, the signal level must be lowered, resulting in poorer sensor measurement accuracy. Tian et al. [19] reported the successful implementation of the idea and very high resolutions for displacement measurement of multiple light beams.

The number of light beams the PSD can measure is only limited by two factors: signal loss from recombination and the bandwidth of the PSD.

REFERENCES

1. S Middelhoek, AC Hoogerwerf. Classifying solid-state sensors: the sensor effect cube. Sensors and Actuators 10:1,1986.
2. DC Karnopp, DL Margolis, RC Rosenberg. *System Dynamics.* 2nd. ed. New York: Wiley & Sons, 1990.
3. IJ Busch-Vishniac. *Electromechanical Sensors and Actuators.* New York: Springer-Verlag, in press.
4. Polaroid Ultrasonic Ranging Experimenter's Kit. Cambridge, Mass: Polaroid Corp., 1980.
5. RC Luo. Sensor technologies and microsensor issue for mechatronics systems. IEEE/ASME Transactions on Mechatronics 1(1):39–49, 1996.
6. YS Lee, KD Wise. A batch-fabricated silicon capacitive pressure transducer with low temperature sensitivity. IEEE Transactions on Electron Devices 29(1):42–48, 1982.
7. D Bosch, B Heimhofer, G Muck, H Seidel, U Thumser, W Welser. A silicon microvalve with combined electromagnetic/electrostatic actuation. Sensors and Actuators A 37–38: 684–692, 1993.
8. J Krautkramer, H Krautkramer. *Ultrasonic Testing of Materials.* 3rd ed. Berlin: Springer-Verlag, 1983.
9. BJ Kuo. *Automatic Control Systems*, *4th ed.* New York: Prentice-Hall, 1982.
10. W Schottky. Ueber den Entsstehungsort der Photoelektronen in Kupfer-Kupferoxydul-photozellen. Phys. Z. 31:913–925, 1930.
11. T. Wallmark. A new semiconductor photocell using lateral photoeffect. Proc. IRE 45:474–483, 1957.
12. G Lucovsky. Photo-effects in nonuniformly irradiated p-n junctions. Journal of Applied Physics 33:1088–1095, 1960.
13. P Connors. Lateral photodetector operating in the fully reverse-biased mode. IEEE Transaction on Electron Devices 18:591–596, 1971.
14. HJ Woltring. Single- and dual-axis lateral photodetectors of rectangular shape. IEEE Transactions on Electron Devices 22:581–586, 1975.
15. W Wang, IJ Busch-Vishniac. The linearity and sensitivity of lateral effect position sensitive devices—an improved geometry. IEEE Transactions on Electron Devices 36:2475–2480, 1989.
16. Hamamatsu Photonics Corp. Position Sensitive Detectors. Product Catalogue, 1987.
17. YZ Xing, CPW Boeder. A new angular-position detector utilizing the latreal photoeffect in Si. Sensors and Actuators 7:153–166, 1985.
18. W Wang, IJ Busch-Vishniac. A four-dimensional non-contact sensing system for micro-automation machines. The Sensors and Instrumentation for In-Process Monitoring of Manufacturing Technical Session, ASME Winter Annual Meeting, Dallas, Texas, November 25–30, 1990.

19. D Qian, W Wang, IJ Busch-Vishniac, AB Buckman. A method for measurement of multiple light spot positions on one position-sensitive detector (PSD). IEEE Transactions on Instrumentation and Measurements 42(1):14–18, 1993. The preliminary results were also presented at the IMTC'92, New York City, New York, May 1992.

2
Microsensors and Microactuators

Michael Goldfarb
Vanderbilt University, Nashville, Tennessee

I. INTRODUCTION

Two major factors distinguish the existence, effectiveness, and development of micro-scale transducers from that of conventional-scale ones. The first is the physics of scaling and the second is the suitability of manufacturing techniques and processes. The former is governed by the laws of physics and thus a fundamental factor, while the latter is related to the development of manufacturing technology, which is a significant, though not fundamental, factor. Due to the combination of these factors, effective micro-scale transducers can often not be constructed as geometrically scaled-down versions of conventional-scale transducers.

Microactuators can be generally classified as either multi-component or solid-state. Multi-component microactuators incorporate two (or more) bodies which perform work by exerting a mutual force between them. The forces typically utilized in these types of actuators are electrostatic or electromagnetic in nature. Solid-state actuators perform work by utilizing material deformation phenomena within a single body. Examples of this type are piezoelectric ceramics, shape memory alloys, magnetostrictive alloys, electrostrictive ceramics and photostrictive materials.

Microsensors are primarily classified by function. Physical microsensors have been developed for the purpose of measuring force, pressure, acceleration, and angular rate (i.e., gyroscopes). Microsensors can additionally be classified as a resonant or analog type. An analog sensor is one in which the measured quantity results in an electrically measurable change in behavior, typically in a proportional manner. A resonant sensor is one in which the measured quantity alters the structural resonant frequency of a member, and is thus acquired by exciting the structure and measuring its resonant frequency. These sensors require a mechanism of excitation in addition to a mechanism of measurement, and thus are typically more complex than the analog type. Since resonant sensors measure frequency instead of amplitude, however, they are generally less susceptible to noise and typically provide a higher resolution measurement than those that operate in an analog fashion.

A. Micromechanics and the Physics of Scaling

The types of forces that influence micro-scale devices are different from those that influence conventional-scale ones. This is because the size of a physical system bears a significant influence on the physical phenomena that dictate the dynamic behavior of that system. For example, larger-scale systems are influenced by inertial effects to a much greater extent than smaller-scale systems, while smaller systems are influenced more by surface effects. As an example, consider small insects which can stand on the surface of still water, supported only by surface tension. The same surface tension is present when humans come into contact with water, but on a human scale the associated forces are typically insignificant. The world in which humans live is governed by the same forces as the world in which these insects live, but the forces are present in very different proportions. This is due in general to the fact that inertial forces typically act in proportion to volume, and surface forces typically in proportion to surface area. Since volume varies with the third power of length and area with the second power of length, geometrically similar but smaller objects have proportionally more area than their larger counterparts. Exact scaling relations for various types of forces can be obtained by incorporating dimensional analysis techniques, as described in references [1–5]. Inertial forces, for example, can be dimensionally represented as:

$$F_i = \rho L^3 \ddot{x}$$

where F_i is a generalized inertia force, ρ is the density of an object, L is a generalized length, and x is a displacement. This relationship forms a single dimensionless group, given by:

$$\Pi = \frac{F_i}{\rho L^3 \ddot{x}}$$

Scaling with geometric and kinematic similarity can be expressed as:

$$\frac{L_s}{L_o} = \frac{x_s}{x_o} = N$$

$$\frac{t_s}{t_o} = 1$$

where L represents the length scale, x the kinematic scale, t the time scale, o the original system and s the scaled system. Since physical similarity requires that the dimensionless group (Π) remain invariant between scales, the force relationship is given by:

$$\frac{F_s}{F_o} = N^4$$

assuming that the intensive property (density) remains invariant (i.e., $\rho_s = \rho_o$). An inertial force thus scales as N^4, where N is the geometric scaling factor. Alternatively stated, for an inertial system that is geometrically smaller by a factor of N, the force required to produce an equivalent acceleration is smaller by a factor of N^4. A similar analysis shows that viscous forces, dimensionally represented by:

$$F_v = \mu L \dot{x}$$

scale as N^2, assuming the viscosity μ remains invariant, and elastic forces, dimensionally represented by:

$$F_e = ELx$$

scale as N^2, assuming the elastic modulus E remains invariant. Thus, for a geometrically similar but smaller system, inertial forces will become considerably less significant with respect to viscous and elastic forces.

B. General Mechanisms of Electromechanical Transduction

The fundamental mechanism for both sensing and actuation is energy transduction. The primary forms of physical electromechanical transduction can be grouped into two categories. The first is multi-component transduction, which utilizes "action at a distance" behavior between multiple bodies, and the second is deformation-based or solid-state transduction, which utilizes mechanics-of-material phenomena such as crystalline phase changes or molecular dipole alignment. The former category includes electromagnetic transduction, which is typically based upon the Lorentz equation and Faraday's law, and electrostatic interaction, which is typically based upon Coulomb's law. The latter category includes piezoelectric effects, shape memory alloys, and magnetostrictive, electrostrictive, and photostrictive materials. Table 1 summarizes these various forms of electromechanical transduction. Each form of transduction is treated separately in the sections that follow.

C. Sensor and Actuator Transduction Characteristics

Characteristics of concern for both microactuator and microsensor technology are repeatability, the ability to fabricate at a small scale, immunity to extraneous influences, sufficient bandwidth and, if possible, linearity. Characteristics typically of concern for microactuators are achievable force, displacement, power, bandwidth (or speed of response) and efficiency. Characteristics typically of concern specifically for microsensors are high resolution and the absence of drift and hysteresis.

II. MULTI-COMPONENT MICROACTUATORS

A. Electrostatic Actuation

The most widely utilized multi-component microactuators are those based upon electrostatic transduction. These actuators can also be regarded as a variable capacitance type, since they operate in an analogous mode to variable reluctance type electromagnetic actuators (e.g., variable reluctance stepper motors). Electrostatic actuators have been developed in both linear and rotary forms. The two most common configurations of the linear type of electrostatic actuators are the *normal drive* and tangential or *comb drive* types, which are illustrated in Figures 1 and 2 respectively. Note that both actuators are suspended by flexures, and thus the output force is equal to the *electrostatic actuation* force minus the elastic force required to deflect the flexure suspension. The normal drive type of electrostatic microactuator operates in a similar fashion to a condenser microphone. In this type of drive configuration, the actuation force is given by:

$$F_x = \frac{\varepsilon A v^2}{2x^2}$$

where A is the total area of the parallel plates, ε is the permittivity of air, v is the voltage across the plates, and x is the plate separation. The actuation force of the comb drive configuration is given by:

Table 1 Summary of Various Forms of Actuation[a]

Mechanism of Transduction	Scaling of Force (Intensive Variable)	Relative Speed	Efficiency
Electrostatic	N^0 (electric field)	fast	high
	N^2 (temperature in wire)		
	$N^{5/2}$ (power flow through wire)		
	N^3 (current density in wire)		
Electromagnetic		fast	fair
Piezoelectric	N^2 (electric field)	fast	fair–high
Shape memory alloy	N^2 (temperature in wire)	slow	low
Magnetostrictive	N^2 (magnetic field)	fast	fair
Electrostrictive	N^2 (electric field)	fast	fair–high
Photostrictive	N^2 (light intensity)	slow	fair–high

[a]Scaling of force column refers to the amount force will decrease for a geometrically similar actuator smaller by a factor of N.

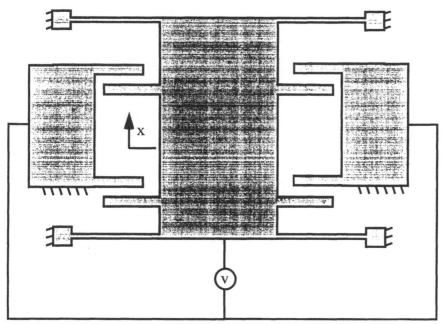

Figure 1 Schematic of a normal-drive electrostatic actuator.

$$F_x = \frac{\varepsilon w v^2}{2d}$$

where w is the width of the plates, ε is the permittivity of air, v is the voltage across the plates, and d is the plate separation. Dimensional examination of both relations indicates that force is independent of geometric and kinematic scaling. That is, for an electrostatic actuator that is geometrically and kinematically reduced by a factor of N, the force produced by that actuator will be the same. Since forces associated with most other physical phenomena are significantly reduced at small scales, micro-scale electrostatic forces become significant relative to other forces. Such an observation is clearly demonstrated by the fact that all intermolecular forces are electrostatic in origin, and thus the strength of all materials is a result of electrostatic forces, see Israelachvili [6].

The maximum achievable force of multi-component electrostatic actuators is limited by the dielectric breakdown of air, which occurs in dry air at about 0.8×10^6 V/m. Fearing [7] estimates that the upper limit for force generation in electrostatic actuation is approximately 10 N/cm^2. Since electrostatic drives do not have any significant actuation dynamics, and since the inertia of the moving member is usually small, the actuator bandwidth is typically quite large, on the order of a kilohertz.

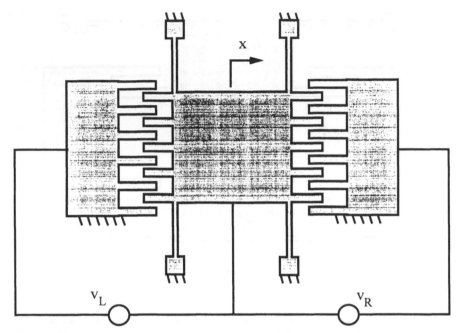

Figure 2   Comb-drive electrostatic actuator. Energizing an electrode provides motion toward that electrode.

The maximum achievable stroke for normal configuration actuators is limited by the elastic region of the flexure suspension and additionally by the dependence of actuation force on plate separation, as given by the above stated equations. According to Fearing, a typical stroke for a surface micromachined normal configuration actuator is on the order of a couple of microns. The achievable displacement can be increased by forming a stack of normal-configuration electrostatic actuators in series, as proposed by Bobbio et al. [8] and Jacobson et al. [9].

The typical stroke of a surface micromachined comb actuator is on the order of a few microns, though sometimes less. The maximum achievable stroke in a comb drive is limited primarily by the mechanics of the flexure suspension. The suspension should be compliant along the direction of actuation to enable increased displacement, but must be stiff orthogonal to this direction to avoid parallel plate contact due to misalignment. These modes of behavior are unfortunately coupled, so that increased compliance along the direction of motion entails a corresponding increase in the orthogonal direction. The net effect is that increased displacement requires increased plate separation, which results in decreased overall force.

The most common configurations of rotary electrostatic actuators are the *variable capacitance motor* and the *wobble or harmonic drive motor*, which are illustrated in Figures 3 and 4 respectively. Both motors operate in a similar manner

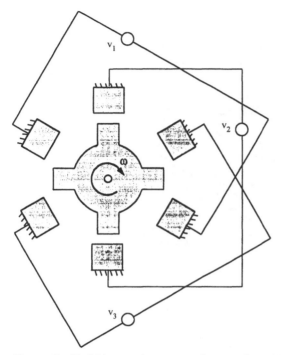

Figure 3 Variable capacitance type electrostatic motor. Opposing pairs of electrodes are energized sequentially to rotate the rotor.

to the comb-drive linear actuator. The variable capacitance motor is characterized by high-speed low-torque operation. Useful levels of torque for most applications therefore require some form of significant micromechanical transmission which does not presently exist. The rotor of the wobble motor operates by rolling along the stator, which provides an inherent harmonic-drive-type transmission, and thus a significant transmission ratio (on the order of several hundred times). Note that the rotor must be well-insulated to roll along the stator without electrical contact. The drawback to this approach is that the rotor motion is not concentric with respect to the stator, which makes the already difficult problem of coupling a load to a micro-shaft even more difficult.

Examples of normal type linear electrostatic actuators are those by Bobbio et al. [8], Jacobson et al. [9] and Yamaguchi et al. [10]. Examples of comb drive electrostatic actuators are those by Kim et al. [11] and Matsubara et al. [12], and a larger-scale variation by Niino et al. [13]. Examples of variable capacitance rotary electrostatic motors are those by Huang et al. [14], Mehragany et al. [15], and Trimmer and Gabriel [16]. Examples of harmonic-drive motors are those by Mehragany et al. [17], Dhuler et al. [18], Price et al. [19], Trimmer and Jebens [20], [21], and Furuhata et al. [22].

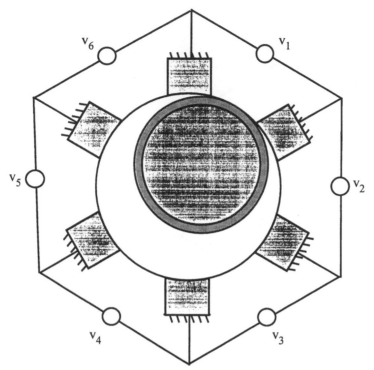

Figure 4 Harmonic drive type electrostatic motor. Opposing pairs of electrodes are energized sequentially to rotate the stator.

Electrostatic microactuators remain a subject of research interest and development, and as such are not yet available on the general commercial market.

B. Electromagnetic Actuation

Electromagnetic actuation is not as omnipresent at the micro-scale as at the conventional-scale. This probably is due in part to early skepticism regarding the scaling of magnetic forces, and in part to the fabrication difficulty in replicating conventional-scale designs. Most electromagnetic transduction is based upon a current carrying conductor in a magnetic field, which is described by the Lorentz equation:

$$dF = I\,dl \times B$$

where F is the force on the conductor, I is the current in the conductor, l is the length of the conductor, and B is the magnetic flux density. In this relation, the magnetic flux density is an intensive variable and thus (for a given material) does not change with scale. Scaling of current, however, is not as simple. The resistance of wire is given by:

$$R = \frac{\rho l}{A}$$

where ρ is the resistivity of the wire (an intensive variable), l is the length, and A the cross-sectional area. If a wire is geometrically decreased in size by a factor of N, then its resistance will increase by a factor of N. Since the power dissipated in the wire is I^2R, assuming the current remains constant implies that the power dissipated in the geometrically smaller wire will increase by a factor of N. Assuming the maximum power dissipation for a given wire is determined by the surface area of the wire, a wire that is smaller by a factor of N will be able to dissipate a factor of N^2 *less* power. Constant current is therefore a poor assumption. A better assumption is that maximum current is limited by maximum power dissipation, which is assumed to depend upon the surface area of the wire. Since a wire smaller by a factor of N can dissipate a factor of N^2 less power, the current in the smaller conductor would have to be reduced by a factor of $N^{3/2}$. Incorporating this into the scaling of the Lorentz equation, an electromagnetic actuator that is geometrically smaller by a factor of N would exert a force that is smaller by a factor of $N^{5/2}$. Trimmer and Jebens [23] and Trimmer [24] have conducted a similar analysis, and demonstrated that electromagnetic forces scale as N^2 when assuming constant temperature rise in the wire, $N^{5/2}$ *when* assuming constant heat (power) flow (as previously described), and N^3 when assuming constant current density. In any of these cases, the scaling of electromagnetic forces is not nearly as favorable as the scaling of electrostatic forces. Despite this, electromagnetic actuation still offers utility in microactuation, and most likely scales more favorably than do inertial or gravitational forces.

Lorentz-type approaches to microactuation utilize surface micromachined microcoils, such as the one illustrated in Figure 5. One configuration of this approach is represented by the actuator of Inoue et al. [25], which utilizes current control in an array of microcoils to position a permanent micro-magnet in a plane, as illustrated in Figure 6. Another Lorentz approach is illustrated by the actuator of Liu et al. [26], which utilizes current control of a cantilevered microcoil flap in a fixed external magnetic field to effect deflection of the flap, as shown in Figure 7. Liu [26] reported deflections up to 500 μm and a bandwidth of approximately 1000 Hz. Other examples of Lorentz-type nonrotary actuators are those by Shinozawa et al. [27], Wagner and Benecke [28], and Yanagisawa et al. [29]. A purely magnetic approach (i.e., not fundamentally electromagnetic) is the work of Judy et al. [30], which in essence manipulates a flexure-suspended permanent micro-magnet by controlling an external magnetic field.

Ahn et al. [31] and Guckel et al. [32] have both demonstrated planar rotary *variable-reluctance type electromagnetic micromotors*. A variable reluctance approach is advantageous because the rotor does not require commutation and need not be magnetic. The motor of Ahn et al. incorporates a 12-pole stator and 10-pole rotor, while the motor of Guckel et al. utilizes a 6-pole stator and 4-pole rotor. Both incorporate rotors of approximately 500 μm diameter. Guckel reports (no

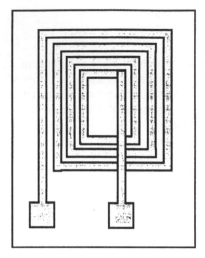

Figure 5   Schematic of surface micromachined microcoil for electromagnetic actuation.

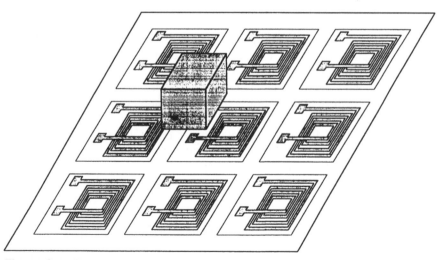

Figure 6   Microcoil array for planar positioning of a permanent micro-magnet, as described by Inoue et al. [25]. Each coil produces a field which can either attract or repel the
permanent magnet, as determined by the direction of a current. The magnet does not levi-
tate, but rather slides on the insulated surface.

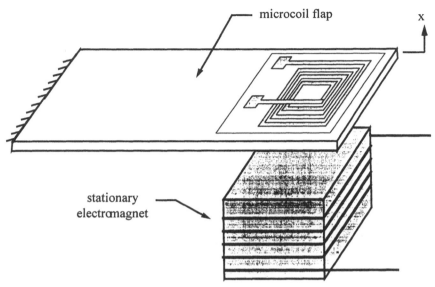

Figure 7 Cantilevered microcoil flap as described by Liu et al [26]. The Lorentz interaction between the energized coil and the stationary electromagnet deflects the flap upward or downward, depending on the direction of current through the microcoil.

load) rotor speeds above 30,000 rev/min, and Ahn estimates maximum stall torque at 1.2 µN-m.

Microfabricated electromagnetic actuators remain a subject of research interest and development, and as such are not yet available on the general commercial market.

III. DEFORMATION-BASED MICROACTUATORS

A. Piezoelectric Actuation

Piezoelectricity (PZT) is a property of certain materials in which application of a mechanical stress on the material induces an electrical charge, and in a converse manner, application of an electric field across the material creates a mechanical strain. These behaviors are called the direct piezoelectric effect and the inverse piezoelectric effect, respectively. The most widely recognized analytical description of the direct and inverse piezoelectric effects was published by a standards committee of the Institute of Electrical and Electronics Engineers (IEEE) [33]. The committee formulated linearized constitutive relations describing piezoelectric continua which form the basis for the model of piezoelectric behavior that is presently in general use. The linearized constitutive relations are typically represented in a compressed matrix notation as follows:

$$S_p = s_{pq}^E T_q + d_{kp} E_k$$

$$D_i = d_{iq} T_q + \varepsilon_{ik}^T E_k$$

where S_p represents an element of the (compressed) strain tensor, s_{pq}^E an element of the elastic compliance matrix when subjected to a constant electrical field, T_q an element of the (compressed) stress tensor, d_{kp} an element of a matrix of piezoelectric material constants, E_k an element of the electric field vector, D_i an element of the electric displacement vector, and ε_{ik}^T is an element of the permittivity matrix measured at a constant stress. The compressed notation eliminates redundant terms by representing the symmetric stress and strain tensors with single column vectors that incorporate elements representing both the diagonal and off-diagonal tensor terms. These equations essentially state that the material strain and electrical displacement (charge per unit area) exhibited by a piezoelectric ceramic are both linearly affected by the mechanical stress and electrical field to which the ceramic is subjected. The actuation force is thus a linear combination of an elastic component and a piezoelectric component, and can be represented dimensionally as:

$$F = k_p E L^2 - k_e x L$$

where F is the actuation force, E the electric field strength, L represents actuator size, x is the actuator displacement, and k_e and k_p are intensive constants associated with the elastic and piezoelectric material properties, respectively. Dimensional analysis indicates that a piezoelectric actuator smaller by a factor of N will exhibit both a piezoelectric component of force and an elastic component smaller by a factor of N^2. Piezoelectric actuation force thus scales as N^2.

Two significant and in general undesirable behaviors of *piezoelectric ceramic* that are not described by the linearized constitutive equations are hysteresis and creep. Piezoelectric materials exhibit a static hysteresis between the applied electrical voltage and the actuator displacement of roughly 15%. The hysteretic behavior of the piezoelectric material arises from the behavior of the molecular dipoles that constitute the ceramic, which have a tendency to flip from one preferred orientation to another as the applied electric field is increased. Piezoelectricity also exhibits significant creep (undesired deformation over time), which can be as much as 15% of the original deformation. Both the hysteresis and creep behaviors make precise open-loop positioning difficult, but can generally be compensated with closed-loop position feedback.

Piezoelectric behavior is established in suitable materials by poling the material with a large electric field. Subsequent application of an electric field in the direction of poling (positive field) induces a tensile mechanical stress that elongates the material, while application of a field opposite that direction (negative field) results in a compressive stress that contracts the material. The material is therefore capable of positive and negative strains. In order to prevent de-poling of the ceramic, however, the magnitude of the negative field is limited to about 30% of the positive, and thus the magnitude of realizable negative (compressive) strains is also about 30% of the positive.

The most common piezoelectric material is a lead-zirconate-titanate ceramic, which is typically referred to as PZT, an acronym of shortened atomic symbols. Piezoelectric ceramics can produce blocked actuation stresses of 40 N/mm^2 (MPa) and tensile strains of about 0.1%. The two most common configurations of PZT actuators are the stack configuration and the bender configuration.

Piezoelectric Stack Actuators

The fundamental component of a *piezoelectric stack actuator* is a wafer of piezoelectric material sandwiched between two electrodes. The piezoelectric material is manufactured so that the direct and inverse piezoelectric effect occurs only along the thickness of each wafer. A typical piezoelectric stack actuator is formed by assembling several of the wafer elements in series mechanically and connecting the electrodes so that the wafers are in parallel electrically, as illustrated in Figure 8. Application of an electrical charge to the stack will generate a mechanical displacement. If the stack is blocked to prevent displacement, it will generate a (compressive) mechanical force in proportion to the applied voltage.

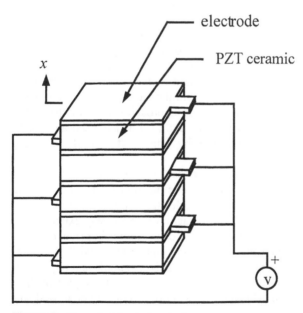

Figure 8 Piezoelectric stack actuator.

A PZT stack actuator is characterized by large force, small displacement, high bandwidth output. A typical PZT stack provides a zero-displacement force of several hundred newtons, can perform step movements with a resolution on the order of a nanometer, and has a bandwidth on the order of a kilohertz. The primary inadequacy of the PZT stack as an actuator is that the strain-based deformations that it provides are limited to approximately 0.1%. A stack that is twenty millimeters in length can therefore provide approximately 20 microns of unloaded displacement. Aside from the limited displacement, another drawback of a PZT stack actuator is that, due to the elasticity of the material, the maximum output force decreases linearly with increasing stroke. Still another limitation of the PZT stack is that it cannot exert (significant) tensile force, and thus can exert significant power in a single direction only. This characteristic is often compensated by preloading the stack in compression and utilizing the preloading mechanism to exert force along the compressive direction.

In order to increase the displacement output of a PZT stack, many devices that incorporate stack actuators utilize a flexure-based mechanism that provides amplification of the stack motion. Examples of PZT stacks that utilize flexure mechanisms are those by Arai et al. [34], Goldfarb and Celanovic [35], Nishimura [36], and Yang and Jouaneh [37].

Piezoelectric stacks have also been incorporated into the design of linear and rotary motors. The two primary types of PZT stack-type motors are the inchworm stepper type and the inertial stepper type. The inchworm type stepper motor consists of two stack-actuated clamps separated by an orthogonal set of stacks, as shown in Figure 9. In order to move the motor along a shaft, the stacks are actuated in a sequential manner so that the rear set of stacks clamp the shaft, the orthogonal stacks expand and push the front set along the shaft, the front set clamp, the rear set unclamp, the orthogonal set contract and pull the rear set forward, and the sequence repeats. Minimum step sizes on the order of nanometers can be achieved with position feedback. Maximum motor speeds are on the order of several millimeters per second, since the maximum step size is on the order of several microns and the excitation frequency can approach several kilohertz. Examples of linear motors are those by Burleigh, Inc. [38] and Pandell and Garcia [39], and examples of rotary motors are those by Ohnishi et al. [40] and Duong and Garcia [41].

Another type of PZT stack-actuated motor is the inertial stepper type, as illustrated in Figure 10. These motors utilize the combined characteristics of Coulomb (i.e., stick-slip) friction and inertia to provide steps on the order of a nanometer to a micron over a large range. A step is achieved by slowly extending the PZT stack (and the proof mass) so that friction holds the motor in place, then rapidly contracting the stack to overcome friction and enable slip, and thus motion, between the motor and ground. These motors are relatively simple and require few actuators, but are limited to relatively small holding forces. Examples of these types of motors are those by Higuchi and Yamagata [42], Higuchi et al. [43], and Isobe et al. [44].

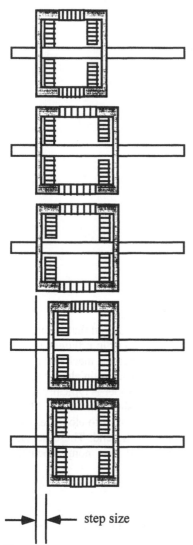

step size

Figure 9 Piezoelectric stack driven inchworm type stepper motor.

Piezoelectric stack actuators are widely available as commercial products from several manufacturers and distributors, and are available in several sizes and configurations.

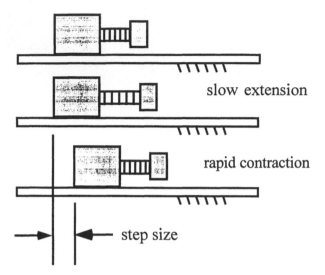

slow extension

rapid contraction

step size

Figure 10 Piezoelectric stack driven inertial-type stepper motor.

Piezoelectric Bending Actuators

A piezoelectric bender is constructed from piezoelectric material that is manufactured so that the direct and inverse piezoelectric effects occur in orthogonal directions. Application of an electric field in one direction generates mechanical stress in the orthogonal directions, and application of a mechanical strain along the latter direction generates an electrical charge in the former. A bender is typically formed by bonding a thin sheet of this material to a substrate, as shown in Figure 11. Application of an electrical charge to the PZT will cause the sheet to elongate, which will generate bending in the beam and displacement at the end. These actuators typically utilize either a single piezoelectric sheet bonded to a nonpiezoelectric

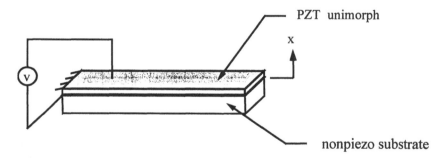

Figure 11 Piezoelectric unimorph-type actuatuor bonded to a nonpiezoelectric substrate.

substrate, which is often called a unimorph, or two piezoelectric sheets bonded together, which is called a bimorph. A bimorph is generally configured in either a parallel or serial electrical configuration, as illustrated in Figure 12. In both configurations, the polarity of the bimorph is arranged so that application of a voltage generates tension in one sheet and compression in the other. Both configurations are mechanically equivalent, though the parallel requires half the electrical voltage and twice the charge for the same mechanical output.

The characteristics of a PZT unimorph are determined in large part by the nonpiezoelectric substrate to which the PZT sheet is bonded, which is typically a metal. A thick metal substrate will typically provide high-bandwidth, low-displacement movement, while a thin substrate will provide the opposite.

One application of unimorph-type actuation is the ultrasonic-type PZT linear motor, which is used to provide focusing motion in several auto-focus cameras. The principle of operation of this type of motor is illustrated in Figure 13. The PZT unimorph actuators are epoxied in specific locations to the bottom of an elastic bar (e.g., a low modulus metal, such as copper). The actuators are excited at high frequencies (e.g., tens of kilohertz, hence the name) with a (typically 90 degrees) phase difference between, which establishes a traveling wave in the elastic bar. A slider then rides on the crests of the traveling wave, driven by the friction

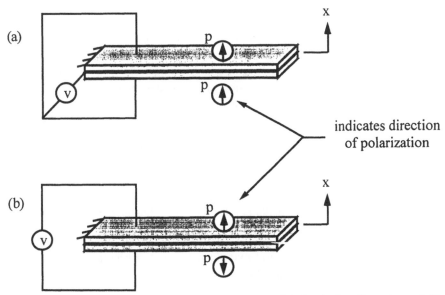

Figure 12 Piezoelectric bimorph-type actuator (a) parallel electrical configuration and (b) series electrical configuration.

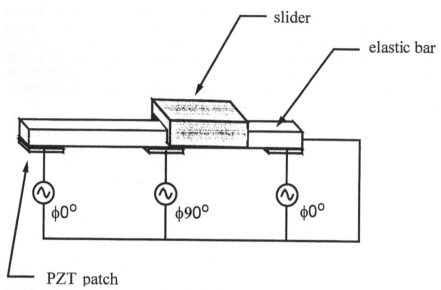

Figure 13   Piezoelectric unimorph-driven ultrasonic-type motor.

between the slider and elastic bar, in a somewhat analogous fashion as a surfer would ride an ocean wave. Examples and descriptions of these devices are those by Hirata and Ueha [45], Seemann [46], Toyoda and Murano [47], Funakubo et al. [48], and Wallaschek [49].

A typical PZT bimorph actuator generates significantly larger displacements than a PZT stack, though at the expense of smaller forces and lower bandwidth output. Specific output characteristics are highly dependent upon actuator geometry. A PZT bender actuator that measures approximately 5 cm in length by 2.5 cm wide by 0.5 mm thick (a typical geometry) can provide a zero-displacement force of a couple of hundred millinewtons, displacements on the order of a couple of millimeters, and has a (no load) bandwidth on the order of tens of Hertz. As is the case with stack actuators, the bimorph force decreases linearly with increasing stroke due to the elasticity of the material. Also, bimorphs tend to exhibit more pronounced static hysteresis than do stack actuators. A comparison of several piezoelectric bending actuators is given in Kugel et al. [50]. Examples of devices that utilize piezoelectric bimorph actuation are those by Chonan et al. [51], Ellis [52], Gaucher [53], Kanno and Okabe [54] and Seki [55].

Piezoelectric unimorph and bimorph actuators are widely available as commercial products from several manufacturers and distributors, and are available in several sizes and configurations.

B. Shape Memory Alloy

Shape memory alloys (SMAs) provide actuation via the shape memory effect (SME), which is due to a phase transformation between austenitic and martensitic crystalline structures. The alloy assumes a martensitic crystalline structure at low temperatures, and once heated to a transition temperature, the crystalline structure transforms to austenite. As the austenitic structure is more compact than the martensite, the material contracts (up to several percent strain) during the transformation, and thus provides actuation. The alloy can be reset to its expanded martensitic form by cooling to a temperature below the transition temperature while also exerting a tensile load. The extent to which the alloy will regain its original form depends upon the amount of tensile load imposed while cooling. Also, the transformation between crystal states is not instantaneous, but rather occurs across a temperature band. The transformation can thus be controlled proportionally so the actuator can be utilized in a proportional manner.

The most common SMA actuator is a nickel-titanium alloy, referred to as NiTi, which is most commonly utilized in a wire form. Heating the alloy is typically achieved by running a current through the wire. Cooling is typically achieved by conduction and/or convection, and the tensile load is applied by preloading the wire against an elastic structure, or by pulling with an antagonist SMA wire. Relative to a PZT stack, these actuators exert slightly greater actuation stress and an order of magnitude more displacement, but are a couple orders of magnitude lower in bandwidth. A typical SMA (tensile) actuation stress is about 150 N/mm^2 (MPa), typical strain is about 2%, and typical bandwidth less than 1 Hz. For example, Neukom et al. [56] reports a NiTi wire of 0.156 mm diameter subjected to 300 mA of current will contract about 4% in 2 seconds while pulling a load of 2.5 N. When the current is discontinued, the wire will cool sufficiently in 2 seconds to re-extend under a tensile load of 0.25 N. It should be noted, however, that the bandwidth can be increased significantly by actively cooling the wire. Peine et al. [57] report that bandwidths of up to 50 Hz can be achieved when incorporating liquid cooling of SMA wire.

With respect to scaling, the phase transformation from martensite to austenite is characterized by a strain, which is a dimensionless quantity. The achievable displacement is therefore linearly related to the size of the actuator. Since the material is linearly elastic, the stress generated (which is an intensive property) is proportional to the strain. The actuation force is proportional to this stress and inversely proportional to the area. Since area diminishes with the second power of size, so does the actuation force. Thus, like the PZT, an SMA actuator smaller by a factor of N will exert an actuation force that is smaller by a factor of N^2.

Though the actuation force becomes proportionately smaller at small scales, the bandwidth (i.e., speed of response) becomes greater. The limiting factor in the speed of SMA actuators is typically the cooling time. Since conduction is proportional to surface area and heat is proportional to material volume (for a given temperature), the cooling rate is proportional to the ratio, which increases linearly with decreasing scale. The bandwidth of these actuators therefore will also increase linearly with decreasing scale. Even for very small scales, however, the bandwidth may be insufficient for a given application, depending on the specific actuation requirements. Recall that the 0.5 Hz bandwidth reported by Nuekom is for a 0.156 mm diameter wire, which is quite small, even with respect to present micromanu-facturing standards. Besides the limited bandwidth, shape memory alloys are also characterized by extremely low efficiency. The typical SMA actuator is about 1% efficient. Most of the inefficiency is a result of using electrical current in the SMA to generate heat, which is not directly utilized for mechanical work.

Some studies of SMA for purposes of microactuation are those by Bergamasco et al. [58], Hunter et al. [59], Lin et al. [60], and Neukom et al. [56]. Examples of devices that utilize SMA actuators are those by Krulevitch et al. [61], Kuribayashi et al. [62], and Russell [63].

SMA, and in particular NiTi thin-wire actuators, are widely available as commercial products from several manufacturers and distributors, and are available in several wire diameters.

C. Magnetostrictive Actuation

A *magnetostrictive alloy* is a material that generates a mechanical strain when subjected to a magnetic field and conversely generates a magnetic flux when subjected to a mechanical stress. Magnetostrictive behavior is closely analogous to piezoelectric behavior. As with piezoelectric materials, an IEEE standards committee [64] formulated linearized constitutive equations for magnetostrictive materials, which are expressed in a similar compressed matrix notation as:

$$S_p = s_{pq}^H T_q + d_{kp} H_k$$
$$B_i = d_{iq} T_q + \mu_{ik}^T H_k$$

where S_p represents an element of the (compressed) strain tensor, s_{pq}^H an element of the elastic compliance matrix when subjected to a constant magnetic field, T_q an element of the (compressed) stress tensor, d_{kp} an element of a matrix of magnetostrictive material constants, H_k an element of the magnetic field vector, B_i an element of the magnetic flux vector, and μ_{ik}^T an element of the magnetic permeability matrix measured at a constant stress. As in the piezoelectric constitutive equations, the compressed notation eliminates redundant terms by representing the symmetric stress and strain tensors with single column vectors that incorporate elements representing both the diagonal and off-diagonal tensor terms. These equations essentially state that the material strain and magnetic flux exhibited by a magnetostrictive alloy are both linearly affected by the mechanical stress and magnetic field to which the alloy is subjected. The actuation force is thus a linear combination of an elastic component and a magnetostrictive component. Assuming a constant magnetic field, a dimensional analysis indicates that a magnetostrictive actuator smaller by a factor of N will exhibit both a magnetostrictive component of force and an elastic component smaller by a factor of N^2, and thus the total actuation force will scale as N^2. Like piezoelectric ceramics, a magnetostrictive alloy will exhibit a static hysteresis between the magnetic field and the material strain. The magnetic field versus mechanical strain relationship of magnetostrictive alloys is typically S-shaped and considerably less linear than the electric field versus mechanical strain relation of piezoelectrics. Another difference is that magnetostriction in the absence of a magnetic bias generates tensile strain only (i.e., no compressive strain), regardless of the direction of applied magnetic field. This effect can be compensated, however, by applying a biasing magnetic field, either with a permanent magnet or with a bias current in the electromagnet, and thus both tensile and compressive strains can be achieved. As with PZT, the maximum compressive strains are typically a fraction of the maximum tensile strains. Also, like PZT, magnetostrictive materials exhibit significant hysteresis between applied magnetic field and displacement. Accurate positioning therefore requires closed-loop position feedback.

The material actuation properties of magnetostrictive materials compare favorably to those of piezoelectric materials. Magnetostrictives exhibit comparable force and similar bandwidth to piezoelectrics, and significantly, twice the strain. In addition to the increased strain, one possible advantage of magnetostrictive actuators is that no wires need be attached to the actuator. This property was utilized by Fukuda et al. [65] in the design of a mobile microrobot that travels inside a pipe. The drawback to a magnetostrictive actuator, however, is that the magnetic field required for magnetostriction typically requires an additional device, usually a coil driven by an electrical current. Actuation of such coils is typically not efficient, does not scale favorably, and involves inductive effects that limit bandwidth. Conversely, the electric field required for piezoelectric actuation requires only electrodes, is highly efficient, and does not entail any significant dynamics.

Magnetostrictive alloys, and in particular one called Terfenol-D, are available as commercial products from a few manufacturers.

D. Electrostrictive Actuation

Electrostrictive materials are also quite similar to piezoelectric ceramics. As with piezoelectric materials, electrostrictive materials exhibit mechanical strain when subjected to an electrical field. The mechanism of transduction, however, is of a fundamentally different nature, and as such, the strain in an electrostrictive material is proportional to the square of the applied electric field, rather than linearly related as in the piezoelectric material. Like magnetostrictive materials, the electrostrictives generate tensile strain only (i.e., no compressive strain), regardless of the direction of applied electric field. The most common *electrostrictive material* is a lead-magnesium-niobate ceramic, which is typically referred to as PMN. These materials exhibit strain, force and bandwidth characteristics comparable to PZT, and also scale similarly (i.e., force decreases as square of geometric scaling).

The most significant advantages of the electrostrictive PMN is that it exhibits significantly less hysteresis and creep than the PZT. PMN typically exhibits about 2% hysteresis and 3% creep, as opposed to PZT, which exhibits about 15% of each. The PMN may therefore be preferable to PZT when utilizing the actuator for open-loop positioning. The most significant weakness, however, is that unlike piezoceramics, PMN exhibits a strong temperature-strain dependence. In fact, Galvagni [66] reports that a variation of 50°C can result in a 50% change in material strain, and as such recommends limiting the range of operating temperatures between zero and 40°C. Such operation is most likely only possible at low actuation frequencies, since higher frequency operation would generate increased mechanical dissipation, which would in turn heat the ceramic.

Some studies of PMN for purposes of actuation are those by Galvagni [66] and Uchino [67]. In addition to ceramic-type electrostrictive materials, Pelrine et al. [68] developed a polymeric electrostrictive material that produces strains on the order of 10%, actuation stresses on the order of 2 N/mm^2 (MPa) and a bandwidth on the order of 10 Hz.

Electrostrictive materials, and in particular PMN ceramics, are available on the commercial market from several manufacturers and distributors.

E. Photostrictive Actuation

A *photostrictive material* is one in which incident light generates a mechanical strain. The most common of these materials is a lead-lanthanum-zirconium-titanium ceramic, referred to as PLZT. The photostrictive effect is a combination of a photovoltaic effect and a piezoelectric effect. Rather than light being converted directly to mechanical strain, the incident light generates a voltage in the material (the photovoltaic effect), which in turn generates a mechanical strain (the piezoelectric effect). The most significant advantage of the PLZT is that no wires are necessary for actuation. The disadvantages, however, are that the PLZT is capable of only 0.01% strain (versus 0.1% of PZT), has a speed of response of a couple of seconds (versus milliseconds for the PZT) and is not capable of significant actuation forces. Additionally, since strain excitation requires light, the geometry is restricted to thin sheets.

These materials are still a topic of research and thus are not yet available on the commercial market. For a study of actuation characteristics and example devices using PLZT, refer to Chu and Uchino [69].

IV. MICROSENSORS

Since microsensors do not transmit power, the scaling of force is not typically significant. As with conventional-scale sensing, the qualities of interest are high resolution, absence of drift and hysteresis, achieving a sufficient bandwidth, and immunity to extraneous effects not being measured.

Microsensors are typically based on either measurement of mechanical strain, measurement of mechanical displacement, or on frequency measurement of a structural resonance. The former two types are in essence analog measurements, while the latter is in essence a binary-type measurement, since the sensed quantity is typically the frequency of vibration. Since the resonant-type sensors measure frequency instead of amplitude, they are generally less susceptible to noise and thus typically provide a higher resolution measurement. According to Guckel et al. [70], resonant sensors provide as much as one hundred times the resolution of analog sensors. They are also, however, more complex and are typically more difficult to fabricate.

The primary form of strain-based measurement is piezoresistive, while the primary means of displacement measurement is capacitive. The resonant sensors require both a means of structural excitation as well as a means of resonant frequency detection. Many combinations of transduction are utilized for these purposes, including electrostatic excitation, capacitive detection, magnetic excitation and detection, thermal excitation, and optical detection.

A. Strain

Many microsensors are based upon strain measurement. The primary means of measuring strain is via piezoresistive strain gauges, which is an analog form of measurement. *Piezoresistive* strain gauges, also known as semiconductor gauges, change resistance in response to a mechanical strain. Note that piezoelectric materials can also be utilized to measure strain. Recall that mechanical strain will induce an electrical charge in a piezoelectric ceramic. The primary problem with using a piezoelectric material, however, is that since measurement circuitry has limited impedance, the charge generated from a mechanical strain will gradually leak through the measurement impedance. A piezoelectric material therefore cannot provide reliable steady-state signal measurement. In contrast, the change in resistance of a piezoresistive material is stable and easily measurable for steady-state signals. One problem with piezoresistive materials, however, is that they exhibit a strong strain-temperature dependence, and so must typically be thermally compensated.

An interesting variation on the silicon piezoresistor is the resonant strain gauge proposed by Ikeda et al. [71], which provides a frequency-based form of measurement that is less susceptible to noise. The resonant strain gage is a beam that is suspended slightly above the strain member and attached to it at both ends. The strain gage beam is magnetically excited with pulses, and the frequency of vibration is detected by a magnetic detection circuit. As the beam is stretched by mechanical strain, the frequency of vibration increases. These sensors provide higher resolution than typical piezoresistors and have a lower temperature coefficient. The resonant sensors, however, require a complex three-dimensional fabrication technique, unlike the typical piezoresistors that require only planar techniques.

B. Pressure

One of the most commercially successful microsensor technologies is the pressure sensor. Silicon micromachined *pressure sensors* are available that measure pressure ranges from around one to several thousand kPa, with resolutions as fine as one part in ten thousand. These sensors incorporate a silicon micromachined diaphragm that is subjected to fluid (i.e., liquid or gas) pressure which causes dilation of the diaphragm. The simplest of these utilize piezoresistors mounted on the back of the diaphragm to measure deformation, which is a function of the pressure. Examples of these devices are those by Fujii et al. [72] and Mallon et al. [73]. A variation of this configuration is the device by Ikeda et al. [74], which utilizes, instead of a piezoresistor to measure strain, an electromagnetically driven and sensed resonant strain gage, as discussed in the previous section. Still another variation on the same theme is the capacitive measurement approach, which measures the capacitance between the diaphragm and an electrode that is rigidly mounted and parallel to the diaphragm. An example of this approach is by Nagata et al. [75]. A more complex approach to pressure measurement is that by Stemme and Stemme [76], which utilizes resonance of the diaphragm to detect pressure. In this device, the diaphragm is capacitively excited and optically detected. The pressure imposes a mechanical load on the diaphragm, which increases the stiffness and in turn the resonant frequency.

C. Acceleration

Another commercially successful microsensor is the silicon microfabricated *accelerometer*, which in various forms can measure acceleration ranges from well below one to around a thousand m/s^2 (i.e., sub-g to several hundred g's), with resolutions of one part in ten thousand. These sensors incorporate a micromachined suspended proof mass that is subjected to an inertial force in response to an acceleration, which causes deflection of the supporting flexures. One means of measuring the deflection is by utilizing piezoresistive strain gages mounted on the flexures. The primary disadvantage to this approach is the temperature sensitivity of the piezoresistive gages. An alternative to measuring the deflection of the proof mass is via capacitive sensing. In these devices, the capacitance is measured between the proof mass and an electrode that is rigidly mounted and parallel. Examples of this approach are those by Boxenhorn and Greiff [77], Leuthold and Rudolf [78], and Seidel et al. [79]. Still another means of measuring the inertial force on the proof mass is by measuring the resonant frequency of the supporting flexures. The inertial force due to acceleration will load the flexure, which will alter its resonant frequency. The frequency of vibration is therefore a measure of the acceleration. These types of devices utilize some form of transduction to excite the structural resonance of the supporting flexures, and then utilize some other measurement technique to detect the frequency of vibration. Examples of this type of device are those by Chang et al. [80], which utilizes electrostatic excitation and capacitive detection, and by Satchell and Greenwood [81], which utilizes thermal excitation and piezoresistive detection. These types of accelerometers entail additional complexity, but typically offer improved measurement resolution. Still another variation of the micro-accelerometer is the force-balanced type. This type of device measures the position of the proof mass (typically by capacitive means) and utilizes a feedback loop and electrostatic or electromagnetic actuation to maintain zero deflection of the mass. The acceleration is then a function of the actuation effort. These devices are characterized by a wide bandwidth and high sensitivity, but are typically more complex and more expensive than other types. Examples of force-balanced devices are those by Chau et al. [82] and Kuehnel and Sherman [83], both of which utilize capacitive sensing and electrostatic actuation.

D. Force

Silicon microfabricated force sensors incorporate measurement approaches much like the microfabricated pressure sensors and accelerometers. Various forms of these force sensors can measure force ranges on the order of milliNewtons to newtons, with resolutions of one part in ten thousand. Mechanical sensing typically utilizes a beam or a flexure support which is elastically deflected by an applied force, thereby transforming force measurement into measurement of strain or displacement, which can be accomplished by piezoresistive or capacitive means. An example of this type of device is that of Despont et al. [84], which utilizes capacitive measurement. Higher resolution devices are typically of the resonating beam type, in which the applied force loads a resonating beam in tension. Increasing the applied tensile load results in an increase in resonant frequency. An example of this type of device is that of Blom et al. [85].

E. Angular Rate Sensing (Gyroscopes)

A conventional-scale *gyroscope* utilizes the spatial coupling of the angular momentum-based gyroscopic effect to measure angular rate. In these devices, a disk is spun at a constant high rate about its primary axis, so that when the disk is rotated about an axis not colinear with the primary (or spin) axis, a torque results in an orthogonal direction that is proportional to the angular velocity. These devices are typically mounted in gimbals with low-friction bearings, incorporate motors that maintain the spin velocity and utilize strain gages to measure the gyroscopic torque (and thus angular velocity). Such a design would not be appropriate for a microsensor due to several factors, some of which include the diminishing effect of inertia (and thus momentum) at small scales, the lack of adequate bearings, the lack of appropriate micromotors and the lack of an adequate three-dimensional microfabrication processes. Instead, micro-scale angular rate sensors are of the vibratory type, which incorporate Coriolis-type effects rather than the angular momentum-based gyroscopic mechanics of conventional-scale devices. A Coriolis acceleration results from linear translation within a coordinate frame that is rotating with respect to an inertial reference frame. In particular, if the particle in Figure 14 is moving with a velocity v within the frame xyz, and if the frame xyz is

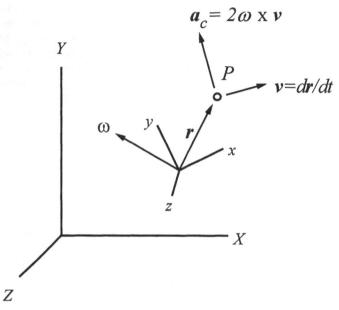

Figure 14   Illustration of Coriolis acceleration, which results from translation within a reference frame that is rotating with respect to an inertial reference frame.

rotating with an angular velocity of ω with respect to the inertial reference frame XYZ, then a Coriolis acceleration will result equal to $a_c = 2\omega \times v$. If the object has a mass m, a Coriolis inertial force will result equal to $F_c = -2m\omega \times v$ (minus sign because direction is opposite a_c). A vibratory gyroscope utilizes this effect as illustrated in Figure 15. A flexure-suspended inertial mass is vibrated in the x-direction, typically with an electrostatic comb-drive. An angular velocity about z-axis will generate a Coriolis acceleration, and thus force, in the y-direction. If the "external" angular velocity is constant and the velocity in the x-direction is sinusoidal, then the resulting Coriolis force will be sinusiodal, and the suspended inertial mass will vibrate in the y-direction with an amplitude proportional to the angular velocity. The motion in the y-direction, which is typically measured capacitively, is thus a measure of the angular rate. Examples of these types of devices are those by Bernstein et al. [86] and Oh et al. [87]. Note that though vibration is an essential component of these devices, they are not technically resonant sensors, since they measure amplitude of vibration rather than frequency.

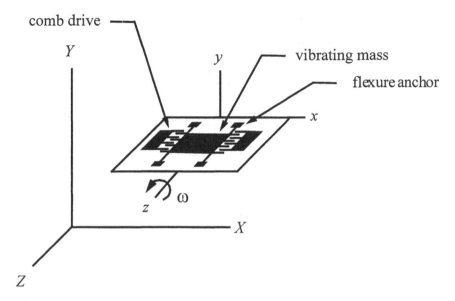

Figure 15 Schematic of a vibratory gyroscope.

REFERENCES

1. PW Bridgman. Dimensional Analysis. 2nd ed. New Haven: Yale University Press, 1931.
2. E. Buckingham. On physically similar systems: Illustrations of the use of dimensional equations. Physical Review 4(4):345–376, 1914.
3. HE Huntley. *Dimensional Analysis*. New York: Dover Publications, 1967.
4. HL Langhaar. *Dimensional Analysis and Theory of Models*. New York: Wiley, 1951.
5. ES Taylor. *Dimensional Analysis of Engineers*. Oxford: Clarendon Press, 1974.
6. JN Israelachvili. *Intermolecular and Surface Forces*. London: Academic Press, 1985, 9–10.
7. RS Fearing. Microactuators for microrobots: electric and magnetic. Workshop on Micromechatronics, IEEE International Conference on Robotics and Automation, 1997.
8. SM Bobbio, MD Keelam, BW Dudley, S Goodwin-Hohansson, SK Jones, JD Jacobson, FM Tranjan, TD Dubois. Integrated force arrays. In: Proceedings of the IEEE Micro Electro Mechanical Systems, 149–154, 1993.
9. JD Jacobson, SH Goodwin-Johansson, SM Bobbio, CA Bartlett, LN Yadon. Integrated force arrays: theory and modeling of static operation. Journal of Microelectromechanical Systems 4(3):139–150, 1995.
10. M Yamaguchi, S Kawamura, K Minami, M Esashi. Distributed electrostatic micro actuators. In: Proceedings of the IEEE Micro Electro Mechanical Systems, 18–23, 1993.
11. CJ Kim, AP Pisano, RS Muller. Silicon-processed overhanging microgripper. Journal of Microelectromechanical Systems 1(1):31–36, 1992.

12. T Matsubara, M Yamaguchi, K Minami, M Esashi. Stepping electrostatic microactuator. International Conference on Solid-State Sensor and Actuators, 50–53, 1991.
13. T Niino, S Egawa, H Kimura, T Higuchi. Electrostatic artificial muscle: compact, high-power linear actuators with multiple-layer structures. In: Proceedings of the IEEE Conference on Micro Electro Mechanical Systems, 130–135, 1994.
14. JB Huang, PS Mao, QY Tong, RQ Zhang. Study on silicon electrostatic and electroquasistatic micromotors. Sensors and Actuators 35:171–174, 1993.
15. M Mehragany, SF Burt, LS Tavrow, JH Lang, SD Senturia, MF Schlecht. A study of three microfabricated variable-capacitance motors. Sensors and Actuators, 173–179, 1990.
16. W Trimmer, K Gabriel. Design considerations for a practical electrostatic micromotor. Sensors and Actuators 11:189–206, 1987.
17. M Mehregany, P Nagarkar, SD Senturia, JH Lang. Operation of microfabricated harmonic and ordinary side-drive motors. In: Proceedings of the IEEE Conference on Micro Electro Mechanical Systems, 1–8, 1990.
18. VR Dhuler, M Mehregany, SM Phillips. A comparative study of bearing designs and operational environments for harmonic side-drive micromotors. IEEE Transactions on Electron Devices 40(11):1985–1989, 1993.
19. RH Price, JE Wood, SC Jacobsen. Modeling considerations for electrostatic forces in electrostatic microactuators. Sensors and Actuators 20:107–114, 1989.
20. W Trimmer, R Jebens. An operational harmonic electrostatic motor. In: Proceedings of the IEEE Conference on Micro Electro Mechanical Systems, 13–16, 1989.
21. W Trimmer, R Jebens. Harmonic electrostatic motors. Sensors and Actuators 20:17–24, 1989.
22. T Furuhata, T Hirano, LH Lane, RE Fontanta, LS Fan, H Fujita. Outer rotor surface micromachined wobble micromotor. In: Proceedings of the IEEE Conference on Micro Electro Mechanical Systems, 161–166, 1993.
23. W Trimmer, R Jebens. Actuators for microrobots. IEEE Conference on Robotics and Automation, 1547–1552, 1989.
24. W Trimmer. Microrobots and micromechanical systems. Sensors and Actuators 19:267–287, 1989.
25. T Inoue, Y Hamaskai, I Shimoyama, H Miura. Micromanipulation using a microcoil array. In: Proceedings of the IEEE International Conference on Robotics and Automation, 2208–2213, 1996.
26. C Liu, T Tsao, Y Tai, C Ho. Surface micromachined magnetic actuators. In: Proceedings of the IEEE Conference on Micro Electro Mechanical Systems, 57–62, 1994.
27. Y Shinozawa, T Abe, T Kondo. A proportional microvalve using a bi-stable magnetic actuator. In: Proceedings of the IEEE Conference on Micro Electro Mechanical Systems, 233–237, 1997.
28. B Wagner, W Benecke. Microfabricated actuator with moving permanent magnet. In: Proceedings of the IEEE Conference on Micro Electro Mechanical Systems, 27–32, 1991.
29. K Yanagisawa, A Tago, T Ohkubo, H Kuwano. Magnetic micro-actuator. In: Proceedings of the IEEE Conference on Micro Electro Mechanical Systems, 120–124, 1991.
30. J Judy, RS Muller, HH Zappe. Magnetic microactuation of polysilicon flexure structures. Journal of Microelectromechanical Systems 4(4):162–169, 1995.
31. CH Ahn, YJ Kim, MG Allen. A planar variable reluctance magnetic micromotor with fully integrated stator and wrapped coils. In: Proceedings of the IEEE Conference on Micro Electro Mechanical Systems, 1–6, 1993.

32. H Guckel, TR Christenson, KJ Skrobis, TS Jung, J Klein, KV Hartojo, I Widjaja. A first functional current excited planar rotational magnetic micromotor. In: Proceedings of the IEEE Conference on Micro Electro Mechanical Systems, 7–11, 1993.
33. Standards Committee of the IEEE Ultrasonics, Ferroelectrics, and Frequency Control Society, 1987. An American National Standard: IEEE Standard on Piezoelectricity, The Institute of Electrical and Electronics Engineers, ANSI/IEEE Std, 176, 1987, New York.
34. F Arai, D Andou, T Fukuda. Adhesion forces reduction for micro manipulation based on micro physics. In: Proceedings of the IEEE Conference on Micro Electro Mechanical Systems, 354–359, 1996.
35. M Goldfarb, N Celanovic. Minimum surface effect microgripper design for force-reflective telemanipulation of a microscopic environment. In: Proceedings of the ASME International Mechanical Engineering Conference and Exposition, DSC 58:477–482, 1996.
36. K Nishimura. A spring-guided micropositioner with linearized subnanometer resolution. Rev Sci Instrum 62(8):2004–2007, 1991.
37. R Yang, M Jouaneh. Design and analysis of a low profile micro-positioning stage. In: Proceedings of the ASME Winter Annual Meeting, PED 58:131–142, 1992.
38. Burleigh, Inc. The Piezo Book. Product Literature.
39. T Pandell, E Garcia. Design of a piezoelectric caterpillar motor. Proceedings of 1996 ASME International Mechanical Engineering Congress and Exposition ARE 52:627–648, 1996.
40. K Ohnishi, M Umeda, M Kurosawa, S Ueha. Rotary inchworm-type piezoelectric actuator. Electrical Engineering in Japan 110(3):107–114, 1990.
41. K Duong, E Garcia. Development of a rotary inchworm piezoelectric motor. Proceedings of SPIE—The International Society for Optical Engineering Smart Structures and Materials 2443:782–788, 1995.
42. T Higuchi, Y Yamagata. Precise positioning mechanism utilizing rapid deformations of piezoelectric elements (2nd report)—motion characteristics with enhanced friction. Journal of the Japan Society of Precision Engineering 58(10):1759–1764, 1992.
43. T Higuchi, Y Yamagata, K Furutani, K Kudoh. Precise positioning mechanism utilizing rapid deformations of piezoelectric elements. In: Proceedings of the IEEE Micro Electro Mechanical Systems, 222–226, 1990.
44. H Isobe, T Moriguchi, A Kyusojin. Development of piezoelectric xy positioning device using impulsive force. Journal of the Japan Society for Precision Engineering 62(4):574–578, 1996.
45. H Hirata, S Ueha. Design of a traveling wave type ultrasonic motor. IEEE Transactions on Ultrasonics, Ferroelectrics, and Frequency Control 42(2):225–231, 1995.
46. W. Seemann. Ultrasonic traveling wave linear motor with improved efficiency. Proceedings of SPIE—The International Society for Optical Engineering 2717:554–564, 1996.
47. J Toyoda, K Murano. Small-size ultrasonic linear motor. Meeting of Ferroelectric Materials and Their Applications 30(9B):2274–2276, 1991.
48. T Funakubo, T Tsubata, Y Taniguchi, K Kumei, T Fujimura, C Abe. Ultrasonic linear motor using multilayer piezoelectric actuators. Proceedings of the 15th Symposium on Ultrasonic Electronics 34(5B):2756–2759, 1994.
49. J Wallaschek. Piezoelectric ultrasonic motors. Journal of Intelligent Material Systems and Structures 6(1):71–83, 1995.
50. VD Kugel, S Chandran, LE Cross. A comparative analysis of piezoelectric bending-mode actuators. Proceedings of the SPIE 3040:70–80, 1997.

51. S Chonan, Z Jiang, S Sakuma. Force control of a miniature gripper driven by piezoelectric bimorph cells. Transactions of the Japan Society of Mechanical Engineers 59(557):150–157, 1993.
52. GW Ellis. Piezoelectric micromanipulators. Science 138:84–91, 1962.
53. P Gaucher. Piezoelectric and electrostrictive bimorph actuators and sensors for smart microsystems. Proceedings of SPIE Conference on Smart Structures and Materials 2779:610–615, 1996.
54. M Kanno, H Okabe. X-Y stage based on piezo-bimorph-elements. Ferroelectrics Proceedings of the European Meeting on Ferroelectricity 186(4):269–272, 1996.
55. H Seki. Piezoelectric bimorph microgripper capable of force sensing and compliance adjustment. Proceedings of the ASME Japan//USA Symposium on Flexible Automation, 707–713, 1992.
56. PA Neukom, HP Bornhauser, T Hochuli, R Paravicini, G Schwarz. Characteristics of thin-wire shape memory actuators. Sensors and Actuators 247–252, 1990.
57. WJ Peine, PS Wellman, RD Howe. Temporal bandwidth requirements for tactile shape displays. Proceedings of the ASME International Mechanical Engineering Congress and Exposition 61:107–114, 1997.
58. M Bergamasco, P Dario, F Salsedo. Shape memory alloy microactuators. Sensors and Actuators 253–257, 1990.
59. IW Hunter, S Lafontaine, JM Hollerbach, PJ Hunter. Fast reversible NiTi fibers for use in microrobotics. IEEE Conference on Robotics and Automation, 166–170, 1991.
60. G Lin, D Yang, D Yu, RO Warrington. A feasibility study of thermal controlled shape memory alloy for application to micro robots. Proceedings of the ASME Winter Annual Meeting, 1–10, 1993.
61. P Krulevitch, AP Lee, PB Ramsey, JC Trevino, J Hamilton, MA Northrup. Thin film shape memory alloy microactuators. Journal of Microelectromechanical Systems 5(4):270–282, 1996.
62. K Kuribayashi, S Shimizu, T Nishinohara, and T Taniguchi. Trial fabrication of micron seized arm using reversible TiNi alloy film actuators. Proceedings of the IEEE/RSJ Conference on Intelligent Robotics and Systems, 1697–1702.
63. RA Russell. A robotic system for performing sub-millimeter grasping and manipulation tasks. Robotics and Autonomous Systems 13:209–218, 1994.
64. IEEE Standard on Magnetostrictive Materials: Piezomagnetic Nomenclature, IEEE Std., p. 319, 1990.
65. Fukuda et al. Giant magnetostrictive alloy (GMA) applications to micro mobile robot as a micro actuator without power supply cables. Proceedings of the IEEE Micro Electro Mechanical Systems, 210–215, 1991.
66. J Galvagni. Electrostrictive actuators and their use in optical applications. Optical Engineering 29(11):1389–1391, 1990.
67. K Uchino. Electrostatic actuators: materials and applications. Ceramic Bulletin 65(4):647–652, 1986.
68. R Pelrine, R Kornbluh, J Joseph, S Chiba. Electrostriction of polymer films for microactuators. Proceedings of the IEEE Micro Electrical Mechanical Systems, 238–243, 1997.
69. SY Chu, K Uchino. Photostrictive effect in PLZT-based ceramics and its application. Ferroelectric 174(1–2):185–196, 1995.
70. H Guckel, JJ Sneigowski, TR Christenson, F Raissi. The application of fine grained, tensile polysilicon to mechanically resonant transducers. Sensors and Actuators A21–A23:346–351, 1990.

71. K Ikeda, H Kuwayama, T Kobayashi, T Watanabe, T Nishikawa, T Yoshida, K Harada. Silicon pressure sensor integrates resonant strain gauge on diaphragm. Sensors and Actuators A21–A23:146–150, 1990.
72. T Fujii, Y Gotoh, S Kuroyanagi. Fabrication of microdiaphragm pressure sensor utilizing micromachining. Sensors and Actuators A34:217–224, 1992.
73. J Mallon, F Pourahmadi, K Petersen, P Barth, T Vermeulen, J Bryzek. Low-pressure sensors employing bossed diaphragms and precision etch-stopping. Sensors and Actuators A21–A23:89–95, 1990.
74. K Ikeda, H Kuwayama, T Kobayashi, T Watanabe, T Nishikawa, T Yoshida, K Harada. Three-dimensional micromachining of silicon pressure sensor integrating resonant strain gauge on diaphragm. Sensors and Actuators A21–A23:1007–1009, 1990.
75. T Nagata, H Terabe, S Kuwahara, S Sakurai, O Tabata, S Sugiyama, M Esashi. Digital compensated capacitive pressure sensor using cmos technology for low-pressure measurements. Sensors and Actuators A34:173–177, 1992.
76. E Stemme, G Stemme. A balanced resonant pressure sensor. Sensors and Actuators, A21–A23:336–341, 1990.
77. B. Boxenhorn, P Greiff. Monolithic silicon accelerometer. Sensors and Actuators A21–A23:273–277, 1990.
78,. H Leuthold, F Rudolf. An ASIC for high-resolution capacitive microaccelerometers. Sensors and Actuators A21–A23:278–281, 1990.
79. H Seidel, H Riedel, R Kolbeck, G Muck, W Kupke, M Koniger. Capacitive silicon accelerometer with highly symmetrical design. Sensors and Actuators A21–A23:312–315, 1990.
80. SC Chang, MW Putty, DB Hicks, CH Li, RT Howe. Resonant-bridge two-axis microaccelerometer. Sensors and Actuators A21–A23:342–345, 1990.
81. DW Satchell, JC Greenwood. A thermally-excited silicon accelerometer. Sensors and Actuators A17:241–45, 1989.
82. KHL Chau, SR Lewis, Y Zhao, RT Howe, SF Bart, RG Marchesilli. An integrated force-balanced capacitive accelerometer for low-g applications. Sensors and Actuators A54:472–476, 1996.
83. W Kuehnel, S Sherman. A surface micromachined silicon accelerometer with on-chip detection circuitry. Sensors and Actuators A45:7–16, 1994.
84. Despont, GA Racine, P Renaud, NF de Rooij. New design of micromachined capacitive force sensor. Journal of Micromechanics and Microengineering 3:239–242, 1993.
85. FR Blom, S Bouwstra, JHJ Fluitman, M Elwenspoek. Resonating silicon beam force sensor. Sensors and Actuators 17:513–519, 1989.
86. J Bernstein, S Cho, AT King, A Kourepenis, P Maciel, M Weinberg. A micromachined comb-drive tuning fork rate gyroscope. IEEE Conference on Micro Electro Mechanical Systems, 143–148, 1993.
87. Y Oh, B Lee, S Baek, H Kim, J Kim, S Kang, C Song. A surface-micromachined tunable vibratory gyroscope. IEEE Conference on Micro Electro Mechanical Systems, 272–277, 1997.

3
Microcomputer Technology

Brendon Lilly and Ljubo Vlacic

Griffith University, Brisbane, Queensland, Australia

In everyday life, microcomputers surround us, in our homes, our cars and at work. We rely on microcomputers in every aspect of our lives. We use them to cook our food (microwave), wash our dishes (dishwasher) and clean our clothes (microcontrolled washing machine).

The desktop personal computer is common today. This form of microcomputer is one that most people can relate to. Most people today come in contact with this type of computer either in the workplace or at home. This machine can perform tasks as complicated as controlling a car welding robot or as trivial as keeping the time. Microcomputers help us perform mundane tasks that bore us human beings.

Computational problems were the driving force in the design of early computers. Today, computers have become smaller and more powerful. What are these microcomputers and how are they used to control mechanical machines? This chapter explores these ideas and describes the design philosophy of the microcontroller, a form of microcomputer, the intelligent core of a mechatronic system.

This chapter is broken up into five sections. The first section describes the history and the structure of a microcomputer. The second section explains the main differences between the microprocessors found in our desktop personal computers and the microcontroller used in most mechatronic systems. We also look at what the digital signal processor offers to mechatronic systems.

The third section deals with designing a microcontroller system for the outside world. The fourth section looks at some design considerations when prototyping a mechatronic system. Finally, the last section contains case studies of popular microcontrollers used in today's control systems.

I. MICROCOMPUTER ARCHITECTURES

This first section looks at the development of the microcomputer from its conception to what we know of it today. The basic architecture found in most microcomputers is then presented as an introduction before describing design specifics in the following sections.

A. A Brief History of Microcomputers

The first electronic computer was invented in 1946 and used vacuum tubes and relays to perform its operations. Called the ENIAC, its programs were hardwired to execute commands of up to 5000 additions and subtractions in one second. Von Neumann then conceptualized an architecture that could store programs instead of having to hardwire them. From this technology M.V. Wikes implemented the EDSAC, the first operational stored-memory computer.

Based on this idea of a stored memory concept, Von Neumann designed an architecture which is still used in the most powerful computers today. This is known as the Von Neumann architecture [1]. The Von Neumann architecture is based on the idea that the arithmetic and logic operations such as ADD, SUBTRACT, AND, OR etc. are stored as commands in memory.

Early computers based on this architecture, were primarily used for scientific computations. Since they used vacuum tubes and electrostatic storage tubes for memory, they were large and took up a lot of space. In 1953, IBM developed a desktop calculator IBM 701 based on the Von Neumann architecture which contained 2K of electrostatic memory. The second generation computer was first conceptualized in 1948 corresponding to the development of the transistor. This allowed designers to develop smaller and higher speed machines. However, it wasn't until 1954 that the first transistorized computer developed by Bell Laboratories arrived on the scene. It contained 800 transistors and was called the TRADIC.

All these computers at this stage were not programmable and they did not contain an *operating system*. All programs were built in by the manufacturer. In the mid-1950s, assembly languages appeared which allow users to develop their own programs. In 1956, a high level language called FORTRAN (FORmula TRANslation) was developed which made programming of these machines somewhat easier.

The invention of integrated circuits sparked the beginning of the third generation computer. Operating systems became realizable and in the mid-1960s, IBM released the IBM 360 which began the evolution of what we know as a computer today. Printed circuit boards (PCBs) were used for the first time and core memory was replaced by faster and more reliable solid state memories.

In 1971, Intel introduced the first microprocessor (see Figure 1). This 4 bit device contained some 2300 transistors using 10 micron technology. Eight bit microprocessors such as the Intel 8008 and 8080; the Motorola 6800 and 6809; Zilog Z80; and the MOS 6502 were released shortly afterwards.

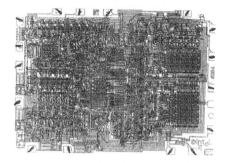

Figure 1 Die photo of the Intel 4004 microprocessor. (Courtesy of Intel Corporation.)

The release of the Intel 8080 in 1974 paved the way for modern day microcomputer architectures found in most personal computers today. This device had 10 times the performance of its predecessor the 8008, and contained only 6000 transistors.

Fourth generation computers were realized with the introduction of large scale integration (LSI). This enabled the designers to pack transistors at higher densities allowing for more features and higher speeds. Different technologies also evolved during this time. P-type metal oxide semiconductor (PMOS) saw the introduction of 16 bit microprocessors with higher speeds and more transistors.

Microcontrollers did not appear on the scene until the 1980s. Intel's 8051 was introduced in 1980 and in 1985, Motorola released the MC68HC11 and Microchip introduced the PIC microcontroller.

Since this time many technologies such as NMOS, HMOS and HCMOS have reduced the size of a transistor and lowered the power consumption of microprocessors and microcontrollers. The number of transistors in a microprocessor has increased from just 800 in the first transistor microprocessor to 7.5 million found in the Intel Pentium II processor.

More recently, digital signal processors (DSP) have been introduced as a high speed, product orientated microprocessor. These devices are revolutionary processors in today's digital age. As technologies improve even further, these devices will become common place just as microcontrollers have for the past 10 or 15 years. Evidence of this is shown by the recent introduction of a microcontroller with a DSP architecture, the Motorola DSP56800 family [2].

The future holds the prospect of higher speed, more functions and better development tools for designers. The next generation microcontroller may be a DSP processor with built-in A/D converters for applications such as remote voice control over the telephone network or portable language translators. Although at this stage, the cost of the processing power required for this sort of application far out-weighs its feasibility.

B. Microcomputer Structure

The architecture of today's modern computers can vary from one design to another. However, all designs contain the same basic elements. This section describes the main parts of the microcomputer architecture. As discussed in Section I.A, today's microcomputer architecture is based on a stored-memory concept. Figure 2 shows a diagram of the basic structure of a stored-memory computer. Here we see that a basic microcomputer consists of a central processing unit (CPU), memory, input/output circuitry and the address, data and control buses. Each of these components are described in more detail in the following sections.

Communication Through Buses

The address, data and control buses in a microcomputer allow each of the different sections to communicate with each other. The address bus is used to set the unique memory location of the data that it wishes to access. This is typically performed by the CPU enabling it to retrieve and store values in memory or peripheral devices. The amount of memory that a processor can access depends directly on the size of the address bus. For example, a 16 bit address bus found on the Intel 8051, can access 64 kBytes (65536 bytes) of data. It is however possible to access more memory than this by using additional I/O pins and performing page switching.

The data bus is used to transfer or copy data from one address location to another as "pointed" to by the address bus. Its width (size) is the figure we talk about when classing the type of microprocessor being used. A 16 bit processor has an internal data bus that is 16 bits wide.

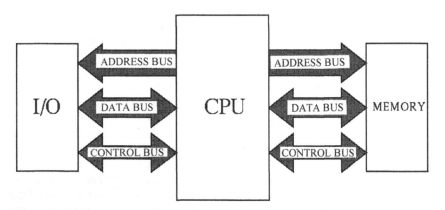

Figure 2   Basic structure of a stored memory computer.

The control bus contains the control signals or control lines that determine the type of transfer that is going to take place. It is usually made up of signals to enable read and write access, and a clock signal that determines when these events happen.

The clock rates in most microcontrollers are in the MHz region and the address and data buses are typically clocked at the same rate. This means that the devices connected to these buses must have small enough access times so that the circuitry timing is valid.

Central Processing Unit

The CPU is the brain of the computer system. This is where instructions are fetched, decoded, administered and executed. The CPU is divided into three main sections: the arithmetic logic unit (ALU), the control unit and the registers as shown in Figure 3.

The registers in a CPU are fast memory locations used for temporary storage of data and status flags of the CPU. General purpose registers are also used at the input of the ALU to hold data ready for execution and/or manipulation. The common registers found in a CPU are:

- ACC   Accumulator
- PC   Program counter
- IR   Instruction register
- SP   Stack pointer
- R0-R7   General registers
- SF   Status flag register
- EAR   Effective address register

The ALU is where all the mathematical and logic operations are executed. It takes the data out of the general registers, performs the required operation and then puts the result back into another or the same register. After execution, a number of status flags in the status flag register are set, cleared or left unchanged depending on the instruction executed. For example, the zero flag will be set if the result from the ALU is a zero.

The control unit controls the fetching of an instruction or op-code from the memory and decodes it. It decodes the instruction into a microprogram which explains the events that have to occur in order to execute that instruction.

Instructions may contain more complex steps or simple ones such as moving data from one address location to another. Once a series of instructions are sequentially added together, we call it a *program*. A program is written for a particular task and can contain millions of different variations of instructions to perform its goal. A more detailed explanation of how a CPU works can be found in [3].

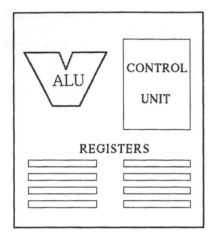

Figure 3   Structure of a CPU.

Memory

Now that we have examined the smaller, inner workings of a microcomputer, we can now extended our scope and look at the devices that contain the programs to be executed in the CPU which are called memory. Table 1 shows the different types of memory used with microprocessors and microcontrollers.

ROM-based memory (masked ROM) is hard coded into a microcomputer during manufacture and is contained in the same circuit as the CPU. As the name suggests, it can only be written once but read many times.

RAM on the other hand can be written to and read from many times. It can be found in the same circuit with the CPU or externally to the microcomputer. This type of memory can be found in a variety of different speeds (access times). High speed RAM is used internally on the chip for various functions including registers. Slower speed RAM is used externally to the microcomputer to hold data such as temporary variables and for temporary storage of programs during prototyping as will be discussed in Section IV.A.

EPROM is the most common memory type for long term storage of programs. It is found in two basic varieties: a plastic package and a ceramic package. The difference between the two is that the ceramic package contains a small, clear window. The EPROM is erased when it is placed under UV light of a particular wavelength for a period of time. Figure 4 shows an example of a Microchip PIC microcontroller in a one time programmable (OTP), plastic package (foreground) and a reprogrammable, ceramic package containing a window (background). This type of memory is usually programmed by a separate programming device before insertion into the circuit.

Table 1 Different Types of Memory

Acronym	Memory Type	Reprogrammable	Programmed in-circuit?	Speed		Data Interface	
				Read	Write	Serial	Parallel
ROM^a (masked)	Read only memory	NO	N/A	FAST	N/A	NO	YES
ROM^b (fused)	Read only memory	NO	NO	FAST	SLOW (once)	NO	YES
RAM	Random access memory	YES	YES	FAST	FAST	YES	YES
EPROM	Electrically programmed ROM	YES	NO	FAST	SLOW	YES	YES
E^{2}PROM	Electrically erased PROM	YES	YES	FAST	SLOW	YES	YES
Flash E^{2}PROM	Flash E^{2}PROM	YES	YES	FAST	FAST	NO	YES

^{a}This type of memory is found onboard some microcontrollers and microprocessors.
^{b}Fused ROM is not commonly used anymore. Fuseable links inside the device are blown during the programming of the chip.

E^{2}PROM or EEPROM is used for the long term storage of data such as the system configuration of a circuit or for logging events of a monitoring system. E^{2}PROM does not require UV light to erase it, instead it can be erased electrically. This memory is not used as RAM as writing the data is very slow. It is used where long-term storage of data is needed and a fast write time is not a major concern, e. g., the BIOS setup in an IBM compatible PC.

An alternative solution that has a relatively fast write time is the flash E^{2}PROM. Flash E^{2}PROM can be used in place of EPROM for the storage of programs and has the facility of being programmed while contained in the circuit.

Figure 4 Microchip PIC microcontrollers with and without a window. (Courtesy of Microchip Technology Inc.)

Input/Output Interface

In general terms, the I/O interface allows the inner workings of the computer system to communicate to the outside world. Such peripherals include monitors, printers, LCD screens, status indicators, etc. which display messages to the user. Keyboards, switches, push buttons and key pads allow the user to instruct the computer system to perform operations. These devices are known as *human interface devices* and are usually used as a form of feedback and control to the computer system.

For mechatronic based circuits, I/O interfaces collect or distribute data from the analog world with its associated problems (noise, etc.) to the digital world found in the microcomputer or system. Such interfaces include digital parallel ports, serial communication ports, analog-to-digital (ADC) converters and digital-to-analog converters (DAC).

These types of interfaces are used in a mechatronic system to display warning signals, measure sensor inputs, control motor driving circuits, and so on. The interfacing of a microcomputer system to the outside world is one of the biggest design problems an engineer is faced with. Section III discusses some I/O devices that can be used for the design of a mechatronic system.

Memory Maps

A memory map describes the position of various memory devices, I/O interfaces and registers in terms of their addressable memory location. Figure 5 shows a typical memory map of the Motorola MC68HC11A8.

The purpose of the memory map is to designate where each of the devices are located so that they do not overlap each other and cause conflicts. The hexadecimal number down the side is the physical[1] memory address location. For example, the interrupt vector table occupies a memory segment at the top of the memory map between FFC0hex and FFFFhex.

After deciding what the system contains and the devices that are needed, a memory map has to be designed so that all components addressable by the memory of the system are accessible. This step is the transition between pure hardware design and the software or programs that access the hardware. Further details of applying memory maps to a design are found in Section III.

II. MICROPROCESSORS TO MICROCONTROLLERS

In the previous section we introduced the basic architecture of modern microcomputers. In this section we are going to extend this to microprocessors and micro-

---

[1]Physical memory is linear memory space as opposed to memory paging or segmented program and data memory which is called *logical* memory. More detail can be found in [3].

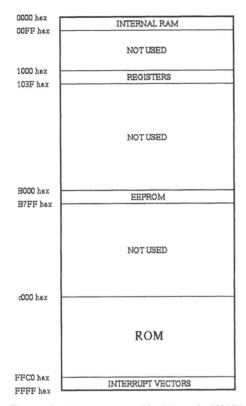

0000 hex	INTERNAL RAM
00FF hex	
	NOT USED
1000 hex	REGISTERS
103F hex	
	NOT USED
B000 hex	EEPROM
B7FF hex	
	NOT USED
c000 hex	
	ROM
FFC0 hex	INTERRUPT VECTORS
FFFF hex	

Figure 5   Memory map of the Motorola 68HC11A8.

controllers and see how they differ. We also examine how digital signal processors (DSP) differ from microprocessors in architecture and application.

A. Difference Between Microprocessors and Microcontrollers

The main differences between a microprocessor and a microcontroller can be described for two areas: architecture and applications. The architecture of a microprocessor can be described as a CPU on a single chip. Therefore all memory, input/output and other peripheral devices are wired externally to a microprocessor. Figure 6 shows a detailed structure of a microcomputer system.

In this diagram, only the CPU with some interrupt control and timers are found on the microprocessor. Due to all the peripheral devices being external to the microprocessor, a high external component count is found in circuits that use them. This is a problem in applications that are required to be small or compact.

This led to the development of the microcontroller, which has a higher degree of integration. Figure 7 shows the structure of most microcontrollers. All

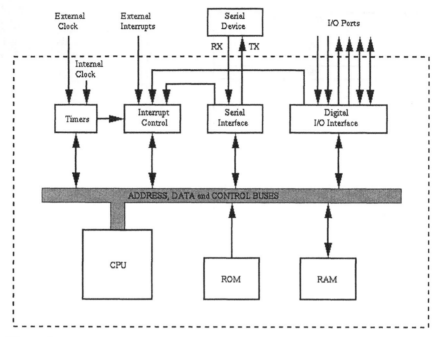

Figure 6   Detailed structure of a microcomputer system.

these devices are integrated onto the microchip allowing for a lower external chip count and a smaller printed circuit board.

The number of bytes of ROM and RAM onboard these microcontrollers is relatively small compared to normal microcomputer systems. This is not considered a disadvantage, as most of the applications that use microcontrollers do not require large amounts of memory. Some microcontrollers allow the use of external memory which means the microcontroller can be utilized for larger applications.

Microcontrollers also utilize a comprehensive and diverse interrupt control system. This is useful for applications where the circuit needs to be operated in real time. The timers and control circuitry associated with interrupt systems are somewhat more diverse on a microcontroller than in microprocessors. As we will see in later sections, timers and interrupts can be used to control circuitry to measure sensor information.

Another important built-in microcontroller feature not found in microprocessors is the I/O interface. Some microcontrollers contain analog to digital converters (ADC) as most sensors measured by microcontrollers are of an analog form.

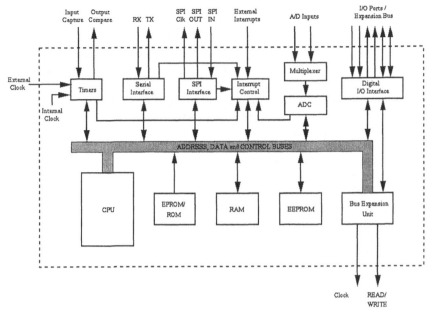

Figure 7   Structure of a microcontroller.

One advantage microprocessors have over microcontrollers is processing power. High processing power requires very large scale integration. This amount of circuitry occupies too much microchip real estate to include the rest of the peripherals that are required for a microcontroller. Therefore, microcontrollers tend to contain smaller, slower speed CPUs.

This makes microcontrollers useful for all mechatronic systems where sensoring and control are required. We will expand on the utilization and design of microcontrollers in later sections.

B. Digital Signal Processors

A digital signal processor (DSP) is a widely used processor for the real-time control and manipulation of signals and can be considered as being a cross between a microprocessor and a microcontroller. A DSP chip contains a high speed processing core as found in some microprocessors with some of the capabilities of microcontrollers. Generally a DSP chip contains some internal RAM and a high speed serial port for communication to other peripherals. They do not contain (at the time this was written) an internal I/O interface such as analog-to-digital convertors as found in microcontrollers.

As for microcontrollers and microprocessors, DSPs have two categories: fixed point and floating point. Fixed point DSP chips are equivalent to the integer-based microprocessors, where a fixed point arithmetic is employed. Floating point devices have a floating point ALU and multiplier that allows the processor to calculate fractional numbers.

Common DSP chips available on the market include such companies as Motorola [2,4], Texas Instruments [5] and Analog Devices [6]. One of the most common places we find DSP chips these days is in our digital mobile phones. The DSP chip in the digital mobile phone performs all the relevant communications as well as the compression and decompression of digitally recorded speech all in real time. Recently, some mobile phones have been released that also contain command recognition for voice dialing.

Microprocessors and microcontrollers based on the Von Neumann architecture were not designed for the execution and manipulation of real world signals in real time. Therefore, the microprocessor was redesigned to include the parallelism and pipelining of the instruction set, so that the performance could be increased. The main application for DSP chips is in the area of signal processing; therefore the parallelism of instructions is optimized for this purpose. More specifically, the design of the instruction set is optimized to perform convolution.

The instruction set of the DSP chip is where most designers see the difference compared to microcontrollers. Some commands on DSP chips have the ability to perform multiple instructions in one, two or three clock cycles. For example, the MUL instruction on the 8051 microcontroller takes 4 clock cycles to execute [1,7]. On the TMS320C25 DSP chip, this instruction (called MPY) can be executed in a single cycle [8]. Also the instruction MACD on the TMS320C25 performs a multiplication, accumates the previous result with a register and shifts some data from one location to another all in 3 clock cycles. Even though the programming of a DSP chip can be a daunting task due to the number of parallel instructions, the performance means that it is suited to a wide variety of real time applications. However, high level languages exist for these devices as will be discussed in Section IV.A under Code Generation and Implementation.

Two large applications of the DSP processor are in the area of speech processing and communications. Speech compression, real time speech recognition, video compression, high speed modems and tracking systems are a few of the applications that use a DSP processor. Although most of these systems are purely signal processing based, control of mechanical machines is also possible. More and more applications that require the analysis of a signal to operate machines are being developed. One such application is the recognition of features in an image for mobile transportation. That is, a mechanical machine that can "see" where it is going. This requires much processing power and signal processing techniques to be feasible, which can be realizable using a DSP chip.

For the rest of this chapter, we consider the DSP chip as being equivalent to a microcontroller in terms of its design process.

III. PRODUCT DESIGN USING A MICROCONTROLLER

In previous sections, we saw how microcontrollers are structured and how they differ from microprocessors. This section looks at designing a microcontroller and associated circuits for a target system. First, the process of designing a circuit for a system is discussed. We then consider the base circuit which shows the minimum components required to get a microcontroller up and running. Finally, the various types of interfacing and devices that can be connected to a microcontroller are discussed. These are illustrated using example interface circuits that can be connected to various microcontrollers.

A. Design Methodology

The design process of a microcontroller based system is presented in this section from its initial design concept to a working prototype. The specific stages for the design are discussed and are depicted as a flow chart shown in Figure 8.

The first thing to consider when designing a microcontroller based circuit is describing the function that the final system must perform.[2] That is, we need to know what the goal is. A good way to specify this is to draw a functional diagram that describes the system such as the one shown in Figure 9. This shows the inter action the circuit you are going to design has with the outside world, without that describes the system such as the one shown in Figure 9. This shows the interaction the circuit you are going to design has with the outside world, without mentioning circuit design specifics. Also the specifications of each part of the system must be decided at this stage. It is possible however, that some specifications of the system may change due to component supply or customer demand.

Once the functional diagram has been drawn, we can then choose a microcontroller that is suited to our application. The base circuit can be designed and we can also identify from our system diagram the appropriate interfacing that needs to be considered and what steps are required to implement this on the microcontroller in the software. This includes whether the circuit contains external memory devices or can operate on the internal devices found in the microcontroller. Consideration as to whether the input circuits need to be robust against adverse environmental conditions should also be addressed at this stage.

Finally, does the design meet the specifications identified in the system diagram? If not, then some adjustments need to be made. Once our design has met our specifications, Section IV discusses other considerations appropriate to our design before it is prototyped.

---

[2]See Chapter 6 for a detailed description of functional specifications and their role in developing a product.

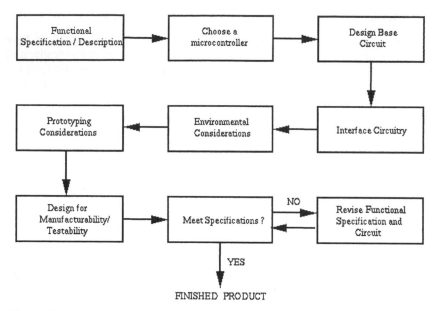

Figure 8 Design methodology flow chart.

B. Choosing a Microcontroller

The hardest decision in a design can be what microcontroller to use. It is vitally important since the decision that you make, influences the rest of the design. Firstly, consider what expertise your company has in microcontroller design and what equipment and software they have that support a particular microcontroller. This can be a cost saving exercise as some development tools for microcontrollers can be quite expensive.

Design/specification influences include the following (not in any particular order):

- processing power
- onboard memory
- amount and type of I/O devices
- ADC resolution (8/10 bit etc.) number and size of internal counters
- temperature range
- cost
- expandability
- ease of design
- usefulness in other products
- design for manufacturability

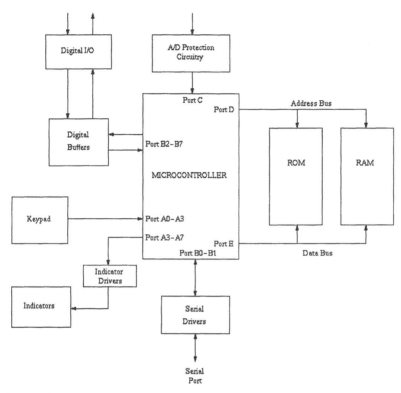

Figure 9 Functional diagram.

- pre-existing company standards
- compatibility with already existing systems

Most microcontrollers have a substantial amount of processing power, however not as much as a dedicated microprocessor. Powerful microcontrollers do exist such as the Motorola MC6833X series based on a MC68020 microprocessing core and the Motorola MPC500 based on the Power PC core, but these are generally more expensive for most basic microcontroller applications. Data sheets for these micrcontrollers are available on the Motorola World Wide Web (WWW) site [9].

One of the most important influences is the amount of onboard memory the device has. The size of the RAM and ROM is application specific and you will find a lot of variations of the same microcontroller with different types and sizes of memory. A microcontroller usually does not contain any ROM memory. It contains what is termed OTP, One Time Programmable memory which is in fact just EPROM. The difference between this and reprogrammable EPROM devices was discussed in Section I.B under Memory.

Other factors include whether the device contains any other types of memory, such as E^2PROM. This is useful if the system has any setup or configuration parameters as they can be stored is this *nonvolatile* memory (does not get erased when power supply is disconnected).

Probably some other important considerations are the amount and type of I/O devices the microcontroller has. These include:

- digital inputs and outputs
- analog to digital converters
- digital to analog converters
- pulse width modulation (PWM) support
- serial ports, high and low speed
- timing capability
- number of external interrupts

The type of I/O depends on what you want the microcontroller to measure from the outside world. This can be decided using a functional diagram such as the one in Figure 9. If external I/O devices are required then a microcontroller that supports such devices is needed. These I/O devices are covered in more detail in the following sections.

If an application has been found that requires the device to operate at very high or very low temperatures, then its operating temperature will also be a deciding factor. Most standard devices have a temperature range of –40°C to +80°C. However, military devices are available which extend the operating range to +125°C.

The final and probably the most decisive factor is cost. The cost of the device itself is not always the best benchmark. To develop the system using the tools available is generally the deciding factor. This is why many companies utilize the tools they already have and develop products based around only a handful of different devices. Also, a decision can be made on the ease with which the device can be designed and programmed as this will save expensive development time. This may be one of the main influences in choosing a certain chip.

No one microcontroller can be utilized for all applications. Microcontrollers are application specific and the choice can change the design of the rest of the system. Consultation of the system diagram such as the system in Figure 9, is vital throughout the design so that a designer can focus on the purpose of the system. The following sections describe the process by which the design takes place once a microcontroller has been chosen.

C. Base Circuit Design

Before we consider interfacing our microcontroller, we need to look at the base circuit design. This is the minimum number of components required to get the microcontroller running. Most databooks have this in the first chapter [7,10,11]. Figure 10 shows a schematic for a typical base circuit.

The first thing to consider when designing the base circuit, is to work out what power supply the circuit requires. From here, we can tailor a power supply circuit to meet the requirements of the microcontroller and associated circuitry.

Most microcontrollers operate on a simple +5 volt DC power supply at relatively low currents (less than 50mA). A +5 volt regulator, such as a 7805, is sufficient for an input supply between 7.5 and 30 volts. The databook for the appropriate microcontroller will specify the operating supply range given a few conditions.

For example, the power supply specifications for the PIC 16C74 microcontroller has an operating voltage (supply voltage) range between 2.0 and 6.5V, given that the supplies must have a minimum regulation of 0.25V. Some microcontrollers also specify a power supply rating given a certain clock speed. Normally, a microcontroller will operate at a lower supply voltage for a reduction in clock speed. The technical specification for the chosen microcontroller will need to be checked before committing to a design.

Microcontrollers and other peripheral circuits in the system also need decoupling. This is normally performed by placing a 0.1μF capacitor across the power and ground power supply pins of the device. This decreases the amount of noise in the peripheral circuits due to the AC being shorted to ground. It is also common to have a large filter capacitor somewhere in the circuit to avoid fast supply changes.

The clock frequency the microcontroller operates at is the next decision. Microcontrollers can run at speeds from a few Hz to several MHz. Given the design specifications of the power supply, this may limit your options.

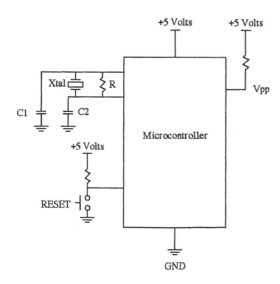

Figure 10 Base circuit design.

Most microcontrollers can be clocked in three different ways. The first and most popular way is to use the internal clock driver in the microcontroller. The external components needed are a crystal and a couple of minor components to correctly load the crystal so that it will oscillate. Figure 10 shows a typical crystal circuit. This is the best way of guaranteeing a stable clock given the temperature specifications of the crystal. Again, the databook for the microcontroller will show typical values to use.

A second way of clocking a microcontroller is through an RC circuit. This is generally a cheaper alternative to the crystal version. However, this is at the cost of a clock rate that may not be accurate and will vary considerably with temperature. Therefore if the timers on the microcontroller are a critical part of the design, an RC circuit is not recommended.

The third and final way is to clock it from an external source. This is normally done when more than one microcontroller exists in the same circuit. Both microcontrollers run off the one crystal circuit thus allowing for synchronization of their operation.

It is often the case that the microcontroller also has some programming supply pins. These also need to be addressed at this stage. For normal operating conditions a Vpp pin would be pulled-up to the supply rail using a resistor (typically 10 kΩ) as shown in Figure 10. During the programming of the device, a higher voltage is placed on this pin. Consult the appropriate microcontroller databook for details.

D. Memory Interfacing

Many applications require external ROM and RAM. External ROM enables us to control program updates and allows us to exchange the ROM on a system without having to replace a comparatively expensive microcontroller. However, size is normally the reason that we require these external devices.

As discussed in Section III.B, some microcontrollers such as the 68HC11 and the Intel 8051, have the ability to access external memory (see Figure 11). There is a cost however involved in adding these peripherals. In these microcontrollers we lose two digital ports. The 16 bits that make up the two data ports are multiplexed to accommodate both the address and data buses. Figures 12 and 13 show example circuits that multiplex the data and address buses to access an external ROM and RAM using the 68HC11 and 8051 respectively. This is also shown in the technical data books for these microcontrollers [7,10].

To enable the two ports to access this memory, these microcontrollers are placed into a different mode. These modes are described in more detail in the appropriate technical manual [7,10]. A commonly used mode is the single chip mode, which operates with no external address and data buses. The second which is used to address external memory and other peripherals is called the expanded multiplexed mode. These modes are selected using external pin(s) on the microcontroller and must be set during power up. Normally these pins are hardwired to the appropriate setting on the circuit board.

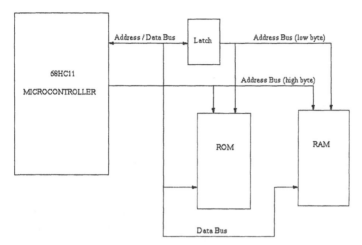

Figure 11 Expanded multiplexed mode of the 68HC11.

During this stage of the design, the memory map as discussed in Section I.B needs to be consulted. We need to decode the memory so that only one device is accessed for one given address space. If no other devices are to be addressed on the external bus, then a simple solution is shown in Figure 14. This decodes the memory so that the entire 64k address space is used with the ROM and RAM occupying half each. However there is a conflict with the internal address spaces. The microcontroller fixes this by accessing the internal devices before any external device. Therefore, where an internal device occupies a given address space, the external devices are not used. Figure 15 shows the new memory map for the 68HC11 with the internal ROM disabled (the 68HC11A1 has this disabled by default, see the databook [10,11] for more details).

E. Peripheral Interfacing

Sometimes other external devices such as additional I/O is needed. This means that the address decoding required can be somewhat more complicated. Logic circuits or PAL/PLA (Programmable Array Logic) devices will be required in this case. Adding additional I/O using the external buses will be described in the next section.

Another common system peripheral is serial communications. This is used commonly to communicate to other devices in the system or to a terminal to change configuration parameters. Most microcontrollers contain TTL serial devices already. However, most devices such as terminals, communicate with specific communications standards such as RS232. Therefore in microcontroller systems, additional circuits are required to convert TTL into the communications line voltages that are needed. "Off the shelf" components are available for this purpose.

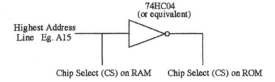

Figure 12 Address decoding for 32k RAM and 32k ROM.

Figure 12 contains a typical circuit using a MAX232 (or equivalent circuit) to convert from the TTL I/O of the microcontroller to the format required for the communication lines.

High speed serial lines also exist on some microcontrollers including the Motorola 68HC11, some 8051 derivatives and some Microchip PIC controllers. These type of serial lines are known as SPI (Serial Peripheral Interface) or I^2C buses and can usually run at half the clock rate. However, there will be restrictions. Usually these high speed serial communications must be contained on the system board the microcontroller is on. This type of communications is not designed to run over a long cable like RS232 or RS422. It is now possible to get a wide range of peripherals designed for these high speed serial buses which include analog-to-digital converters and additional I/O. It is also a popular method for communicating between multiple processors.

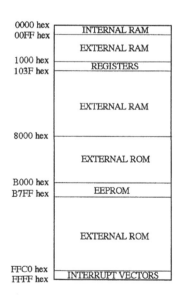

Figure 13 Memory map when adding external ROM and RAM on the 68HC11.

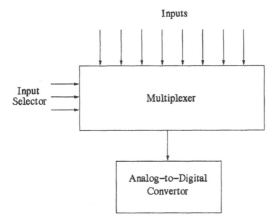

Figure 14   Multiplexed input for an analog-to-digital convertor.

F. Sensor Interfacing

Most mechatronic systems require the use of interface circuits to provide and gather information to and from the outside world. These interface circuits are used to convert a signal from a sensor or provide a signal to control a mechanical device to an electrical signal compatible with the microcontroller.

There are several interface techniques that convert signals to and from microcontrollers:

- Digital inputs and outputs (I/O)
- Analog-to digital convertors (A/D)

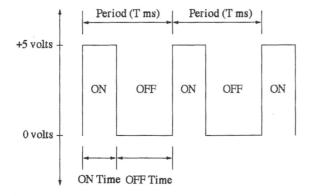

Figure 15   Pulse width modulation (PWM) waveform.

- Digital-to-analog convertors (D/A)
- Pulse width modulation (PWM)
- Isolation devices

Most microcontrollers available today have some of these interface circuits built in. This again enhances the philosophy of the single chip solution. Microcontrollers come in many forms and a major selection criteria may be based on the type of interface circuits that it contains. The most common of all the interface circuits on a microcontroller is digital I/O. Microcontrollers can have anything between 4 and 60 lines of digital I/O which can have a wide range of applications.

The first thing to consider when designing a system using digital I/O is to allocate the number of lines that are required in the circuit. This was done when we designed our functional diagram describing the structure of our circuit for which an example is shown in Figure 9. Here we need to decide which digital I/O line from the microcontroller is connected to which part of the circuit.

Our decision is based first on whether it is an input or an output. Not all the digital I/O lines on a microcontroller may be bidirectional. Some digital lines may be multipurpose allowing them to be configured to behave differently. For example as seen in Section III.D, it is possible to use two of the 8 bit digital ports as a 16 bit address bus and an 8 bit data bus to enable external peripherals to be added.

Other specifications of the digital I/O lines such as the current output that can be delivered and timing aspects also need to be considered. Another important procedure is to ground any digital input lines that aren't being used. This will reduce the amount of noise on surrounding digital lines and make the internal circuits of the microcontroller more stable.

In some cases, the amount of onboard digital I/O may not be enough. Alternative methods are therefore required. One such method is to add an additional circuit called a general purpose input-output (GPIO) device. This requires an external address and data bus and will also require to be address decoded as was discussed in Section III.E. Common GPIO devices include the Motorola 6821 and the Intel 8255 chips. Figure 13 shows an example circuit containing an 8255 device. This may not be a good solution especially if the microcontroller does not support an external address/data bus and it also takes up valuable printed circuit board (PCB) real estate.

If the microcontroller supports high speed serial such as SPI, we can use this to obtain as much I/O as we want. The SPI serial bus works by transmitting a byte out the serial port. Upon sending the byte, a byte is received into a serial register. More detail can be found in the appropriate references [10,11]. A good example of the use of the I^2C bus can be found in [12].

Analog-to Digital/Digital-to-Analog Converters

It is often the case that a system needs to sample a sensor that has an analog voltage or current value. In order for us to read this value, we need to convert it into a form that a microcontroller can read. This requires a device called an analog-to-digital converter or ADC. Conversely, we may want to output an analog value to a circuit. For this we can use a device called a digital-to-analog converter or DAC. Together, these devices make it possible to read and write analog values to and from a system or sensor using the microcontroller.

ADC devices are commonly found on-board microcontrollers. On-board DAC devices are rare as there are generally other ways of solving the problem. For example, when controlling the speed of a motor, we don't necessarily need to supply a variable voltage to change the speed. We can use Pulse Width Modulation (PWM) which is in digital form. This is discussed in the following section.

The ADC and DAC devices are commonly found in circuits using DSP processors. This is because the main use of the DSP processor is to manipulate analog signals. However, ADCs and DACs are not commonly found on-board DSP processors.

DAC and ADC devices are commonly available and have a wide range of specifications. The most important of all and the first to consider is the sampling speed. ADCs need to operate at twice the speed of the maximum bandwidth (Nyquist bandwidth) that needs to be sampled, in order for the signal to be correctly represented. For example, say we have a sensor that outputs a voltage that represents the speed of some motor, the bandwidth will be the maximum acceleration and deceleration of the motor. In design practice, we make the sampling frequency greater than the twice the maximum bandwidth so that we get a better representation and so that we do not get aliasing (where twice the maximum signal bandwidth exceeds the sampling frequency).

To avoid aliasing, an anti-aliasing filter is used. An anti-aliasing filter is just a low pass filter (LPF) with a cutoff frequency (half-power frequency) just below half the sampling frequency. Thus, any signal that exceeds this frequency is discarded.

After determining the frequency of the incoming or outgoing signal we wish to sample, we need to determine the resolution or how many levels of representation we require. For example, a 12 bit ADC has 4096 levels and a resolution of (voltage range/4096). Therefore if our input or output signal has a voltage range of 5 volts, our resolution would be (5/4096) = 1.22mV. Most control applications suffice with only 8 or 10 bit ADCs and DACs. The Motorola 68HC11 and Intel 8051 are available with 8 bit and 10 bit ADC devices respectively. If a higher resolution is needed, then an external device is required.

An external ADC or DAC device can be connected to a microcontroller in several ways. By far the easiest way is to use one or more of the digital I/O ports. The disadvantage of this is that digital ports are used for communication with the ADC or DAC and thus cannot be used for any other purpose. The only real advantage of this design is that it is relatively easy to program.

Single chip ADC and DAC devices are available that utilize the high speed serial bus (SPI, I^2C). These devices have internal logic to latch a serial bit stream from the microcontroller. Also these devices can coexist with other devices on the same serial line. Additional digital lines may be required to select the active device on the serial line (SPI bus).

rules

Internal ADC devices are available in most microcontrollers. A single ADC device is normally buffered by some sort of multiplexing to select the required input signal. This enables the microcontroller to sample multiple signals with one AD converter as shown in Figure 16. This type of architecture is what is found on microcontrollers such as the 68HC11.

As mentioned earlier, DSP processors are commonly used with ADC and DAC devices. However, on-board devices are not common but could become common-place in the near future. A DSP microprocessor usually uses a CODEC which contains both an ADC and a DAC device. A CODEC is usually wired to one of the serial ports on the DSP chip. A CODEC and DSP microprocessor make a simple standalone system useful for evaluation purposes [13]. Such evaluation systems are available commercially from Texas Instruments [13], Motorola and Analog Devices [14] for a relatively small cost. However, these systems contain relatively small amounts of memory which is unsuitable for medium to large applications, however they are a good start for new users.

Pulse Width Modulation

An efficient way of transferring information to and from a microcontroller is using pulse width modulation (PWM). Pulse width modulation has the advantage that the waveform is in a digital form but represents an analog signal. Figure 17 shows the PWM waveform. Adjusting the width of the ON pulse enables us to represent a value of an incoming sensor or an output to a circuit.

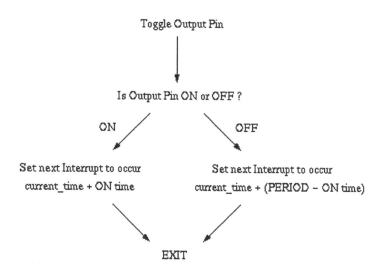

Figure 16   Flow chart of the PWM interrupt routine.

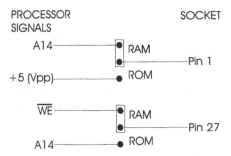

Figure 17 ROM/RAM selector jumpers.

Most microcontrollers contain timer controlled pins that allow PWM waveforms to be generated. They have output control or output compare pins that are logically connected to timer registers. They also have timer input or input capture pins that save the timer value when a particular edge of a digital waveform is detected. This is useful for reading in PWM waveforms.

Although the timer functions are available, a PWM waveform is generally not generated by the hardware in a microcontroller. It is normally controlled using the timer interrupts in software. Figure 18 shows the operation of the software interrupt routine to control a PWM output. Here you will notice that the routine calculates when the next interrupt is to happen based on the frequency (period) of the PWM signal and the width of the pulse.

```
0000 hex  ┌──────────────────────┐
          │     INTERNAL RAM     │
00FF hex  ├──────────────────────┤
          │      Jump Table      │
          │       NOT USED       │
1000 hex  ├──────────────────────┤
          │      REGISTERS       │
103F hex  ├──────────────────────┤
          │                      │
          │                      │
          │       NOT USED       │
          │                      │
          │                      │
B000 hex  ├──────────────────────┤
          │       EEPROM         │
B7FF hex  ├──────────────────────┤
          │                      │
          │       NOT USED       │
          │                      │
c000 hex  ├──────────────────────┤
          │                      │
          │                      │
          │       MONITOR        │
          │                      │
FFC0 hex  ├──────────────────────┤
          │   INTERRUPT VECTORS  │
FFFF hex  └──────────────────────┘
```

Figure 18 Memory with the BUFFALO™ monitor program.

An ADC can be realized by constructing a circuit that converts an analog voltage into a PWM signal. Such circuits can be designed using op-amps or using a 555 timer circuit [15]. The basis behind this circuit is a slow scan, single ramp ADC that is controlled by the microcontroller in software.

First, a constant ramp signal is produced which is reset by the microcontroller. For slower rates and high accuracy, the software is modified for a different scan rate and the components controlling the slope are recalculated. This circuit is useful when low cost is required and slow conversion rates are tolerable.

The DAC device is nowhere near as involved. All that is required is that the PWM output signal is low-pass filtered to obtain the value it represents. The voltage we want to output is in fact the average of the PWM waveform. Therefore if we low-pass filter the PWM signal such that all the harmonics in the PWM signal are removed, we will be left with just the DC voltage or the average voltage. However, the low pass filter is also the output reconstruction filter of this DAC and the cutoff frequency has to be carefully designed so that the output bandwidth is considered.

This DAC design has the disadvantage in that there will be ripples in the output signal if the low pass filter is not sharp enough. This is generally not a problem with some applications such as the speed control of electric motors, as the motor does not have a quick enough response time to be affected by the ripples.

Isolated Input/Output

Some specialized applications require a microcontroller system to be isolated from a sensor or from the rest of the system. Many types of isolation can be designed into a system depending on the application.

Due to the advancements in switch-mode power supply design efficient isolated power supplies can be produced [16]. Parts of the circuit or system may be powered by a suitable regulated switch mode supply while others can derive their power from an isolated power supply. The isolated supply can be generated from the first supply by adding an additional winding to the inductor making a transformer.

Another device which is typically used in conjunction with an isolated supply is the opto-coupler. An opto-coupler is used to transfer signals in a digital or PWM form across the isolation barrier.

One important design factor needs to be considered when designing an isolated circuit. The breakdown voltage of the isolated supply and across the opto-coupler needs to be determined. This is important especially in medical applications where the patient needs to be isolated from the mains power socket.

For more details concerning this topic see [16].

G. Design for Real World Environments

A circuit or system that has been designed for a practical environment requires some form of protection. The level of protection depends heavily upon the type of use.

First, protection of inputs (including power supplies), needs to be considered for the testing and installation of the system. Such protection will resist incorrect wiring or connection. For example, a simple diode on the input of a DC power supply will avoid the system failing due to the power supply being wired backwards.

Protection for the inputs and outputs of a microcontroller can be realized by using buffers. In the event of a digital I/O line being "stressed," the buffer will be destroyed and not the microcontroller that it is protecting. A buffer IC is much cheaper to replace than another microcontroller.

Protection for when the system is placed in the field is the most important factor. Any inputs or outputs to the system are vulnerable. Excess voltage protection may need to be considered. A simple and effective solution is a zener diode which clamps an input to a specified voltage and draws sufficient current to blow a fuse or fusible resistor.

Noise can also be a problem with some inputs, especially power supply "hum" from the mains. A simple low pass or notch filter at the input can usually solve this problem. Another simple solution for digital lines is called debouncing. This is usually used for measuring the state of a switch or keypad. A simple implementation is performed in software and samples the state of the switch every 100ms. The state of the switch can be determined when 2 or 3 successive samples are "measured" as being the same state.

Another important form of protection for a microcontroller is under voltage protection of the power supply. If for some reason a voltage drop occurs in the power supply and drops below the operating range of the microcontroller, the microcontroller may not continue to operate even though the power supply returns to its regulation voltage (called brown-out). Such a device exists that provides a reset to the microcontroller when the supply voltage drops below a certain limit. Such devices include the Seiko 8054 and the Motorola MC34064 [17]. These are three pin devices that look just like a transistor.

Most microcontrollers also have a "watch dog" circuit that also provides this function. If a write to a particular register is not performed within a specified time frame, then the microcontroller will reset. This is a software implementation and can be used in conjunction with the under voltage circuit.

Another form of protection that may also need considering is radiation (such as RF transmitters and nearby lightning strikes). This can be minimized by the inclusion of a ground plane into the printed circuit board and protective shields over sensitive circuits. However, phenomena such as lightning is impossible to completely protect electronics against.

IV. IMPLEMENTATION TECHNIQUES

The previous section detailed the considerations needed to design a system given a simple structure and specifications. Before we can finish the electronic design of our mechatronic system, we need to consider some implementation issues. These include: How we are going to construct and test our prototype circuit? How we are going to develop the firmware for the system? And how can we make the circuit easy to manufacture and test? The first section deals with some important issues when prototyping the circuit and developing the software/firmware. The second section shows some simple rules for the design of the PCB layout and looks at what is required to make the system easy to manufacture and test.

A. Prototyping

Prototyping a microcontroller-based circuit can be a costly exercise. Common practice is to design a circuit based on the system functionality as in Section III. Considerations into how this design can be manufactured, tested and prototyped are then decided before the prototype is actually made. During the construction of the prototypes, problems in manufacturing can be found and addressed ready for the next revision. Being a prototype circuit, tests can be made on its suitability for environmental conditions using an environmental chamber for example.

Once the software is almost completed, the schematic circuit can be finalized and the production run of the hardware can be manufactured. The advantage of developing a circuit this way is that you can trial some new techniques and test the circuit against the required specifications before placing it in the field. One of the biggest disadvantages is time. It takes time to revise a circuit and then manufacture a revised batch. Thus, for systems with short development times, this is generally not practical (however generally good practice).

Development time can be reduced by merging the productions of the prototype and production of the PCB into one. Simple partial prototype circuits using strip board (also known as Veroboard) can be tested before finalization of the circuit. Also, problems that have already been solved by the company in previous designs can and should be used. This also has an advantage to the person who has to test and repair the circuit as they will be familiar with the concept. The disadvantage with this concept is that a high number of failures may occur if the design is not properly thought through during the design and prototyping stage.

The following two sections describe these development techniques in greater detail.

Prototyping Hardware

In microcontroller-based circuit design, hardware design needs to be considered at a software level. Also, hardware design needs to be considered for the prototyping stage of the development. Such considerations include how the software is going to be tested on the prototype. One easy and common solution is known as ROM/RAM switching.

Generally a microcontroller with external memory will have an external ROM (EPROM) and RAM. It is possible that you can change the pins on the ROM socket so that a RAM IC can be put in its place. Figure 19 shows a simple circuit that incorporates a switch that switches between the two types of memory ICs.

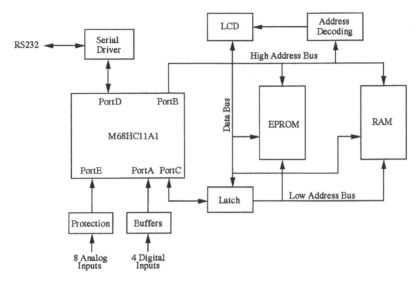

Figure 19 Function description for the 68HC11A1 development board.

By doing this, we can start the initial part of the software development by downloading the program into RAM using a *monitor program*. The concept of monitors will be discussed in more detail in the next section. Using RAM instead of an EPROM speeds up the software development. This is because for every small change in the software, another EPROM does not need to be programmed. The program only has to be downloaded to the prototype board again.

A small problem using this method, is that the monitor program itself needs to occupy some memory. In the old versions of the Motorola 68HC11A8, the monitor was programmed into the internal EPROM of the device. This makes the previous method easy to implement. Another solution, is the monitor can be programmed into the external EPROM placed in the ROM socket. This way the RAM can be used as both program and data memory for software development.

For example, the case study using the 68HC11A1 (Section V.A) was designed as a development system. The ROM is a monitor and the RAM is used for both program and data memory. Once the program is downloaded into the RAM, the monitor allows the user to execute the program, setting break points and to look at memory locations. Once the program has been developed, the EPROM containing the monitor can be removed and an EPROM containing the newly developed software can be put in its place.

A problem occurs using this structure when the program and data exceed the amount of RAM on the board. Once at this stage, a large amount of the software has been written and tested, so we can revert back to the method of programming EPROMs.

Software Development

As discussed in the previous section, the simplest way to develop software in a system that has external ROM and RAM, is to program EPROM every time you want to test a new part of the software. For a system utilizing a single chip solution (that is with no external ROM and RAM), this is the only method that can be used. Software is developed using reprogrammable microcontrollers with windows to allow erasing of the EPROM as shown in Figure 4.

However, due to the time it takes to switch between EPROMs and program them, an alternative solution is to replace the program ROM with an EPROM containing a monitor. Therefore, if the RAM is large enough, we can use it as both our program and data memory (depending on the architecture of the microcontroller). As discussed in the previous section, if the monitor program exists in the internal EPROM of the microcontroller, then two RAM chips can be utilized.

An example monitor program that is available for 68HC11 called BUF-FALO™, is available on the Internet [18]. This monitor program occupies 8k Bytes of address space at the top of the memory map as shown in Figure 20. As this overlaps the interrupt vector table, the monitor uses the internal RAM on the microcontroller as a *jump table*. This change in addressing needs to be considered when developing software as the location of the interrupt routine vectors has changed.

Therefore, instead of changing the address at the top of the address space, an address location in the internal RAM is changed instead. Other monitor programs exist for other microcontrollers; however it is possible to write one that suits your own application or needs.

One of the most important considerations when developing a system that requires both hardware and software, is how is the software going to access the hardware devices. This can make the decision on which hardware is chosen and which method is the best. The most common hardware changes during the prototyping stage is due to the lack of software control. Therefore, when designing a particular part of the hardware, an engineer should be able to "picture" the software routine that is going to access it. That is, no software is written, but the method in which the software accesses the hardware should be thoroughly thought through and a simple flowchart should be drawn. This is useful in the later development of the software.

Code Generation and Implementation

Once the decision has been made about how the software accesses the hardware, code can be written. Programs are normally written in an assembly code or in a high level language such as C. In order to convert these languages into the binary opcodes that a microcontroller requires, a cross-assembler or cross-compiler is used.

The cross-assembler or cross-compiler is executed on a host system such as a PC or workstation. No communication between the host system (PC) and the target system is required during the compilation. The assembly code or C program is cross-assembled/compiled on the host system and produces a binary file or a file in a format suitable for downloading to the target system (a format which a particular monitor program accepts).

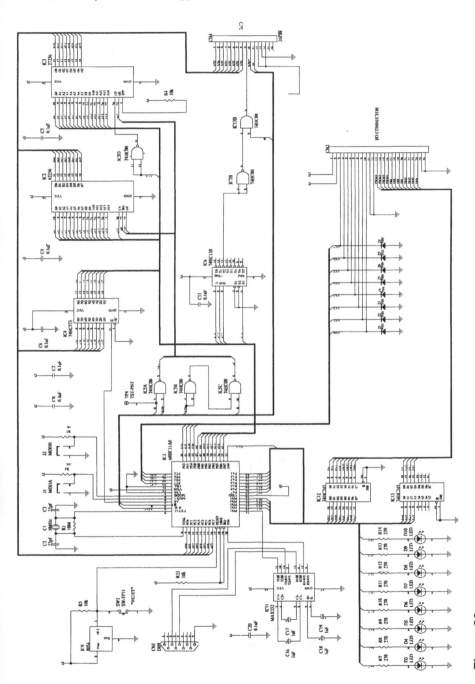

Figure 20 Schematic diagram of system.

This is where the on-board monitor of the microcontroller can be used (see Section IV.A under Prototyping Hardware). By connecting the target system to the host system using a serial or parallel cable, the target system can download a file or "look" at registers and memory using the monitor program. When downloading the binary file output by the assembler or compiler, the monitor is setup such that it is ready to receive the file (performed by the user). The file is then downloaded to the target system and the monitor places the program into the specified memory location. This memory location is specified using assembler specific instructions in the assembly code or in a setup or configuration file in the case of the compiler (see appropriate documentation of assembler/compiler for more details). Once the program has been downloaded, the monitor can be told to execute the program.

This is the only way software can be tested. As mentioned in Section IV.A (Prototyping Hardware), another method is to program an EPROM or the microcontroller itself. This is a common method once the software has been tested and is ready for field testing. Again, the software is assembled/compiled as above. The EPROM programmer software can then read the binary file output by the assembler or compiler and program the EPROM or internal EPROM in the microcontroller.

V. CASE STUDIES

A.   Development Board using the Motorola 68HC11 Microcontroller

This case study was designed for use as a development board and educational tool for developing software for the 68HC11A1 micrcontroller. The reason that this microcontroller was chosen, was for its popularity in industry and its application diversity. There is no shortage of documentation for this microcontroller. References include the Motorola databooks [10,11] and "after-market" publications [19]. Also many world wide web (WWW) pages exist with appropriate data and software [2,9,18,20,21,22,23].

The major features of this microcontroller include [10]:

- 8K Bytes of On-Chip EPROM (OTP ROM)[3]
- 256 Bytes of On-Chip RAM
- 512 Bytes of On-Chip EEPROM
- 16 Bit Timer System
- 8 Bit Pulse Accumulator
- Real-Time Interrupt Circuit
- COP Watchdog System
- Synchronous Serial Peripheral Interface (SPI)

---

[3] The On-Chip EPROM is not used in this case study.

- Serial Communications Interface (SCI)
- 8 Channel 8 Bit Analog to Digital Converter
- 38 General Purpose I/O Pins

Additional features that this development board offers include:

- 8K Bytes of External EPROM (BUFFALO Monitor Program)
- 32K Bytes of External RAM
- Buffered I/O lines
- RS232 Compatible SCI System

The following sections provide the functionality of the platform, describe the hardware of the platform and show some example programs that can be executed on the board.

Functional Description

As stated above, the major objective of this design was as a development and education tool. Therefore, we wanted to utilize as many functions as possible while providing adequate protection to the I/O of the microcontroller. This meant compromising between flexibility and limiting the functionality of the board by adding protection. As the primary function of this platform was for educational use, we added protection to all the I/O while providing as much functionality as possible. Figure 20 shows the functional diagram for this system. Firstly, a serial port was added to communicate to a host PC system for the downloading and debugging of embedded software. We could therefore use simple terminal software which is available on most PC workstations.

0000 hex	
00FF hex	INTERNAL RAM
	EXTERNAL RAM
1000 hex	
103F hex	REGISTERS
	EXTERNAL RAM
7FFF hex	
	LCD SCREEN
B000 hex	
B7FF hex	EEPROM
	NOT USED
C000 hex	
	ROM
FFC0 hex	
FFFF hex	INTERRUPT VECTORS

Figure 21 Memory map for the 68HC11A1 development board.

Secondly, we needed to interface the microcontroller to an unknown system that may contain keypads, LCD screens, switches, digital sensors such as push buttons or reed switches and analog sensors such as temperature sensors. This microcontroller has the capability to access all these. Therefore, we designed a platform that utilized as many of these types of I/O as possible.

The digital I/O in this design are controlled directly from the micrcontroller via buffer chips, which are included for protection. In some circumstances, the buffer chips are not required; however this design opted for robustness.

The analog inputs are usually the core of most mechatronic systems. They represent signals from the analog world in a digital form for use by microcontrollers. In this design, we utilize all 8 analog inputs. These analog inputs can alternatively be used as digital inputs. This gives the user some added flexibility.

PWM waveforms (see Section III.F, Pulse Width Modulation) can also be generated using this platform. The digital lines associated with the Output Compare timers (see [10,11,19]) are buffered to the output connector so the user is able to control mechatronic systems using this type of waveform. Alternatively, these digital lines can be used as general purpose digital outputs. Section V.A (Memory Map) shows some example code written in the C language that generates a PWM waveform for this platform as suggested in Section III.F (Pulse Width Modulation) and Figure 14.

Another useful tool to teach newcomers which can be useful for debugging, is an LCD screen. In this design, we placed the LCD screen directly on the address and data buses and decoded the address so that it could be accessed without conflicts to any of the other devices. The LCD screen connector is capable with the 2 line, 16 character Seiko alphanumeric type.

Hardware Description

Figure 12 in Section III.D shows the schematic circuit diagram of the system. First we see that this system has external EPROM (IC3) and RAM (IC2) which requires that the microcontroller runs in an expanded multiplexed mode (see Section III.D). On the 68HC11, the address and data buses are multiplexed on the digital ports B and C. Using a latch (IC4) to demultiplex the data and low 8 bits of the address bus, the microcontroller generates an external 16 bit address bus and an 8 bit data bus [10,11,19].

This platform provides a variety of digital and analog I/O. Table 2 shows a summary of the I/O available on this system. Eight analog lines are provided to the

Table 2 Summary of the I/O on the 68HC11A1 Development Board

I/O	Number of lines	68HC11A1 port pins
Digital inputs	4	PA0, PA1, PA2, PA7
Digital outputs	4	PD2, PD3, PD4, PD5
Analog inputs	8	PORT E
PWM outputs	4	PA3, PA4, PA5, PA6

microcontroller directly from the outside with zener diodes to clamp the input voltage. This is to prevent the input voltage rising above the maximum rating of the microcontroller. Optionally, fuses can be added to the inputs (before the zeners) for complete protection.

Four general purpose digital inputs and four general purpose digital outputs are also provided. These are protected using buffer chips (IC12, IC13). Four other digital outputs also exist which are connected to the internal timers of the 68HC11 which enable the user to configure them as an output compare function.

The SCI serial system of the microcontroller is used to communicate to a host system (such as a PC) for the downloading and debugging of the user's software. An LCD screen port is also provided

Memory Map

Before we can design and write software for this platform, we need to know the memory map of the board. Figure 22 shows the memory map for this platform. Some of the internal RAM is used by the monitor program for monitor variables and a jump table. Usually the user would not require the 256 Bytes of RAM and therefore it is left for the monitor program to use. However, it is possible to use the first

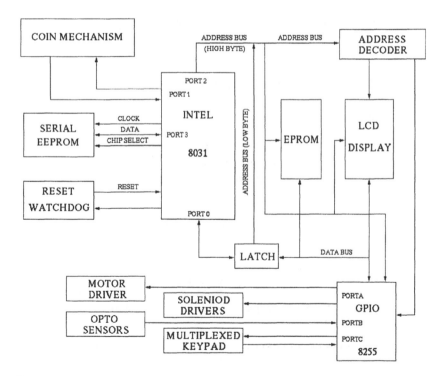

Figure 22 Functional description for the automatic coin sorter/counter.

100 or so bytes. The best solution is to use the external RAM for both program space and data space. The external RAM occupies the lower 32k Bytes of address space. However as can be seen by the memory map in Figure 21, the internal registers from 1000h to 103Fh and the internal RAM, overwrite the external RAM. This reduces the size of the external RAM from 32768 Bytes to 32448 Bytes. Not a major problem!

The LCD screen contains only two registers. Instead of completely decoding the address space to access these registers, we can decode the LCD screen using incomplete address decoding. This means that the registers will appear multiple times throughout the decoded address space. A good example of how to do this is presented in the next case study (see Section V.B, Hardware Description). Using incomplete addressing, we see that the LCD screen actually occupies 4k Bytes instead of only 2 Bytes if we completely decoded it.

The EPROM is decoded to the top 16k Bytes of address space. This is because in the 68HC11 the reset vector (points to the start of the user's program) is located at addresses FFFEh and FFFFh. We have placed the monitor program BUFFALO™ into this EPROM which is available on the Internet [22]. Users can use their own programmed EPROM with their own monitor program or software they have developed.

Software Description

As there is no specific software for this platform, this section instead presents some practical examples mentioned throughout this chapter. The software presented here is written in the C language. The examples include:

- Reading and writing digital I/O and using the serial port
- Generating PWM waveforms
- Using the analog inputs

Program 1 shows an example program that reads in the digital inputs and reflects the bits onto the digital outputs. It also sends a string out the serial port showing the status of the four digital input lines.

Program 1   Digital I/O and serial output routine written in C

```
/******************************************************   *
 * TESTIO.C
 *
 * Program that reflects the digital inputs on PORT
 * A to the digital outputs on PORT D. It also sends
 * a string to the serial port showing the status of
 * the input lines.
 *
 *****************************************************
 */
```

```
#define      PORTA   (*(unsigned char *)0x1000)
#define      PORTD   (*(unsigned char *)0x1009)
#define      DDRD    (*(unsigned char *)0x1008)
#define      PACTL   (*(unsigned char *)0x1026)
#define      SCCR2   (*(unsigned char *)0x102D)
#define      SCSR    (*(unsigned char *)0x102E)
#define      SCDR    (*(unsigned char *)0x102F)

#define      D0      0x01
#define      D1      0x02
#define      D2      0x04
#define      D3      0x08
#define      D4      0x10
#define      D5      0x20
#define      D6      0x40
#define      D7      0x80

#include <stdio.h>
#include <intrpt.h>

static char counter, SerialBuffer[50];

/* serial interrupt routine */
void interrupt SerialRoutine(void)
{
   SCDR = SerialBuffer[counter];
   if (SerialBuffer[counter] == '\r')
   {
     counter = 0;      /* reset counter */
     SCSR &= ~D7;      /* disable serial interrupts */
   }
   else
     counter++;                /* increment counter */
}

/* MAIN routine */
void main(void)
{
   char data0, data1, data2, data3;

   /* set the direction of the ports */
   DDRD = 0x3F;               /* set PORTD to output */
   PACTL &= ~D7;              /* set PORTA bit 7 to input*/

   /* set the serial interupt routine */
   set_vector(0xFFD6, SerialRoutine);


   while (1)          /* loop forever */
   {
```

```
    /* clear the serial status lines */
    data0 = data1 = data2 = data3 = 0;

  /* reflect the digital lines */
  if (PORTA & D0)
    {
    PORTD |= D2;   /* set D2 of PORTD if PORTA*/
data0 = 1;         /* D0 is high */
    }
  else
    PORTD &= ~D2; /* else clear the bit. */

  if (PORTA & D1)
    {
    PORTD |= D3;   /* set D3 of PORTD if PORTA*/
    data1 = 1;          /* D1 is high */
    }
  else
    PORTD &= ~D3; /* else clear the bit. */

  if (PORTA & D2)
    {
    PORTD |= D4;   /* set D4 of PORTD if PORTA*/
    data2 = 1;          /* D2 is high */
    }
  else
    PORTD &= ~D4; /* else clear the bit. */

  if (PORTA & D3)
    {
    PORTD |= D5;   /* set D5 of PORTD if PORTA*/
data3 = 1;         /*D3 is high */
    }
  else
    PORTD &= ~D5; /* else clear the bit. */

  sprintf(SerialBuffer, "0:%d  1:%d  2:%d  3:%d\r"
    ,data0, data1, data2, data3);
  counter = 0;
  SCCR2 |= D7;               /* enable serial interrupt */
  while (SCSR & D7);  /* wait until transmitted */
  }
}
```

The program begins by initializing the digital ports by assigning the direction of each bit on the I/O port. The digital lines on Port A need to be configured as inputs and Port D as outputs. Port A on the 68HC11A1 only has one bit that can be configured as an input and output. This is different for other versions of the 68HC11, such as the 68HC11E9 which has 2 configurable lines [11]. To set the digital line on port A to an input, bit 7 of the internal register PACTL (1026h) needs to be clear (see documentation [10,19]). Port D has a corresponding data direction register (DDRD) which sets the direction of the port. Writing all ones to all the bits in this register sets the port to an output port.

The program then sets up to loop indefinitely and reads in port A and places the corresponding bits on port D. The loop then places a string in a buffer and enables the serial interrupt. The routine SerialRoutine sends a serial string of the status of the bits on port A.

Program 2 shows the outline of a program that generates a PWM waveform on the PORT A pin that corresponds to output compare timer 2 (OC2). The main loop sets up the interrupt routine and enables the PORT A pin to be toggled every time the interrupt routine is accessed. It then goes into an infinite loop. The value that corresponds to the width of the ON pulse, can be changed in the main loop and the interrupt routine will change the pulse in the background.

To scan in all 8 analog lines into the 68HC11 requires the user to scan in four lines at a time. Program 3 shows a simple routine that scans in the first four analog input lines and transmits the results out the serial port using the routine SerialRoutine in Program 1. The first part of the main loop, starts the conversion. On this microcontroller, the four conversions take 128 clock cycles (given that no interrupt routine has been in service in that time). Once the conversion is initiated, the program waits until the conversion is completed. It then constructs a string into the serial buffer and sends it.

Overall, this platform has many uses. As a development tool, engineers can debug software before implementing it in the final hardware platform (useful if the target hardware platform is not available at the time). As an educational tool, this platform teaches newcomers the basics of using a microcontroller.

Program 2 Generating PWM waveforms

```
#define      PWM_PERIOD    10000  /* clock cycles */

static int PWM_value;

void interrupt PWM(void)
{
  if (PORTA & D6)
    TOC2 += PWM_value;
  else
    TOC2 += PWM_PERIOD - PWM_value;

  TFLG1 &= D6;      /* clear interrupt flag */
}
```

```
void main(void)
{
    .
    .
```

Program 3 Scanning the analog input lines

    .
    .

```
void main(void)
{
    .
    .
    .

  while (1)
  {
    ADCTL = 0x10;          /* start 4 channel conversion */
    while (ADCTL & CCF);   /* wait until converion is com-
plete */
    sprintf(SerialBuffer, "analogs:  1:%d  2:%d  3:%d
4:%d\r", ADR1, ADR2, ADR3, ADR4);
    SCCR2 |= D7;
    while (!(SCSR & D7));   /* wait until transmitted */
  }
}
```

B. Coin Counter Using the Intel 8051/8031 Microcontroller

The 8051 has many derivatives which have been produced by many companies.
Some derivatives have dual serial ports, PWM outputs, analog inputs and so on. In
the following case study, we present the implementation of a coin sorting machine
based around the 8051/8031 microcontroller [1]. The difference between the 8031
and the 8051 microcontroller is that the 8031 does not contain an internal
EPROM. The standard 8031 has the following specifications:

- 0 bytes EPROM
- 128 bytes RAM
- Four of 8 bit I/O Ports
- Two of 16 bit Timers/Counters
- Full duplex serial UART
- Five maskable interrupts

The complete system provides some additions features. These include:

- Up to 64k EPROM
- 1k Serial EEPROM
- Three of 8 bit I/O ports (supplied by the 8255)
- RESET WATCHDOG control chip
- Two line by 16 character, LCD display
- 12 key keypad, of which only nine can be accessed by the user

Most vending machines store the coins they collect in a cash box where all the denominations are mixed together. To sort these coins manually, requires a significant amount of time and human error can introduce incorrect coin counts and coin mixing. The automatic coin sorter/counter (ACSC) presented in this case study was designed to overcome these problems. The ACSC takes the coins from a vending machine, sorts them into their denominations and counts them so they can be easily banked. The rest of this case study outlines the functional design and implementation of an ACSC based around the 8031 microcontroller.

Functional Description

Before we look at the electronic design, we need to know how the mechanics of the coin sorter operate. The mechanical aspects of the ACSC consists of a hopper which holds the unsorted coins, a coin discriminator which identifies the coin, individual coin tubes which hold the sorted coins and rejected coins, various sensors which detect the current coin position and the microcontroller.

The coin discriminator used in the ACSC contains a channel through which the coins travel. Along this channel there are multiple sensors which perform checks on the properties of the coin. These properties include coin composition, diameter and surface features. These are measured by using three oscillators. The inductors for these oscillators are located adjacent to the coin channel. As the coin passes each inductor, the frequency of the oscillator changes by a consistent value for each coin denomination. This check is performed with three different frequency oscillators, using three different inductors. For a coin to be identified, the frequency shifts of each of the oscillators must fall in the range of one of the coin's frequency shifts stored in the discriminator memory. If one of the frequency shifts is outside the range for that coin, the coin is rejected.

The coin discriminator also contains its own microcontroller. This microcontroller performs checks on the coin as it passes through the discriminator, to determine its denomination. If the discriminator cannot determine the coin denomination, it is rejected. Otherwise the coin value is outputted onto one of its 6 output lines.

The coin collecting tubes have solenoid-operated gates which open under processor control. These tubes are positioned in such a way that each coin rolls past each gate, until it reaches the tube that matches its denomination (which will have its gate open).

To operate the ACSC, the unsorted coins are placed into the hopper and the start button is pressed. This starts the hopper rotating which feeds one coin at a time into the coin discriminator via the discriminator gate. If the coin denomination is identified, the coin is then passed to the channel which feeds the coin tubes. If a coin does not get identified, it sends it to the reject tube. As the coin is identified by the discriminator, the microcontroller opens the gate to the coin tube associated with its value and the coin falls into the tube. If the microcontroller does not detect the coin going into the tube, the unit stops and displays an error message.

Figure 23 shows the functional description of the electronics of the ACSC. This mechatronics platform is made up of several functional units. These include:

- Serial EEPROM. The purpose of the EEPROM is to store configuration data, such as the currency unit, value of each coin, number of coins per tube, etc. Access to the EEPROM is via a 4 wire bus (clock, data in, data out and chip select). [24]
- Coin gates and motor control. The coin gates and motor control are controlled via the 8255 (GPIO) which has its outputs buffered by an open collector transistor array. These provide the required current to operate the solenoids.
- OPTO PHOTO INTERRUPTER Coin Sensors - A photo interrupter is a nontactile sensor, which uses the presence or absence of infrared light on a sensor to detect a moving object. A common example of this simple device is the "magic eye" that prevents people from walking out the entrance in some stores by closing a gate.

  The purpose of these sensors is to check if the coin has gone into the right tube. Also on startup to check if there are any coins in any of the tubes. In total, there are 13 sensors: 6 just after the coin gates; 6 at the bottom of the tubes and one at the gate to the coin discriminator. The 6 sensors at the bottom of the tube are ANDed together so that only one digital line to the microcontroller is required to read the empty status.
- RESET/WATCHDOG - The majority of the microcontrollers include a Watchdog as part of the internal workings. The standard 8031 does not include this feature. Therefore, we have added a Reset/Watchdog chip to provide a Watchdog timer for the microcontroller. A Watchdog is a timer that needs to be reset before it times out. If it times out, the Watchdog will reset the microcontroller. To prevent the Watchdog from timing out the microcontroller toggles the STROBE pin of the Watchdog chip.
- LCD Display and Keypad - To display messages to the user we have also included an LCD display. This enables us to display error messages and current coin counts/value. It also allows us to display the responses from the users instructions from the keypad.

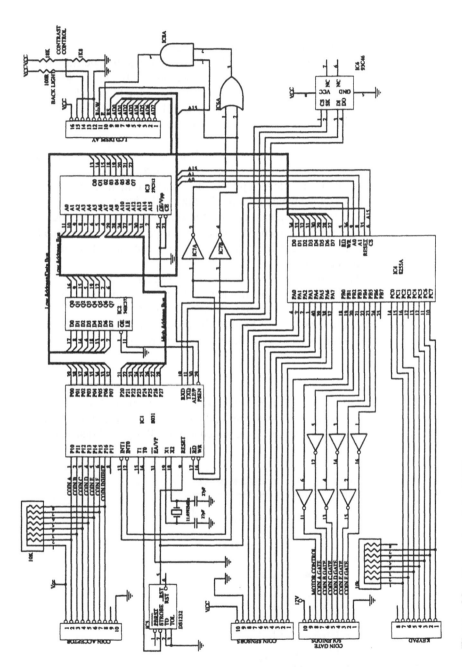

Figure 23 Electronics of the ACSC.

Hardware Description

Figure 13 in Section III.D shows the schematic diagram of the ACSC. First we see that the system has an external EPROM (IC3) which requires the microcontroller to run in expanded multiplexed mode (see Section III.D). On the 8031, the low address and data buses are multiplexed on ports 0 and the high address is on port 2. Using a latch (IC2) to de-multiplex the data and low 8 bits of the address bus, the microcontroller generates an external 16-bit address bus and 8-bit data bus.

The 8255 (IC4) provides the extra digital data lines required to control the motor, solenoids, sensors and the keypad. Port C of the 8255 can be operated in a nibble wise format, with one nibble as input and the other as output. This is perfect for the scanning of the keypad which requires 3 output and 4 input lines.

The Watchdog chip (IC5) can be set for different timeout periods by changing the logic states of its TD pin. The TOL pin is used to set the tolerance for the power supply. For more in-depth explanation of the operation of this chip refer to Dallas Semiconductor [25].

The EEPROM (IC6) is used to store the system setting during power down. It has 1k (1024) bits of memory space, arranged as 64 memory locations of 16 bit words. More information can be obtained from Microchip [26,27].

A more detailed explanation of how the hardware operates, is presented in the software description in Section V.B (Software Description).

Memory Map

The 8031 does not have continuous program/data memory space. Instead it contains a separate memory space for program and data as shown in Figure 24. This has the ability to address 64k bytes of program code and 64k bytes of data using a

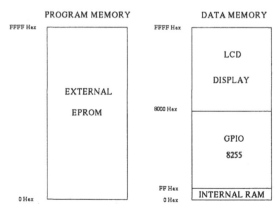

Figure 24 Memory map for the automatic coin sorter/counter.

16-bit address bus. The program memory can still be used to store constant data, such as strings and lookup tables. However, it cannot be used for random access of variables or external device addressing.

The program code for the operation of this unit is stored in an external EPROM. This allows for easier program updates to be made without having to replace the microcontroller. Which would be the case with an OTP microcontroller (See Section I.B, Memory). The maximum addressable EPROM size in this design is 64k bytes. This will be more than enough program space to perform the function required. The amount of program code should be only several kilobytes, which would also fit onto an OTP version of the 8031. This option was not chosen because:

- Printed circuit board space was not a consideration.
- The cost of an OTP can be more than a standard 8031 with additional EPROM.
- The use of an OTP microcontroller would require a special OTP programmer.
- If an OTP microcontroller were programmed incorrectly it would be useless and would have to be discarded. An EPROM can be reprogrammed (see Section I.B, Memory).

The 8031 contains 128 bytes of internal RAM. As the memory content is lost when power is removed, the internal RAM was used for volatile variable storage. The data address space in this design is used to access the 8255 and the LCD display. As a result of this, these devices were connected in such a way that the 8255 occupies half of the entire address space (the upper 32k bytes) and the LCD occupies the lower half. The benefit of this is that you do not require complicated address decoding. This is called incomplete or redundant address decoding. This means several memory addresses can be used to access the same physical memory location.

To include full address decoding, we would require a circuit functional equivalent to the one presented in Figure 25. Equivalently Figure 26 gives the circuit using incomplete address decoding. Although there is only one logic gate difference between the two circuits, there would be more due to the unavailability of a 13 input AND gate which makes this circuit not practical. To realize this circuit, two 8 input AND gates would have to be used with their outputs ANDed together with a 2 input AND gate. Alternatively the entire circuit could be placed in a gate array, PAL or similar device.

The LCD used in this design is a 2 line, 16-character alphanumeric type, with backlighting. It can be used in a 4 bit or 8 bit mode. These two modes differ only in that the commands are sent to the display as 8 bit byte or as two *nibbles*. In this design, we connect the LCD directly onto the address/data bus and therefore program it in the 8 bit mode. The LCD contains 80 bytes of RAM of which

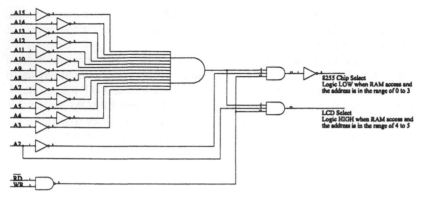

Figure 25   Complete address decoding example.

only 16 bytes are displayed on the screen at any one time. The other 64 bytes can be used as general purpose RAM or for additional screen data.

Software Description

The software flow chart in Figure 27 shows the flow of the software for the Automatic Coin Sorter/Counter. On startup, the ACSC checks the sensors to see if any coins are left in the unit and waits until the operator removes them if there is. The reason for this is to start the ACSC in a known condition (i.e., in case someone took some coins out of the unit while it was switched off).

The keypad and sensors are scanned by one of the timer interrupts of the 8031 and the debounced states are stored as flags. This allows the ACSC to respond to the STOP key at any time in the program. It also allows the program to have ready access to key and sensor data without having to call a function and wait for the response after debounce.

As the keys on the keypad bounce (the contacts open and close, repetitively) over several milliseconds, the microcontroller can be confused as to whether the key is pressed or not. To overcome this problem, the key states are sampled at a set interval, and the state of the key is averaged over the last 3 samples (see Section III.G). An example of the software code to perform this function is shown in

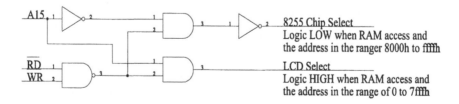

Figure 26   Incomplete address decoding example.

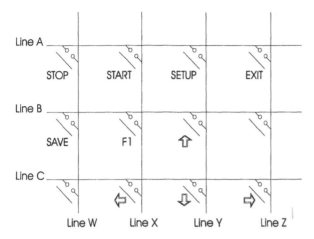

Figure 27 Keypad matrix.

Program 4. The sample code is called as part of the 1ms software timer interrupt in the microcontroller. The keys are not scanned at 1ms intervals, but at 10ms intervals (every 10 calls, the interrupt scans the keys). So for a key to be in the pressed state, 2 or 3 of the last 3 samples need to read that the key is pressed.

Though it is not shown in the software flow chart, the Watchdog's Strobe input is toggled whenever the microcontroller is waiting for a user or sensor input. The keypad on the ΛCSC is arranged in a matrix format. This allows 7 digital data lines to monitor 12 separate keys. This is accomplished by having the keys connected to the digital lines as shown in Figure 28.

Program 4   1ms interrupt routine which calls the keypad routine

```
;*******************************************************
;timr0 -     Timer interrupt which is call every 1ms.
;            This timer is divided down to give a
;            10ms virtual interrupt

timr0:       push   acc
             push   psw

             clr    TR0     ;disable interrupts
             clr    ie.7    ;while we reset the
                            ;timeout

             ;reset 1kHz counters, allow correction
             ;of 7 clock cycles
             mov    a,#low (-922+7)      ;110592/12/100
                                         ;+ 7 cycles
```

```
                add     a,TL0
                mov     TL0,a
                mov     a,#high (-922+7)        ;ideally 921.6
                addc    a,TH0
                mov     TH0,a
                setb    TR0
                setb    ie.7
                setb    psw.4           ;select register bank2
                setb    psw.3
```

{ do all 1ms functions (make sure that it does not
take over 1ms, otherwise we can have the timer being
Corrupted by having the interrupt being called while
still in the interrupt }

```
                djnz    timr1c,no100Hz          ;check divider
                                                ;for slower
                                                        ;interupts
                mov     timr1c,#10
                ljmp    timr1                   ;at 10mS do timr1
no100Hz:        lcall   scanopt                 ;scan optos and
                                                ;switches
                                                ;(every 1ms)

                pop     psw                     ;return from 1ms
                pop     acc                     ;timer interrupt
                reti
```

To scan the keypad, line A is set to a LOW state, then the state of the lines W, X, Y and Z are read. If any of these 3 lines are LOW, then we know that the switch that shorts that line to line A is pressed. Next line A is set back to a HIGH level. The process of setting a line LOW and scanning W, X, Y and Z is repeated for each of the other lines (B AND C). This type of key scanning is used in many applications. One common example is the keyboard of an IBM compatible computer, which has 101+ keys to scan using an 8 bit microcontroller.

The overall operation of the ACSC therefore requires that the keypad is periodically scanned while performing the foreground task which is to look after the coin sorter mechanism.

C. Microchip PIC Microcontroller[4]

The following case study will deal with the design of a motor speed controller based around Microchip's PIC16C74 microcontroller. This microcontroller is versatile and cost effective, suitable for a range of applications - particularly mechatronic control systems. The reader is asked to consult the Microchip Technology

---

[4]Prepared by Mr. Mark Hitchings, R&D Engineer, the Intelligent Control Systems Laboratory, Griffith University, Nathan, Australia.

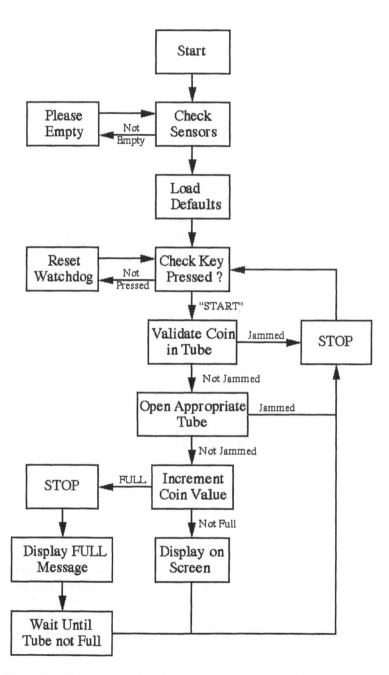

Figure 28 Software flow chart for the automatic coin sorter/counter.

data book for the PIC16C74 [26] for a comprehensive discussion of the various hardware aspects of this microcontroller. Among some of the major features of the PIC16C74 are:

- 4 kByte ROM
- 196 byte RAM
- 2 PWM modules
- 2 timer-counter modules
- 8 sources of interrupt
- 5 Input/output ports (I/O)
- 8 channels of 8 bit analog to digital converters (A/D)
- Harvard architecture with RISC-like instruction set

The aim of this case study is to present the complete design process from conception through to realization. Firstly, a functional description of the motor speed controller will be given. Sections detailing hardware and software design will follow this. Finally, a discussion on the implementation of the system will be given which will draw conclusions on its overall effectiveness.

Functional Description

The goal of this case study is to design an automatic DC servomotor speed controller that will maintain the motor shaft speed irrespective of a reasonable mechanical load placed on the shaft. The magnitude of this load will depend on the motor's no load speed and rated torque. An effective means of achieving this goal is to use a proportional integral (PI) controller. The PI controller is commonly known as a feedback controller. This simply means that the parameter that is to be controlled (in this case motor speed) is passed back to the input of the controller. The difference (or error) between the desired value and the actual value of the parameter is used to vary the controller output until a state of equilibrium is reached. Typically mechatronic control systems are comprised of the following major functional items:

- the plant or actuator (output)
- sensor/s to measure the controlled parameter
- the controller
- the input

The plant in this case is a DC servomotor. It is desired that the velocity, or rate of motor shaft spin, is maintained at a constant set level. This level will be set via an input source or an electrical signal fed to the controller indicating the desired motor shaft velocity. In control theory this signal is referred to as the setpoint. The speed of the motor must be translated from a rotational force into an electrical signal that the controller can use to compare it to the set-point value. This conversion is achieved by way of a sensor. The sensor we will be using is an optical shaft encoder. More detail relating to optical shaft encoders will be given in coming sections. The controller is a generic term used for any device that automatically controls connected plant according to a desired set-point condition. In our case we will be using a computing device called a microcontroller, which performs the control tasks according to software pre-programmed into its memory.

It is perhaps prudent at this stage to draw up a block diagram of the system that will make up our motor speed controller. This is usually one of the first steps of the design process after we have decided how we will achieve the desired system. Figure 29 shows the system block diagram and the following section will investigate in detail each of the major functional blocks.

Hardware Description

From Figure 29 it can be seen that there are a number of major sections in the design. These sections will be detailed in the following discussion and include:

- Input
- Controller
- Sensor
- Output

The input section is fairly simple and consists of a variable resistor and fixed value resistor forming a voltage divider function between the supply voltages of 0 volts and 5 volts. The purpose of the input section to provide signal that is representative of the desired amount of motor speed. This is referred to as the set-point value. Our system will read the input voltage level set by the variable resistor

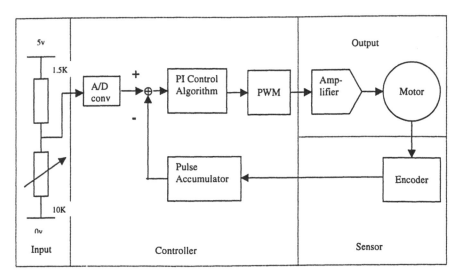

Figure 29 System block diagram of the motor speed controller.

into an analog to digital (A/D) converter. This will convert the input analog voltage level into a digital word, which can be used by the PI control algorithm as a set-point value. Using Kirchoff's voltage laws, it can be found that when the potentiometer is turned fully anti-clockwise (0 ohm resistance) an input of 0 volts is presented to the controller. We will use this value as an indicator of the slowest speed desired of the motor. It can also be found that when the potentiometer is turned fully clockwise (10 Kohm resistance) an input of 4.35 volts is presented to the controller. We will use this value as an indicator of the fastest speed desired of the motor.

Next on Figure 29 we see the controller section. It can be seen that the controller encompasses the A/D converter, pulse accumulator, summer, PI control algorithm and pulse width modulation (PWM). This perhaps represents a good example of the great advantage of microcontrollers in mechatronic systems. Microcontrollers combine a number of peripheral functions onto the same chip as the CPU and memory. The A/D converter on the PIC16C74 will be used to convert the input voltage set-point to an 8 bit digital word. This converter uses the supply voltage as a reference to it's input and produces values between 00_{16} and FF_{16}. An A/D output of 00_{16} corresponds to an input of 0 volts and FF_{16} corresponds to an input of 5 volts. All inputs in between are assigned a value in steps of 0.02 volts. For more details on A/D conversion, please refer to one of the many books dealing with the subject. Proakis and Manolakis [27] is one good example. The summer and PI control algorithm are implemented in our system as software and will be dealt with in the following section.

The motor speed is controlled using PWM. Pulse width modulation works on the principle of applying an electrical signal to the motor consisting of a repeating on/off pulse pattern. The motor speed is dependent on the amount of time the signal pulse is "on" compared to the amount of time that the signal is "off." It can be seen from Figure 30 below that there are two parameters to control in PWM—the *period* and the *duty cycle*.

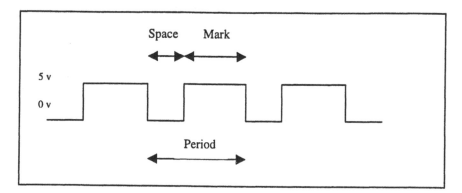

Figure 30 Typical PWM signal components.

As it can be seen the waveform is made up of a series of marks and spaces. The marks are simply the regions where the signal is "on" (5 volts). The spaces are the regions where the signal is "off" (0 volts). The period is defined as the amount of time that the signal takes to repeat itself (one mark and one space). The *duty cycle* is defined as the ratio or percentage of mark time to space time or mark to-space ratio. Let us fix the period for now at say 1 millisecond. If we set the mark time to 0.5 millisecond, we have a repeating signal that is 5 volts for 0.5 millisecond and 0 volts for 0.5 millisecond. This corresponds to a 50% duty cycle. It may also be concluded that the *average* voltage of this signal over time is 2.5 volts. This is how PWM is used to vary motor speed - by varying the duty cycle, the average (or DC) voltage to the motor is varied. The motor speed is dependent on the voltage applied to its terminals. A 100% duty cycle corresponds to the PWM signal being 5 volts all the time giving the maximum average voltage to the motor of 5 volts and thus the maximum motor speed. The PWM signal is generated in our system using the on-board PWM module of the PIC16C74. Once this module has been set up, it runs independently of the CPU and need only be accessed when changing the PWM period or duty cycle.

The output signal to the motor from the microcontroller must be buffered. This simply means that the microcontroller is unable to supply sufficient current to the motor via direct application of the PWM signal and so that signal must be amplified using a peripheral device. One good method is to use what is known as a H-bridge. This is a diode network that is configured so that current may be applied to the motor in different polarities for different motor rotation directions. A good example of a H-bridge amplifier chip is the National Semiconductor LM293.

So how do we measure the motor speed? There are numerous methods. However in our case we shall use an optical shaft encoder. An optical shaft encoder comprises a disk mounted on the shaft of the motor and a light emitting diode transmitter and photodiode receiver. The disc is comprised of alternate white and black segments such as shown below in Figure 31.

The light emitting diode (LED) and receiver are manufactured into one package and are fixed near the motor, facing the disc. The LED emits light onto the disc and if a white segment is lined up with the incoming light, the light will be reflected back to the receiver. If a black segment is lined up with the transmitter, then the incoming light is absorbed and not reflected. If the reflected light is collected by the phototransistor and used to develop an electrical signal, a pulse train similar to that shown in Figure 31 will be developed. It can be seen from Figure 31 that our encoder wheel has 32 segments and so it can be expected that there will be 32 pulses per motor shaft revolution. We have only briefly discussed optical wheel encoders. For a more complete discussion please refer to one of the many excellent books available such as Jones, 1993 [28].

If we count the electrical pulses produced by the shaft encoder over time, a measure of the motor rotational velocity is obtained. This is done using pulse accumulation techniques on the microcontroller. The accumulated value may be read at any time in the software running on the microcontroller, which can be used to

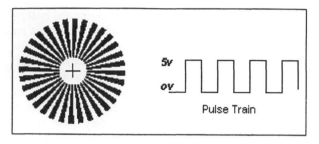

Figure 31 Optical wheel encoder disc and resulting signal.

calculate the motor shaft speed. All that remains in our hardware design is to draw up the schematic diagram of the hardware. From this schematic the circuit may be implemented on a prototyping board, or a printed circuit board (PCB) may be designed and manufactured. The complete schematic diagram is presented as Figure 32.

Software Description

The general functionality of the software algorithm is as follows:

1. Read the set-point value (A/D conversion).
2. Read the current motor speed (pulse counting).
3. Compare the set-point to the actual motor speed value to obtain the *error value.*
4. Multiply the error value by a *proportional constant* - this is the proportional part of the algorithm.
5. Integrate (accumulate the error value over time) and multiply by an *integral action constant* - this is the integral part of the algorithm.
6. Add the proportional and integral errors together.
7. Add this error value to the duty cycle value of the PWM.

Repeat steps 1 through 7 after a set period of time.

From this list of functionality it can be seen that there are four significant functions of the software algorithm and we will discuss each of them separately in turn. These are:

*   PI control algorithm
*   Pulse width modulation (PWM)
*   Analog to digital (A/D) conversion
*   Pulse counting

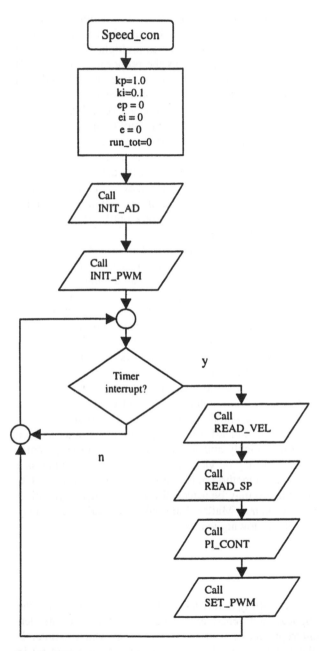

Figure 32   Schematic diagram.

Let us first discuss the proportional part of the PI control algorithm. From Figure 29 it can be seen that the error signal is derived from subtracting the actual velocity from the set-point value. This error signal is then multiplied by a constant value called the *proportional constant*, k_p, giving us the proportional error e_p. If the actual motor speed is less than the set-point, e_p is positive and this value is added to the PWM duty cycle signal fed to the motors to increase the motor speed. If the actual motor speed is more than the set-point, e_p is negative and this value is subtracted from the signal fed to the motors and so the motor slows down. By varying the value of k_p we increase the amount of proportional control and hence the speed of response of the controller. Care must be exercised however, not to increase k_p by too much, otherwise the system becomes unstable. An effect of instability is that the controller overcompensates for changes in motor speed and the motor tends to hunt (increase and decrease speed in an oscillatory fashion). The controller continuously adjusts the speed of the motor until a steady state is reached and no further speed corrections are required.

Now let us look at the integral part of the PI control algorithm. The integral part works by adding the error signal to a running total over time. This running total is multiplied by an *integral action constant* k_i giving us the integral error e_i. This resulting error signal is then added to the proportional error e_p to achieve the overall controller output. As with k_p, the value of k_i may be varied in order to achieve the best trade-off between speed of controller response and system stability. The question remains as to what are the optimum values of k_p and k_i. The simple answer is that they are found by experimentation. The system must be implemented and its speed of response and stability characteristics must be measured for different values of k_p and k_i. This method is haphazard, time consuming and may not necessarily lead us to the optimum values. It is far wiser to use one of the many modeling and simulation tools available on the market. The motor and controller characteristics may be simulated in software on a personal computer and the values of k_p and k_i may be varied quickly and easily. The various output parameters such as motor speed and so on may be viewed in graphical form, giving us a much more time efficient and certain method of determining optimum values of k_p and k_i. The modeling and simulation may be achieved very easily using graphical and matrix based tools available such as Matlab, Simulink and Matrix$_x$, and many more. For now let us settle on the following values:

$k_p = 1.0$

$k_i = 0.1$

The pulse counting is very straight forward. All we need to do is use the pulses generated by the optical shaft encoder to interrupt the microcontroller. When this happens, an interrupt service routine is invoked in which counter is incremented. The accumulated count is read periodically—the time period set in the software of the microcontroller. We will call this our sample time. Every time this period is finished the program:

- reads the value from the pulse counter
- resets this value to zero for the next time period
- reads the value present at the A/D converter - our set-point
- restarts the time period in the software for the next sample time

The amount of time between samples can be worked out as follows. Let us say the maximum speed of the motor is 4500 revolutions per minute, giving us 75 revolutions per second. The optical wheel encoder generates 32 pulses per revolution which will mean that there will be 2400 pulses per second. The PIC16C74 microcontroller has an 8 bit data bus. This gives a maximum value that may be held in a data register of 255 (FF_{16}). We should therefore read the pulse counter register before it overflows at a value of 255. This equates to a maximum sample time of approximately 0.1 seconds. It can be concluded that after every sample time of 0.1 seconds, we can read the pulse counter and its value will be somewhere between 0 and 240 depending on the current motor speed.

Let us use the value read from the analog to digital converter as the set-point value and it is limited to 240. An important point to note at this stage concerns the choice of potentiometer to be used for the input. Potentiometers typically are either linear or logarithmic in scale. This means that the terminal resistance given by the potentiometer changes either linearly (increasing in evenly spaced steps) or logarithmically (logarithmically increasing steps). For smooth control of our motor speed over the range of control we must choose a linear control type potentiometer. Based on a set-point range of 0 to 240 and a pulse counter range of 0 to 240, the error value will have a range of –240 to +240. As an example, if the motor is stopped and we adjust the potentiometer value for full speed of 240, the error at that instant will be a maximum of 240. This error will decrease over time as the controller endeavors to reduce this error by increasing the motor speed. Let us fix the period of the PWM signal at 240 microseconds. The duty cycle value can therefore take values between 0 and 240 microseconds. The adjustment of the speed is applied by adding the value of the error from the PI controller to the value of the duty-cycle. Care must be taken to ensure that the duty cycle value does not fall outside its possible range - obviously we can't have a negative duty-cycle value or one that exceeds the period. The algorithm must limit the value of the duty cycle to these maximum and minimum values.

So let us put all of these thoughts together into one logical format from which the software code may be derived. There are many ways to express an algorithm; however in this case let us use a flowchart. A flowchart expresses all of the logical steps of a program in a graphical format, where the flow of a program may be traced quite easily. Our flowchart appears as Figure 33. Due to space restrictions, we won't expand on the subroutine calls, but as they are fairly straightforward, this shouldn't be a major limitation to understanding the software.

Now all that remains in the design of our system is the writing of the software itself and downloading it into our hardware. There are a few methods of producing the software for a microcontroller systems such as: using higher level lan-

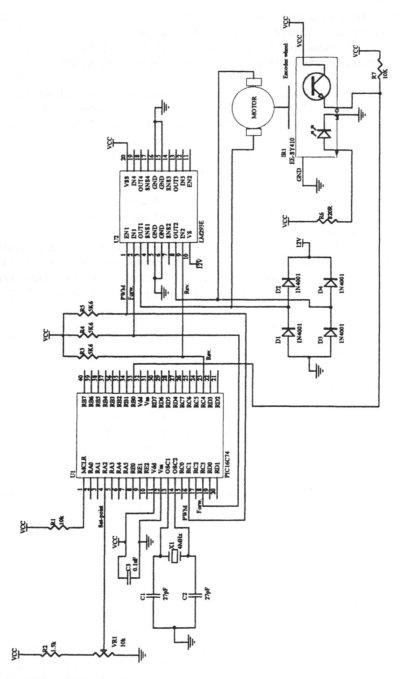

Figure 33 Flowchart of the motor speed controller software.

guages such as C and cross-compiling this code into the machine language format suitable for downloading into the microcontroller system; and writing an assembler language program and compiling this into machine language format suitable for downloading into the microcontroller system.

We will use assembler language programming here because it is easy to understand and there is no prerequisite purchase of a C cross-compiler. The assembler language program is first written on a personal computer using an ASCII text editor. This code is then assembled using a third party assembler available free of charge from Microchip Technology Inc. This assembler may be downloaded from Microchip's Internet website or is available from your local Microchip retailer. The resulting binary file produced from the assembler is then down-loaded into the EPROM of the PIC16C74 using a commercially available programmer. The microcontroller "plugs into" the programmer which is connected to a personal computer to download the binary file from the computer into the microcontroller. There are many of these programmers available on the market and it is probably wise to contact your nearest Microchip retailer for a list of currently available models. The assembler code that forms the algorithm for our motor speed controller is shown as Program 5.

It is not the intention of this case study to discuss line by line the code for operation of each of the microcontroller modules such as A/D converter, interrupts or PWM. The reader is encouraged to consult the relevant Microchip application notes for full details on the programming of each microcontroller module. Some useful application notes may be found in their embedded control handbook [27] or from their Internet website.

Conclusions

A case study involving the design of a motor speed controller using the Microchip PIC16C74 was presented. Mechatronic design typically requires the understanding of many individual components and the combination of these components into one cohesive system. Microcontrollers such as the Microchip PIC16C74 offer the mechatronic designers a flexible and yet powerful platform on which to base their designs. It is hoped that this case study provided the reader with a little insight into the design process - both hardware and software of a typical mechatronic system. It is also hoped that an appreciation of the factors to consider while designing a mechatronic system will also be gained.

Program 5 Software for the motor speed controller

```
;TITLE:          "SPEED_CON.ASM"
;AUTHOR:         Mark Hitchings
;DATE:           20/7/97
;REVISION:       1.0
;PLATFORM:       Microchip PIC16C74 assembly language
;DESCRIPTION:
;                This program implements a motor speed
                 controller using a PI control algorithm
```

```
            list    P=16C74, F=INHX8M, n=66
            ;8 BIT HEX., 66 LINES TO A PAGE
            include "p16c74.inc"
            __config 0X3F31        ;configuration   word   CP0=1,
PWRTE=0, WDTE=0, XT
            ERRORLEVEL 1,-302      ; Turn OFF message ID 302

;Declare RAM locations
E           EQU     0X20
EP          EQU     0X21
EI          EQU     0X22
RUN_TOT     EQU     0X23
SET_PT      EQU     0X24
VELOCITY    EQU     0X25
VEL_CTR     EQU     0X26
CTR3        EQU     0X27
CTR2        EQU     0X28
CTR1        EQU     0X29

            ORG     0X00
RESET
            BSF     STATUS,RP0     ;RAM Bank 1
            GOTO    START          ;Reset, then start the pro-
gram
;-----------------------------------------------------------
-
;interrupt Service Routines go here
;-----------------------------------------------------------
-
            ORG     0X04
PER_INT_V
            CALL    PUSH           ;Save accumulator and status
            BTFSC   INTCON,INTF    ;Is   it   an   PORTB(0)   inter-
rupt?
            CALL    INC_CTR        ;YES,  then  increment  veloc-
ity
                                   counter
            CALL    POP            ;Restore accumulator and
                                   status
            RETFIE
;-----------------------------------------------------------
-
;Start of main code
;-----------------------------------------------------------
-
START
            BCF     STATUS, RP0    ;Cleared for RAM bank 0
            CLRF    STATUS         ;Clear status flags
            CLRF    PIR1           ;Clear peripheral interrupt
                                   flags
```

```
;-----------------------------------------------------------
-
;Initialize the Peripheral interrupts
;-----------------------------------------------------------
-
          BSF     STATUS,RP0     ;Bank 1
          CLRF    INTCON         ;Clear    interrupt    control
flags
;-----------------------------------------------------------
-
;Set up I/O Port directions and clear the data registers
;-----------------------------------------------------------
-
          MOVLW   0XFF
          BSF     STATUS, RP0    ;Set for bank 1

;CLEAR PORTS' DATA REGISTERS
          BCF     STATUS,RP0     ;Cleared for bank 0
          CLRF    PORTA
          CLRF    PORTB
          CLRF    PORTC
          CLRF    PORTD
          CLRF    PORTE

;SET UP PORTS FOR I/O DIRECTION
          BSF     STATUS,RP0     ;Set for bank 1
          MOVLW   0X3F
          MOVWF   TRISA          ;Set PORTA to all inputs
          MOVLW   0XF1           ;
          MOVWF   TRISB          ;Set PORTB MSBs AND RB0 to
                                 input and REST output
          CLRF    TRISC          ;Set PORTC to all outputs
;-----------------------------------------------------------
-
          CALL    INIT_AD
          CALL    SET_PWM
          CLRF    E
          CLRF    EP
          CLRF    EI
          CLRF    RUN_TOT
          CLRF    VEL_CTR

          BSF     PORTC,RC4      ;Forward control of motor
                                 enabled
          BCF     PORTC,RC3      ;Reverse control of motor
                                 disabled

;-----------------------------------------------------------
-
;CODE MAIN LOOP
```

```
;-------------------------------------------------------------
-
LOOP1
        CALL    WAIT_S1         ;Wait 0.1 seconds
        CALL    READ_VEL        ;Read the current velocity
        CALL    READ_SP         ;Read the set-point value
        CALL    PI_CONT         ;PI controller
        CALL    SET_PWM         ;Adjust the speed
        GOTO    LOOP1           ;do it all again
;-------------------------------------------------------------
-
;SUBROUTINES
;-------------------------------------------------------------
-
;INIT_PWM
;This subroutine initialises the PWM Period and Duty Cycle
variable values
;-------------------------------------------------------------
-
INIT_PWM
        BCF     STATUS,RP0      ;Register Bank 0
        MOVLW   0X30            ;
        MOVWF   DC_HI           ;Most sig. 8 bits of duty
                                cycle (default)

        MOVLW   0X00            ;
        MOVWF   DC_LO           ;Least sig. 2 bits of duty
                                cycle (default)

        MOVLW   0X50            ;Period of the PWM
        MOVWF   T2_PERIOD       ;Set the period (default)
        CALL    SET_PWM
        RETURN
;-------------------------------------------------------------
-
;INIT_PWM
;This subroutine initializes the PWM Period and Duty Cycle
variable values
;-------------------------------------------------------------
-
INIT_AD
        BSF     STATUS,RP0      ;RAM BANK 0
        MOVLW   B'00000010'     ;
        MOVWF   ADCON1          ;Make PORTA analog
        BCF     STATUS,RP0      ;RAM BANK 1
        MOVLW   B'11010001'     ;
        MOVWF   ADCON0          ;A/D RC osc., RA2, ADON
        CLRF    ADRES           ;Clear the result register
        RETURN
;-------------------------------------------------------------
```

```
-
;DELAY
;A subroutine which implements a delay of 514 clock cycles =
0.5 ms
;----------------------------------------------------------------
-
DELAY
          BCF    STATUS,RP0     ;Cleared for bank 0
          MOVLW  0XFF           ;
          MOVWF  CTR1           ;Load 255 into loop counter
DELAY2
          DECFSZ CTR1,1         ;Decrements   counter   until
zero
          GOTO   DELAY2
          RETURN                ;Returns when count is zero
;----------------------------------------------------------------
-
;WAIT_S1
;A subroutine which implements a delay of 111,157 clock cy-
cles = 0.1 SECOND
;----------------------------------------------------------------
-
WAIT_S1
          BCF    STATUS,RP0     ;Cleared for bank 0
          MOVLW  0X01           ;
          MOVWF  CTR3           ;Load 3 into outside loop
                                counter
CONT_S1O
          MOVLW  0XD7           ;Inside loop = 111,157 clock
                                cycles
          MOVWF  CTR2           ;Load 215 into inside loop
                                counter
CONT_S1I
          CALL   DELAY          ;Waste 515 microseconds
          DECFSZ CTR2,F         ;Decrements   counter   until
zero
          GOTO   CONT_S1I       ;Not zero then continue in-
side
                                loop
          DECFSZ CTR3,F
          GOTO   CONT_S1O       ;Not zero then continue
                                outside loop
          RETURN                ;Returns when count is zero
;----------------------------------------------------------------
-
;READ_VEL
;----------------------------------------------------------------
-
READ_VEL
          MOVF   VEL_CTR,W
```

```
            MOVWF   VELOCITY
            CLRF    VEL_CTR
            RETURN
;------------------------------------------------------------
-
;READ_SP
;------------------------------------------------------------
-
READ_SP
            MOVLW   0X10            ;
            MOVWF   CTR1            ;Load 16 into loop counter
SH_1
            DECFSZ CTR1,F           ;Decrements    counter    until
zero
            GOTO    SH_1
            BSF     ADCON0,GO       ;Start AD process
WAIT_AD
            BTFSC   ADCON0,DONE     ;Conversion done?
            GOTO    WAIT_AD         ;NO, keep waiting
            MOVF    ADRES,W         ;YES, move AD result to
            accumulator
            MOVWF   SET_PT          ;Save the current SS in temp
                                    var.
            RETURN
;------------------------------------------------------------
-
;PI_CONT
; PI control algorithm
;------------------------------------------------------------
-
PI_CONT
            MOVF    VELOCITY,W
            SUBWF   SET_PT,W
            BTFSC   STATUS,C
            GOTO    SLOW_IT
SPEED_IT
            MOVWF   E
            ;KP MULTIPLY HERE
            MOVF    RUN_TOT,W
            ADDWF   E,W
            MOVWF   RUN_TOT
            ;KI MULTIPLY HERE
            MOVF    P,W
            ADDWF   I,W
            MOVWF   E
            ADDWF   DC_HI,W
            BTFSS   STATUS,C
            GOTO    PI_DONE
            MOVLW   0XF0
            MOVWF   DC_HI
```

```
            GOTO    PI_DONE
SLOW_IT
            MOVF    SET_PT,W
            SUBWF   VELOCITY,W
            MOVWF   E
            ;KP MULTIPLY HERE
            SUBWF   RUN_TOT,W
            MOVWF   RUN_TOT
            ;KI MULTIPLY HERE
            MOVWF   I
            ADDWF   P,W
            MOVWF   E
            SUBWF   DC_HI,W
            BTFSS   STATUS,C
            GOTO    PI_DONE
            MOVLW   0X00
            MOVWF   DC_HI
            GOTO    PI_DONE
PI_DONE
            CALL    SET_PWM
            RETURN
;-------------------------------------------------------------
-
;PWM SERVICING
;Wait for the Duty Cycle value to roll over then reload the
D/C and period
;values
;-------------------------------------------------------------
-
;Set up the CCP modules for PWM operations
;-------------------------------------------------------------
-
SET_PWM
            CLRF    PR2             ; Clear the period register
                                    of tmr2
            BCF     STATUS,RP0      ;Cleared for bank 0
            MOVLW   0X0C
            MOVWF   CCP1CON         ;Set up CCP1 module for PWM
                                    mode
            CLRF    PIR1            ;Clear the peripheral
                                    interrupt reg.
            CLRF    T2CON           ;Clear the Timer 2 control
                                    register
            BCF     T2CON,0         ;Make TMR2 prescaler 1
            BSF     STATUS, RP0     ;Set for bank 1
            BSF     INTCON,PEIE     ;Enable   peripheral   inter-
                                    rupts
            BSF     INTCON,GIE      ;Enable Global Interrupts
            BCF     STATUS,RP0      ;Cleared for bank 0
            BSF     T2CON,TMR2ON    ;Turn on Timer 2
```

```
              CALL    WAIT_DC
              RETURN
WAIT_DC
              BCF     STATUS,RP0    ;Cleared for bank 0
              MOVF    TMR2,W        ;Read current timer 2 value
              SUBWF   PR2,W         ;How close is the timer to
                                       rolling over?
              ANDLW   0X0F          ;Does this make it zero?
              BTFSC   STATUS,Z      ;If it is zero, then
                                       rollover is near
              GOTO    WAIT_DC       ;If it is not zero repeat
                                       above until it is
              MOVF    DC_HI,W       ;If rolled over then load
                                       the duty cycle msb's
              MOVWF   CCPR1L        ;Set the DC high bits - CCP1
              MOVLW   0X0F          ;
   ANDWF      CCP1CON,F             ;Clear the lsb's of D/C -
CCP1
              BTFSC   DC_LO,1       ;Set the DC low bits - CCP1
              BSF     CCP1CON,CCP1X ;Set the DC low bits - CCP1
              BTFSC   DC_LO,0       ;Set the DC low bits - CCP1
              BSF     CCP1CON,CCP1Y ;Set the DC low bits - CCP1
              BCF     PIR1          ;Clear the TMR2=PR2 Flag
;------------------------------------------------------------
-
;Wait until TMR2 value is equal to the preset period value,
then calculate
;if any offset the period value needs to be added.
;------------------------------------------------------------
-
WAIT_PR
              BTFSS   PIR1,TMR2IF
              GOTO    WAIT_PR       ;Wait until TMR2=PR2 so that
                                     the DC latches

              BSF     STATUS,RP0    ;Bank 1
              MOVLW   0X0F          ;Load  timer  2  period  with
the
                                     min. value
              MOVWF   PR2

              MOVLW   0XF0          ;
              ANDWF   T2_PERIOD, W  ;
              BTFSC   STATUS,Z      ;Determines if TMR2 period
                                     needs to increase
              GOTO    NO_OFFSET     ;No, it's fine

PR_OFFSET
              MOVLW   0X0F          ;Yes... calculate the
                                     additional offset
```

```
            SUBWF  T2_PERIOD,W
            ADDWF  PR2,F          ;Add the period offset

NO_OFFSET
            BCF    STATUS,RP0     ;Bank 0
            RETURN
;-----------------------------------------------------------
-
;INC_CTR
;-----------------------------------------------------------
-
INC_CTR
            MOVF   VEL_CTR,W
            ADDLW  0X01
            MOVWF  VEL_CTR
            RETURN
;-----------------------------------------------------------
-
;PUSH
; This routine pushes the accumulator and STATUS register
into
;   temporary storage variables during an interrupt service
routine.
;-----------------------------------------------------------
-
PUSH
            MOVWF  TEMP_W         ;Save W register
            SWAPF  STATUS,W       ;Get STATUS register
            BCF    STATUS,RP0     ;Bank 0
            MOVWF  TEMP_STAT      ;Save STATUS register
            RETURN
;-----------------------------------------------------------
-
;POP
; This routine pops the accumulator and STATUS register val-
ues from
;   temporary storage variables at the completion of  an in-
terrupt
;   service routine.
;-----------------------------------------------------------
-
POP
            SWAPF  TEMP_STAT,W    ;Restore STATUS register
            MOVWF  STATUS
            SWAPF  TEMP_W,1
            SWAPF  TEMP_W,W       ;Restore W register
            RETURN
;-----------------------------------------------------------
-
;END OF PROGRAM
```

```
;-----------------------------------------------------------
-
            ORG     PMEM_END          ;End of program memory
            END
```

ACKNOWLEDGMENTS

The authors like to thank David Rowlands for his comments about the content of this chapter. Also, thanks must go to Bill Lindbergs and Mark Hitchings for their contribution of the case studies. Thanks also to our families you have supported us throughout this project.

REFERENCES

1. IS MacKenzie. The 8051 Micrcontroller, USA: Prentice-Hall, 1995.
2. http://www.motorola-dsp.com/.
3. M Rafiquzzaman, R Chandra. *Modern Computer Architecture*, West Publishing Company, 1988.
4. http//www.mot.com/SPS/DSP/products/.
5. http://www.ti.com/.
6. http://www.analog.com/.
7. INTEL 8051 DATABOOK.
8. TMS320C2X Users Guide, Texas Instruments Incorporated, 1992.
9. http://www.mcu.motsps.com/mc.html.
10. Advance Information MC68HC11A8. HCMOS Single Chip Microcontroller, Motorola. Order Code: MC68HC11A8/D, 1988.
11. Advance Information MC68HC11E9. HCMOS Single Chip Microcontroller, Motorola. Order Code: MC68HC11E9/D, 1988.
12. http://www.microchip2.com/products/micros/.
13. http://www.ti.com/sc/docs/dsps/univprog/teachkit.htm.
14. http://www.analog.com/products/sheets/ADSP2100.html.
15. *Switchmode. A Designer's Guide for Switching Power Supply Circuits and Components*. Motorola, 1992.
16. Linear and Interface Integrated Circuits, Printed in U.S., Motorola, 1988.
17. http://mot-sps.com/books/dl128/pdf/mc34064rev2f.pdf.
18. http://www.mcu.motsps.com/freeweb/areas.amcu.html.
19. P. Spasov, Microcontroller Technology: The 68HC11., New York: Prentice-Hall, 1993.
20. http://www.mcu.motsps.com/hc11/home.html.
21. http://www.mcu.motsps.com/lit/fam_11.html.
22. http://www.mcu.motsps.com/freeweb/.
23. http://freeware.mcu.motsps.com:80/lit/index.html.
24. http://www.microchip2.com/products/memory/allmem.htm#I2C.
25. http://www.dalsemi.com?DocControl/PDFs/pdfinger.html.
26. Microchip Technology Inc., *PIC16C7X 8-bit CMOS Microcontrollers with A/D Converter- Product Data Book*. Microchip Technology Inc., 1996.

27. J Proakis, D Manolakis. *Digital Signal Processing - Principles, Algorithms and Applications*. Upper Saddle River, New Jersey: Prentice-Hall, 1996.
28. J Jones, A Flynn. *Mobile Robots: Inspiration to Implementation*. Wellesley: A.K. Peters Ltd., 1993.
29. Microchip Technology Inc. *Embedded Control Handbook*. Microchip Technology Inc., 1996.

4
Intelligent Controllers

Vojislav Kecman
The University of Auckland, Auckland, New Zealand

I. INTRODUCTION

Recently many new products equipped with intelligent controllers have been launched on the market, a lot of effort has been made in R & D departments around the world, and numerous papers have been written on how to apply neural networks (NN), fuzzy logic models (FLM), and related ideas of learning (in classic control terms of adaptation) for solving control problems of both linear and non-linear systems. NN based algorithms have been recognized as attractive alternatives to the standard and well established adaptive control schemes. Due to their well known ability to be a universal approximator of multivariate functions, NN and FLM are of particular interest for controlling highly nonlinear, partially known and complex plants or processes. Many good and promising results have been reported and the whole field is developing rapidly but it is still in the initial phase. This is mainly due to a lack of firm theory despite the very good and seemingly far reaching and important experimental results.

Let us first indicate what we are going to discuss in this chapter. The title of this chapter only describes the subject partially, mainly because the meaning of intelligence is very variable and indeterminate. The first thing is the question of what intelligent control means *today*. Recent presentation of intelligent control techniques is of interest. To facilitate the understanding of the permanently changing concept of intelligent control systems let us first review the last 220 years of industrial control engineering.

There is no doubt that the first two control devices, namely Polzunov's float regulator (1765) for controlling a steam boiler's water level and Watt's flyball governor (1769) for controlling the speed of a steam engine, were highly intelligent at the time. Before the introduction of these two regulators, humans had to solve such control problems. They did it with human intelligence by observing the changes in water level (engine speed) and by closing or opening a water (steam) valve to the steam generator (engine). Thus, these very first simple intelligent mechanical devices had successfully replaced man. These two controllers were regulators of so-called proportional or P type. P action is the simplest of all the feedback laws and usually results in an unavoidable offset of the controlled variable from its set point.

Almost a half century of trial and error experiments was devoted to the quest for better controllers. Finally, by introducing integral or I action in the speed governor system, the problem of offset was solved. The PI controller was a much more intelligent device than simple P action. It was able to eliminate offset without human intervention.

A further improvement and refinement in control system performance was achieved just after the turn of 20th century by introducing derivative or D action into control laws. This highly intelligent PID controller was the best solution for the majority of control problems for the next 60 years. P action responded quickly and strongly enough to the actual control error, I action forced the error to zero value by integrating (accumulating or, in modern terms, by memorizing) all of the past errors, and D action predicted the future behavior of the control system by carefully observing trends (derivatives) of the actual error.

Hence, almost all but one of the important constitutive elements of what is meant by intelligence are embodied in the PID control law. The good old PID controller can both *memorize* and *predict*. The important part and the fundamental ability of any intelligent system (biological brain or artificial device), to *learn* and/or to *adapt*, was badly missing in this early PID controllers. Once tuned, these early regulators were not able to cope with the changes of plant dynamics. Faced with time-varying systems, the PID controller did not behave intelligently any longer - it was not able to adapt to the changing environment.

Before plunging into a description of basic adaptive control schemes for linear systems, and into presentation of NN and FLM based adaptive control structures, let us clarify the differences and similarities between the concepts of *learning* and *adaptation*. These two concepts form the basis of intelligent control systems today. This might be important because the present intelligent control is a mixture and an overlapping of classic *adaptive* control and standard NN based *learning* techniques.

In the simplest of colloquial terms we might say that learning is an adaptation from scratch, where scratch means ignorance of the system. Thus, learning usually means acquiring knowledge of a previously unknown system or concept, and adaptation connotes change or refinement of the existing knowledge of the system. This process of knowledge acquisition is naturally achieved through learning, which is also in the field of control is usually called adjustment or tuning. Historically, adaptive techniques originated from attempts to control *time-varying* and *linear* systems. NN and FLM are by their very nature nonlinear modeling systems. Thus an alternative way to think about the learning may also be that it represents an adaptation of the nonlinear NN and/or FLM.

Let us continue our journey from the first rigid and intelligent (though not considered so today) mechanical regulators towards flexible and intelligent (possibly not considered so tomorrow) microprocessors based adaptive control schemes. Thus, the early PID controller with fixed parameters was not able to provide acceptable system performance when coping with time-varying linear or, even worse, with an unknown nonlinear plant. In the late 1950s this fact led to a deep interest in adaptive algorithms starting with the development of the *gain-scheduling* techniques. In the meantime many different adaptive control structures for controlling linear systems have been developed. The two most useful and popular configurations in classic adaptive control theory and practice are self-tuning regulators (STR) and model reference adaptive control (MRAC). These two basic tools for adaptive control of *linear* systems are represented in Figure 1. However, it should be stressed that due to the adaptation algorithm both schemes, while trying to control linear plant, will represent highly *nonlinear* control systems. Because of this fact it is usually not an easy task to study adaptive control systems analytically. The analytical part of the problem will doubtless be much more difficult for the neuro-fuzzy based adaptive control of a nonlinear plant.

From Figure 1 it follows that STR and MRAC have many similarities. Both systems have two feedback loops. The inner one is a standard feedback loop containing a plant and a controller. The controller parameters are set by the outer loop and this tuning is based on feedback from the plant inputs and outputs. When the design algorithms of the inner loop and the parameter adjustment techniques are the same, STR and MRAC are identical. This is not usually the case and both methods have been developed independently in their own way.

Standard adaptive control of a linear system can be done in two different ways. In so-called *direct* control the adjustment of the controller parameters is based on the error between plant and reference model output. An alternative method is to estimate the plant parameters and to adjust the controller parameters on the basis of such estimates. This approach involves an identification step (resulting in intermediate parameters of the plant model) and is called *indirect* control. Hence, in applying the indirect approach, the identification error between plant and plant model is used to estimate plant model parameters, which in turn are used to adjust controller parameters. These indirect schemes are also sometimes called *explicit* since the design is based on an explicit plant model. By its very nature STR uses the indirect control approach while MRAC can be achieved by using either of these two approaches. (A direct MRAC scheme is presented in Figure 1). NN and FLM based control schemes for controlling nonlinear plants belong to indirect adaptive control algorithms.

In the linear domain there are numerous different identification and controller design algorithms. Depending on which parameter estimation scheme is chosen, and which control design algorithm is used, we can generate a wide spectrum of adaptive control algorithms. Some combinations may be better than others; some approaches converge in a stable manner; and some may not work at all. Some adaptive schemes work well in practice but without analytically proven

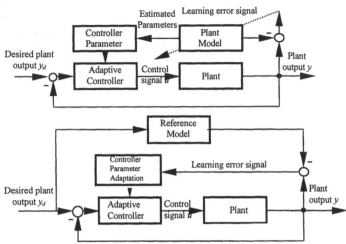

Figure 1   Self-tuning regulator (STR) scheme, indirect control (above); model reference adaptive control (MRAC) scheme, direct control (below).

convergence and stability. This is exactly the case with the majority of present-day NN and FLM based adaptive control approaches. These two crucial problems of convergence and stability of different adaptive control systems, as well as many other interesting questions in adaptive control of linear systems, are well covered in mathematical and engineering literature. The more interested reader is referred to the references at the end of this chapter for their detailed study, [1–7].

Today, STR and MRAC aimed at and developed for control of linear systems are no longer considered highly intelligent. In this respect they share the destiny of the PID controller with fixed structure and operating algorithm (which is still very efficient). Faced with partially known and nonlinear systems, these two classic adaptive algorithms do not behave intelligently. They cannot cope with nonlinearities and basically they do not possess the ability to incorporate existing human knowledge into their algorithm. NN and FLM seem to be promising tools for solving such problems.

After this short introduction to the basic ideas of adaptive control we can conclude with the statement that "learning is only part of intelligence." We will not try to give a definition of intelligence here. This issue has been and always will be investigated and addressed by other disciplines (notably philosophy, psychology, biology and artificial intelligence). Staying firmly in the engineering domain, some comments on the *contemporary* term *intelligent control* are now in order. Without any doubt, human mental faculties of *memorizing, prediction, adaptation* and *learning* should be the *foundation* of any *intelligent* artificial device or *control system.* STR and MRAC already exhibit these properties when the plant to be controlled is a linear one.

We will now, in the main, discuss two aspects of intelligent controllers: one, learning in a nonlinear, uncertain and time-variant environment and also the techniques for embedding human structured knowledge (experience, heuristics) into a workable control algorithm. Speaking broadly, the paradigm of neural networks will cover the first aspect of our intelligent controller and the second part will be processed by the fuzzy logic modeling tools. Clearly, knowing the laws of learning as well as of understanding and transferring human expertise into usable algorithms would not automatically make our systems highly intelligent or human-like, and it would not immediately tell us a lot of specific things that are part of human intelligence. To be more precise, part of a controller's intelligence should be the ability to cope with a large amount of noisy data coming from different sensors. They must also be able to plan under large uncertainties, to set the hierarchy of priorities and to coordinate many control loops simultaneously. Besides, the tasks of today's intelligent controllers will increasingly be the detection of faults, or their early diagnosis, in order to have enough time for the reconfiguration of control strategies, maintenance or repair. These tasks will only be a small part of the decision making capacities of the next generation of intelligent systems.

Today's NN and/or FLM based intelligent control systems presented below are the next hierarchical level after classic adaptive control schemes for linear systems.

A. Neural and/or Fuzzy Logic Approach

In recent years neural networks and fuzzy logic models have been used in many different fields. There is practically no area of human activity left untouched by NN and/or FLM. These two primary *modeling tools* are the two sides of the same coin. When and why one should try to solve a given problem by NN, and when the more appropriate tool would be FL model, depends upon availability of previous knowledge (expertise) about the system to be controlled, as well as upon the amount of measured process data. More about the answer to this question may be found in Table 1.

Table 1 is self explanatory. In short, the less previous knowledge there is, the more neural and less fuzzy will our model be and vice versa. When applied in a system control area, neural networks can be regarded as *nonlinear identification* tools. This is the closest connection with the standard and well developed field of estimation or identification of linear control systems. If the problem at hand is a linear one, NN would degenerate into one single linear neuron and in that case the weights of the neuron correspond to the parameter of the plant's discrete transfer function $G(z)$.

In order to avoid too high an expectation of these new concepts in computing (and particularly after they have been connected with intelligence), as well as avoiding underestimating these computing paradigms, it might be of use to

Table 1   Neural Networks and Fuzzy Logic Modeling

Neural Networks	**Fuzzy Logic Models**
	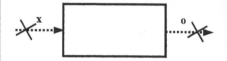
Black-Box	**White-Box**
No previous knowledge, but there are measurements, observations, records, data	Structured knowledge (*experience, expertise or heuristics*) IF - THEN rules are the most typical examples of the structured knowledge
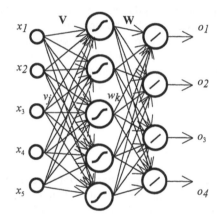	Example: Distance between two cars on the road
	R1: IF the speed is *low* AND the distance is *small* THEN the force on brake should be *small*
	R2: IF the speed is *medium* AND the distance is *small* THEN the force on brake should be *big*
	R3: IF the speed is *high* AND the distance is *small* THEN the force on brake should be *very big*
Behind NN stands the idea of **learning** from the data	Behind FLM stands the idea of **embedding human knowledge** into workable algorithms.

At many instances we do have both *some knowledge* and *some data.*

This is the most common **gray box** situation covered by the paradigm of
Neuro-Fuzzy *or* Fuzzy-Neuro models.
If we do not have any prior knowledge AND we do not have any measurements (by all accounts very hopeless situation indeed) it may be hard to expect or believe that the problem at hand may be approached and solved easily. This is a **no-color -box** situation.

Source: Ref. 8.

mention the basic advantages claimed for NN and FLM (Table 2) and the counterclaims advanced by their discouraged users or cautious beginners in the field (Table 3). Because of the wide range of applications of these two new modeling tools it is hard to disprove or object to some of these claims and counterclaims. It is certain that everyone working with NN and FLM will find his or her own answers to at least some or most of these claims.

But the fast growing number of companies and products (software and hardware) in the NN and FLM fields with applications in almost all areas of human activity and the growing number of new neural and fuzzy computing theories and paradigms are proofs that despite the many still open questions and problems NN and FLM are already well established engineering tools which are overcoming their teething problems and are becoming common computational means for many everyday real life tasks and problems.

II. NEURAL NETWORKS AND FUZZY LOGIC SYSTEMS

This section provides the principal ideas and approaches as well as all basic tools needed for NN based adaptive control of nonlinear and time-varying systems. First, in Section A, we will present the basic theoretical foundation and rationale for the wide application of both NN and FLM, namely their capability to model, or approximate, any multivariate function to any desired degree of accuracy. In Section B, the neural networks will be described, starting with the basic building block, the neuron. These neurons will be connected to create *feedforward* neural networks, and it will be shown how neural networks can be used as

Table 2 Claimed Advantages of NN and FLM

Neural networks	Fuzzy logic models
i) have the property of learning from the data mimicking human learning ability.	i) are an efficient tool for embedding human (structured) knowledge into useful algorithms.
ii) can approximate any multivariate nonlinear function.	ii) are efficient for modeling of substantially nonlinear processes.
iii) do not require deep understanding of the process or the problem being studied.	iii) are applicable when a mathematical model is unknown or impossible to obtain.
iv) are robust to the presence of noisy data.	iv) operate successfully despite a lack of precise sensor information.
v) have parallel structure and can be easily implemented in hardware.	v) are useful at the higher levels of hierarchical control systems.
vi) the same NN can cover a broad range of tasks.	vi) are the appropriate tool in generic decision making processes.

Table 3 Claimed Disadvantages of NN and FLM

Neural networks	Fuzzy logic models
i) need extremely long training or learning time (problems with local minima or multiple solutions) with little hope of many real time applications.	i) There must be human solutions to the problem and this knowledge must be structured. Experts may have problems with structuring the knowledge.
ii) do not uncover basic internal relations of physical variables and do not increase our knowledge of the process under consideration.	ii) Experts sway between extreme poles - too much aware of their expertise or tend to hide their knowledge.
iii) are prone to bad generalization. (When a large number of weights are present, there is a tendency to overfit the data with poor performance on previously unseen data during the test phase.)	iii) The number of rules increases exponentially with the increase in the number of inputs, as well as the number of fuzzy subsets per input variable.
iv) There is very little or no guidance at all about the NN structure, optimization procedure or even the type of NN for a particular problem.	iv) Learning (changing of membership functions' shapes and positions and/or the rules) is highly constrained and typically more complex task than in NN.

approximators of nonlinear functions. Recurrent NN or NN with dynamic neurons will not be discussed. The elements and basic ideas of fuzzy logic will be given in Section C.

There are remarkable mathematical similarities, and sometimes even mathematical equality, between these two seemingly different modeling tools—NN and FLM. This equality will be presented in Section D.

After familiarizing the reader with the mathematical equality of these two conceptually different modeling tools, our presentation will be based on neural networks only. Following Section D, neural networks stand strictly for, or may be understood as neural networks, fuzzy logic, neuro-fuzzy or fuzzy-neuro models. The principal idea of the whole of Chapter 4 is that of learning as a prerequisite for the control of nonlinear systems. Whether the initial model of both the controller and the plant being controlled resulted from a black, white or gray box is of less importance. When adaptation starts, the mathematics behind such adaptation is the same for differently labeled models.

A. Approximation Capabilities of NN and FLM

This section presents the basics of the theoretical approximation ability of both neural networks and fuzzy logic models. Without any doubt these faculties of approximating any nonlinear function to any degree of accuracy are the foundation of modeling by NN and FLM. We will present the classic Weierstrass theorem for the approximation by polynomials, Cybenko's theorem which states identical abilities of sigmoidal functions, as well as the theorem of the universal approximation properties of the Gaussian (radial basis function).[1]

Classical Weierstrass' Theorem

The set $P_{[a, b]}$ of all polynomials is dense in C $[a, b]$. In other words, given an $f \in C$ $[a, b]$ and $\varepsilon > 0$, there is polynomial p for which: $|p(x) - f(x)| < \varepsilon$ for all $x \in [a, b]$.

Cybenko's Theorem

Let σ be any sigmoidal function and I_d the d-dimensional cube $[0, 1]^d$. Then finite sums of the form,

$$F(\mathbf{x}) = \sum_{j=1}^{n} w_j \sigma_j (\mathbf{v}_j^T \mathbf{x} + b_j),$$ (1)

are dense in C $[I_d]$. In other words, given an $f \in C$ $[I_d]$ and $\varepsilon > 0$, there is a sum, $F(\mathbf{x})$ of the above form for which: $|F(\mathbf{x}) - f(\mathbf{x})| < \varepsilon$ for all $x \in I_d$. (Where w_i, $\mathbf{v}$ and b_i represent output layer weights, hidden layer ones and bias weights of the hidden layer respectively.)

Theorem for the Density of Gaussian Functions

Let G be a Gaussian function and I_d the d-dimensional cube $[0, 1]^d$. Then finite sums of the form,

$$F(\mathbf{x}) = \sum_{j=1}^{n} w_j G(\|\mathbf{x} - \mathbf{c}_j\|),$$ (2)

are dense in C $[I_d]$. In other words, given an $f \in C$ $[I_d]$ and $\varepsilon > 0$, there is a sum, $F(\mathbf{x})$ of the above form for which: $|F(\mathbf{x}) - f(\mathbf{x})| < \varepsilon$ for all $x \in I_d$. (Where w_i, and $\mathbf{c}_i$ represent output layer (OL) weights and centers of hidden layer (HL) multivariate Gaussian functions respectively.)[2]

The same results of universal approximation properties do exist for fuzzy models, too. Results of this type can also be stated for many other different

---

[1]The reader less interested in these theoretical issues may, without loss of continuity, skip the first part of this Subsection and continue the reading at the first paragraph after Eq. (4).
[2]In passing, let us note that Eqs. (1) and (2) represent the two most popular feedforward neural networks used today—multilayer perceptron and radial basis function NN. Their graphical representation will be given in Section II.B. Multilayer perceptron stands for NN with one or more hidden layers comprising neurons with sigmoidal activation functions. The standard representative of such fucntions is a tangent hyperbolic function. Structure of RBF networks is the same, but the activation functions are radially symmetric.

functions. They are very common in approximation theory and hold under very weak assumptions. Density in the space of continuous functions is a *necessary* condition that every approximation scheme should satisfy.

In control tasks (as well as in other engineering problems where NN and FLM are being successfully applied) we are generally not faced with the problem of approximating a continuous univariate $f(x)$ or a multivariate functions $f(\mathbf{x})$.Typical engineering problems, that we will have to solve, are *interpolation/approximation* of sets of *sparse and noisy training data points*. Let us be more specific. Usually, NN has to model mapping of a finite set of P n-dimensional input training patterns to the corresponding P m-dimensional output or target patterns. In other words NN should model the underlying function or hypersurface $H: \Re^n \to \Re^m$. The learning or training phase of NN adaptation corresponds to the (non)linear optimization of a fitting procedure based on knowledge of the training data pairs. This is hypersurface fitting in the generally high dimensional space $\Re^n \otimes \Re^m$. Radial basis functions (RBF) networks have a nice property in that they can interpolate any set of P data points. The same is also valid for fuzzy logic models or multilayer perceptrons.

The interpolating problem is now stated as follows: Given is a set of P measured data: $X = \{\mathbf{x}_p, \mathbf{d}_p, p = 1, \ldots, P\}$ consisting of the *input pattern* vectors $\mathbf{x} \in \Re^n$ and output *desired responses* or targets $\mathbf{d} \in \Re^m$. An interpolating function is such that,

$$F(x_p) = d_p, \quad p = 1, \ldots, P. \tag{3}$$

The *cost* or *error function E* that measures the quality of modeling (here, we use the sum of error squares) in the case of interpolation must equal zero,

$$E = \sum_{p=1}^{P} e_p^2 = \sum_{p=1}^{P} [F(\mathbf{x}_p) - \mathbf{d}_p]^2 = 0. \tag{4}$$

(Strictly speaking F and E are parameterized by approximation coefficients named here as the weights, and more proper notation would be $F(\mathbf{x}, \mathbf{w}, \mathbf{V})$ and $E(\mathbf{w}, \mathbf{V})$. Writing this dependency explicitly stresses the fact that the weights will be subjected to the optimization procedure that should result in a "good" F and a "small" E).

Note that in order to interpolate a data set X, the RBF network should have, in a hidden layer, exactly P neurons. In the NN field we typically work with a set of the thousand patterns (measurements) and this would mean that the size of such an interpolating network would have to be too big. There is another important reason why the idea of data interpolation is usually not a good one. Real data are corrupted by noise, and interpolation of noisy data leads to the problem of *overfitting*. What we basically want our NN to do, is to model the underlying function and to filter out the noise contained in the training data. There are many different techniques by which this can be done.

The number of neurons in the hidden layer is the most important design parameter concerned in the approximation abilities of the neural network. Remember that both the number of input components (features) and the number of output neurons is in general determined by the very nature of the problem. Thus the real representation power of NN and its generalization capacity are primarily determined by the number of HL neurons. In the case of general nonlinear regression, which is basically done in NN based control, the main task is to model the underlying function between the given inputs and outputs by filtering out the disturbances contained in the noisy training data set. By changing the number of HL nodes two extreme solutions should be avoided: first filtering out the underlying function (not enough HL neurons) and then modeling of the noise or overfitting the data (too many HL neurons.)

Thus, during the optimization of the network's size, one of the smoothing parameters is certainly the number of HL neurons which should be small enough to filter the noise out and big enough to model the underlying function. Example 4.1 is a simple case of one-dimensional mapping and may be a good presentation of overfitting. It is clear that perfect performance on a training set doesn't guarantee a good model (see left graph where the interpolating function passes through the training data denoted by crosses). Same phenomena will be observed while fitting multivariate hypersurfaces.

Example 1
Dependency (plant or system to be identified) between two variables is given by $y = x + \sin(2x)$. Using highly corrupted (25% Gaussian noise with zero mean) training data set containing 36 measured patterns (x, d) interpolate and approximate this data by RBF network. A graphical presentation of this example is given in Figure 2.

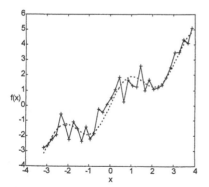 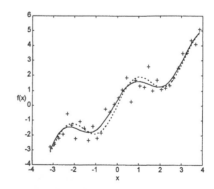

Figure 2   Interpolation and overfitting noisy data (36 HL neurons, left). Approximation and smoothing noisy data (8 HL neurons, right). Underlying function $y = x + \sin(2x)$ - dashed. Number of training patterns $P = 36$.

In order to avoid overfitting we must relax the interpolation requirements (3) while fitting noisy data, and instead of strict interpolation we typically do an *approximation* of the training data set that can be expressed as,

$$F(x_p) \approx d_p, \quad p = 1, \dots, P. \tag{5}$$

In the case of approximation we no longer want $E = 0$. We only want the error (cost) function $E = \sum_{p=1}^{P} e_p^2 = \sum_{p=1}^{P} [F(\mathbf{x}_p) - \mathbf{d}_p]^2$ to be small and, at the same time, the noise filtered out as much as possible.

The interpolation (Figure 2 left) and approximation (Figure 2 right) in Example 1 were achieved by RBF NN as given in Eq. (2) and having 36 and 8 neurons in the hidden layer respectively. Gaussian functions (here referred to as HL activation functions) were placed symmetrically along the x-axis with the same standard deviation σ equal to double the distance between two adjacent centers. With such a choice of σ we obtain a nice overlapping of the basis (activation) functions. Note that both parameters (c_i, σ_i) of Gaussian bells were fixed during the calculation of the best output layer weights w_i. (In terms of NN, the hidden layer weights, and in terms of FLM, the parameters that define positions and shapes of membership functions, were fixed or frozen during the fitting procedure.)

This was the problem of *linear approximation* because the parameters w_i enter linearly into the expression for the approximating function $F(\mathbf{x})$. In other words, approximation error $e(\mathbf{w})$ depends linearly upon the parameters (here the OL weights w_i). Notice that in this example the resulting function $F(\mathbf{x})$ physically represents the output from the single OL neuron and the approximation error for the p-th data pair (x, d) can be written as,

$$e_p = e_p(w) = F(x_p) - d_p = o(x_p) - d_p. \tag{6}$$

Generally x will be the vector $\mathbf{x}$. When approximation error e depends linearly upon the weights, the error function $E(\mathbf{w})$ defined above as the sum of error squares, is a hyperparaboloidal bowl with a guaranteed single minimum. The weight vector $\mathbf{w}^*$ that gives this minimal point $E_{min} = E(\mathbf{w}^*)$ is the wanted solution and in this case, the approximating function $F(\mathbf{x})$ from Eqs. (2) or (7) has the property of *best approximation*. Note that despite the fact that this *approximation problem is a linear one*, the approximating function $F(\mathbf{x})$ is a nonlinear one resulting from the summation of the *weighted nonlinear basis function* φ_n.

There are a variety of choices of basis functions for approximation all of which have been or may be used in NN or FLM. (In NN area basis functions are typically called activation functions (AF) and in the field of FLM the most common names are membership functions or possibility distributions.) Let us list a few-out-of-many possible basis functions. The basic *univariate* (one-dimensional) *linear* approximation is given as,

$$F(x) = \sum_{i=1}^{n} w_i \varphi_i(x) \tag{7}$$

and some of the basis functions are,

$$\varphi_i(x) = x^i \quad (i = 0, n-1) \quad algebraic\ polynomials, \tag{7a}$$

$$\varphi_i(x) = cos(ix), \quad sin(ix) \quad trigonometric\ polynomials, \tag{7b}$$

$$\varphi_i(x) = e^{-\frac{1}{2}(\frac{x-c_i}{\sigma_i})^2} \qquad radial\ basis\ functions\ e.g.,\ Gaussians. \tag{7c}$$

Unlike the linear approximations above, the one given in Eq. (1) represents the *nonlinear multivariate approximation* (now **x** is a vector and not a scalar),

$$F(\mathbf{x}) = \sum_{j=1}^{n} w_j \sigma_j (\mathbf{v}_j^T \mathbf{x} + b_j) .$$

As before $F(\mathbf{x})$ is a nonlinear function but the name *nonlinear approximation* is due to the fact that $F(\mathbf{x})$ is no longer the weighted sum of *fixed* basis function. The positions and shapes of basis function σ_i defined as a HL weight vectors $\mathbf{v}_i$ are now subjects of optimization procedure too. The approximating function $F(\mathbf{x})$ and error function E depend now on two sets of weights; linearly upon the OL weight vector **w** and nonlinearly upon the HL weight matrix **V**, where the rows of **V** are weight vectors $\mathbf{v}_i$. If we want to stress this fact we may write these dependencies explicitly as $F(\mathbf{x}, \mathbf{w}, \mathbf{V})$ and $E(\mathbf{w}, \mathbf{V})$. Now, the problem of finding the best set of weights is a *nonlinear optimization problem* that is much more involved than a linear one. Basically, the optimization or search for the weights that result in the smallest error function $E(\mathbf{w}, \mathbf{V})$, will now be a lengthy iterative procedure which doesn't guarantee that global minimum will be found.

Let us finish this subsection by introducing the *norm* of approximation, or the measure of how well a specific approximation $F(\mathbf{x})$ of the given form, matches the given set of noisy data. Norms are (positive) scalars used as a measure of error, length, size, distance, etc. (depending on context). The most common class of norms, in the case of the measured discrete data set, is the L_p (Hölder) norm. L_p-norm is the p-norm of $(N, 1)$ vector **e** given as,

$$\|\mathbf{e}\|_p = \|\mathbf{d} - \mathbf{F}(\mathbf{x})\|_p = \|\mathbf{d} - \mathbf{o}\|_p = \left(\sum_{i=1}^{N} |d_i - o_i|^p \right)^{1/p} . \tag{8}$$

d and **o** stand for N-dimensional vectors of desired and actual output of the NN respectively. Note that Eq. (8) is strictly valid for $\Re^d \to \Re^l$ mapping, or for NN with a single ouput layer neuron. This will typically be the case when controlling single-input, single-output (SISO) systems. For a larger number of output layer neurons, norm would be defined as an appropriate matrix norm.

Mostly $p = 1$, 2, or ∞ and these norms are called one-, two- (Euclidean) and infinity (Chebishev) norms respectively. The choice of the appropriate norm or the measure of the approximation's goodness depends primarily on the data and on the (simplicity of) approximation and on the optimization algorithms available. The L_2 norm is the best one for the data corrupted by normally distributed (Gaussian) noise. In this case it is known that the estimated variables obtained in L_2 norm correspond to the maximum-likelihood estimates. The L_1 norm is much better than the Euclidean norm for data which have outliers (small number of isolated, distant or wild points) because L_1 norm tends to ignore such data. The Chebishev norm is ideal in the case of exact data with errors in uniform distribution.

In the field of system identification and control, L_2 or the Euclidean norm is almost universal for the following reasons. First, the assumption that the Gaussian character of noise in the control systems environment is an acceptable and reasonable one. The second reason for the wide application of the L_2 norm is its mathematical simplicity and tractability. (Strictly speaking, the slightly changed version of the Euclidean norm is used here - the sum of error squares as given in Eq. (4).)

B. Basics of Neural Networks

The (artificial) neural networks are software or hardware models inspired by the structure and behavior of biological neurons and the nervous system. There are about 50 different types of NN in use today. Here we will discuss and describe *feedforward* NN with *supervised learning*. This section deals with the *representation* abilities of NN. It shows what NN can model, and why they can represent any underlying function that generated training data. How the best set of weights, that enables NN to be an universal approximator, can be calculated (learned) by using these data is the problem of learning that will be described in Section III.

Feedforward NN are the most used in the present-day intelligent control of nonlinear dynamic systems. Besides, feedforward NN are mathematically very close, and sometimes even equal, to the fuzzy logic models. Both neural and fuzzy logic approximation techniques could be given graphical representation which could be called a neural network or fuzzy logic model. With such an interpretation of NN or FLM tools we did not add anything new to the approximation theory but, from the point of view of implementation (primarily in the sense of parallel and massive computing), this graphical representation is a more desirable property.

The (artificial) neural networks are composed of many computing units called neurons. Strength of the connection, or link, between two neurons is called the weight. These weights are true network parameters and the very subject of the learning procedure in NN. Depending upon the problem, they have different physical meaning and sometimes it is hard to find any physical meaning at all. Their geometric meaning is much clearer. The weights define the positions and the shapes of basis functions in neural and fuzzy models.

The neurons are typically organized into layers in which all of the neurons usually possess the same activation functions. The genuine neural networks are the networks with at least three layers of neurons - an input, a hidden and an output-providing that HL neurons have nonlinear and differentiable AF. Note that such a NN has *two layers of adjustable weights* that are typically organized as the elements of the weights matrices **V** and **W**, where **V** stands for the hidden layer weights and **W** for the output layer weights. For a single OL neuron, **W** degenerates into weight vector **w**.

The nonlinear AF in hidden layer enable the neural network to be a universal approximator. Thus the nonlinearity of the activation functions solves the problem of representation. The differentiability of the HL neurons AF, as it will be seen later, will enable the solution of nonlinear learning. (Today, by using random optimization algorithms, e.g., the genetic algorithm, we may also think of the learning HL weights in the cases when the hidden layer neurons AF are no longer differentiable. The fuzzy logic models are the most typical networks that usually have non-differentiable *activation functions*.)

Here, we don't treat the input layer as a layer of the neural processing units. The input units are merely *fan out* nodes. Generally, there won't be any processing in the input layer and, although in its graphical representation it looks like a layer, the input layer is not a layer of neurons. Rather, it is an input vector, eventually augmented by a bias term, whose components will be fed to the next (hidden or output) layer of neural processing units. The OL neurons may be linear ones (for the regression type of problem) or they can have sigmoidal activation functions (for the classification or pattern recognition tasks). For NN based adaptive control schemes, OL neurons are always linear units. The most elementary, but still powerful feedforward NN is given in Figure 3. This is a graphical representation of the approximation scheme (4.1) repeated below with the notation as in Figure 3,

$$o(\mathbf{x}, \mathbf{V}, \mathbf{w}, \mathbf{b}) = F(\mathbf{x}, \mathbf{V}, \mathbf{w}, \mathbf{b}) = \sum_{j=1}^{J} w_j \sigma_j (\mathbf{v}_j^T \mathbf{x} + b_j). \tag{9}$$

By explicitly writing $o = o(\mathbf{x}, \mathbf{V}, \mathbf{w}, \mathbf{b})$ we have deliberately stressed the fact that the output from NN depends upon the weights contained in **V**, **b** and **w**. Input vector **x**, bias weight vector **b**, HL weight matrix **V** and OL weight vector **w** are given below,

$$\mathbf{x} = \begin{bmatrix} x_1, & x_1, \cdots, x_n \end{bmatrix}^T, \tag{10}$$

$$\mathbf{V} = \begin{bmatrix} v_{11} & \cdots & v_{1i} & \cdots & v_{1n} \\ \vdots & & \vdots & & \vdots \\ v_{j1} & \cdots & v_{ji} & \cdots & v_{jn} \\ \vdots & & \vdots & & \vdots \\ v_{J1} & \cdots & v_{Ji} & \cdots & v_{Jn} \end{bmatrix}, \tag{11}$$

$$\mathbf{b} = \begin{bmatrix} b_1, & b_2, & \cdots, & b_J \end{bmatrix}^T \tag{12}$$

$$\mathbf{w} = \begin{bmatrix} w_1, & w_2, & . & . & ., & w_J, & w_{J+1} \end{bmatrix}^T \tag{13}$$

At this point a few comments may be needed. Figure 3 represents a general structure of both multilayer perceptrons and RBF networks. In the case of multilayer perceptron, x_{n+1} will be the constant term equal to 1 (possibly wrongly) called *bias*. The bias weight vector **b** can simply be integrated into an HL weights matrix **V** as its last column. After such a concatenation, Eq. (9) can be simplified as $o = \mathbf{w}^T\sigma(\mathbf{Vx})$.[3] For RBF networks $x_{n+1} = 0$ or there is no bias.

For both multilayer perceptrons and RBF networks, the HL bias term may be used but is not generally needed. The HL activation functions shown in Figure 3 are sigmoidal ones indicating that this particular network represents a multilayer perceptron. The very structure of multilayer perceptron, RBF network and FLM is the same or very similar. The basic distinction between the sigmoidal and radial basis function network is how the input to each particular neuron is calculated.

The basic computation that takes place in NN is a very simple one. After a specific input vector is presented to NN, the input signals to all HL neurons u_j are computed either as scalar (dot or inner) products between the weight vectors $\mathbf{v}_j$ and **x** (for a linear or sigmoidal AF), or as Euclidean distances between the centers $\mathbf{c}_j$ of the radial basis function and **x** (for RBF activation functions). Radial basis AF is typically parameterized by two sets of parameters. Besides the center **c,** the second set of parameters is the one which determines the shape (width) of the RBF. In the case of a one-dimensional Gaussian function this second set of parameters is the well known standard deviation σ. For the case of a multivariate input vector **x,** parameters that define the shape of the hyper-Gaussian function are elements of a variance matrix Σ. To put it simply, HL bias weight of sigmoidal activation functions loosely corresponds to the centers of RBF functions, while weights v_{ji} which define the slope of sigmoidal functions with respect to each input variable correspond to the width parameter of the RBF.

Thus the inputs to the HL neurons for sigmoidal AF are given as,

$$u_j = \mathbf{v}_j^T\mathbf{x}, \quad j = 1,...,J, \tag{14}$$

and for RBF activation functions the distances are calculated as,

$$r_j = \sqrt{(\mathbf{x}-\mathbf{c}_j)^T(\mathbf{x}-\mathbf{c}_j)}, u = f(r, \sigma). \tag{15}$$

[3]Note the simplicity of this notation. The use of the summation sign is avoided. Product **Vx** is a column vector of the inputs to the HL neurons. After transforming these inputs through the HL activation functions, the NN output is obtained as a scalar product $\mathbf{w}^T\sigma$ between the OL weights **w** and the HL neurons output vector $\sigma = \mathbf{y}$ in Figure 3.

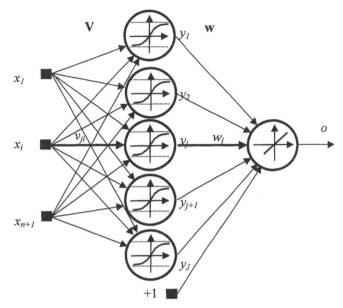

Figure 3 Feedforward neural network that can approximate any $\Re^n \to \Re^l$ nonlinear mapping.

In the case of three dimensional input when $\mathbf{x} = [x_1, x_2, x_3]^T$ and $\mathbf{c} = [c_1, c_2, c_3]^T$, the Euclidean distance r_j can be readily calculated as,

$$r_j = \sqrt{(\mathbf{x} - \mathbf{c}_j)^T (\mathbf{x} - \mathbf{c}_j)} = \sqrt{(x_1 - c_1)^2 + (x_2 - c_2)^2 + (x_3 - c_3)^2} \qquad (16)$$

The output from each HL neuron depends on the type of activation function used. The most common activation functions in multilayer perceptrons are the squashing sigmoidal functions; *unipolar logistic* function (17) and the *bipolar* sigmoidal, or tangent hyperbolic, function (18). Instead of a sigmoidal function, any nonlinear, smooth, differentiable and, preferably, nondecreasing function could be used, too.

$$o = \frac{1}{1 + e^{-u}}, \qquad (17)$$

$$o = \tanh\frac{u}{2} = \frac{2}{1 + e^{-u}} - 1. \qquad (18)$$

Two of the most popular RBF networks activation functions for use in control tasks are the Gaussian (19) and the *inverse* multiquadric function (20),

$$o = e^{-(\frac{r^2}{\beta})} = e^{-(\frac{(x_1-c_1)^2 + \cdots + (x_j-c_j)^2 + \cdots + (x_n-c_n)^2}{2\sigma^2})} , \beta > 0, \qquad (19)$$

$$o = (r^2 + \beta)^{-\frac{1}{2}} , \beta > 0. \qquad (20)$$

Note that the Gaussian in Eq. (19) is normalized with a maximal height equaling 1. At the same time the maximum of the *inverse* multiquadric (20) depends upon the "shape" parameter β. The normalized *inverse* multiquadric activation function with maximal height at center equaling 1, is given below,

$$o = \sqrt{\beta}(r^2 + \beta)^{-\frac{1}{2}} , \beta > 0. \qquad (21)$$

In Figure 6, the normalized two-dimensional radial basis functions, the Gaussian (left) and the *inverse* multiquadric function (right), are presented.

After the outputs from the HL neurons $y_j = o_j = f(u_j)$ are calculated, the output from the whole NN equals the scalar product of the HL output vector and OL weights vector,

$$o = y^T w = \sum_{j=1}^{J+1} w_j y_j . \qquad (22)$$

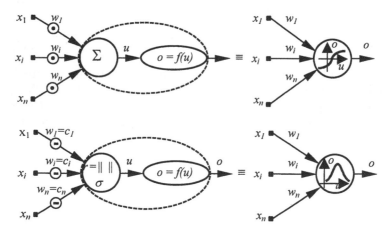

Figure 4   Formation of the input signal to the basic neurons of: sigmoidal type (above), input to the neuron is a *scalar product* $v^T x$,; radial basis function type (below), input to the neuron depends upon the *distance r* between x and the RBF center $c(r = \|x - c\|)$ as well as upon the parameters of *shape* of the RBF (for the Gaussian function these are the elements of covariance matrix Σ).

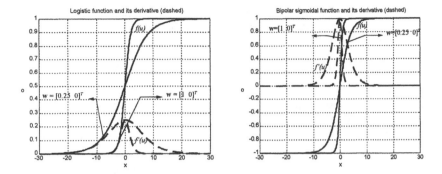

Figure 5 Logistic (17) and bipolar sigmoidal (18) activation functions and their derivatives.

Note that φ_i from Eq (7) equals y_j. Consequently, and in connection with Eqs. (1), (2), and (7), for the multilayer perceptron we can write, $y_j = \varphi_j = \sigma_j$, and for the RBF having Gaussian basis functions, $y_j = \varphi_j = G_j$.

So far as the problem of identification and control of dynamic systems is concerned, the most popular activation functions for the feedforward NN are the linear, the sigmoidal (e.g., tangent hyperbolic) and the Gaussian functions. Typically, we find linear AF in the OL neurons while the latter two are the activation functions of the HL neurons. It is interesting to note that the standard identification of linear systems can be represented and performed by single linear neuron. Example 2 shows a basic neural approach to the identification of linear systems.

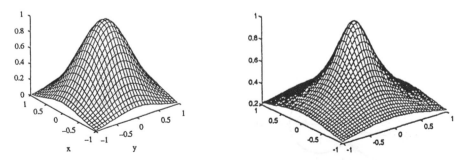

Figure 6 Two-dimensional radial basis functions; Gaussian (left) and inverse multiquadric (right). (Note that both activation functions are normalized, i.e., their maximum here equals 1.)

Example 2 'Neural' representation of a linear system identification.
One way to describe a linear SISO dynamic systems can be the following generic discrete equation,

$$y_k = a_1 y_{k-1} + a_2 y_{k-2} + \cdots + a_n y_{k-n} + b_1 u_{k-1} + b_2 u_{k-2} + \cdots + b_n u_{k-m} + n_k \quad (23)$$

where y_{k-i} and u_{k-i} are past inputs and outputs, and n_k is additive white noise. The system identification problem of determining a-s and b-s can be viewed as a regression (functional approximation) problem in $\Re^{n+m+1}$ space. In case we want to model a second order dynamic system, the input patterns are four dimensional vectors consisting of u_{k-1}, u_{k-2}, y_{k-1} and y_{k-2} for each instant k and $k \in [3, K]$. K denotes the total number of discrete steps. Simulation time equals $K\Delta t$ and Δt is the sampling rate. The output from the linear neuron is y_k. The structure of the linear neuron that can successfully solve this problem is given in Figure 7.

　　　Note that in this type of discrete difference equation, the constant input (bias) term is missing. This stresses the fact that the hyperplane in 5-dimensional space is a homogeneous one. More simply, this hyperplane passes through the origin. As k increases from 0 to K, the points in the $[u_{k-1}, u_{k-2}, y_{k-1}, y_{k-2}, y_k]$ space will trace a trajectory which, in the case where there is no noise, lies on this hyperplane. In the presence of disturbances, this trajectory will lie around it. The use of expressions $w_i \approx a_i$, or $w_i \approx b_i$, in Figure 7 denotes the fact that by the end of the learning, w_i will be the estimates of the parameters a_i and b_i.

　　　The previous example might be an easy to understand one, so let us show the ability of RBF network to model nonlinear dynamic plants. This faculty of NN is a principal one for the NN based adaptive control of nonlinear systems due to

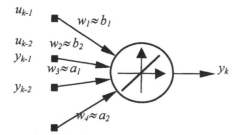

Figure 7   Structure of linear neuron for the identification of a 2nd order system.

the fact that a basic control structure will typically comprise two neural networks; one for modeling the plant and another one for modeling the inverse of the plant. The latter NN will act as a controller. Thus both NN should be able to model any nonlinear mapping correctly. The example below represents the ability of the RBF to model highly nonlinear dynamic systems.

Example 3
Consider the nonlinear 1st order dynamic system,[4]

$$y(k) = \frac{y(k-1)}{1+y^2(k-1)} + u^3(k-1) \qquad (24)$$

The output of this system is corrupted by the additive 5% white noise with zero-mean. Identify the plant by RBF neural network and show the performance of a trained NN on previously unseen step input.

The identification of this plant was done by RBF network having 9 HL neurons with Gaussian AF. Neural network was trained by genetic algorithm (100 generations, population size = 104, training time = 2694 seconds on the Pentium 133 PC) and resulting 3-D surfaces $y_k = f(u_{k-1}, y_{k-1})$ of the plant and its model are represented in Figure 8.

The training input signal was a white noise. Note that without output disturbances, the trajectories for any given input lie on this surface.

---

[4]This plant is adapted from Narendra and Parthasarathy [9]. Results presented are from Löchner [10].

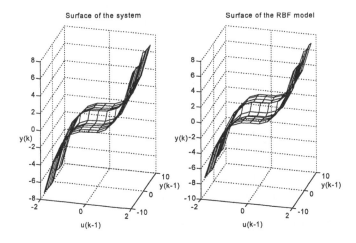

Figure 8 3D surface of the nonlinear dynamic plant (4.24) and that of an RBF model.

After the training phase, the test was done by previously unseen inputs. The plant desired step response y_d, the RBF network step response y_{estim}, the input step signal u and the modeling error $e = y_d - y_{estim}$ are represented in Figure 9. The graphs show a good generalization ability of this RBF network.

C. Basics of Fuzzy Logic Modeling

We can start this section with a short cut definition of what fuzzy logic may be: *Fuzzy logic is a tool for embedding human structured knowledge into workable algorithms.* We feel that a little more should now be said about this novel and powerful tool.

The concept of fuzzy logic is used in many different senses. In the narrow sense FL is considered as a logical system aimed at providing a model for modes of human reasoning which are approximate rather than exact. In a wider sense, FL is treated as a fuzzy set theory of classes with unsharp or fuzzy boundaries. Fuzzy logic methods may be used to design *intelligent systems* on the *basis of knowledge expressed in common language.* This methodology permits the processing of both symbolic and numerical information. Systems designed and developed utilizing fuzzy logic methods have often been shown to be more efficient than those based on conventional approaches. Today, in combination with NN techniques, fuzzy logic models are used to design robust adaptive (intelligent) control systems.

The theory of fuzzy sets (FS) is a theory of graded concepts, "a theory in which everything is a matter of degree, or everything has elasticity" (Zadeh, [1]). It is aimed at dealing with complex phenomena which do not lend themselves to analysis by a classical method based on bivalent logic and probability theory. Many of the systems in real life are too complex and/or too ill-defined to be susceptible to exact analysis. Even in cases where the systems, or some concepts, seem to be unsophisticated, our perception and understanding of them doesn't necessarily have to be simple. The introduction of fuzzy sets, or classes which admit intermediate grades of membership in them, opens the possibility of

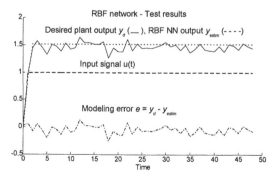

Figure 9   Test of the RBF neural networks modeling performance.

analyzing such systems both qualitatively and quantitatively by allowing the system variables to range over fuzzy sets.

Here we are mostly interested in the role of FL as a technique for mathematical expression of linguistic knowledge and ambiguity. In order to follow the presentation of FL tools it may be useful to understand the relationship between human knowledge and basic concepts such as sets and functions first.

> Human knowledge is structured in the form of IF-THEN rules. We'll soon see that these rules represent mappings, or multivariate functions that map the input variable (IF part variable, antecedents, causes) into output variables (THEN part variables, consequents, effects). It is well known that the functions are the relations defined over product sets, or over Cartesian products. The latter are the sets of all ordered pairs from two (or more) basic sets. Therefore, the best way (and a natural one) to present the ideas of FL as well as to understand the knowledge, is the bottom-up approach. From the sets, operation on them and Cartesian products, to the relations, multivariate functions and to the IF-THEN rules as the linguistic form of the human structured knowledge.

Consequently, we will initially introduce the basics of fuzzy set theory and compare these concepts with classic crisp logic. The important concept of membership function is introduced here. The representation of fuzzy sets by a membership function will be used as an important link with neural networks. Subsequently we will discuss basic set operations (notably intersection and union) and will connect them with proper operators (notably with MIN and MAX operators). Afterwards we will introduce the concepts of (fuzzy) relations and of relational matrix, as well as of composition of fuzzy relations. Formal treatment of fuzzy IF-THEN statements, questions of fuzzification and defuzzification and compositional rule of fuzzy inference with such statements, concludes this section on the basics of fuzzy logic modeling.

Crisp (or Classic) and Fuzzy Sets[5]

Sets or classes in a *universe of discourse U* could be differently defined:
by list of elements:

$$S_1 = \{John,\ Ana,\ Jovo,\ Mark\} \quad S_2 = \{Beer,\ Wine,\ Juice,\ Slivovitz\}$$

$$S_3 = \{Horse,\ Deer,\ Wolf,\ Sheep\} \quad S_4 = \{1,\ 2,\ 3,\ 5,\ 6,\ 7,\ 8,\ 9,\ 11\},$$

---

[5]Notation: Sets are denoted by capital letters and their members (elements) by lower-case letters. So, A denotes the universe of discourse, or universal set, or a collection of objects that contains all the possible elements *a-s* of concern in each particular context. A is assumed to contain a *finite number* of elements a, unless otherwise stated.

by definition of some property:

$$S_5 = \left\{ x \in N \mid x \langle 12 \right\} \qquad S_6 = \left\{ x \in R \mid x^2 \langle 25 \right\}, \text{(Note, } S_4 \in S_5\text{)}$$

$$S_7 = \left\{ x \in R \mid x \rangle 1 \wedge x \langle 7 \right\}, S_8 = \left\{ x \in R \mid \text{'} x \text{ is much smaller then } 10' \right\},$$

(symbol '$\wedge$' stands for logical AND, or for the operation of *intersection*),

by *membership* (in a crisp set also named as characteristic) *function.*

For crisp sets $\mu_S(x) = \begin{cases} 1 \, if \ x \in S \\ 0 \, if \ x \notin S \end{cases}$ (Figure 4.10, left).

For fuzzy sets $\mu_S(x)$ is mapping of X on [0,1]. (Figure 10, right).

In engineering applications, *the universe of discourse U* stands for (linguistic) input and output variables, or for antecedent and consequent variables, or for IF-part and THEN-part variables of the rule. Membership functions (or *possibility distributions*) of two typical *fuzzy* sets are represented in Figures 10 (right) and 11. The last one shows a fuzzy set S of *all real numbers close to* 12.

$$S = \left\{ (x, \mu_S(x)) \mid \mu_S(x) = (1 + (x - 12)^2)^{-1} \right\}.$$

Note the similarities of these last two membership functions with sigmoidal and radial basis activation functions given in the previous section! We will use it to show the equality of neural and fuzzy models in the next subsection.

In human thinking, it is somehow natural that the maximal degree of belonging to some set cannot be higher than one. Related to this is a definition of *normal* and *not-normal* FS. Both sets are shown in Figure 12. Typically, FS of input variables (IF part variables of IF-THEN rules) will be normal sets and the resulting FS, or FS of output variables, will be not-normal fuzzy sets.

Thus there is an important difference between classic and fuzzy sets that is represented in Table 4. In fuzzy logic an element can be a member of two or more sets at the same time. Element x belongs to A AND to B, and not to only one of these two sets. The very same x is just *more* or *less* a member of A and/or B.

Another notation for *finite* FS (sets comprising finite number of elements) is when set S is given as a set of pairs μ/x (see below and in Figure 13). Note that μ is a function of x,

$$\mu = \mu(x): S = \{\mu/x \mid x < 3\}, \text{e.g.,} S = \{(1/1), (0.5/2), (0/3)\}.$$

Usually, human reasoning is very approximate. Our statements depend on the content and we describe our physical and spiritual world in a rather vague manner. Imprecisely defined classes are an important part of human thinking. Let us finish this introduction of fuzzy sets with two more real and everyday examples.

Modeling of the concept *young man* is both imprecise and subjective. Three different membership functions of this fuzzy set, or class, depending on the person who is using it, are given in Figure 14. (Obviously, the two dashed membership functions would be defined by the persons who are in their thirties or eventually forties.)

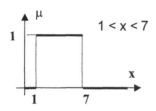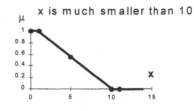

Figure 10 Membership functions of crisp (left) and fuzzy set (right).

Similarly, the order given in a pub "Bring me a *cold beer*[6] please" may have different meanings in different parts of the world. The author's definition of this fuzzy class is shown in Figure 15.

The membership functions may have different shapes. The most common ones in engineering applications are represented in Figure 16.

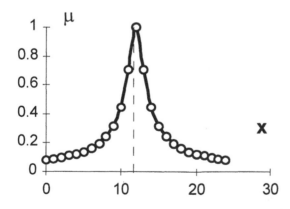

Figure 11 Membership function of fuzzy set *S* of *all real numbers close to* 12.

---

[6]*We would rather order* hot slivovitz *as a nice* rememorari a patria mea, *instead.*

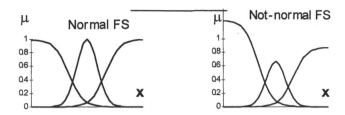

Figure 12 Membership functions of *normal* and *not-normal* fuzzy sets.

Table 4 Differences Between
Crisp and Fuzzy Approach

Crisp sets	Fuzzy sets
Either-or	And
Bivalent	Multivalent
Yes or no	More or less

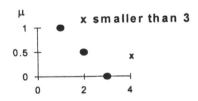

Figure 13 Membership functions as discrete pairs μ/x.

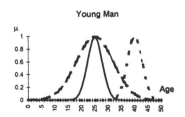

Figure 14 Three different membership functions $\mu(x)$ of the class *young man*.

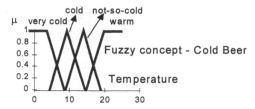

Figure 15   Membership functions $\mu(x)$ for S (*cold beer*) = {*very cold, cold, not-so-cold, warm*}.

Basic Set Operations

Out of many set operations the three most important ones are: complement S^C (or *not-S*), intersection and union. Figure 17 shows the complement S^C of crisp and fuzzy set in Venn diagrams and by using the membership functions.

The other two most common operations with sets are the *intersection* and the *union*. Figure 48 shows these operations by using membership functions. Figure 18 is obtained by using MIN operator for *intersection* (interpreted as logical AND) and MAX operator for *union* (interpreted as logical OR), i.e.,

$$\mu_{A \wedge B} = MIN\left(\mu_A, \mu_B\right), \mu_{A \vee B} = MAX\left(\mu_A, \mu_B\right).$$

(In fuzzy logic theory the operators of intersection are called *T-norms,* and the ones of union are called *T-conorms* which are also known as *S-norms.*)

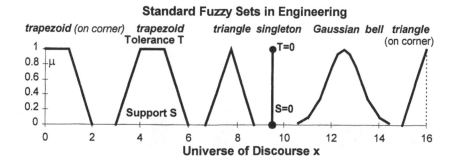

Figure 16   The most common fuzzy sets in engineering applications.

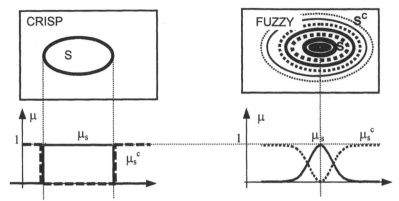

Figure 17 The two different ways of representing crisp (left) and fuzzy set S (right) and corresponding complement set S^C: the Venn diagrams (above) and the membership functions (below). The thickness of the lines in upper-right graph denotes the degree of belonging, or membership degree μ, of the elements of U to the fuzzy set S. For the complement set the following is true: $\mu_{not\text{-}S} = 1 - \mu_S$.

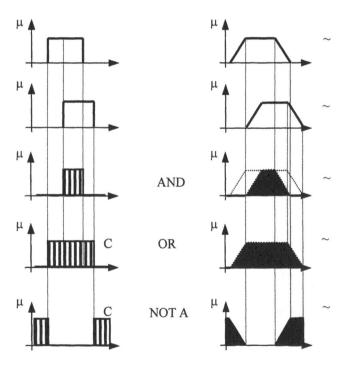

Figure 18 Operations of *intersection* and *union* (as well as complement of A) for crisp (left) and fuzzy sets (right) represented by the corresponding membership functions.

These are not the only operators that could be chosen to model the intersection and union of FS but they are the most commonly used ones in engineering applications. For the intersection, a popular alternative to the MIN operator is the *algebraic product*

$$\mu_{A \wedge B} = \mu_A \cdot \mu_B \ ,$$

that typically gives much smoother approximations. Besides, a kind of compensatorial operator could be the best alternative for certain tasks.

Before closing this introductory part on basic logical operators it may be interesting to show some basic differences between crisp and fuzzy set calculus. Namely, it is well known that the intersection between crisp set S and its complement S^C is an empty set, and that the union between these two sets is a universal set. In fuzzy logic calculus is different. Expressed by membership degrees, these facts look like:

Crisp set calculus		*Fuzzy set calculus*	
$\mu \wedge \mu^c = 0$	$\mu \vee \mu^c = 1$	$\mu \wedge \mu^c \neq 0$	$\mu \vee \mu^c \neq 1$

For fuzzy sets this can be verified readily as shown in Figure 19.

Fuzzy Relations

Let us first introduce the notion of an *ordered pair*. When we make pairs of anything the order of elements is usually of great importance. (Remember, for example, that the points (2, 3) and (3, 2) in an *x-y* plane are different. The order matters.) A pair of elements that occur in a specified order is called an ordered pair. *A relation is a set of ordered pairs.*

The relations express the connections between different sets. A crisp relation represents the presence or absence of association, interaction or interconnectedness between the elements (objects) of two or more sets.

If we generalize this concept, allowing the various degrees or strength of relation between elements, we get fuzzy relation. Due to the fact that *the relation itself is a set,* all set operations can be applied to it without any modifications. Also, the relations are subsets of a Cartesian product, or simply of the product set.

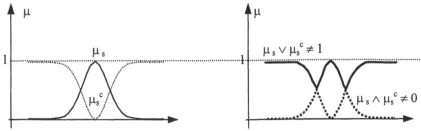

Figure 19 Interesting properties of the fuzzy set calculus.

In other words the relations are defined over *Cartesian products* or *product sets*. Thus we should introduce the concept of the Cartesian product.

Cartesian product of two crisp sets X and Y, denoted by $X \times Y$, is the *crisp set* of *all* ordered pairs such that the first element in each pair is a member of X and the second element is a member of Y,

$$X \times Y = \{(x, y) \mid x \in X \text{ and } y \in Y\}$$

Let $X = \{1, 2\}$ and $Y = \{a, b, c\}$ be two crisp sets. Cartesian product is given as,

$$X \times Y = \{(1, a), (1, b), (1, c), (2, a), (2, b), (2, c)\}.$$

Now, we can choose some subsets at random, or we may choose those that satisfy specific conditions in two variables. In both cases these subsets are relations. We typically assume that in one relation variables are somehow connected, but this random choice of, say, three ordered pairs $\{(1, b), (2, a), (2, c)\}$ being a subset of the product set $X \times Y$ is also the relation.

Cartesian product can be generalized for n sets in which case elements of Cartesian product are *n-tuples* $(x_1, x_2, \ldots, x_n)$. Here we will concentrate our presentation to relations between *two sets* known as a *binary relation* and denoted as $R(X, Y)$ or simply as R. Thus the binary relation R is defined over a Cartesian product $X \times Y$.

If the elements of the latter come from *discrete* universes of discourse, R can be presented in a form of the *relational matrix* or, graphically, as a discrete set of points in a 3-dimensional space $(X, Y, \mu_R(x, y))$.

Example 4
Let X and Y be the two sets given below. Present the relation R: *x is smaller than y* graphically, and in the form of a relational matrix.

$$X = \{1, 2, 3\}, Y = \{2, 3, 4\}, \qquad R; x \langle y$$
$$R = \{(1, 2), (1, 3), (1, 4), (2, 3), (2, 4), (3, 4)\}$$

Note that R is a set of pairs, and that R is a binary relation. The relational matrix, or membership array, in this crisp case comprises only of ones and zeros.

x/y	2	3	4
1	1	1	1
2	0	1	1
3	0	0	1

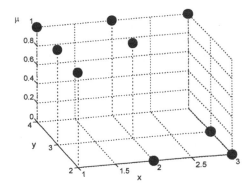

Figure 20 Discrete membership function $\mu_R(x, y)$ of the relation R: *x is smaller than y*.

Elements of the relational matrix are degrees of memberships $\mu_R(x, y)$, or possibilities, or degrees of belonging of a specific pair *(x, y)* to the given relation R, e.g., the pair (3, 1) belongs with a degree 0 to the relation *x is smaller than y*, or the possibility that $3 < 1$ is 0. The relation above is a typical example of a crisp relation. The condition involved in this relation is a precise one and one which is either fulfilled or not fulfilled.

The common mathematical expression '*x is approximately equal to y*' or the relation R: $x \approx y$ is different. It is a typical and yet a simple example of an imprecise, or a fuzzy, relation. Example 5 is very similar to Example 4, with the difference being that the degree of belonging of some pairs *(x, y)* from the Cartesian product to this relation may be any number between 0 and 1.

Example 5

Let X and Y be the two sets given below. Present the relation R: *x is approximately equal to y* in the form of a relational matrix.

$$X = \{1, 2, 3\}, \quad Y = \{2, 3, 4\}, \qquad R \; ; x \approx y$$

$$R = \{(1, 2), (1, 3), (1, 4), (2,2), (2,3), (2,4), (3,2), (3,3), (3,4)\}$$

x/y	2	3	4
1	0.66	0.33	0
2	1	0.66	0.33
3	0.66	1	0.66

Now, the discrete membership function $\mu_R(x, y)$ is again a set of discrete points in a 3-dimensional space *(X, Y, $\mu_R(x, y)$)* but with membership degrees that can have any value between 0 and 1.

When the universes of discourse are *continuous sets* comprised of an infinite number of elements, the *membership function* $\mu_R(x, y)$ *is a surface* over the Cartesian product $X \times Y$, and not a curve as in the case of the fuzzy sets.[7] Thus, the relational matrix is an (∞, ∞) matrix and it doesn't have any practical meaning. This is a common situation, which is resolved by proper discretization of the universes of discourse. Let us illustrate this by the following example.

Example 6

Let X and Y be the two sets given below. Show the membership function of the relation R: *x is approximately equal to y*, and present the relational matrix after discretization.

$$X = \{x \in R \mid 1 \le x \le 3\}, \; Y = \{y \in R \mid 2 \le x \le 4\}, \qquad R : x \approx y$$

Relational matrix after the discretization by a step of 0.5, is given below.

x/y	2	2.5	3	3.5	4
1	0.6667	0.5000	0.3333	0.1667	0.0000
1.5	0.8333	0.6667	0.5000	0.3333	0.1667
2	1.0000	0.8333	0.6667	0.5000	0.3333
2.5	0.8333	1.0000	0.8333	0.6667	0.5000
3	0.6667	0.8333	1.0000	0.8333	0.6667

In the examples above the sets are defined on the same universes of discourse. But, the relations may be defined on rather different linguistic variables expressing a variety of different associations or interconnections.

Example 7

Relation R is given as an association or interconnection between the fruit *color* and *state*. Present R as a *crisp* relational matrix.

$$X = \{green, yellow, red\}, \qquad Y = \{unripe, semiripe, ripe\}$$

R	Unripe	Semiripe	Ripe
Green	1	0	0
Yellow	0	1	0
Red	0	0	1

---

[7]Note also that in the case that there are n independent universes of discourse (n linguistic or input variables) membership is a hypersurface over an n-dimensional Cartesian product.

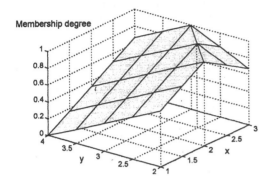

Figure 21 Membership function $\mu_R(x, y)$ of the relation R: *x is approximately equal to y* over the Cartesian product of two continuous sets *X* and *Y*.

It may be interesting to realize that we can interpret this *relational matrix as a notation*, or model, of an existing empirical set *of* IF-THEN *rules:*

R1:   IF (the tomato is) green, THEN (it is) unripe
R2:   IF   yellow, THEN   semiripe
R3:   IF   red, THEN   ripe

In fact, *relations are a convenient tool to model* IF-THEN *rules.* Before plunging into this important part of FL modeling, let us mention that the relational matrix given above is a crisp one, and not in total agreement with our experience. Eventually better interconnection between the fruit color and the state of the fruit may be given by the *fuzzy relational matrix* below,

R	Unripe	Semiripe	Ripe
Green	1	0.5	0
Yellow	0.3	1	0.4
Red	0	0.2	1

Example 8
Present the fuzzy relational matrix for the relation *R* that represents the concept of being *very far* in geography. Two crisp sets are given as,

$X = \{Auckland, Tokyo, Belgrade\}$
$Y = \{Sydney, Athens, Belgrade, Paris, New York\}$

R: very far	Sydney	Athens	Belgrade	Paris	New York
Auckland	0.2	0.8	0.85	0.90	0.55
Tokyo	0.5	0.5	0.5	0.55	0.4
Belgrade	0.8	0.1	0	0.15	0.5

Hence the relational matrix doesn't necessarily have to be a rectangular one. We can model many other concepts using the relations on different universes of discourse. Note that, similarly as for crisp relations, fuzzy relations are fuzzy sets in product spaces. As an example, let us analyze the meaning of the linguistic expression: *a young tall man*.

Example 9
Find the relational matrix of the concept: *a young tall man*. Implicitly, the concept *a young tall man* means both: *young* AND *tall man*. Therefore, we have to define two fuzzy sets *young man* and *tall man* first, and afterwards to apply the *intersection* operator to these two sets defined on different universes of discourse, *Age* and *Height* respectively.

One-out-of-many operators for modeling fuzzy intersection is the MIN operator. (The second and the most commonly used operator is the algebraic product). Thus, we can write,

$$\mu_R\big(Age,\ Heights\big) = \text{MIN}\big(\mu_1\,(Age),\mu_2(Heights)\big)$$

After the discretization as in Figure 22 is done, the relational matrix follows as,

$$\mu_1 = \begin{bmatrix} 0 \\ 0.5 \\ 1 \\ 0.5 \\ 0 \end{bmatrix}, \quad \mu_2 = \begin{bmatrix} 0 \\ 0.5 \\ 1 \\ 1 \\ 1 \end{bmatrix}, \quad R = \mu_1 \times \mu_2^T = \begin{bmatrix} 0 \\ 0.5 \\ 1 \\ 0.5 \\ 0 \end{bmatrix} \times \begin{bmatrix} 0 & 0.5 & 1 & 1 & 1 \end{bmatrix} =$$

R	170	175	180	185	190
15	0	0	0	0	0
20	0	0.5	0.5	0.5	0.5
25	0	0.5	1	1	
30	0	0.5	0.5	0.5	0.5
35	0	0	0	0	0

$$S_1 = \big\{15,\ 20,\ 25,\ 30,\ 35\big\}, \quad S_2 = \big\{170,\ 175,\ 180,\ 185,\ 190\big\}$$

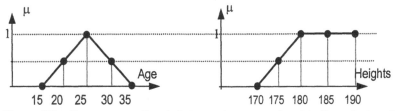

Figure 22 Fuzzy sets, or linguistic terms, *'young man'* and *'tall man'* given by corresponding membership functions.

The last relational matrix is actually a *'surface'* over the Cartesian product *Age x Heights*, which represents the membership function, or a *possibility distribution* of a given relation. In a general case, given the different universes of discourse, one can graphically get this surface utilizing the *EXTENSION PRINCIPLE*. (Cylindrical extension, in particular).

Composition of Fuzzy Relations

Fuzzy relations in different product spaces can be combined with each other by the *COMPOSITION*. (Fuzzy sets could also be combined with any fuzzy relation in the same way.) Many different versions are possible. The most common one and the best known one is a MAX-MIN composition. MAX-PROD and MAX-AVERAGE can also be used. MAX-PROD composition might often be the best alternative. Note that the composition is the fuzzy set, too. The three most important compositions are given below.

Let $R_1(x, y)$, $(x, y) \in X \times Y$ and $R_2(y, z)$, $(y, z) \in Y \times Z$, be two fuzzy relations.

MAX-MIN composition $R_1 \ max-min \ R_2$ is the fuzzy set:

$$R_1 \circ R_2(x,z) = \left\{ \left[(x,z), \ \max_y \left\{ \min \left\{ \mu_{R_1}(x,y), \mu_{R_2}(y,z) \right\} \right\} \right] \mid x \in X, \ y \in Y, \ z \in Z \right\}$$

and $\mu_{R_1 \circ R_2}$ is the membership function of a fuzzy composition on fuzzy sets.

MAX-PROD composition

$$R_1 \underset{\otimes}{\circ} R_2(x,z) = \left\{ \left[(x,z), \ \max_y \left\{ \mu_{R_1}(x,y) * \mu_{R_2}(y,z) \right\} \right] \mid x \in X, \ y \in Y, \ z \in Z \right\}$$

MAX-AVE composition

$$R_1 \underset{ave}{\circ} R_2(x,z) = \left\{ \left[(x,z), \ \frac{1}{2} \max_y \left\{ \mu_{R_1}(x,y) + \mu_{R_2}(y,z) \right\} \right] \mid x \in X, \ y \in Y, \ z \in Z \right\}$$

Later, trying to make *fuzzy inference,* a composition of fuzzy set and fuzzy relation (and not the one between the two fuzzy relations as above) will be of practical importance. The following example should facilitate the understanding of a rather mathematical presentation of these three different compositions.

Example 10

R_1 is a relation that describes an interconnection between *Color x* and *Ripeness y* of a tomato, and R_2 represents an interconnection between the *Ripeness y* and *Taste z* of a tomato:

R_1 (x - y connection) is given by following relational matrix,

$R_1(x,y)$	Unripe	Semiripe	Ripe
Green	1	0.5	0
Yellow	0.3	1	0.4
Red	0.	0.2	1

Relational matrix R_2 (y - z connection) is given as,

$R_1(x,y)$	Unripe	Semiripe	Ripe
Green	1	0.5	0
Yellow	0.3	1	0.4
Red	0	0.2	1

Relational matrix R_2(y-z connection) is given as,

R (x, z)	Sour	Sweet-sour	Sweet
Green	1	0.5	0.3
Yellow	0.7	1	0.4
Red	0.2	0.7	1

The entries of relational matrix R above were calculated as follows,

$$r_{11} = \text{MAX}(\text{MIN}(1,1), \text{MIN}(0.5,0.7), \text{MIN}(0,0)) = \text{MAX}(1, 0.5, 0) = 1$$

$$r_{23} = \text{MAX}(\text{MIN}(0.3,0), \text{MIN}(1,0.3), \text{MIN}(0.4,1)) = \text{MAX}(0, 0.3, 0.4) = 0.4$$

MAX-PROD composition would give a slightly different relational matrix,

$$R_1 \underset{\otimes}{\circ} R_2 = MAX \begin{bmatrix} (1*1,05*07,0*0) & (1*02,05*1,0*07) & (1*0,05*03,0*1) \\ (03*1,1*07,04*0) & (03*02,1*1,04*07) & (03*0,1*03,04*1) \\ (0*1,02*07,1*0) & (0*02,02*1,1*07) & (0*0,02*03,1*1) \end{bmatrix}$$

$$MAX \begin{bmatrix} (1,0.35,0) & (0.2,0.5,0) & (0,0.15,0) \\ (0.3,0.7,0) & (0.06,1,0.28) & (0,0.3,0.4) \\ (0,0.14,0) & (0,0.2,0.7) & (0,0.06,1) \end{bmatrix} = \begin{bmatrix} 1 & 0.5 & 0.15 \\ 0.7 & 1 & 0.4 \\ 0.14 & 0.7 & 1 \end{bmatrix}$$

Note that the resulting MAX-MIN and the MAX-PROD relational matrices differ a little only in two elements; R_{13} and R_{31}. It might also be interesting to compare the result of the MAX-PROD composition with the classical multiplication of the two matrices R_1*R_2. Note that in a standard matrix multiplication, after the multiplication of the specific elements in corresponding rows and columns, instead of the MAX operator the SUM operator is used.

$$R_1 * R_2 = \begin{bmatrix} (1*1+05*07+0*0) & (1*02+05*1+0*07) & (1*0+05*03+0*1) \\ (03*1+1*07+04*0) & (03*02+1*1+04*07) & (03*0+1*03+04*1) \\ (0*1+02*07+1*0) & (0*02+02*1+1*07) & (0*0+02*03+1*1) \end{bmatrix} = \begin{bmatrix} 1.35 & 0.7 & 0.15 \\ 1 & 1.34 & 0.7 \\ 0.14 & 0.9 & 1.06 \end{bmatrix}$$

The linguistic interpretation of the resulting relational matrix R is a straight one, corresponding to our experience, and it could be given in the form of IF-THEN rules. Example 4.10 shows clearly that *fuzzy relations are suitable means for the expression of fuzzy (uncertain, vague) implications*. Let us give such a linguistic interpretation in the form of the rules of the above given resulting relational matrices:

R1: IF the tomato is *green*, THEN it is *sour*, it is less likely to be *sweet-sour* and it is most unlikely to be *sweet*.

R2: IF the tomato is *yellow*, THEN it is *sweet-sour*, it is possible for it to be *sour* and it is the least likely to be *sweet*.

R3: IF the tomato is *red*, THEN it is *sweet*, it is eventually *sweet-sour* and it is unlikely to be *sour*.

Fuzzy sets, or linguistic variables, are given in *italics*. Note the multivaluedness of the fuzzy implications. Compare the crisp relational matrix from Example 7 with the R_1 given here, as well as compare their corresponding *crisp and fuzzy implications*.

Fuzzy Inference

In classical propositional calculus there are two basic inference rules—*The Modus Ponens* and *The Modus Tollens*. Modus Ponens is associated with the implication *A implies B* or *B follows from A* and it is much more important for engineering applications of logic than Modus Tollens is.

Modus Ponens can typically be represented by the following inference scheme: *Fact or Premise*: *x* is *A*, *Implication*: IF *x* is *A* THEN *y* is *B*, *Consequence or Conclusion*: *y* is *B*. (In Modus Tollens inference, the roles are interchanged; *Fact or Premise*: *y* is not *B*, *Implication*: IF *x* is *A* THEN *y* is *B*, *Consequence or Conclusion*: *x* is not *A*.

The Modus Ponens from standard logical propositional calculus cannot be used in the fuzzy logic environment due to the fact that such an inference can take place if and only if the fact or premise is exactly the same as the antecedent of the IF-THEN rule. In fuzzy logic the *Generalized Modus Ponens* (which allows inference when the antecedent is only partially known, or when the fact is just similar, and not equal, to it) is used. A typical problem in *fuzzy approximate reasoning* looks like:

Implication: IF the tomato is *red*, THEN it is *sweet*, it is eventually *sweet-sour* and it is unlikely to be *sour*.

Premise or Fact: The tomato is *more or less* red ($\mu_{Red} = 0.8$).

Conclusion: **Taste** = ?

The question now is, having *the state of the nature* (premise or fact) that is not exactly equal to the antecedent, and the IF-THEN rule (implication) what is *the conclusion*?

In traditional logic (classical propositional calculus, conditional statements) an expression as such, IF *A* - THEN *B*, is written as $A \Rightarrow B$ (A implies B, B follows from A). Such an *implication* is defined by the following truth table,

A	B	$A \Rightarrow B$
T	T	F
T	F	F
F	T	T
F	F	T

Note that the following identity is used in calculating the above,

$$A \Rightarrow B \equiv A^C \vee B,$$

as well as the "strange" character of the last two rows. Conditional statements, or implications, sound paradoxical when the components are not related. In everyday *human reasoning*, *implications* are given to *combine* the somehow *related statements* but, in the classical two-valued logic's use of the conditional there is no requirement of relatedness. Thus, "unusual" but correct results could be produced using the operator given above. Example 11 illustrates this curious character of standard Boolean logic.

Example 11
The statement

IF 2 * 2 = 5, THEN cows are horses

is **true** (row # 4), but this one

IF 2 * 2 = 4, THEN cows are horses

is **false**(row # 2).

Note that in Boolean logic there doesn't have to be any real causality between the antecedent (IF part) and the consequent (THEN part). It is different in
human reasoning. Our rules express *cause-effect relations*, and fuzzy logic is a
tool for transferring such a structured knowledge into workable algorithms. Thus,
fuzzy logic cannot be and it is not Boolean logic. It must go beyond crisp logic!
This statement is connected with the fact that, in engineering as well as in many
other fields, there is no effect (output) without cause (input).

Therefore, which operator is to be used for *fuzzy* conditional statements (implications) or for *fuzzy* IF-THEN rules? In order to give an answer to this question,
let's see what would be the result of the everyday (fuzzy) reasoning, if the *crisp
implication* algorithm was used. Thus, starting with the crisp implication rule,

$$A \Rightarrow B \equiv A^C \vee B \text{, and with } A^C = 1 - \mu_A(x) \text{, as well as with}$$
$$A \vee B = \text{MAX}(\mu_A(x), \mu_B(y)) \text{, for fuzzy OR operator,}$$

fuzzy implication would be,

$$A \Rightarrow B \equiv A^C \vee B = \text{MAX}(1 - \mu_A(x), \mu_B(y)) \ .$$

This result is definitely not an acceptable one for *related* fuzzy sets which are
subjects of everyday human reasoning, because in the cases when the premise is
not fulfilled ($\mu_A(x) = 0$), the result would be the truth value of the conclusion
$\mu_B(y) = 1$. This doesn't make really much sense in engineering, where a system
input (cause) produces a system output (effect). Or, in other words, if there is no
cause, there will be no effect. Thus, for $\mu_A(x) = 0$, $\mu_B(y)$ must equal 0. For
fuzzy implication the implication rule states that:

> The truth value of the conclusion must not be bigger than that of the
> premise.

There are many different ways by which one can find the truth value of the premise or how one can calculate the relational matrix that describes a given implication. The minimum and product implications are the two most widely used today.
(They were proposed by Mamdani and Larsen respectively).

$$\mu_{A \Rightarrow B}(x,y) = \text{MIN}(\mu_A(x), \mu_B(y))$$

$$\mu_{A \Rightarrow B}(x,y) = \mu_A(x)\mu_B(y)$$

Zadeh's Compositional Rule of Inference

If R is a fuzzy relation from the universe of discourse X to the universe of discourse Y, and x is a fuzzy subset of X, then the fuzzy subset y of Y which is induced by x is given by the *composition*,

$$y = x \circ R$$

As mentioned earlier, the very operator of the composition is MAX-MIN, with alternatives in MAX-PROD or MAX-AVE. Let us show a compositional rule of inference by using MAX-MIN operator in the following example.

Example 12
R represents a relation between *Color* x and *Taste* z of the tomato as given in Example 4.10, and the *state of the nature (premise, fact* or *input x)* is: *The tomato is red.* This premise can be expressed as the input vector x.

Premise or Fact:[8]

$$x = \begin{bmatrix} 0 & 0 & 1 \end{bmatrix}$$

Implication R:

x/z	Sour	Sweet-sour	Sweet
Green	1	0.5	0.3
Yellow	0.7	1	0.4
Red	0.2	0.7	1

Note that the linguistic interpretation of this implication (or of the relational matrix above) is in the form of IF-THEN rules in Example 10.
The conclusion, is a result of the following composition,

---

[8]Note that X has three possible liguistic values; *green, yellow,* and *red.* Thus, the fact that tomato is red is expressed with the vector $x = [0\ 0\ 1]$. This is a *fuzzification step* that *transforms a crisp value* into *a vector of membership degrees.*

$$y = x \circ R = \begin{bmatrix} 0 & 0 & 1 \end{bmatrix} \circ \begin{bmatrix} 1 & 0.5 & 0.3 \\ 0.7 & 1 & 0.4 \\ 0.2 & 0.7 & 1 \end{bmatrix} =$$

$$\text{MAX}\big[m(0,1),\, m(0,0.7),\, m(1,0.2)\ \ m(0,0.5),\, m(0,1),\, m(1,0.7)\ \ m(0,0.3),\, m(0,0.4),\, m(1,1) \big] =$$

$$\begin{bmatrix} 0.2 & 0.7 & 1 \end{bmatrix} \ \ \text{(m denotes a MIN operator)}$$

In solving the previous example we used a composition between x and a given relational matrix R.

Thus when modeling human structured knowledge, the IF-THEN *rules* (as its most common form) *should first be transformed into relational matrices*. Only after the proper relational matrix R of the rule is calculated, can fuzzy inference take place. How to find this relational matrix R of the IF-THEN rules is shown below by using a simple yet generic example.

Example 13
Find the relational matrix R of the following rule (implication),

IF $x = $ small, THEN $y = $ high.

First fuzzy sets *small* and *high* should be defined. These are shown in Figure 23 by their membership functions. In order to obtain a matrix R of finite dimension each membership function must be discretized. Discrete points shown in Figure 23, but not the straight lines of the triangle, represent now fuzzy sets *small* and *high* (see also Figure 13). Thus, the universes of discourse X and Y have now five (or, a finite number of) elements each,

$X = \{-40, -20, 0, 20, 40\}$, $Y = \{-4, -2, 0, 2, 4\}$

In order to calculate the entries of the relational matrix R, let us remind ourselves that *the truth value of the conclusion must be $\leq$ the truth value of the premise*. This will be ensured by using, for example, the MIN or PROD operator. (There are also many others). The result obtained by the MIN operator is,

$$\mu_R (x,y) = \text{MIN}(\mu_S (x), \mu_H (y)).$$

The relational matrix R can be calculated by a vector product using the same procedure as in Example 9,

$$R = \text{MIN}\{\mu_S (x)^T \mu_H (y)\} = \text{MIN}\{[0\ 0.5\ 1\ 0.5]^T [0\ 0.5\ 1\ 0.5\ 0]\}.$$

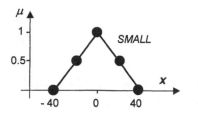 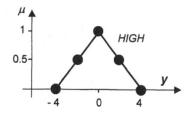

Figure 23 Fuzzy sets, or linguistic terms *small* and *high* given by corresponding membership functions in different universes of discourse.

e.g., for x = -20 and y = 0 the membership degree of the relational matrix will be,

$$\mu_R(x=-20, y=0) = \text{MIN}\{\mu_S(-20)\mu_H(0)\} = \text{MIN}\{0.5, 1\} = 0.5.$$

The whole R looks like this,

R	−40	−20	0	20	40
−40	0	0	0	0	0
−20	0	0.5	0.5	0.5	0
0	0	0.5	1	0.5	0
20	0	0.5	0.5	0.5	0
40	0	0	0	0	0

The fuzzy inference for x' = -20 (the shaded row of R) is the result of the following *composition,*

$$\mu_{S' \circ R}(y) = \mu_{H'}(y) = \underset{x \in X}{\text{MAX}}\ \text{MIN}(\mu_{S'}(x), \mu_R(x,y))$$

$$\mu_{S' \circ R}(y) = \mu_{H'}(y) = [0\ 1\ 0\ 0\ 0] \begin{bmatrix} 0 & 0 & 0 & 0 & 0 \\ 0 & 0.5 & 0.5 & 0.5 & 0 \\ 0 & 0.5 & 1 & 0.5 & 0 \\ 0 & 0.5 & 0.5 & 0.5 & 0 \\ 0 & 0 & 0 & 0 & 0 \end{bmatrix} = [0\ 0.5\ 0.5\ 0.5\ 0]$$

Note that a crisp value x' = -20 was fuzzified or transformed into a membership vector $\mu_{S'}$ = (0 1 0 0 0) first. This is due to the fact that x is a singleton at x', and is shown in Figure 24.

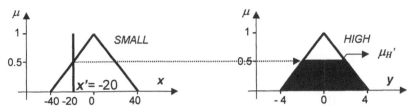

Figure 24 MAX-MIN fuzzy inference. The *conclusion is* not a crisp value but *not-normal fuzzy set.*

Another popular fuzzy inference scheme is MAX-PROD which typically results in a smoother model. The relational matrix R as well as the graphical result of the MAX-PROD inference are given below.

R	−40	−20	0	20	40
−40	0	0	0	0	0
−20	0	0.25	0.5	0.25	0
0	0	0.5	1	0.5	0
20	0	0.25	0.5	0.25	0
40	0	0	0	0	0

Typical control problems are the ones with more input variables, and corresponding rules are given in the form of a rule table,

R1:   IF $x_1 = small$ AND $x_2 = medium$ THEN $y = high$
R2:   IF $x_1 = small$ AND $x_2 = high$ THEN $y = very\ high$, etc.

Now, the rules R are 3-*tuple* fuzzy-relations having membership functions that are hyper-surfaces over the 3-dimensional space spanned by x_1, x_2 and y, e.g., for R_1,

$$\mu_R\,(x_1, x_2, y) = MIN(\mu_S\,(x_1), \mu_M\,(x_2), \mu_H\,(y)).$$

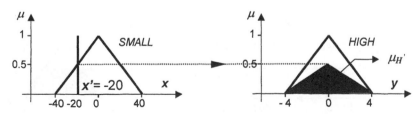

Figure 25 MAX-PROD fuzzy inference. The *conclusion is* not a crisp value but a *not-normal fuzzy set.*

The relational matrix has got a third dimension now! It is a cubic array.

Example 14

Membership functions of two fuzzy sets *small* and *medium* ones, as well as the rule R_1, are,

R1:   IF $x_1 = small$ AND $x_2 = medium$ THEN $y = high$

Figure 26 shows the results of fuzzy inference for the two crisp values, $x_1' = 2$ and $x_2' = 600$.

We actually want to find the output for the two given input values, or $y(x_1' = 2, x_2' = 600) = ?$. At this point we are not sufficiently equipped to say anything about the *crisp* value of y. A part of the tool called *defuzzification* method, is missing at the moment. (It will be introduced later.) But, we can find the *consequent* of the rule R_1. First note that the antecedents 1 and 2 (*small* and *medium*) are connected with an AND operator, meaning that the fulfillment degree of the rule R_1 will be calculated by using a MIN operator, $H = \text{MIN}(\mu_S(2), \mu_M(600)) = 0.5$. Thus, the resulting consequent is a not-normal fuzzy set $\mu_{H'}$ as shown in Figure 26.

Example 15

For $x' = 20$ find the output fuzzy set of the single-input, single-output system described by the two following rules,

R1: IF $x = S$,   THEN $y = H$
R2: IF $x = M$,   THEN $y = M$.
$y(x' = 20) = ?$ $\mu_{x=S}(20) = 0.5$, $\mu_{x=M}(20) = 0.75$
$\mu(y) = MAX \ (\text{MIN}(0.5, \mu_{y=H}(y)), \text{MIN}(0.75, \mu_{y=M}(y))$
$= MAX(\mu_{yH'}(y), \mu_{yM'}(y))$

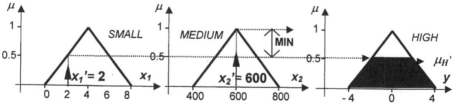

Figure 26   Construction of the consequent membership function $\mu_{H'}$ for the rule R_1.

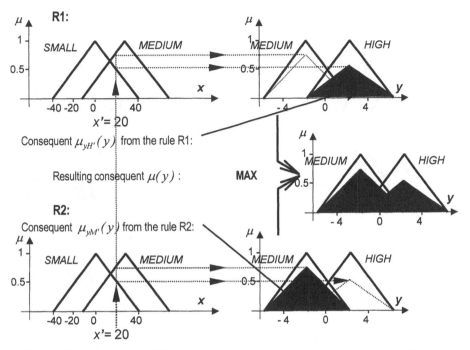

Figure 27 Construction of the consequent membership function from the two active rules for the single-input, single-output system.

Typically, in actual engineering problems we will have both more input variables and more fuzzy sets (linguistic terms) for each variable. In such a situation there will be $N_R = n_{FS1} * n_{FS2} * ... * n_{FSu}$ rules, where n_{FSi} represents the number of fuzzy sets for i-th input variable x_i, and u is the number of input variables. For example, when there are three ($u = 3$) inputs with 2, 3 and 5 fuzzy sets respectively, there will be $N_R = 2*3*5 = 30$ rules. During the operation of the fuzzy model, more rules will generally be active simultaneously. Note that all the rules make a *union* of rules. In other words, the rules are implicitly connected by an OR operator.

Example 15 above shows the fuzzy inference in a simple case when there is one input and one output only ($\mathcal{R}^1 \rightarrow \mathcal{R}^1$ mapping). A slightly more complex and generic situation when there are two inputs and one output ($\mathcal{R}^2 \rightarrow \mathcal{R}^1$ mapping) is given in the Example 16.

Example 16
Find the output fuzzy set for the system with two inputs (having two fuzzy sets for each input) and one output, described by the following four rules:

R1:

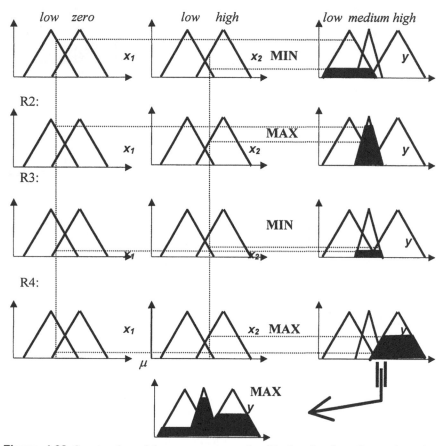

Figure 4.28 Construction of the consequent membership function from four active rules for the system with two inputs and one output.

R1: IF $x_1 = low$ AND $x_2 = low$, THEN $y = low$
R2: IF $x_2 = low$ OR $x_2 = high$, THEN $y = medium$
R3: IF $x_1 = zero$ OR $x_2 = high$, THEN $y = medium$
R4: IF $x_1 = zero$ OR $x_2 = high$, THEN $y = high$

Finally, how to find the crisp output value y from $\mu(y)$, or how to *defuzzify* $\mu(y)$, will be introduced below.

Defuzzification

In the last few examples, we saw that the conclusions happened to be not-normal fuzzy sets! For practical purposes, a crisp output signal to the actuator is needed.

The procedure for obtaining a crisp output value from the resulting fuzzy set, is called a *defuzzification*. Note the subtle difference between the *fuzzification* (which we met in Examples 12 and 13) and the defuzzification:

The fuzzification represents the transformation of the crisp input into a vector of membership degrees, and the defuzzification transforms (typically a not-normal) fuzzy set into a crisp value.

Which method is to be used to find this crisp output value? There are several methods used today. Here we will present the four most popular. Before doing this, it may be useful to get an intuitive insight into defuzzification first.

What do you think, what would be a crisp output value for the resulting not-normal fuzzy set from the last example repeated below? Just by observing the geometry of the resulting fuzzy set, we would conclude that the resulting output value y might be between $y = 50$ and 100, and that the right value could be $y = 58$. At this value it is actually the *center-of-area* (or the *center-of-gravity*) of the resulting consequent from the four rules given above. This is one of the many methods of defuzzification. Figure 4.30 shows three of the most common defuzzification methods.

Each of these methods possesses some advantages in terms of, for example, complexity, computing speed and smoothness of resulting approximating hyper-surface. Thus the first-of-maxima method is the fastest one and is of interest for real time application but the resulting surface is rough. The center-of-gravity for singletons is eventually the most practical method as it has similar smoothness properties to the center-of-area method, but is a simpler and faster method. When the membership functions of the output variables are singletons and when the PROD operator is used in inference, it is relatively easy to show equality of the NN and the FL models (see next subsection). The resulting crisp output in this particular case is calculated as,

$$y' = \frac{\sum_{i=1}^{m} y_i H_i}{\sum_{i=1}^{m} H_i}. \tag{25}$$

CENTER-OF-AREA or
CENTER-OF-GRAVITY

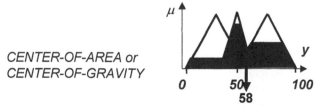

Figure 29 Defuzzification, or obtaining the crisp value from the fuzzy set: Center-of-area (or Center-of-gravity) method.

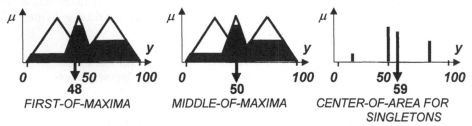

Figure 30 Graphical representation of three popular defuzzification methods.

At the beginning of this Section II.C on fuzzy logic, we stated that human knowledge structured in the form of IF-THEN rules represented mapping, or a multivariate function, which maps the input variables (IF part variables, antecedents, causes) into the output ones (THEN part variables, consequents, effects). Now, in closing this part on FLM, we will illustrate this fact by showing how our common knowledge in controlling the distance between our car and the vehicle in front of us while driving, is indeed a function. In order to show this function graphically, we restrict the input variables to the two most relevant to this control task: the *distance* between the two vehicles and the *speed*.

We call the surfaces represented in Figure 31 the *surfaces of knowledge* because all our control actions (producing the braking force in this task) are the results of sliding on this surface. Normally, we are totally unaware of the very existence of this surface, but it is stored in our mind and all our decisions concerning the braking are in accordance with this two-dimensional function. In reality, this surface is a projection of one hypersurface of knowledge in a 3-dimensional space. In other words, there are additional input variables involved in this control task, for example, visibility, wetness, or the state of the road, our mood and our estimation of the quality of the driver in the car in front. Taking into account all these input variables there is a mapping of the 6 above mentioned input variables into 1 output variable (the braking force). Thus, in this real life situation, this function is a *hypersurface of knowledge* in 7-dimensional space (it is actually $\mathfrak{R}^6 \rightarrow \mathfrak{R}^1$ mapping). Let us stay in our 3-dimensional world and let us analyze a fuzzy model for controlling the distance between the two cars on the road.

Example 17
A fuzzy model for controlling the distance between the two cars traveling on the road should be developed. The resulting surface of knowledge should be shown graphically.

We now have two input variables (the *distance* and the *speed*) and one output variable (the *braking force*). We have chosen 5 fuzzy subsets (membership functions, *attributes*) for each linguistic variable.

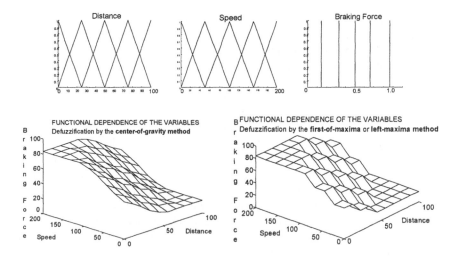

Figure 31 Fuzzy model for controlling the distance between two cars, showing the fuzzy subsets of the two input variables (distance and speed) and one output variable (braking force) and the two *surfaces of knowledge* obtained by different defuzzification methods (below).

The membership functions for the input (or the IF part) variables are the triangles. The fuzzy subsets (the attributes) of the output variable are singletons.

Fuzzy subsets (attributes) of distance: [very small, small, moderate, large, very large]
Fuzzy subsets (attributes) of speed: [very low, low, moderate, high, very high]
Fuzzy subsets (attributes) of braking force: [zero, one fourth, half, three fourth, full]

Now, the rule basis comprises 25 rules of the following type;

R1: IF distance *very small* AND speed *very low* THEN braking force *half.*

The inference was made by using a MAX-MIN operator. Two surfaces of knowledge are represented below in Figure 31. The left and smooth one is obtained by the center-of-gravity defuzzification and the right and rough one is obtained by using the first-of-maxima defuzzification method.

There are many interesting comments and conclusions which can now be made. Let us mention a few of the most important ones. Firstly, by using an FL model, we tried to model structured human knowledge. This knowledge is highly imprecise. We all drive a car differently. Already at the very first step, defining the universes of discourse, each of us would define them differently. The younger, or less cautious, of us would definitely consider distances of 100 yards intolerably large. They drive *'very close'*, meaning that the maximal value of the distance's fuzzy subsets [*very large*] would be perhaps 50 yards. On the other hand, more cautious drivers would never even think of driving at velocities higher than 75 mph. Secondly, the choice (shapes and positions) of the membership functions is highly individual. Thirdly, the inference mechanism and the defuzzification method applied will also have an impact on the final result. Despite all these *fuzzy* factors, the resulting surface of knowledge which represents our knowledge with regard to the solution of the given problem is usually an acceptable one. If there is good knowledge, then the fuzzy logic provides the tool to transfer it into an efficient algorithm. Compare the two surfaces given above. Both surfaces model very well known facts: that a decrease in the distance and/or an increase in driving speed, demand a larger braking force and vice versa.

Note that in the case where there are several input and output variables, nothing but computing time and memory changes. If the resulting surfaces reside in four, and higher dimensional space, visualization is not possible any longer but the algorithms remain the same. (Yes, we rename these surfaces into *hypersurfaces.*)

D. Mathematical Similarities and the Equality of NN and FLM

In the previous sections we have mentioned that NN and FLM are based on very similar, or sometimes even equal, underlying mathematics. This fact has been reinvented and shown by different authors independently. Here, in order to show this very important result, we follow the Kecman and Pfeiffer paper, [12]. This presentation shows when and how the *Learning of Fuzzy Rules* (LFR) from numerical data is mathematically equal to the training of *Radial Basis Function* (RBF), or regularization, networks. Although both approaches have their origin in different paradigms of intelligent information processing, it is demonstrated that the mathematical structure is the same. They also share the property of being 'universal approximators' of any real continuous function on a compact set to arbitrary accuracy. In the LFR algorithm proposed here, the subjects of learning are the rule conclusions, i.e., the positions of the membership functions of output fuzzy sets (also called attributes) which are in form of singletons. For fixed number, location and shape of input membership functions in LFR or of basis functions in RBF, LFR and RBF training becomes a least squares optimization problem which is linear in the unknown parameters. (These parameters are the OL weights **w** for RBF or the rules **r** for LFR). In this case the solution boils down to the pseudoinversion of a rectangular matrix. The presentation below can be readily extended to the other, not necessarily radial, activation functions.

Using these two approaches, the general problem of approximating or learning, a mapping f from an n-dimensional input space to an m-dimensional output space, given a finite training set of K examples of f $\{(x_1, y_1), (x_2, y_2), \dots, (x_K, y_K), y_k = f(x_k)\}$ will be exemplified by the one to one mapping presented in Figure 32.

First, it should be stated that for real world problems with noisy data we will never try to find the function F which interpolates a training set passing through each point in the set, i.e., we do not demand $F(x_k) = y_k \ \forall k \in \{1, \dots, K\}$. The approximating function F of the underlying function f will be obtained with the relaxed conditions that F does not have to go through all the training set points, but we do want to get our function F as close as possible to the data.

Usually, the criterion of such a closeness is least squares. It is clear that in the case of the noisy free data the interpolation is the better solution. This will be true as long as the size of the data set is not too large. Generally, in the case of large data sets (say, more than 1000 data patterns), and due to the numerical problems, one is forced to find an approximation solution.

In the RBF approximation techniques (that resulted from the regularization theory by Tikhonov and Arsenin [13], which is a good theoretical framework for the treatment of the approximation or interpolation problems) under some mild assumptions the expression for an approximating function F has the following simple form,

$$y = F(x) = \sum_{i=1}^{N} w_i \varphi_i(x, c_i).\tag{26}$$

where w_i are weights to be learned, c_i are centers of the radial basis functions φ_i, which can have different explicit forms (e.g., spline, Gaussian, multiquadrics).

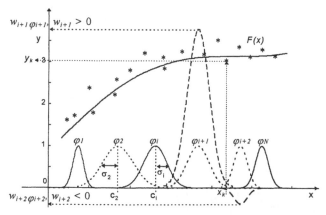

Figure 32 Training data set (*), basis φ, or membership functions μ, and nonlinear approximating function $F(x)$. Note that this approximation is *linear in parameters*.

It is important to realize that in the case when the number N, positions c_i and shapes (which are defined by the parameter σ and by the covariance matrix Σ for the 1-dimensional and higher-dimensional Gaussian basis functions respectively) are fixed before the learning, the problem of approximation is linear in the parameters (weights w_i) that are the very subject of learning. Thus, the solution boils down to the pseudoinversion of matrix $\Phi(K, N)$. This matrix is obtained by using Eq. (26) for the whole training set. If any of the parameters c_i or σ_i, that are "hidden" behind the nonlinear function φ_i, become for any reason part of the training, the problem of learning will have to be solved by the nonlinear optimization. And certainly, it will be much more involved.

In order to show the equality of NN and FLM we consider, without loss of generality, a scalar output variable. The RBF network modeling data set is given as,

$$y = F(\mathbf{x}) = \sum_{i=1}^{N} w_i \varphi_i(\mathbf{x}, \mathbf{c}_i). \tag{27}$$

For the case that φ is a Gaussian function (and usually the normalized Gaussian with the amplitude $G(c_i, c_i)=1$ is used) one can write,

$$y = F(\mathbf{x}) = \sum_{i=1}^{N} w_i G_i(\mathbf{x}, \mathbf{c}_i). \tag{28}$$

Figure 32 presents Eq. (28) graphically. For $N = K$ an interpolation and for $N < K$ an approximation will take place.

The same approximation problem can be considered as the problem of learning fuzzy rules (LFR) from examples. Figure 32 still represents our problem setup, but now the Gaussian bumps are interpreted as membership functions μ_i of the linguistic attributes (fuzzy subsets) of the input variable x (input is now a 1-dimensional variable or scalar).

For reasons of computational efficiency, the attributes of the linguistic *output variable* are defuzzified off-line, by replacing the fuzzy set of each attribute with a singleton at the center of gravity of the individual fuzzy set (as in Figure 31, braking force graph). The parameters to be learned are the positions r_i of the singletons describing the linguistic rule conclusions. The corresponding continuous universes of discourse for linguistic input and output variables $Input_1, ..., Input_n$, and *Output,* are called $X_1, ..., X_n, Y$ respectively. Rule premises are formulated as fuzzy AND relations on the Cartesian product set $X = X_1 \times X_2 \times ... X_n$, and several rules are connected by logical OR. Fuzzification of a crisp scalar input value x_1 produces a column vector of membership grades to all the attributes of $Input_1$, and similarly for all other input dimensions, e.g.,

$$\mu_1 = \begin{bmatrix} \mu_{1,"attr1"}(x_1) \\ \mu_{1,"attr2"}(x_1) \\ \vdots \end{bmatrix}, \mu_2 = \begin{bmatrix} \mu_{2,"attr1"}(x_2) \\ \mu_{2,"attr2"}(x_2) \\ \vdots \end{bmatrix}. \tag{29}$$

The degrees of fulfillment of all possible AND combinations of rule premises are calculated and written into a matrix $\mathbf{M}$. For ease of notation, the following considerations are formulated for the case of only two input variables, but they can be extended straightforward to the higher dimensional input spaces. If the algebraic product is used as an AND operator, this matrix can be directly obtained by multiplication of a column and a row vector,

$$\mathbf{M}(\mathbf{x}) = \mu_1(x_1)\, \mu_2^T(x_2). \tag{30}$$

Otherwise, the minimum operator can be applied to all pairs of membership values. As the attributes of the linguistic output variable are singletons, they appear as crisp numbers in the fuzzy rule base. The first rule, for example, reads,

IF Input$_1$ is *attribute$_1$* AND Input$_2$ is *attribute$_2$*, THEN Output is r_{11},

and its conclusion is displayed as the element r_{11} in the relational (rule) matrix,

$$\mathbf{R} = \begin{bmatrix} r_{11} & r_{12} & \cdots \\ r_{21} & r_{22} & \cdots \\ \vdots & \vdots & \vdots \end{bmatrix}. \tag{31}$$

$\mathbf{R}$ has the same dimensions as $\mathbf{M}$. IF THEN rules are interpreted as AND relations on $X \times Y$ i.e., the degree of membership to the output fuzzy set of a rule is limited to the degree up to which the premise is fulfilled. A crisp output value y is computed by the Center of Singletons (CoS) algorithm (25) as a weighted mean value

$$y = F(\mathbf{x}) = \frac{\sum_{jl} \mu_{jl}(\mathbf{x})\, r_{jl}}{\sum_{jl} \mu_{jl}(\mathbf{x})} \tag{32}$$

where $\mu_{jl} = H_i$ and $r_{jl} = y_i$. The sum covers all elements of the two matrices $\mathbf{M}$ and $\mathbf{R}$. If the membership functions of the input attributes are the Gaussians, the μ_{jl} are space bumps $G_i(\mathbf{x}, \mathbf{c}_i)$ representing the joint possibility distribution of each rule. If, moreover, the elements of matrix $\mathbf{R}$ are collected in a column vector,

$$\mathbf{r} = (r_{11},\ r_{12},\ \ldots,\ r_{21},\ r_{22},\ \ldots)^T = (r_1,\ r_2,\ \ldots,\ r_N)^T, \tag{33}$$

the approximation formula becomes,

$$y = F(\mathbf{x}) = \frac{\sum_{i=1}^{N} G(\mathbf{x}, \mathbf{c}_i)\, r_i}{\sum_{i=1}^{N} G(\mathbf{x}, \mathbf{c}_i)}. \tag{34}$$

The structural similarity of (28) and (34) is clearly visible, where the rule conclusions r_i correspond to the trainable weights w_i. These two equations could be given graphical representation in the form of a "neural" networks. For the bivariate functions $y = f(x_1, x_2)$ this is done as in Figure 33.

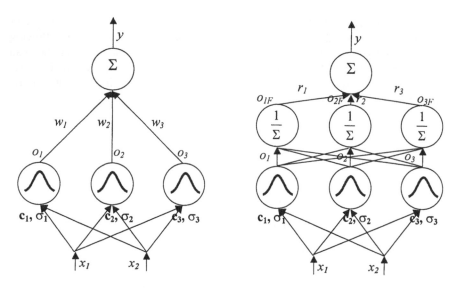

Figure 33   Networks for interpolation ($N=K$) or approximation ($N<K$) of a bivariate function $y = f(x_1, x_2)$. RBF (left) and Fuzzy Network or Soft RBF (right). Here $N=3$.

The structure of both networks is the same in the sense that both have just one hidden layer, and that in both networks the connections between the input and the hidden layer are fixed and not the subject of learning. The subjects of learning are the connections w or r between the hidden layer and the output layer, respectively. It should be stressed that the seemingly second hidden layer in Fuzzy (or Soft RBF) network is not an additional hidden layer, but the normalization part of the only hidden layer. Due to this normalization the sum of the outputs from the hidden layer in Soft RBF is equal to one, i.e., $\Sigma\ o_{iF} = 1$. This is not the case in classical RBF. (The meaning of the words "soft" and "normalization" are explained below.)

The equality of these two approximation schemes is obvious if (28) and (34) are compared. The only difference is that in the so-called fuzzy approximation the output value from the HL y is *"normalized."* The word normalized in quotation marks is used because y is calculated using the normalized output signals o_{iF} in Figure 33 from the neurons whose sum is equal to 1. This is not the case with standard RBF network. The fuzzy approximation, due to the effect of "normalization," is doing some kind of soft approximation with the approximating function always going through the middle point between the two training data. In analogy with the *"softmax"* function introduced into neural network community for the sigmoidal type of activation functions by John Bridle in [14], we name the *fuzzy approximation* as a *soft RBF approximation scheme.*

More on the very mathematics of the learning fuzzy rules in the FLM or, on the OL weights adaptation in the RBF networks, can be found in the Appendix A. For the fixed HL weights this learning rules are linear. The genuine NN learning problem is an nonlinear optimization task in which we want to find optimal values of both HL and OL weights. We present the basics of the learning in the NN in the next section.

III. LEARNING TECHNIQUES

Without any doubt the capacity to learn the underlying dependencies, or functions, between the input and output variables by using sets of measured data, is the most important ability of NN models. This learning was, and still is, an intriguing concept that has been developed and used independently in different scientific and engineering areas. A solution of regression problems in statistics and an identification of system parameters in linear control engineering, are the two most important fields which rely heavily on the learning techniques. The solution to these problems is an easy, and relatively straightforward, task as long as the approximating models are linear in parameters.

The equation (A.7) is a standard, and the best in the least squares sense, solution when the model output (or the error) depends linearly upon unknown parameters. In the case of NN and FL models, the problem is for fixed number, locations, and shapes of the hidden layer activation (basis, membership) functions as a linear one. This means from a practical point of view that the HL weights (matrix V in Figure 3 or c_i-s and σ_i-s in Figure 33) are fixed and not the subject of learning.

This dramatically changes if the HL weights are trained. Unfortunately, this is almost always the case. The solution to the problem becomes one of nonlinear optimization, which is difficult. There are plenty of books on nonlinear optimization and practically all presented techniques can be used for HL and OL weights adaptation. However, NN and FLM have a specific structure and only a small part of the established optimization techniques have passed the test. By far, the most popular learning method in the NN field is the *error back propagation (EBP)* algorithm. Basically, EBP is *a gradient based* learning approach developed for the NN. Below we will devise the EBP for one single nonlinear neuron only. In the case that we should apply the EBP to the NN with hidden layer(s) the basic approach is the same but the resulting models are more complex ones.

The basic adaptation algorithm developed in 1960s for linear neuron by Widrow and Hoff (see in [15] and [16]) is known as the *delta learning rule*. The delta learning rule is an on-line adaptation algorithm that learns the weights by using data pairs only. The variable (or signal) δ designates an *error signal*, and not the error itself as defined in (6). Thus, δ will generally not be equal to the error $e = d-o$. Interestingly, as will be shown below, the equality $\delta = e$ will hold for a linear activation function. In the world of neural computing the error signal δ used to have the highest importance. The adapting rule which in 1986 had made a breakthrough (after the stall in development of learning rules for the multilayer networks of almost 20 years), was called *the generalized delta learning* rule. Today this rule is known as the *error back propagation (EBP)* algorithm, too. The basics of this *error correction delta rule* which uses a gradient descent strategy for adapting the weights in order to reduce the error or cost function is presented here. This presentation will be exact for the neurons that use *scalar product* [see (14)] between the input and weight vector as an input to the neuron. (These are typical neurons in the so-called multilayer perceptrons. For the RBF neurons the final result is only slightly different.)

The EBP algorithm will be derived for a *single neuron,* having *any* differentiable activation function presented in Figure 34, which forms the input signal u as a scalar product. After devising the EBP for any activation function the *least mean square (LMS)* rule for the linear neuron will be treated as only one of the possible cases when the activation function is a linear one. The gradient based methods optimize the weights according to the simple learning rule,

$$\mathbf{w}_{p+1} = \mathbf{w}_p + \Delta\mathbf{w}_p = \mathbf{w}_p - \eta \frac{\partial E}{\partial \mathbf{w}}\bigg|_p , \tag{35}$$

where index p stands for the pattern presented, or for the iteration step, and η *is the* learning rate that controls the stability and rate of adaptation. Thus the problem to solve is to find the expression for, and to calculate, the gradient $\nabla_{\mathbf{w}} E\big|_p$ given in (35), by using a training set of pairs of input and output patterns. Recall that we want to learn in the on-line mode so let us first define our error function for this case,

$$E_p = \frac{1}{2}e_p^2 = E(\mathbf{w}_p). \tag{36}$$

Note that this is an on-line version of the error function (4). The constant ½ is used for computational convenience only. It will be canceled out, by the following required differentiation. Note also that $E(\mathbf{w})$ is a nonlinear function of the weight vector. Fortunately, in this simple case, the calculation of gradient is straightforward. For this purpose all we have to do is to use the *chain rule* as follows,

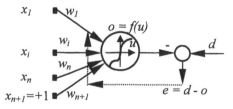

Figure 34 Single neural unit with any differentiable activation function.

$$\frac{\partial E}{\partial \mathbf{w}}\Big|_p = \underbrace{\frac{\partial E_p}{\partial u_p}}_{-\delta} \frac{\partial u_p}{\partial \mathbf{w}_p},\tag{37}$$

where the first term on the right hand side is called the *error signal* δ. This delta signal is a measure of how the error changes due to the input to the neuron u when the p-th pattern is presented. The second term shows the influence of the weight vector $\mathbf{w}$ on that particular input. By applying the chain rule again we obtain,

$$\frac{\partial E}{\partial \mathbf{w}}\Big|_p = \frac{\partial E_p}{\partial e_p} \frac{\partial e_p}{\partial o_p} \frac{\partial o_p}{\partial u_p} \frac{\partial u_p}{\partial \mathbf{w}_p},\tag{38}$$

$$\frac{\partial E}{\partial \mathbf{w}}\Big|_p = e_p(-1)f'(u_p)\,\mathbf{x}_p.\tag{39}$$

The last term in (39) follows from the fact that $u_p = \mathbf{w}^T\mathbf{x}_p$, and the δ learning rule can be written as,

$$\mathbf{w}_{p+1} = \mathbf{w}_p - \eta\frac{\partial E}{\partial \mathbf{w}}\Big|_p = \mathbf{w}_p + \Delta\mathbf{w}_p = \mathbf{w}_p + \eta(d_p - o_p)f'(u_p)\,\mathbf{x}_p.\tag{40}$$

This is the most general learning rule which is valid for a single neuron having *any nonlinear and differentiable* activation function and whose *input u* is formed as a *scalar product* of the pattern and weight vector. Because of the important role which this delta signal plays in the development of the general EBP algorithm for the multilayer neural network, it might be useful to present separately the expression for the error signal δ_p as,

$$\delta_p = (d_p - o_p)f'(u_p).\tag{41}$$

The calculation of the particular change of the single weight vector component w_j is straightforward and becomes after rewriting the vector equation (40) in terms of vectors' components,

$$w_{j,p+1} = w_{j,p} + \eta(d_p - o_p)f'(u_p)\,x_{j,p} = w_{j,p} + \eta\,\delta_p\,x_{j,p}.\tag{42}$$

Note the fact that the error signal δ_p is the same for each component of the weight vector. Thus, the change of w_j is determined by, and proportional to, its corresponding component of the input vector x_j. The famous LMS learning rule for a linear neuron taking into account that $f'(u_p) = 1$ is given as,

$$\mathbf{w}_{p+1} = \mathbf{w}_p + \Delta\mathbf{w}_p = \mathbf{w}_p + \eta(d_p - o_p)\,\mathbf{x}_p = \mathbf{w}_p + \eta\,\delta_p\,\mathbf{x}_p.\tag{43}$$

The LMS is an error correction type of the rule, in the sense that the weight change $\Delta\mathbf{w}_p$ is proportional to the error $e_p = (d_p - o_p)$.

The EBP, and its original variant for linear neuron LMS, are the iterative learning schemes that use the gradient, or the first derivative of the error function in respect to the weights. Thus, they belong to the group of so-called "first order" optimization algorithms. It is well known that by using the information about the curvature of the error function to be optimized we can considerably speed up the learning procedure. Information about the curvature is contained in the second derivative of the error function in respect to the weights, or in the Hessian matrix. More about the second order algorithms can be found in specialized references. In the general case, the NN comprises many layers and devising of the EBP rule is much more involved. The final algorithm as well as its application are simple ones though. The on-line EBP learning algorithm, for the NN as given in Figure 3, but with K OL neurons, is presented in Table 5. Example 18 shows how the given algorithm can be readily used.

Example 18
For the network below calculate the expressions for the weight changes by the EBP algorithm in an on-line learning mode.
After presenting input vector $\mathbf{x} = [x_1\, x_2]^{\mathrm{T}}$, the output vector $\mathbf{o} = [o_1\, o_2]^{\mathrm{T}}$ is calculated first. Knowing activation functions in neurons their derivatives can be readily calculated and, by using given desired vector $\mathbf{d} = [d_1\, d_2]^{\mathrm{T}}$, we can calculate delta signals for the OL neurons

$$\delta_{ok} = \underbrace{(d_k - o_k)}_{e_k} f_{ok}'(u_k),\ (k = 1, 2).$$

Having δ_{ok} we can find the hidden layer neurons' deltas (or error signals) δ_{yj} as follows $\delta_{yj} = f_{hj}'(u_j)\sum_{k=1}^{K}\delta_{ok}w_{kj}$, $(j = 1, 2, 3, k = 1, 2)$. Only now can we calculate the weight changes for specific weights. Thus, for example, $\Delta v_{12} = \eta\delta_{h1}x_2$, $\Delta v_{32} = \eta\delta_{h3}x_2$, $\Delta w_{23} = \eta\delta_{o2}y_3$, $\Delta w_{13} = \eta\delta_{o1}y_3$, etc.

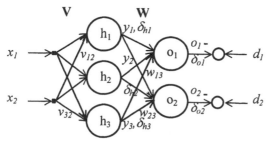

Figure 35   Scheme of the variables for the EBP learning in a multilayer NN.

Table 5 Summary of the EBP Algorithm[a] (On-Line Version)

Given is a set of P measured data that will be used for training: $X = \{x_p, d_p, p = 1, ... , P\}$ consisting of the input pattern vectors x and the output desired responses d:

$$\mathbf{x} = \begin{bmatrix} x_1, & x_2, & . & . & ., & x_n, & +1 \end{bmatrix}^T, \mathbf{d} = \begin{bmatrix} d_1, & d_2, & . & . & . & d_K \end{bmatrix}^T.$$

Note that there are K output layer neurons.

FEEDFORWARD PART

STEP 1: Choose the learning rate η and predefine the maximally allowed, or desired, error E_{des}.

STEP 2: Initialize weight matrices $V_p(J, I)$ and $W_p(K, J+1)$.

STEP 3: Perform the online training (weights are adjusted after each training pattern). p=1 ,..., P. Apply the new training pair (x_p, d_p) in sequence, or randomly drawn, to the hidden layer neurons.

STEP 4: Calculate consecutively the outputs from the hidden neurons and from the output layer ones

$$y_{jp} = f_h(u_{jp}), o_{kp} = f_o(u_{kp})$$

STEP 5: Find the value of the sum of errors square cost function E_p for the data pair applied and given weight matrices V_p and W_p, (in the first step of an epoch initialize $E_p = [ ]$)

$$E_p = \frac{1}{2} \sum_{k=1}^{K} (d_{kp} - o_{kp})^2 + E_p$$

Note that the value of the cost function is accumulated over all the data pairs.

BACK PROPAGATION PART

STEP 6: Calculate the output layer neurons' error signals δ_{okp} as follows,

$$\delta_{okp} = \underbrace{(d_{kp} - o_{kp})}_{e_{kp}} f_{ok}'(u_{kp}), (k = 1, ..., K)$$

STEP 7: Calculate the hidden layer neurons' error signal δ_{yjp},

$$\delta_{yjp} = f_{hj}'(u_{jp}) \sum_{k=1}^{K} \delta_{okp} w_{kjp}, (j = 1, ..., J).$$

STEP 8: Calculate the updated output layer weights $w_{kj, p+1}$,

$$w_{kj, p+1} = w_{kj p} + \eta \delta_{okp} y_{jp}$$

STEP 9: Calculate the updated hidden layer weights $v_{ji, p+1}$

$$v_{ji, p+1} = v_{ji p} + \eta \delta_{yjp} x_{ip}$$

STEP 10: If p < P go to STEP 3, otherwise go to STEP 11

STEP 11: The learning epoch (the sweep through all the training patterns) is completed, (p = P). For $E_p < E_{des}$ terminate the learning, otherwise go to STEP 3 and start a new learning epoch, p=1.

[a]Algorithm is given for the multilayer perceptron networks (NN in which the input to the neurons is defined as a scalar product $u = w^T x$). Also, the dimension of the matrix W is given for the case when there is a bias term in hidden layer as shown in Figure 3.

Despite the sound theoretical foundations that concern the representational capabilities of NN and FLM (as being general multivariate function approximators in the sense that they can uniformly approximate any continuous function to within an arbitrary accuracy), and the apparent success of the EBP, there are many practical aspects of the back propagation which does not guarantee that this algorithm will not have "serious" difficulties. The most troublesome is the unusually long training process that does not guarantee that the absolute minimum of the cost function, meaning the best performance of the network on the test data set, will be achieved. The algorithm may become stuck at some local minimum and such a termination with a suboptimal solution will require the repetition of the whole learning process by changing the structure and/or some of the learning parameters that influence this iterative scheme.

As in many other scientific and engineering disciplines, in the artificial neural networks field, the theory (or at least some part of it) was only established after a number of practical neural network applications had been implemented. Many practical questions are still open and for the broad range of engineering tasks the design of NN, as well as the learning procedures and corresponding training parameters, are still in the realm of the empirical art. In this respect the EBP algorithm is a genuine representative of the nonlinear optimization schemes.

This chapter will not discuss the practical aspects of the EBP learning. Instead it will mention only the most important design parameters which should basically be determined by simulational experiments. These are: the number of hidden layers (for control purposes one is actually enough), the number of neurons in the hidden layer (highly problem dependent), the type of activation functions (the most common ones are mentioned before), weight initialization (some previous knowledge and data preprocessing can be a lot of help), the magnitude of the learning rate (can be determined in a few simulational runs) and introduction of the momentum term (a highly recommended).

IV. NEURAL NETWORKS BASED ADAPTIVE CONTROL

This final section focuses on the NN and FLM based adaptive control.[9] In particular, after presenting the basic guiding ideas of the NN based control approaches, we will introduce the *Adaptive Backthrough Control (ABC)* scheme as one of the most serious candidate for the future control of the large class of nonlinear, partially known and time-varying systems (Kecman [8]; Kecman and Rommel [17] and Rommel [18]). Recently, the area of NN control has been exhaustively investigated and there is a large number of different NN based control methods. (Rigorous comparisons show that the NN based controllers perform far better than the well established conventional options when the plant characteristics are poorly known (Boskovic and Narendra [19]). A systematic classification of the very different NN control structures is a formidable task indeed (Agarwal [20]). Here, we will describe the approach based on feed-forward networks having "static" neurons as given in Figures 3 and 33.

---

[9]In what follows NN will stand for either neural, fuzzy, neuro-fuzzy, or fuzzy-neuro models.

The standard control task and the basic problem in controlling an unknown dynamic plant is to find the proper, or desired, control (actuation) value u_d as an input to the plant which should ensure that,

$$y(t) = y_d(t), \quad \forall t, \tag{44}$$

where the subscript d stands for *desired*. $y(t)$ and $y_d(t)$ denote the plant output and desired (reference) plant output respectively. A controller that could produce this value u_d would be the best controller, and the output of the plant would exactly follow the desired input y_d. In *linear* control, (44) will be ensured when,

$$G_{ci}(s) = G_p^{-1}(s). \tag{45}$$

Hence, the ideal controller transfer function $G_{ci}(s)$ should be the inverse of the plant transfer function $G_p(s)$. Because of the many practical constraints, this is an idealized control structure (Kecman [21]). However, we can try to get as close as possible to this ideal controller solution $G_{ci}(s)$. The ABC approach that is presented in this section, can achieve a great deal (and sometimes even nearly all) of this ideal controller. The block diagram of the *ideal control* of any nonlinear system is given in Figure 36.

$\mathbf{f(u, y)}$ represented in Figure 36 stands for any nonlinear mapping between an input $\mathbf{u}(t)$ and an output $\mathbf{y}(t)$. In the general case of a dynamic system, $\mathbf{f(u, y)}$ represents a system of nonlinear differential equations. Here we will primarily be concerned with discrete-time systems, and the model of the plant in the discrete-time domain will be in the form of a nonlinear discrete equation $\mathbf{y}(k+1) = \mathbf{f(u}(k), \mathbf{y}(k))$. Now, the basic problem is how to learn (design or obtain) the inverse model of the unknown dynamic plant by using NN.

The wide application of NN in control is based on the universal approximation capacity of neural networks and fuzzy models. Thus learning (identification, adaptation, training) of the plant and inverse plant dynamics represents both the basic mathematical tool and the basic problem to be solved. Therefore the analysis presented below assumes the stability as well as the complete controllability and observability of the plant. So far as the representation of the dynamic system is concerned, we use a so-called NARMAX model here. In the extensive literature on modeling dynamic plants, it was proved that under some mild assumptions any nonlinear, discrete and time invariant system can always be represented by the following NARMAX model,

$$y(k+1) = f\{y(k), \cdots, y(k-n); u(k), \cdots, u(k-m)\}, \text{ or}$$

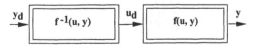

Figure 36 The ideal (feedforward) control structure for any plant.

$$y_{k+1} = f(y_k, y_{k-1}, y_{k-2}, \ldots, y_{k-n}, u_k, u_{k-1}, u_{k-2}, \ldots, u_{k-m}), \quad (46)$$

where y_k and u_k are the input and output signals at instant k, and y_{k-i} and u_{k-j} ($i = 1$, ..., n and $j = 1, \ldots, m$) represent the past values of these signals. Typically one can work with $n = m$. (46) is a simplified deterministic version of the NARMAX model (there is no noise terms in it), and it is valid for dynamic systems with K outputs and L inputs. For $K = L = 1$ we obtain the so-called SISO (single-input single-output) system which is studied here.

In reality, the nonlinear function f from (46) is very complex, and generally unknown. The whole idea in the application of NN is to try to *approximate* f by using some known and simple functions which, in the case of the application of NN and FLM, are their activation and membership functions respectively.

This identification phase of the mathematical model (46) can be represented graphically (Figure 37). Note that two different identification schemes are presented in Figure 37 - a *series-parallel* and a *parallel one*. (The names are introduced by Landau, [4]). Identification can be done by using either of the two schemes,

$$y(k+1) = f\{y(k), \cdots, y(k-n); u(k), \cdots, u(k-n)\} \quad \text{Series-Parallel,} \quad (47)$$

$$y(k+1) = \{\hat{y}(k), \cdots, \hat{y}(k-n); u(k), \cdots, u(k-n)\} \quad \text{Parallel.} \quad (48)$$

It is hard to say which scheme is the better one. Narendra and Annaswamy in [5] showed (for linear systems) the series-parallel method to be globally stable, but similar results are not available for the parallel model yet. The parallel method has the advantage of avoiding noises which exist in real plant output signals. On the other hand, the series-parallel scheme uses actual (meaning correct) plant outputs, and this is what generally enforces identification. It should be mentioned that the questions regarding better performance, advantages and shortcomings of the

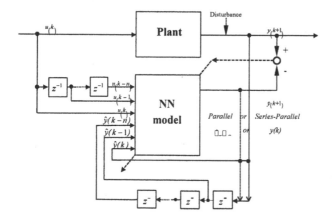

Figure 37 Identification scheme using NN.

series-parallel model on one side (advanced and used by Narendra & Parthasarathy in [9], for example) and the parallel model on the other, are still unanswered.

The following outlines the "history" of, seemingly the strongest stream of the NN based control strategies. Inside this stream of a *feedforward control*, there were a few relatively independent and partly dissimilar basic directions in the search for a good control strategy. However, the leading idea remained the same in these otherwise different control schemes. The final goal was always the determination of a good inverse model of the plant dynamics $\mathbf{f}^{-1}(\mathbf{u}, \mathbf{y})$ as required in ideal feedforward control structure shown in Figure 36. We will show the promising new ABC approach and its performance. The presentation that follows is based on (Kecman [8]).

A. General Learning Architecture or Direct Inverse Modeling

Figure 38 shows how the inverse plant model of a *stable* plant can be trained by using the *general learning architecture* introduced by Psaltis et al [22]. Independently, the same approach developed in Jordan and Rumelhart [23] is called *direct inverse modeling*. This is basically an off-line procedure and for nonlinear plants it will usually precede the on-line phase. (If the plant is unstable, a stabilization with a feedback loop is necessary. This can be done with any standard control algorithm.) To learn the inverse plant model, an input signal u is chosen and applied to the input of the plant to obtain a corresponding output y. In the following step the neural model is trained to reproduce this value u at its output.

After this training phase, the structure of the on-line operation looks like that given in Figure 36. Thus, the NN representing the inverse of the plant precedes the plant. The trained NN should be able to take a desired input value y_d and produce the appropriate $u = u_d$ as an input to the plant. This architecture is unfortunately not "goal-directed." Note that we normally do not know what output u_d of the controller corresponds to the desired output y_d of the plant. Therefore, this learning scheme should cover the large operational regime of the plant, with the limitation that a control system cannot be selectively trained to respond accurately in the region of interest. Thus, one of the important parts of learning within the general learning architecture is the selection of adequate training signals u, which

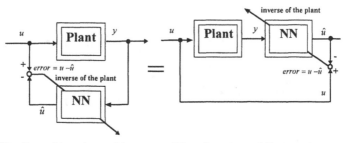

Figure 38 General learning architecture or Direct inverse modeling.

should cover the whole input range. Also, as this is an off-line approach unsuitable for on-line applications, the controller cannot operate during this learning phase. Besides, due to the use of the EBP algorithm (that minimizes the sum of the error squares cost function) this structure may be unable to find a correct inverse if the plant is characterized by many-to-one mappings from the control inputs u to the plant outputs y. But, despite these drawbacks, in a number of domains (stable systems and one-to-one mapping plants) this general architecture is a viable technique.

B. Indirect Learning Architecture

As a second concept Psaltis et al. [22] introduced *indirect learning architecture*. In this adaptive control structure the controller or network NN_1 (which is a copy of the trained inverse plant model NN_2) produces, from the desired output y_d, a control signal u_d that drives the plant to the desired output $y = y_d$. The aim of the learning is to produce a set of NN_2 weights, which will be copied into the network NN_1, in order to ensure correct mapping $y_d \rightarrow u$ over the range of the *desired* operation.

The positive feature of this arrangement would be the fact that the network can be trained in the region of interest. It is "goal-directed." Also, the fact that this is an on-line learning procedure, is advantage in respect to the general architecture. However, Psaltis et al. in [22] mentioned that this method is unfortunately not a valid training procedure because minimizing the controller error $e_1 = u - \hat{u}$ does not necessarily minimize the performance error $e_3 = y_d - y$. (Actually, the very name of this architecture comes from the fact that the subject of minimization is not *directly* the performance error e_3 between the desired and actual plant output, but rather the controller error e_1.) This structure also uses the EBP algorithm, and it has similar problems as a general learning architecture if the plant performs many-to-one mappings from control inputs u to plant outputs y.

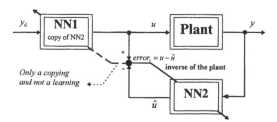

Figure 39   Indirect learning architecture.

C. Specialized Learning Architecture

The third approach as presented in Figure 40 is the specialized learning architecture introduced in Psaltis et al, in [22].

This structure operates in on-line mode and it trains a neural network to act as a controller in the region of interest (meaning that it is "goal directed"). In this way this scheme avoids some of drawbacks of the previous two structures. Here, in the specialized learning architecture, the controller no longer learns from *its* input-output relation, but from a *direct* evaluation of the *system's* performance error $e_3 = y_d - y$. The network is trained in order to find the best control value u that drives the plant to an output $y = y_d$. This is accomplished by using a steepest decent (EBP) learning procedure. Despite the fact that the specialized architecture operates in on-line mode, in the case of nonlinear plant, a pretraining or off-line phase is usually both very useful and highly recommended.

A critical point of the specialized learning architecture is that the EBP learning algorithm requires knowledge of the Jacobian matrix of the plant. (For the SISO systems the Jacobian matrix becomes a scalar which represents the plant's gain.) The emergence of the Jacobian is obvious. The subjects of the learning are the weights of NN and, in order to correct the weights in the right direction, learning algorithm should have information concerning the error caused by inaccurate weights. But, there is no such direct information because the plant intervenes between the unknown NN outputs or control signals u, and the desired plant outputs y. The "teacher" in the EBP algorithm is typically an error (here the performance error $e_3 = y_d - y$) and this teacher is a distal one (Jordan and Rumelhart [23]). More detailed description of "distal teacher's" approaches is given in Appendix B. In the next subsection, the promising Adaptive Backthrough Control method is described.

D. Adaptive Backthrough Control (ABC)

The so-called NN based control is the one that typically uses two neural networks. Figure 41 represents the ABC scheme. We will first present the performance of the ABC structure comprised of the two networks, which is in line with the above

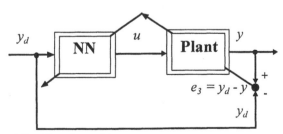

Figure 40   Specialized learning architecture.

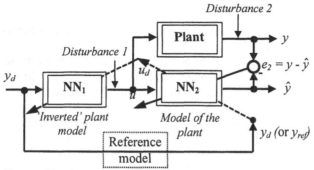

Figure 41 Neural networks based adaptive backthrough control (ABC) scheme.

mentioned approaches. Finally, we will show when, why and how the ABC structure can perform even better with only one NN. Basically, we suggest that there is no need for the NN_1 which acts as an inverse of the plant dynamics. Let us first present the structure which is in line with previous approaches.

The control loop structure proposed here comprises NN_2 representing the (approximate) model of the plant, and NN_1 which acts as a controller. The latter represents (approximate, again) the inverse of NN_2. Note that NN_1 is an "inverse" of the plant model and not of the plant itself.

The structure given in Figure 41 is almost standard one in the field of neuro-fuzzy control. In this respect, the adaptive backthrough control (ABC) proposed here, is in the line with the basic results and approaches in (Psaltis et al [22], Saerens and Soquet [24], Garcia and Morari [25], Jordan [26], Jordan and Rumel-hart [23], Hunt and Sbarbaro [27], Narendra and Parthasarathy [9], Saerens, Ren-ders, and Bersini [28] and Widrow and Walach [16].

While being similar in graphical appearance, there are a few important dis-tinctive features that differentiate the ABC approach from all other standard NN based control methods.

The principal feature of the ABC is that, unlike the other approaches, it does not use standard training errors (e.g., e_3) as the learning signals for adapting the controller (NN_1) weights. Rather, the true desired value y_d (signal that should be tracked or reference signal) is used for the training of NN_1. In this way, the de-sired, but unknown control signal u_d, results from the *backward* transformation of the y_d *through* NN_2. The origin of the label for this approach as a *backthrough* method, lies in this backward step for the calculation of u_d. Thus the ABC basi-cally represents a younger (and apparently more direct and powerful) relative of the distal teacher idea by Jordan [26], or of the approach in Saerens and Soquet [24], as well as in Saerens, Renders, and Bersini [28]. Besides using different error signals for the training (see comments on the next page), they use the steepest de-scent for optimization.

In the ABC, as long as the control problem is linear in parameters (linear dependence of the cost function upon the NN weights) the recursive least squares

(RLS) learning algorithm is strictly adhered to. This is a second and interesting feature of the ABC approach. Note that in many cases, for both NN and fuzzy logic model based controller, this assumption about the linearity in the parameters model is a realistic and acceptable one. This is typically the case when the hidden layer weights (positions and shapes of the basis functions and the membership ones in NN and FLM respectively) are fixed (Kecman and Pfeiffer [12]). Note, however, that the proposed ABC algorithm doesn't strictly depend on the use of the RLS technique and that the standard gradient (EBP) or any other learning procedure can also be used. The RLS based learning of the ABC method behaves much better on a quadratic error surface than any gradient based search procedure. This is another reason why the ABC algorithm seems to be more promising than the first order EBP approach.

As is the case with the adaptive inverse control (AIC) devised by Widrow, the ABC scheme proposed here is effective as long as the plant is a stable one. The ABC solves the problems of tracking and disturbance rejection for any stable plant. The same will be true in the case of unstable plants as long as the unstable plant is first stabilized by some classic control method. It seems as though the proposed ABC algorithm can handle nonminimum phase systems easier than the AIC. The ABC is the adaptive control system design algorithm in discrete domain, and as long as a suitable sampling rate (meaning, not too small) is used, there are no difficulties with discrete zeros outside the unit circle. The control structure in Figure 41 has some of the good characteristics of the idealized control system design with a positive internal feedback that does not require the plant model NN_2 to be a perfect model of the plant (Tsypkin, [29]). The latter control scheme is structurally equal to the internal model control (IMC) approach. Besides a structural resemblance the learning (feedback) part will cause a different (better) behavior of the ABC system. Besides, it seems as though the ABC method uses much fewer weights than both the AIC and IMC approach. Also, there is no need for explicit design of the first order filters that is the typical design practice in IMC. (Unless some control of the actuator signal variable u is needed, the reference block shown in Figure 4.41 is not required. All results below were obtained by using $G_{ref}(s) = 1$.)

The basic idea of the ABC is to design an adaptive controller which acts as the inverse of the plant. In order to learn the characteristics of the plant and to adapt the controller to the plant's changes, the neural network that works as a controller must be told what the desired control value should be. In general this value u_d is not available. We will show that by using the ABC approach we can find desired control values u_d that will usually be very close to the ideal ones.

During the operation of the whole system (meaning, during the adaptation or learning of both the plant model and controller parameters) there are several error signals that may be used for the adjustment of these parameters. Similarly to Jordan and Rumelhart [23], the following errors are defined in Table 6. (If the reference model is used the value y_d should be replaced with the output value of the reference model y_{ref}.)

Table 6 Definition of the Errors

$e_1 = \hat{u}_d - \hat{u}$	Controller error
$e_2 = y - \hat{y}$	Prediction error
$e_3 = y_d - y$	Performance error
$e_4 = y_d - \hat{y}$	Predicted performance error

Other methods such as Psaltis et al. [22], Widrow and Walach [16], Widrow and Plett [30], Saerens and Soquet [24] and Jordan and Rumelhart [23] use different approaches in order to find the error signal term that can be used to train the controller. Psaltis et al in [22] make use of the performance error e_3 modified by the plant Jacobian to train the controller. Saerens and Soquet [24] use a similar approach when using the performance error e_3 but unlike Psaltis et al, they multiply e_3 by the sign of the plant Jacobian only.

Jordan and Rumelhart [23] with their distal teacher idea, differ significantly from these two approaches in using the Jacobian of the plant model and not the one of the real plant. They discuss the application of three errors in training the plant model and controller. For the learning of the plant forward model, they use the prediction error e_2 (which is the usual practice for identification of unknown systems) and for learning of the controller, they proposed either the use of the performance error e_3 or the predicted performance error e_4.

In the approaches proposed by Widrow and his collaborators (Widrow and Walach [16]) and (Widrow and Plett [30]) the performance error e_3 for controller training is used. As far as the structure of the whole control system is concerned, they use different structures depending upon whether the plant is a minimum or non-minimum phase, and whether there is a need for noise cancelling. The adaptive neural networks in Widrow's approach are primarily of finite impulse response (FIR) filter structure. In the ABC approach presented below, we typically work with the infinite impulse response (IIR) structure.

The ABC structure originated from these previous approaches with a few basic and important differences.

First, the estimate of the desired control signal u_d can be calculated, and an error (delta) signal as in the distal teacher approach is not needed. For the ABC of linear systems, the calculation of u_d is straightforward. The forward model NN_2 is given as,

$$\hat{y}(k+1) = \sum_{i=1}^{N} w_{2j} \cdot x_{2j} = \mathbf{w}_2^T \cdot \mathbf{x}_2 \tag{49}$$

where $N = 2n$, n is the order of the model and x_2 is an input vector to NN_2 comprised of present and previous values of u and y. For the calculation of the desired value $\hat{u}_d$ equation (4.49) should, and can, be rearranged with respect to the input of the neural network NN_2,

$$\hat{u}_d(k) = \left(\frac{y_d(k+1) - w_{2,1}y_d(k) -, \cdots, -w_{2,n}y_d(k-n+1) - w_{2,n+2}\hat{u}(k-1) -, \cdots, -w_{2,2n}\hat{u}(k-n+1)}{w_{2,n+1}} \right). \quad (50)$$

Therefore, when applied to the control of linear systems the calculation of the control signal u_d by using (40) is similar to the predictive (deadbeat) controller approach. Note that in the calculation of the best estimates of the desired control signal $\hat{u}_d(k)$ to the plant and to the NN$_2$, the desired output values of the system $y_d(k+1)$, $y_d(k), \cdots, y_d(k-n)$ are used. It is interesting to note that instead of using the present and previous desired values, one can use the present and previous actual plant outputs $y(k), \cdots, y(k-n)$. It seems as though this second choice of the variables is a better one. This question should be investigated in detail. All results below were obtained by using (50).

In the case of *nonlinear system control*, the calculation of the desired control signal u_d which corresponds to the desired output from the plant y_d is a much more involved task. In Kecman [8] it was shown that for the *monotonic* nonlinearities (i.e., for one-to-one-mapping of the plant inputs u into its outputs y) control signal u_d would be calculated by an *iterative algorithm* that guarantees finding the correct u_d for any desired y_d. This is the most important result in the proposed ABC algorithm. Due to lack of space, we cannot go into details here (other details, as well as some other alternative approaches, may be seen in Kecman [8]). Instead, the performance of the ABC will be demonstrated using a number of different examples.

Secondly, the important feature of the ABC is that the output layer weights adaptation is strictly based on the RLS algorithm. (Although any other established NN learning algorithm e.g., the EBP, may be used.)

Thirdly, the ABC uses different error signals for the learning of the forward plant model (NN$_2$) and for the controller adaptation (NN$_1$). The prediction error e_2 is used for the training of NN$_2$, and the controller error e_1 is used for the adaptation of the controller NN$_1$. *All previous methods do not use e_1 in combination with a forward plant model during the learning.* This is an interesting advantage, and it seems to be a powerful novelty, because there is *no direct* influence of the plant output disturbance on the learning of the controller weights as in the case of the distal teacher procedure from Jordan and Rumelhart [23]. It is obvious that in the case of *linear systems*, for any Gaussian disturbance at the output (providing one has an infinitely long learning time; the orders of the plant model and of the real plant are equal; training signal u is rich enough and uncorrelated with the noise) there will be no influence of the noise at all, and the controller will produce the desired u_d. perfectly

Here, we will present the performances of the ABC for a number of different systems: first, in the linear case, when the orders of the plant and plant model (or emulator NN$_2$) are the same and without noise, by using *linear 3rd order non-minimum phase oscillatory system* we show that the ABC results in perfect *adaptive poles-zeros cancelling*. In the presence of the uncorrelated noise, perfect cancelling will be achieved after a longer training time. The bigger the noise the longer the learning should be.

Example 20 presents the abilities of the ABC faced with the *mismatched model orders* of a plant and of an emulator NN_2. Here, the plant is a 7th order linear system and both NN are the 2nd order IIR filters.

Example 21 shows the results of the ABC when coping with a *monotonic nonlinear 1st order plant* (one-to-one mapping of the plant).

Example 19
Consider the 3rd order nonminimum phase linear system given by the following transfer function,

$$G(s) = \frac{s - 0.5}{s^3 + s^2 + 5s + 4} .$$

The sampling rate was 2.25s. The plant model (emulator NN_2) was of the 3rd order too. The resulting adaptive controller (NN_1) perfectly cancels the poles of the system. This example is shown graphically in Figure 42.

Example 20
The ABC of the 7th order plant by using the 2nd order model (NN_2) and controller (NN_1). Both networks are IIR filter.

The plant is a stable linear system without zeros and with poles at [-1, -2, -5, -8, -10, -12, -15]. The plant gain is $K_{plant} = 0.5$. The additive output measurement noise during training is $n_2 = 5\%$. Note the extremely large errors at the beginning of the learning and the very good performance at the end of the learning in Figure 43. After only 750 learning steps, the ABC performs well. It tracks desired y_d with a settling time of ~ 2s (Figure 43 right). Note that the settling time of the 7th order plant is ~7s. The sampling rate was 1.75 s.

At the beginning of training, and due to the fact that the learning actually started from the scratch, there are large discrepancies between the NN model and the real plant output (see the left graph in Figure 43). But, after only a few hundred steps, the whole system is adjusted enough, and shows acceptable behavior. Thus, when the order of the emulator is lower than that of the actual plant (typical real life situation) the ABC scheme performs well. It is robust in respect with the unmodeled plant dynamics as well as to the additive measurement noise.

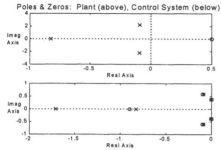

Figure 42 Perfect poles-zeros cancelling by the ABC.

Example 21

Nonlinear 1st order dynamic plant given by the difference equation shown below, should be controlled by the ABC structure.

$$y(k + 1) = 0.1y(k) + \tan(u(k))$$

Both NN were RBF networks with 100 hidden layer neurons each having 2-dimensional Gaussian activation functions. (Concerning the number of HL neurons, it should be mentioned that the ABC worked well with the networks having fewer HL neurons after the optimization by the orthogonal least square method.) All Gaussians were symmetrically placed and had fixed centers and width. In other words hidden layer weights were not a subject of learning.

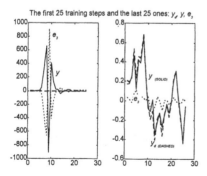

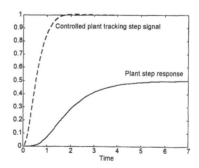

Figure 43 The ABC of the linear 7th order plant by using the 2nd order model. *Left:* Desired output y_d, actual plant output y and error $e_3 = y_d - y$ in the first 25 learning steps and in the last 25. Noise $n_2 = 5\%$. *Right:* Tracking of the unit step input signal without controller (solid) and with controller (dashed).

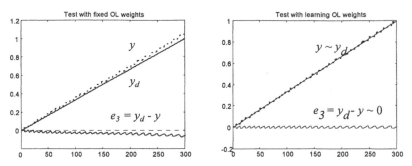

Figure 44 Test results with previously unseen ramp y_d without the on-line training (left) and with it (right).

During learning only the output layer weights have been changed. Pretraining was done by using 1000 random and uniformly distributed input signals y_d. After this off-line learning phase, two tests by a previously unseen ramp signal, were carried out. In both simulations the hidden layer weights were not the subjects of the learning. In the first simulation (Figure 44 left) the output layer weights were fixed and in the second one, both networks operated in a learning mode by adapting the OL weights (Figure 44 right). The graphs in Figure 44 show that the whole ABC structure can be successfully trained in on-line mode as long as the plant surface is a monotonic one. Figure 45 shows that NN is a good model of this nonlinear plant. There is no big difference between the actual plant surface and the one modeled by NN_2. Note that all the trajectories of the controlled plant lie on this surface.

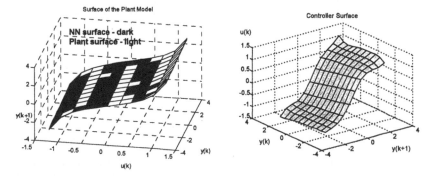

Figure 45 Modeling of the nonlinear plant by NN_2 (left) and its inverse by NN_1, or controller (right).

For nonlinear systems pretraining of both networks is necessary. It is important to realize that the ABC scheme with two networks performs well when faced with monotonic nonlinear plants.

All former results have been obtained by the ABC structure comprised of the two networks as shown in Figure 41. This structure is "inherited" from the previously described approaches and it is directly related to the classic EBP learning. The task of the network NN_1 which acts as a controller, is to learn the inverse dynamics of the controlled plant. When properly trained and after receiving the desired plant output signal y_d, NN_1 should be able to produce the best control signal u_d which would drive the plant to produce the desired y_d. However, the ABC learning is different from the EBP algorithm. Note that in the ABC algorithm the best control signal u_d is calculated for each operating step and it is used for the adaptation of NN_1's weights in order that this controller produces an output signal u, which should be equal or very close to the u_d. Thus, there is a great deal of redundancy and it seems as though both the very structure of the control system and the learning should be able to be halved.

Having the signal u_d calculated, the controller network NN_1 is no longer needed. The ABC structure with only one NN which simultaneously acts as the plant model and controller (inverse plant model) is shown below in Figure 46.

The performance of the ABC scheme having one NN is superior to the structure comprised of the two networks as given in Figure 41. The redundant part of the training and use of NN_1 is avoided here and this contributes to the overall efficiency. This will be demonstrated in examples 22 and 23 below.

First we show that for the time invariant plants the ABC perfectly tracks any desired signal. Next we show that the ABC can also cope with nonlinear fast changing plants. This is one of the toughest problems in the control field.

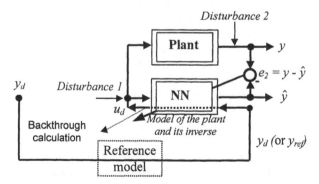

Figure 46 Neural or fuzzy network based Adaptive Backthrough Control (ABC) scheme with one network which simultaneously acts as the plant model and as a controller (inverse plant model).

Finally we show a series of simulation results of ABC performance while controlling the nonlinear plants described by non-monotonic mappings. We have only chosen first order systems for the sake of graphical visualization of the results obtained.

Example 22

Nonlinear 1st order dynamic plant which was identified in the Example 3 should be controlled by the ABC scheme comprised of the single network only. The plant equation (24) is repeated below,

$$y(k) = \frac{y(k-1)}{1 + y^2(k-1)} + u^3(k-1).$$

The neural network that simultaneously acts as a plant model and as its controller is comprised of 39 neurons in a hidden layer. Basis functions in all HL neurons are the 2-dimensional Gaussians with the same covariance matrix $\Sigma = diag(0.2750, 0.0833)$, and with positions determined by an orthogonal least squares selection procedure (Orr [31]). The NN was pretrained by using 1000 data pairs. The training input signal was a uniformly distributed random signal. Note that the ABC control structure is much simpler than the one in Narendra and Parthasarathy [9] which used two NN for the identification and one as a controller. Each network had 200 neurons. Besides, in the off-line training phase, they used 25,000 training pairs.

After the training was done, a number of simulation runs showed a very good performance of the ABC scheme while controlling this *time invariant nonlinear* system. Figure 47 (left) shows the plant response when tracking the input $y_d = sin(2\pi k / 25) + sin(2\pi k / 10)$. The plant response is indistinguishable from the desired trajectory. One can say that the tracking is perfect.

A much more complex task is to control a *time variant nonlinear* plant. There is no *general* theory, approach or method in adaptive control of nonlinear time variant plants. These are the toughest control problems. Here, we only present initial results on how the ABC scheme copes with such plants. We do not pretend to be able to answer all open questions in this field. In particular, we don't discuss the problems of the convergence or of the stability of the ABC, when faced with a nonlinear time variant plant, but rather we try to throw a little light on its performance. (Note that the problems of NN based control of the time variant plant is rarely discussed in the literature.) Figure 47 (right) shows the error when the pretrained but fixed NN tries to control fast changing plant given on p. 207.

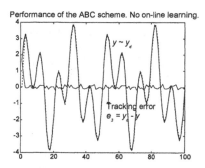

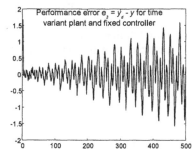

Figure 47 ABC comprising one NN only: Perfect tracking in the case of nonlinear mono-tonic time invariant plant (left). Performance error for fixed pretrained NN controlling of the time variant plant, when the plant gain is halved in 500 steps, (right).

$$y(k) = \frac{y(k-1)}{1 + y^2(k-1)} + (1 - 0.001k) * u^3(k-1).$$

This is a model of the plant which halves the plant gain in 500 steps. Without an adaptation the performance error $e_3 = y_d - y$ increases rapidly (Figure 47, right). Figure 48 shows e_3 in the case of the on-line adaptation of a neural network. Results are obtained by using a forgetting factor $\lambda = 0.985$.

The adaptation and control process is a stable one and, in comparison to the error in Figure 47, the final error in Figure 48 is three times smaller.

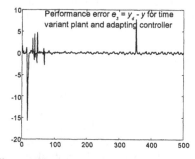

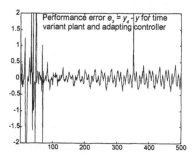

Figure 48 ABC: Performance error in controlling the time variant plant with on-line adaptation of the NN OL weights. Forgetting factor $\lambda=0.985$. (Right graph is drawn using the same scale as Figure 47 right.)

The process is a hairy one and this problem of smoothing the adaptation procedure should be further investigated. (Readers who are more familiar with the identification of the linear systems are well acquainted with the wild character of identification procedures. In the case of nonlinear system identification as the one in this example, one can expect rougher transients.)

There are many open questions in the adaptive control of nonlinear time variant processes. All important questions from linear domain are present here (dual character of adaptive controller, identifiability, persistency of excitation etc.). One specific question in a nonlinear domain is the choice of the input signals for the pretraining phase. The standard binary signals used in linear systems identification are not good enough. During the pretraining, the whole region of plant operation should be covered and the best choice would be to use the uniformly distributed random signals. (See in Figures 49 to 51 (right graphs) which parts of the plant dynamics surface are properly covered by using three different desired signals y_d.) Unfortunately, we cannot go into these important details here. Instead, we will show a few more simulation results of the ABC when controlling a non-monotonic nonlinear plant. In this way the reader will be able to understand at least a part of the important properties and specific features of the NN based control of the nonlinear dynamic systems.

Example 23
Consider the ABC of the following nonlinear dynamic plant,

$$y_{k+1} = sin(y_k)*sin(u_k) - u_k/\pi \qquad (60)$$

The characteristic feature of this plant is that (60) is not a monotonic nonlinear function. In other words, there is one-to-many mapping of the u_k to the y_{k+1}. However, the function $y_{k+1} = f(u_k, y_k)$ represents one-to-one mapping and the ABC can successfully model both the plant dynamics (mapping of u to y) and the plant inverse dynamics (mapping of y to u). The NN was optimized by using a feedforward orthogonal least square method. The basis functions in all neurons are the 2-dimensional Gaussians with the same covariance matrix $\Sigma = diag(0.0735, 0.1815)$. At the beginning of the RBF selection, there were 169 symmetrically placed neurons in a hidden layer and at the end only 47. Such a network models the plant very well. (Note that this structure corresponds to the fuzzy logic model with a rule basis comprised of 47 rules).

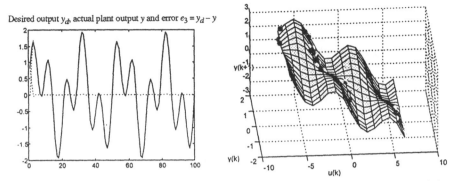

Figure 49 ABC *Left:* Perfect tracking of the desired signal $y_d = sin(2\pi k / 25) + sin(2\pi k / 10)$ for the time invariant plant [Eq. (60)]. Pretrained NN weights are fixed. No adaptation. *Right:* Trajectory shown by dots lies on the surface described by Eq. (60).

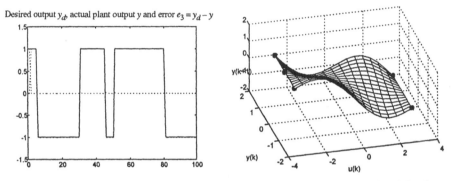

Figure 50 ABC *Left:* Perfect tracking of the desired rectangular random signal for the time invariant plant [Eq. (60)]. Pretrained NN weights are fixed. No adaptation. *Right:* Trajectory shown by dots lies on the surface described by Eq. (60).

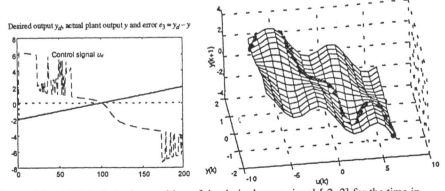

Figure 51 ABC *Left:* Perfect tracking of the desired ramp signal [-2, 2] for the time invariant plant [Eq. (60)]. Pretrained NN weights are fixed. No adaptation. *Right:* Trajectory shown by dots lies on, or sneaks through, the surface described by Eq. (60).

V. CONCLUSIONS

This chapter presents the neural network and fuzzy logic model based control of nonlinear dynamic plants. Due to their well known ability to be universal approximators of multivariate functions, NN and FLM are of particular interest for controlling highly nonlinear, partially known and complex plants or processes. Many simulation examples presented here have shown these capabilities of the neural and fuzzy based controllers. The approach presented here seems to be very good for the feedforward adaptive control of *stable* nonlinear systems.

In the first part of the chapter the basic ideas of classic adaptive control were reviewed. Thereafter, we discussed when and why one should try to solve a given problem by NN, and when the more appropriate tool is FL model. That introductory presentation was followed by a brief description of the neural and fuzzy models approximation capabilities.

After that, the basics of the neural networks and fuzzy logic systems were introduced. More specifically, we presented the two most used feedforward types of NN - multilayer perceptron and radial basis function networks. The latter are very closely related to the FL models. Both modeling tools can be represented by the same structure which comprises input, hidden and output layers. Moreover, we have shown that the mathematical models of these different modeling tools is either the same or very similar. This is a very powerful result which naturally leads to a better understanding of both NN and FLM and to the development of hybrid neuro-fuzzy or fuzzy-neuro systems.

We were chiefly interested in part of the FL as technique for mathematical expression of linguistic knowledge and ambiguity. Therefore, the fuzzy logic was presented as a suitable mathematical tool for embedding human structured knowledge into workable algorithms. This knowledge is structured in the form of IF-THEN rules, and these rules represent mappings, or multivariate functions, which map the input variables (IF part variables, antecedents, causes) into the output variables (THEN part variables, consequents, effects). On the other hand, the roots of the functions theory are in the set theory. Therefore, our presentations of the FL followed this path - from the sets, operation on them and Cartesian products, to the relations, multivariate functions, and to the IF-THEN rules as the linguistic form of the human structured knowledge.

After establishing the mathematical equality of NN and FLM we presented one of the most explored streams in the NN based control which basically uses the control structure comprised of the two neural (fuzzy) networks - plant model (emulator) and the controller (inverse of the emulator). In particular it was shown that the neural network based adaptive backthrough control (ABC) scheme can be successfully applied for control of both linear and nonlinear dynamic plants. The principal idea of the ABC, the use of *controller error signal* e_l instead of an error back propagated delta signal for the learning of the controller, is introduced. Due to this fact ABC performance seems to be superior to the other NN based feedforward adaptive control approaches. For linear plants, the resulting feedforward controller, providing that the order of the plant and plant model are equal, is a perfect adaptive poles-zeros canceller. In this way, the ABC has the character of a predictive controller. Faced with nonlinear plants, the ABC performs as a kind of nonlinear deadbeat controller. Though these claims should be analyzed more rigorously. The neural networks based ABC scheme can also be successfully trained in noise. As is the case in such an environment, both networks nicely approximate the conditional expectation of the corresponding output patterns.

Finally, we introduced the idea of using only one NN, which should simultaneously act as both the plant model and the model of the plant inverse. In this way we avoided a great deal of redundancy while training two networks. The number of simulation results shows the good performance of such a structure. These abilities clearly indicate that the neural network based ABC scheme exhibits intelligent adaptive behavior. The ABC is a very good candidate algorithm as well for the control of nonlinear processes for which no perfect mathematical model is available.

APPENDIX A

The Mathematics of the LFR and RBF NN

The mathematical side of the solution is defined as follows; for fixed number N, positions c_i and width s of the basis f, or membership function m, the problem of approximation is linear in learned parameters w or r, and it will be solved by the simple inversion of the matrix A given in (A.5) below. (In the case of interpolation, i.e., when the number N of basis or membership functions (attributes) for each input variable is equal to the number K of data pairs, A is a square matrix. When there is less basis or membership functions than data pairs, A will be calculated by pseudoinversion. The latter case of approximation will be more common for real world problems.) This property of the algorithm being linear in parameters is not affected by the choice of the algebraic product as fuzzy AND operator, and this algorithm remains the same for the minimum operator. It seems as though the soft RBF is more robust in respect to the choice of the width parameter and that it has better approximation properties in cases when there is no big overlapping of basis f, or membership functions m. In such a situation (for small s in the case of Gaussian functions) the approximation obtained with classical RBF given by (28) will be much more spiky than the one obtained with fuzzy approximation, or soft RBF, given by (34).

There is a significant difference in the "physical" meaning of the learned (or trained) weights w_i or rules r_i in these two paradigms. Approaching the problem from the "fuzzy" side, the rules have from the very start of the problem formulation, a clear "physical" meaning, saying that an output variable should take a certain value under certain conditions of input variables. There is no such analogy in the classical RBF approach to functional approximation. In the latter case, the meaning of weights w_i is more abstract and depends on such small subtleties as, e.g., whether or not we use normalized Gaussians $G(c_i, c_i)=1$. Generally, in both methods, with increased overlapping of the basis or membership functions the absolute values of the parameters w or r will increase. But, in the fuzzy case when the resulting output variables are the rules r, we are aware of their physical limits and these limits will determine actual overlapping of the membership functions in input space. There are no such caution signs in classical RBF because this approach is derived from the "clean" mathematical domain. We hope that by exposing of both methods we will shed much more light on both methodologies.

In order to apply a standard least squares method in the spirit of the parameter estimation schemes, the dedicated fuzzy identification algorithm for the center of singletons defuzzification method should be slightly reformulated by collecting the elements of $\mathbf{M}$ in a column vector,

$$\mu = (\mu_{11}, \ \mu_{12}, \ ..., \ \mu_{21}, \ \mu_{22}, \ ...)^T , \tag{A.1}$$

and by defining a vector of ones with the same dimension $\vec{1} = (1, \ 1, \ ...)^T$. Using these vectors (32) can be written with the numerator and denominator calculated as scalar products,

$$y = \frac{\mu^T \mathbf{r}}{\mu^T \vec{1}} , \tag{A.2}$$

which is equivalent to,

$$\mu^T \mathbf{r} = \mu^T \vec{1} \, y . \tag{A.3}$$

The input data are fuzzified according to the attributes of the linguistic variables $Input_1$ and $Input_2$. For each sample k, and input data set $\mathbf{x}_k$, a corresponding vector μ_k is obtained by applying the formulas (29), (30) and (A.1) successively, and an equation of the form (A.3) is stated,

$$\mu^T_k \mathbf{r} = \mu^T \vec{1} \, y_k . \tag{A.4}$$

From this equation a system of linear equations is constructed for $k = 1, ..., K$,

$$\underbrace{\begin{bmatrix} \mu_1^T \\ \mu_2^T \\ \vdots \end{bmatrix}}_{\mathbf{A}} \mathbf{r} = \underbrace{\begin{bmatrix} \mu_1^T \vec{1} \, x_1 \\ \mu_2^T \vec{1} \, x_2 \\ \vdots \end{bmatrix}}_{\mathbf{b}} \tag{A.5}$$

This system is in a linear form, with a known but rectangular matrix $\mathbf{A}$ and a known vector $\mathbf{b}$.

$$\mathbf{Ar} = \mathbf{b} \tag{A.6}$$

Now, (A.6) can be solved for the unknown vector $\mathbf{r}$ by any suitable numerical algorithm, e.g., by taking the pseudoinverse as an optimal solution in a least squares sense,

$$\mathbf{r} = (\mathbf{A}^T\mathbf{A})^{-1}\mathbf{A}^T\mathbf{b} = \mathbf{A}^+\mathbf{b} \tag{A.7}$$

Finally, the elements of vector $\mathbf{r}$ can be regrouped into the rule matrix $\mathbf{R}$. The matrix $\mathbf{A}$ actually contains the degrees of fulfillment of all rules. For a system with N rules, its dimensions are (K, N). Therefore the matrix $\mathbf{A}^T\mathbf{A}$ is of dimension (N, N) and can be, even for very large numbers of data samples, easily inverted.

This explains the equality of RBF and FL models and it also shows how the weights, or rules, can be adapted (trained). The final learning rule was a relatively simple one due to the fact that the hidden layer weights were fixed. Nevertheless, this one-out-of-many possible learning rules was a good introduction into generic nonlinear learning tasks which was dealt with in the Section 3.

APPENDIX B

General Distal Learning

Let us show the EBP algorithm for the general distal teacher learning situation. In order to apply the EBP algorithm, the NN and the plant are treated as a single neural network in which the plant represents a fixed (unmodifiable) output layer. In this way the real OL of NN becomes the hidden layer. The whole EBP learning now concerns a calculation of proper *deltas*, or error signals δ, associated with each neuron. (See Table 5 and Example 18.) In order to find these signals, the delta signals δ_{ok} for the true OL neurons of NN should be first determined. For the sake of simplicity we will show how this can be done for the SISO plant in this way avoiding matrix notation. Having δ_{ok} enables a relatively straightforward calculation of all other deltas and specific weights changes, see (B.1).

Assume that NN is a network operating in parallel mode by having $2n$ inputs (where n represents the model order), or that NN is given by model $u(k+1) = f\{y_d(k), \cdots, y_d(k-n); u(k), \cdots, u(k-n)\}$. There are enough HL neurons which can provide a good approximation, and there is one *linear* OL neuron having the output u. The plant is given as $y = f(u, y)$. The EBP algorithm for learning the NN weights, as given by Eq. (35), is a steepest descent procedure, and the cost (error) function to be optimized is,

$$E = \frac{1}{2}e^2 = \frac{1}{2}(y_d - y)^2 = E(w_{ij}). \tag{B.1}$$

Note that $y = f(u, y)$ and $u=(f_n(u_n))$ so that $y = f((f_n(u_n)), y)$. f_n and u_n stand for the activation function of, and input signal to, the OL neuron respectively. (For a linear OL neuron, f_n represents identity i.e., $u = u_n$.)

In order to calculate the OL neuron's error signal δ_o we apply a chain rule in calculation of the cost function's gradient,

$$\frac{\partial E}{\partial w_i} = \frac{\partial E}{\partial e} \frac{\partial e}{\partial y} \frac{\partial y}{\partial u} \frac{\partial u}{\partial u_n} \frac{\partial u_n}{\partial w_i} = \underbrace{(y_d - y)(-1) \frac{\partial f(u, y)}{\partial u} f_n'}_{-\delta_o} \frac{\partial u_n}{\partial w_i} = -\delta_o \frac{\partial u_n}{\partial w_i} \quad \text{(B.2)}$$

The error signal of the OL neuron δ_o is determined in accordance with Eq. (37). f_n' stands for the derivative of the OL neuron activation function and in this case for a linear neuron $f_n' = 1$. For the multilayer perceptron network, where the input signal to the neuron is obtained as a scalar product $u_n = \mathbf{w}^T\mathbf{x}$, the derivative $\partial u_n / \partial w_i = x_i$. (Note that in the RBF networks, this expression for the OL error signal δ_o will be identical. There will, however, be a difference in the expressions for the learning of the HL neuron weights.)

It is important to realize that the derivative $\partial f(u,y) / \partial u$ represents the Jacobian of the plant. Here, for SISO plant this is a scalar, or more precisely (1, 1) matrix. Generally, plant dynamics and the Jacobian are unknown and this is a serious shortcoming of this final result that is otherwise useful. There are two basic approaches which overcome this difficulty and these will be commented on below.

Let us first make some final comments concerning the weights adaptation in the specialized learning architecture under the following assumptions: the Jacobian is known; the OL neuron is linear; and the input to the neuron is calculated as a scalar product. With these assumptions we can directly apply Table 5. Note that the calculation of δ_o in (B.2) means that *STEP* 6 in Table 4.5 is completed. Knowing the structure of NN and following Table 5, *STEPS* 7 to 11 result in HL deltas and in the new weights adapted by applying their corresponding weight changes $\Delta w_i = \eta\, \delta x_i$. Hence, in this *backpropagation through the plant algorithm*, the determination of the networks' OL delta signal is the most important step. In order to do this the Jacobian of the plant must be known.

Generally this is not the case, and two alternative approaches on to handling the ignorance of the plant Jacobian are: *the approximation of the plant Jacobian by its sign;* and *the distal teacher approach.*

Approximation of the Plant Jacobian by its Sign

Specialized learning with the EBP through the plant, can be achieved with approximation of partial derivatives of the Jacobian by their signs as presented by Saerens and Soquet in [24]. In principle, they use the same basic equations for the calculation of the deltas, with the difference that they approximate sensitivity derivatives in the Jacobian matrix by their signs. These signs are generally known when qualitative knowledge about the plant is available. This, from a practical point of view means that the entries in the Jacobian matrix are +1 or −1. The main disadvantage of this approach is slower training. This is a consequence of the fact that this approach makes no use of the whole of the information available.

The Distal Teacher Approach

The structure and the concept presented in Jordan and Rumelhart [23] (similar approaches and structures have been proposed and used in many papers from Widrow and his coworkers under the global name of the *adaptive inverse control*) differ significantly from the two methods presented above, by using *the Jacobian of the plant forward model* instead of the real plant's Jacobian, or instead of the signs of the Jacobian derivatives of the real plants. The whole feedforward control system now comprises two neural networks. One is a model of the plant and the second, which will be trained with the help of the first one, acts as a controller. This structure is practically the same as ABC given in Figure 41.

The learning or modeling proceeds in two phases. In the first phase, a *forward model* of the plant mapping from inputs u to outputs y is learned by using the standard supervised learning algorithm (EBP). In the second phase, the inverse model and the forward model are combined, and an identity mapping is learned across the composed network. It is important to note that the whole learning procedure is based on the performance errors e_3 between the desired plant outputs y_d and actual plant outputs y.

The learner (NN_1 or controller) is assumed to be able to observe the states, the inputs and the outputs, and can therefore model the inverse plant dynamics. If the plant is characterized by many-to-one mapping from the input to the output, then there may be a number of possible inverse models. In their paper Jordan and Rumelhart comment how the distal teacher approach resolves this problem of finding a particular solution. (Unfortunately, they don't go into details.) An important feature of this approach is that the feedforward model of the plant (NN_2) can be an approximate model. It is the use of the performance error e_3 that ensures that NN_1 can learn an exact inverse model of the plant even though the forward model is only approximate. Before closing this survey of the basic approaches to NN and/or FLM control, a few comments, concerning the practical aspects of the NN implementation, may be in order.

In the case that the plant is nonlinear the standard approach is to combine generalized and specialized learning. This method combines the advantages of both procedures. A possible method to compound these two approaches is to initially learn (with general architecture) the approximated behavior of the plant. Then, the fine tuning of the network in the operating region of the system should be done by specialized training (Psaltis et al [22]). The advantage is that generalized learning will produce a better set of initial weights for the specialized learning. In this way we will be able to cover a large range of input space as well as making the specialized learning faster. The same approach is used in the ABC scheme. In the case of nonlinear plants pretraining of both the controller NN_1 and the plant model NN_2 is essential. After this pretraining step the on-line ABC adaptation can be started with previously learned weights. In the case of linear plant this pretraining is not essential.

Sometimes it may be useful to introduce a reference model, too. This step is not crucial for the ABC approach but an important result may be that with a reference model tuning of the control effort is possible. This will be necessary for many real existing systems because the actuators usually operate only within a specific range, and leaving this range is either not possible or can harm the system's performance.

ABBREVIATIONS

ABC   Adaptive backthrough control
AF   Activation function
EBP   Error back propagation
FLM   Fuzzy logic models
FS   Fuzzy system
HL   Hidden layer
MIMO   Multi-input multi-output systems
MRAC   Model reference adaptive control
NN   Neural networks
OL   Output layer
SISO   Single-input single-output systems
STR   Self-tuning regulator

REFERENCES

1. KJ Aström, B Wittenmark. *Adaptive Control*. Reading: Addison-Wesley, 1989.
2. P Ioanou, J Sun. Robust *Adaptive Control*. Upper Saddle River, New Jersey: Prentice-Hall, 1996.
3. M Krstic, I Kanellakopoulos, P Kokotovic. *Nonlinear and Adaptive Control Design*. New York: Wiley, 1995.
4. ID Landau. *Adaptive Control*. New York: Marcel Dekker, 1979.
5. KS Narendra, AM Annaswamy. *Stable Adaptive Systems*. Englewood Cliffs, New Jersey: Prentice-Hall, 1989.
6. WL Brogan. *Modern Control Theory, 3rd. ed*. Englewood Cliffs, New Jersey: Prentice-Hall, 1991.
7. Z Gajic, M Lelic. *Modern Control System Engineering*. London: Prentice-Hall Europe, 1996.
8. V Kecman. Neural Networks and Fuzzy Logic Based Control. Report 575. The University of Auckland, 1997.
9. KS Narendra, K Parthasarathy. Identification and control of dynamical systems using neural networks. IEEE Transactions on Neural Networks 1:4–27.
10. J. Löchner. Identification of dynamic systems using neural networks and their optimization through genetic algorithms. Report No. 96-30. The University of Auckland, New Zealand.

11. LA Zadeh. Outline of a new approach to the analysis of complex systems and decision processes. IEEE Transaction Systems, man and Cybernetics, SMC 3:28–44, 1973.
12. V Kecman, B-M Pfeiffer. Exploiting the structural equivalence of learning fuzzy systems and radial basis function neural networks. EUFIT 1994 Proc. 1:58–66, 1994.
13. AN Tikhonov, VY Arsenin. Solutions of Ill Posed Problems. Washington, DC: Winston, 1977.
14. JS Bridle. Probabilistic Interpretation of Feedforward Classification Network Outputs, with Relationships to Statistical Pattern Recognition. In: F Fougelman Soulie, J Herault, eds. Neuro Computing: Algorithms, Architectures and Applications, Springer Verlag, 1989.
15. B Widrow, SD Stearns. *Adaptive Signal Processing*. Englewood Cliffs, New Jersey: Prentice-Hall, 1985.
16. B Widrow, E Walach. *Adaptive Inverse Control*. Upper Saddle River, New Jersey: Prentice-Hall, 1996.
17. V Kecman, T Rommel. Neural networks based adaptive backthrough control. Proceedings of the 5th European Congress on Intelligent Techniques and Soft Computing. EUFIT 1997, Aachen, Germany, 1997.
18. T Rommel. Neural networks based adaptive control. Report No. 97-30. The University of Auckland, New Zealand (Thesis).
19. JD Boskovic, KS Narendra. Comparison of linear, nonlinear and neural-network-based adaptive controllers for a class of fed-batch fermentation processes. Automatica 31(6):817–840, 1995.
20. M Agarwal. A systematic classification of neural-network-based control. IEEE Control Systems 17(2):75–93, 1997.
21. V Kecman. Foundations of Automatic Control (in Serbocroatian). Zabreb: Skolska Knjiga, 1988.
22. DA Psaltis, A Sideris, AA Yamamura. A multilayered neural network controller. IEEE Control Systems Magazine 8:17–21, 1988.
23. MI Jordan, DE Rumelhart. Forward models: supervised learning with a distal teacher. Journal of Cognitive Science 16:307–354, 1992.
24. M Saerens, A Soquet. Neural controllers based on backpropagation algorithm, IEE Proc F 138(1):55–62, 1991.
25. CE Garcia, M Morari. Internal model control. 1. Unifying review and some new results. IEC Proc Des Dev 21:308, 1982.
26. MI Jordan. Connectionists Models of Cognitive Processes. Course 9.641, Massachusetts Institute of Technology, Cambridge, Massachusetts, 1993.
27. KJ Hunt, D Sbarbaro. Neural networks for nonlinear internal model control. IEE Proc D 138(5):431–438, 1991.
28. M Saerens, J-M Renders, H Bersini. Neurocontrollers based on backpropagation algorithm. In: M Gupta, N Sinha, eds. Intelligent Control Systems. IEEE Computer Society Press, 1996.
29. Ja Z Tsypkin. Fundamentals of Automatic Control Theory (in Russian). Moscow: Nauka, 1972.
30. B Widrow, GL Plett. Adaptive Inverse Control Based on Linear and Nonlinear Adaptive Filtering. Proc. Of NICROSP '96 Intern. Workshop on NN for Identif., Control, Robot., and Sign./Image Process., Venice, Italy, IEEE 1996 0-8186-7456-3/96, 1996.
31. MJL Orr. Regularization in the selection of radial basis function centers, Report, Center of Cognitive Science, University of Edinburgh, 1996.

The interested reader may wish to consult several readily available sources for
more material on this broad and interdisciplinary field.
Books on neural networks
CM Bishop Neural Networks for Pattern Recognition. Oxford: Clarendon Press, 1995.
A Cichocki, R Unbehauen. Neural Networks for Optimization and Signal Processing.
    Chichester: BG Teubner, Stuttgart and John Wiley and Sons, 1993.
S Haykin. Neural Networks: A Comprehensive Foundation. New York: MacMillan College
    Publishing Co.
J Zurada. Introduction to Artificial Neural Systems. St. Paul, Minnesota: West Pub. Co.,
    1992.

Books on fuzzy logic
J Kahlert, H Frank. Fuzzy-Logik und Fuzzy-Control, (In German). Wiesbaden: Vieweg-
    Verlag, 2.Auflage, 1994.
JG Klir, TA Folger. Fuzzy Sets, Uncertainty, and Information. Englewood Cliffs, NJ: Pren-
    tice-Hall, 1988.
HJ Zimmermann. Fuzzy Set Theory and Its Applications. Boston: Kluwer, 2nd E, 1991.

Special books devoted to the intelligent control
MM Gupta, NK Sinha, (Eds.). Intelligent Control: Theory and Practice. Piscataway, NJ:
    IEEE Press, 1995.
TW Miller, RS Sutton III, PJ Werbos, (Eds.). Neural Networks for Control. Cambridge,
    Massachusetts: MIT Press, 1990.
DA White, DA Sofge, eds. Handbook of Intelligent Control. New York: Van Nostrand
    Reinhold, 1992.

Worth of consulting on intelligent control would be three
Special Issues on Neural Networks in Control Systems of the IEEE *Control System Maga-*
    zine. 10(3), April 1990; *12(3),* April 1992; and *15(3),* June 1995.

Additional sources on the related topics are
DS Broomhead, D Lowe. Multivariable functional interpolation and adaptive networks.
    Complex Systems 2:321-355, 1988.
G Cybenko. Approximation by superpositions of a sigmoidal function. Mathematics of
    Control, Signals and Systems 2:304-314, 1989.
K Funahashi. On the approximate realization of continuous mappings by neural networks.
    Neural Networks 2(3):183-192, 1989.
S Geman, E Bienenstock, R Doursat. Neural networks and the bias/variance dilemma. Neu-
    ral Computation 4(1):1-58, 1992.
EJ Hartman, JD Keeler, JM Kowalski. Layered neural networks with Gaussian hidden units
    as universal approximations. Neural Computation 2(2):210-215, 1990.
K Hornik, M Stinchcombe, H White. Multilayer feedforward networks are universal
    approximators. Neural Networks 2(5):359-366, 1989.
V Kecman. Application of Artificial Neural Networks for Identification of System Dy-
    namics. Dept. of ME, Tech. Rep., TR 93-YUSA-01, Massachusetts Institute of Tech-
    nology, Cambridge, Massachusetts, 1993.
V Kecman. Connectionists Networks and Fuzzy Logic Systems. Lectures Manuscripts. The
    University of Auckland, New Zealand, 1996.

M Maruyama, F Girosi, T Poggio. A connection between GRBF and MLP. A.I. Memo No. 1291, Massachusetts Institute of Technology, Cambridge, Massachusetts, 1992.

J Park, IW Sandberg. Universal approximation using radial basis function networks. Neural Computation 3(2):246-257, 1991.

T Poggio, F Girosi. A theory of networks for approximation and learning. A.I. Memo No. 1140, Artificial Intelligence Lab, Massachusetts Institute of Technology, Cambridge, Massachusetts, 1989.

T Poggio, F Girosi. Regularization algorithms for learning that are equivalent to multilayer networks. Science 247:978-982, 1990a.

T Poggio, F Girosi. Networks for approximation and learning. Proc. of IEEE 78:1481-1497, 1990b.

T Poggio, F Girosi. Learning, Approximation and Networks. Lectures, Course 9.520, Massachusetts Institute of Technology, Cambridge, Massachusetts, 1993.

BT Polyak. Introduction to Optimization. New York: Optimization Software, Inc., 1987.

DE Rumelhart, GE Hinton, RJ Williams. Learning internal representations by error propagation, 1986. In: DE Rumelhart, JL McClelland, and the PDP Research Group, eds. Parallel Distributed Processing 1:318-362. Cambridge, Massachusetts: Massachusetts Institute of Technology Press. Reprinted in Anderson and Rosenfeld, 1988.

5
Communications Systems

Dobrivoje Popovic
University of Bremen, Bremen, Germany

Data communication systems have become an essential component of industrial plant and manufacturing automation systems. Thus, they are also of great importance for mechatronic systems, where they integrate control and monitoring facilities and support the robots, conveyor systems, and vehicles in data provision for execution of their tasks and help product and production process design by data exchange between the automation system and the CAD/CAM facilities. They also enable the data transfer between individual hierarchical levels of the automation system.

A need for advanced communication links can be particularly found within the flexible manufacturing systems which contain a large number of numerically controlled machine tools, programmable controllers, monitoring and command facilities, computer guided vehicles, etc. mutually linked for information exchange. Here, even a special backbone communication link, the MAP (Manufacturing Automation Protocol) bus system has been developed that, combined with the Field Bus, facilitates solving the communication and data transfer problem at factory floor.

Thus, data communication technology is considered essential for integrating both the mechatronic product and its production plant. This chapter presents the relevant communication concepts and related international standards.

I. COMMUNICATION CONCEPTS

In factory automation, the evolution of communication concepts has been strongly influenced by computer application needs on various plant automation levels. During the early days of computer-based automation single, dedicated computers were installed at the field instrumentation level for sensor data collection for plant monitoring purposes. At that time, there was no direct data transfer between such computers and the central computer facility of the enterprise, where the process and production data have been stored and processed. For this data transfer usually magnetic media have been used. However, the interrelations between the data collected at different automation levels and the growing functional interdependency of tasks to be executed at these levels has called for a *direct* data transfer and for a closer functional interactivity between the individual computers by their electronic interfacing. In this way the individual computers, mutually interconnected, have gradually grown into a *multicomputer system.*

However, for building the high-performance multicomputer systems high-performance data communication systems are required as interconnection links between the computers as well as between the accompanying monitoring facilities scattered over the enterprise. This, at the very beginning, was a difficult task to be implemented because of mutual hardware and software incompatibilities of computers.

Already in early applications, bus oriented interconnections, i.e. bus systems, were seen as an appropriate tool for computer interfacing. Such interconnections were realized by various companies building professional multicomputer systems to be launched on the marketplace. Examples of bus systems used in the early days of multicomputer systems building are: the HIGHWAY of Honeywell, DATA HIGHWAY of Fisher Controls, FOXNET of Foxboro, FEEDER BUS of BBC, etc. Later on, some more popular communication systems have been launched, internationally still widely used, such as Arcnet, Hyper/Net, Omninet, and Ethernet [1].

In the 80s, the progress in standardization of local area networks (LANs) has brought to the market a number of universal communication links for implementation of distributed and hierarchically organized plant automation systems. Token Ring, Token Bus, fiber distributed data interface (FDDI), etc. have been the results of hard, long-time standardization work within the IEEE and the ISO. Standardization efforts of IEEE on LANs within its *Project 102* and ISO efforts in standardization of open systems interconnection model (OSI Model) have laid down the fundamentals of new data communication technology, from which many of further developments in this field have profited, because the resulting IEEE and ISO international standards have immediately been accepted by the industry as adequate means for solving data communication problems in factory automation.

In order to interconnect computers programmed to execute jobs at different automation levels and to organize their interconnections and their jobs as required by the process and/or manufacturing plant to be automated, the computer manufacturers and the users have, in a cooperative work over the years, developed a multilevel model of functions to be implemented within individual computers of a hierarchically organized distributed computer system. For instance, in *process plant automation* the following automation levels have been identified:

- Field instrumentation
- Process control
- Plant control
- Production control

In *manufacturing* a similar hierarchical organization has been worked out that identifies the

- Machinery
- Station
- Cell
- Shop floor

Later on, by defining integrated enterprise automation systems, the levels of Production Planning and Scheduling, Corporate Management, Product and Process Design, as well as some further commercial and personnel management levels have been added.

For computer interconnections at each hierarchical level, as well as between different levels, appropriate communication links are required. For example, for interconnection at the lowest level, i.e. at Field Instrumentation Level or Machinery Level, remote communication links are required for acquiring sensor data. At this automation level, the progress in control instrumentation technology was closely followed by the evolution of data transfer and communication technology. Soon after the instrumentation standard of (0-20) mA for remote signal transfer has been accepted, the work on protocol-based data transfer within the plant automation systems has started. The first initiative in this direction has been the IEC standardization of PROWAY (Process Highway) communication system. However, after the standard proposal documents of PROWAY were completely developed, the PROWAY project was abandoned, because of ISO work on standardization of LANs and of the Open System Interconnection Model. Taking into account these standardization efforts, the IEC has concentrated on development of the *Field Bus Standard*, that is still going on within the WG6 of IEC TC 65.

For computer interconnections at *intermediate automation levels,* i.e. at Process and Plant Control or at Station and Cell levels where the monitoring and supervisory facilities are situated within the factory, high-performance communication links—such as Local Area Networks—are required, whereby at higher automation levels broadband communication systems are preferable. Furthermore, interconnection of different mutually incompatible communication networks within a distributed computer system requires special *internetworking facilities*.

Based on the standardized OSI Model of ISO, the *Network Bridges* and *Network Gateways* have been standardized for this purpose. Using the interfacing facilities, a complex distributed multicomputer system could be built as shown in Fig. 1.

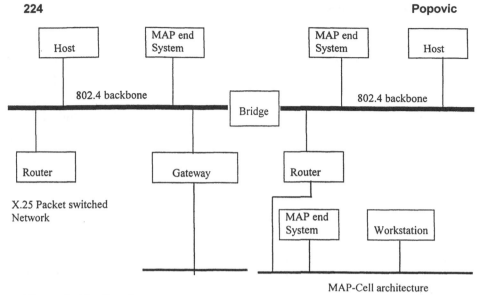

Figure 1   Distributed computer system.

The evolution of computer-based instrumentation devices and material handling facilities—such as PLCs, production planing and monitoring stations, robots,
etc. —has enabled new automation concepts in production engineering and manufacturing. For instance, the originally widely used CNC concept has been replaced
by modern ones such as the CIM and FMS concepts. Here, the problem of interconnection of a large number of programmable logic controllers, robots, and various intelligent devices dispersed within the factory had to be solved using special
communication systems to be developed. This was the main impetus for starting
standardization work in the area of industrial LANs. Major engineering companies, both vendors of mechanical, electrical and electronic equipment as well as
the users, have been working for more than one decade on standardization of factory oriented communication technology. The General Motors' initiative to work
on standardization of a communication system for manufacturing industry was
followed by Boeing, Ford, McDonnell-Douglas, General Electric, DuPont, Eastman Kodak, Hewlett Packard, Texas Instruments, Motorola, Intel, and many others. As a result, two industrial communication standards have been worked out:

- Manufacturing automation protocol (MAP) and
- Technical office protocol (TOP)

It was realized that the MAP implementation as a backbone bus system,
based on a single broadband coaxial cable, could facilitate building integrated
manufacturing systems by interconnecting of existing "automation islands,"
equipped with thousands of programmable controllers and robots. Fig. 2 shows an

example of a MAP-based factory automation system. Furthermore, a MAP/TOP-based communication system can help building enterprise automation systems in manufacturing industry (Fig. 3).

For long-distance interconnections, like for information exchange between different production units and the head office of the company, the *Fiber Distributed Data Interface* (FDDI) standard has later been accepted. It enables information transfer at a rate of 100 MBPS at a distance of up to 60 km.

In production engineering, the automation concept of *Computer Integrated Manufacturing* (CIM) is based on an advanced communication concept, also shown in Fig. 3, in which the Field Bus, MAP, TOP, and various LANs are integrated.

In the mean time, the term *Enterprise Network* has been coined [2]. It is a corporate-wide network that interconnects the communications, processing, and

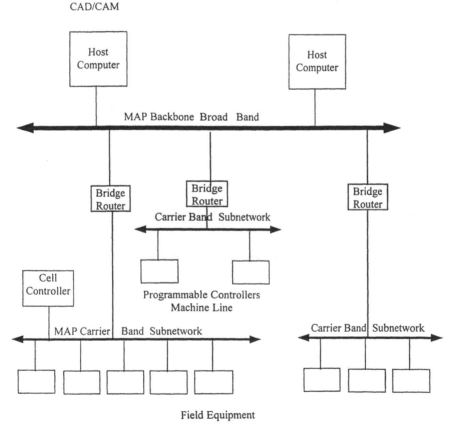

Figure 2 MAP backbone system.

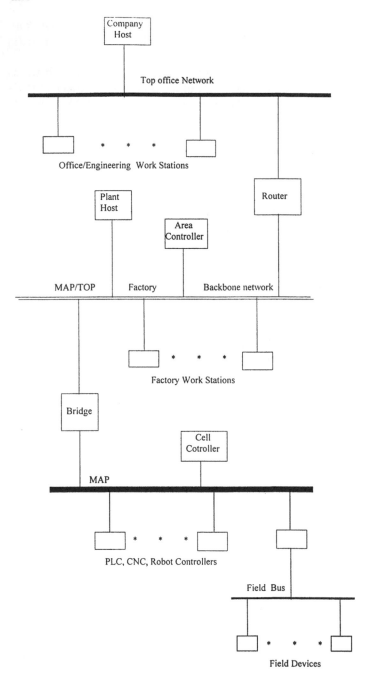

Figure 3 MAP/TOP enterprise networking.

storage resources of a corporation and makes them available to the users distributed within the corporation. The term has been extended to include the *Integrated Enterprise Network*, in which all forms of communications, voice, data, video, and image are integrated. It is a multimedia systems with conferencing possibilities, etc. The component of the networks are processors and distributed processing support software, LAN hubs and switched LANs, virtual LANs and internetworking elements, as well as network management systems.

Recent standardization activities of IEEE are in the area of *Wireless Local Area Networks* [3–5]. The IEEE has installed a committee for standardization of dubbed 802.11 for wireless radio LANs (WLANs) within the 2.4 Ghz band. The future standard should enable data transmission rate of 1 or 2 Mbps using CSMA/CA medium access protocol and error control at frame level.

II. LOCAL AREA NETWORKS

Local area networks are digital communication systems for data exchange between intelligent devices situated within a building or a production plant, or scattered over a factory or a university campus. The networks are characterized by their

- Topology
- Transmission medium
- Transmission signals
- Medium access control methods
- Area of application

A. Network Topology

Network topology is the term used to describe the *configuration* of the network, i.e. its structure, decomposable into *subnetworks* (or *subnets*), defined in terms of their *nodes* and *links*. *Final devices* or *terminals*, such as computers, workstations, monitors, etc., are attached to the subnetworks, representing their nodes.

In industrial automation systems the most preferable network topologies are

- Star topology
- Ring topology
- Bus topology

Star topology is the simplest network configuration in which the *point-to-point* links are established virtually between the terminals of the network via a common, active *central switch* (Fig. 4) called the *network node* or *hub*. Consequently, to enable communication between any two terminals, the hub should be addressed by the sending device, which should also specify the address of the destina-

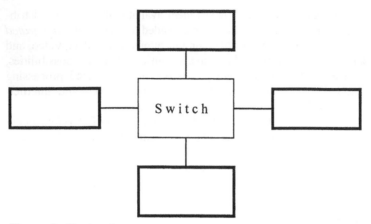

Figure 4 Star topology.

tion terminal selected. The hub is usually a powerful *main frame* computer equipped by appropriate communication software.

Star topology was from the very beginning a favored topology because the star structured networks are easy to implement, as in fiberoptic technology, using the *star coupler* as a star node. The main disadvantage of such networks, however, is the hub itself that, being the central network facility, represents a bottleneck and an unreliability factor, requiring a redundant back-up and some additional implementation investments.

Ring topology, in which network nodes are arranged to form a closed path in the form of a *ring* (Fig. 5), also enables direct point-to-point connections but

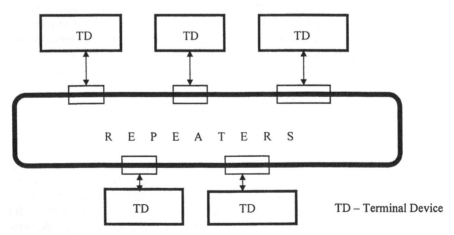

Figure 5 Ring topology.

only between the adjacent nodes. Information is sequentially exchanged between two distant terminals, i.e. it is node-to-node transmitted along the ring. The intermediate nodes, (i.e. the nodes between the sending and the receiving node) are active devices, intelligent enough to identify their own address if present in the data packet destinated to them. If this is not the case, the devices serve as *repeaters* for data transfer to other nodes, addressed by the data packets. Each node therefore operates as a *listener* that, depending on the destination address indicated in the data packet, *accepts* the data under transfer or *bypasses* it. This relatively simple operational mode of the nodes requires also relatively simple devices to serve as *message receivers* or as *message repeaters*.

The transfer of data within the ring is u*nidirectional*, i.e. the data is transmitted (or it circulates) only in one direction. For *bidirectional transfer* the *twin-ring* configuration is required.

Ring topology eliminates the communication dependence on a single central switch and enables a fast data transmission between the terminal devices. Instead of using a costly central hub, that is an expensive computer system, relative simple repeater devices are used in the ring, capable to receive, identity, and transmit the received data. Consequently, the installation costs for a network ring, particularly when twisted-pair lines are used, are relatively low. However, the ring network implementation using fiber optics is relatively expensive because it requires optical-to-electrical and electrical-to-optical converters, optical amplifiers, etc. This increases the complexity of nodes and decreases the network reliability. To this, the powering problem of individual nodes, where the converters are placed, is to be added because the optical fiber cannot transfer electrical power.

Although the ring configuration simplifies the transfer of broadcast messages sent to all ring participants, it is still involved when new participants are to be attached to the ring, because each time a new node is to be added to an existing ring configuration, the ring operation has to be temporarily suspended. The same holds for the on-line ring extension.

Bus topology (Fig. 6) uses a common transmission medium to which all network terminals are connected. Sharing a common transmission medium implies that only one point-to-point interconnection between the two bus participants can be established at one time. Each terminal attached to the bus is provided by its individual, *unique logic address*, that should be identified by the destination terminal when the data transferred along the bus reaches it. This requires that each terminal should continuously monitor the data transferred along the bus and should collect only the data addressed to it. This characterizes the bus as a *time-shared transmission path*, supported by a mechanism that prevents multiple simultaneous transmissions that would cause transmission conflicts.

Bus implementation using twisted-pair lines or coax cable is simple. Considerable difficulties arise when this should be done using optical fiber. Also here, like in the ring topology, the use of opto-electric and electro-optic coupler and optical amplifiers is required.

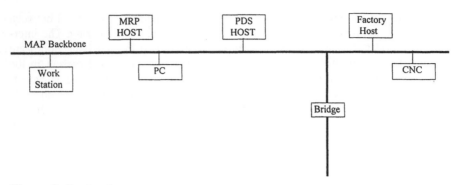

Figure 6 Bus topology.

B. Development of LAN Standards

IEEE initiated the first activities in definition of local area networks by establishing Project 802. The documents worked out within the project have been the subject of standardization work by ANSI (American National Standardization Institute), ECMA (European Computer Manufacturers Association), and ISO. Later on, the IEEE extended its activities to a number of particular aspects of Local Area Networks and published thus far the project documents:

- IEEE 802.1 Addressing and higher layer management
- IEEE 802.2 Logic link control (LLC)
- IEEE 802.3 CSMA/CD Medium access control method
- IEEE 802.4 Token passing bus medium access control method and physical layer specifications
- IEEE 802.5 Token passing ring memory access control method and physical layer specifications
- IEEE 802.6 Metropolitan area networks (MANs)
- IEEE 806.7 Broadband LANs
- IEEE 806.8 Fiber optic LANs
- IEEE 806.9 Integrated voice and data LANs

Two main ISO standardization issues, related to the LANs, the

- Medium access control methods and
- Open systems interconnection model

are discussed below. For details, see [6–7].

C. Medium Access Control Methods

Based on the IEEE documents 802.2, 802.3, 802.4, and 802.5, ISO has standardized the following medium access control methods (MAC methods):

- CSMA/CD medium access control method
- Token passing bus medium access method
- Token passing ring medium access method

Carrier Sense Multiple Access with the Collision Detection

Medium access is an extension of the ALOHA concept, originally known as *listen-while-talking protocol (LWT protocol)*. The improved version of the protocol, worked out in joint efforts of DEC, INTEL, and XEROX, was used for implementation of Ethernet.

The concept underlying carrier sense multiple access with collision detection (CSMA/CD) is as follows: the terminal device, intending to send the data, first listens to the network to check whether the network is *free* or *busy*. If busy, the listing device waits until the network becomes free and then tries to use the network. It could happen however that two devices waiting for the free network, after discovering that the network is free, simultaneously try to use it. This causes the transmission collision. After realizing that the collision is present, both devices resign from the use of the bus and initiate a new trail after a certain time-interval, specific (predetermined or randomly selected) for each device, has elapsed. The collision avoidance considerably increases the traffic volume of the network (up to over 90%) and is frequently used in industrial LANs. The technique does not require specific knowledge about the internal structure of the data packets to be transferred, and is predominantly implemented as hardware, i.e. at signal level. For discovering the current state of the bus it is enough to monitor the waveform of analog signal on the bus. The superposition of two carrier signals, generated by two different sending devices, is detected as an over-voltage and processed in the sense described forehead, both for baseband or broad-band transmission mode.

Token Passing Medium Access Control Method

The token passing medium access control method is used for implementation of token bus and token ring.

Token Bus

Token bus implementation takes into account the fact that the terminal devices attached to the bus are offered services only one at a time, whereby the servicing sequence is firmly predetermined and controlled by token passing. Only the device having the token is authorized to use the bus for data transmission and this only within a predetermined time interval. Thereafter, if the transmission has still not been terminated, the device has to pass the token to the next device in the predetermined servicing sequence. A device, receiving the token, checks its immediate need for using the bus. If there are no immediate needs, the device forwards the token to the device next in the servicing sequence.

Due to the rigid sequential servicing schedule of devices attached to a token bus, the devices form a *logical ring* (Fig. 7). To establish and run the services in the ring, the following basic functions have to be implemented:

- Ring initialization: by very start of the network the list of device actively participating in the token passing process is to be initiated
- Recovery: the same initialization should be carried out after the logical ring has failed or some errors occurs in it
- Adding new devices: the added device should be incorporated into the list of token passing order
- Deletion of a device: the deleted device should be removed from the list and its predecessor and successor spliced together

Token Ring

Token ring is also the implementation of the token passing medium access control method in a communication ring, in which the token is circulated from node to node around the ring and can be used by any device within the ring when addressed. For instance, when the ring is idle, i.e. not used by any device, the *free* token is circulating along the ring until a device, ready to use the ring, captures it. Before using the ring, the device first converts the *free* token into the *busy token*, indicating that the ring is under device control.

Usually, the *free token* is encoded as an eight bit string 11111111, to be converted into the *busy token* encoded as 11111110. The conversion is done by the device intending to use the ring. After completing the data transmission or after the allowed servicing time has elapsed, the device converts back the busy token into the free token.

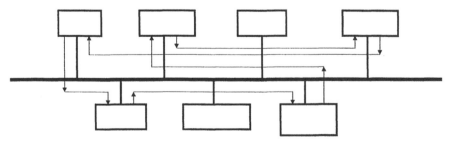

Figure 7 Bus logic ring.

Token Passing

Token passing is used in Ethernet, a bus system enabling in its original version data to be transferred over a distance of up to 2.5 km with the transfer rate of 10Mb/s. The frame format of the net (Fig. 8) includes:

- Preamble for packet synchronization
  Start frame delimiter, a bit pattern similar to the preamble, but ending with the bit 11
- Destination address, a 48 bit field (for up to 1024 participants)
- Source address, also a 48 bit field length, a two byte long pattern supplied by the LLC sublayer of the net
- Data field, a 46 to 1500 bytes field for the placement of user's data
- Frame check sequence, a 4 byte field for cyclic redundancy check

This results in a packet size of 72 to 1526 bytes.

Slotted Ring

Slotted ring is a standardized alternative to the token ring. In a slotted ring, a number of fixed-length *time slots* circulate along the ring instead of a token, each slot containing a *header bit*, that marks the slot status as *empty* or *full*. The device, ready to use the ring, waits for the next empty slot to arrive, marks it full, and inserts in it the destination address and the information to be transmitted (not longer than the slot length). Then, it waits until the slot comes back in order to continue the sending of data, if permitted. The destination device, after identifying its address within the information filled slot frame, reads the information and sets the accept bit. This should be the transmitting device be identified as the *acknowledgment* that now may transmit the next chunk of information.

A considerable advantage of the slotted ring is its simplicity, but the substantial disadvantage is its low transfer efficiency. A typical example of a slotted ring is the *Cambridge ring*.

In the area of fiber optic communication links the fiber distributed data interface (FDDI), developed by X3T9.5 of ANSI and internationally standardized by SC 25 of ISO/IEC JTC-1, is a *dual ring* network using a slightly modified token ring protocol for medium access control, that directly supports the LLC protocol.

Preamble	Start-Frame Delimiter Sequence	Destination Address	Source Address	Length	Information Field	Pad	Frame -Check Sequence

Figure 8 Ethernet protocol frame.

A fiber distributed data interface has been developed to serve as a backbone communication link for 500 to 1000 participants, distributed over a distance of 100km or more. The data transfer rate of 100Mbps is realizable using an 850nm wavelength carrier. Its frame of data transfer consists of a number of coded fields, the most important being [8]:

- Preamble (64 bits max.)
- Initial delimiters
- Control frame (64 bits max.)
- Transmission direction
- User data field
- Frame check sequence
- Final delimiters
- Frame-state field

The physical layer of the system is split into two sublayers:

- Physical medium dependent (PMD) and
- Physical layer protocol (PHY)

D. The OSI Model of ISO

In the late 1970s, ISO started the standardization work on *network architecture*, specifying the interrelations between the network functions and protocols required for implementation of network services. As a result, the *OSI model of ISO*, a standard draft proposal on *open systems interconnection*, was worked out [9–10], in which the *layering concept* was adopted to describe network architecture and related functions. The concept is based on the decomposition of network structure into a number of smaller, well-defined and functionally independent parts (*layers*) in terms of which any complex communication system can be described. The objective of the proposed model was not to *prescribe* any particular network implementation, but rather to specify *how* the implementation should conceptually look like and *what* the individual parts of it should perform. The term *"open"* should only indicate the possibility that two implementations, meeting the standard requirements, can easily be internet worked [10].

The initial task of ISO was to define the layers, required for implementation of an open system interconnection model, the services provided by each layer, and the information to be exchanged between the layers when executing data communication tasks. To each layer a subset of functions has been assigned, required for implementation of its services and for communication with the neighboring layers. The general idea was that each layer provides services to the next higher layer, based on the services provided to it by the next lower layer. The chain of services is however independent of individual, really implemented layers in the system, this in the sense that the removal of an existing layer or the insertion of a new layer does not require functional changes in other system layers. The individual layers should hence be mutually independent functional modules.

The OSI reference model, defined by ISO, when implemented in its full version, consist of seven hierarchical layers (Fig. 9):

- Layer 1: Physical
- Layer 2: Data link
- Layer 3: Network
- Layer 4: Transport
- Layer 5: Session
- Layer 6: Presentation
- Layer 7: Application

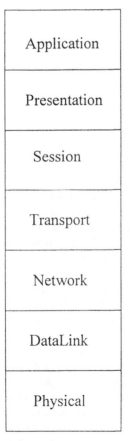

Figure 9 OSI reference model.

The physical layer, the lowest layer of the hierarchical model, is required for direct control of the data transfer medium, and defines the *physical (*electrical and mechanical*), functional*, and *procedural* standard recommendations for *accessing* and *using* the medium, without specifying the type and the parameters of the cable to be used for data transfer. The layer primarily helps establish and release the communication links, supports the synchronization and framing in transmission process, controls the data flow, detects and corrects the transmission errors, etc. It further handles electrical voltages, pulses and pulse duration, and detects—when CSMA/CD medium access control technique is applied—the collisions during the transmission and serves as interface to the transmission medium. Typical standards are RS 232-C, RS 449/422A/423A, X.21.

The data link layer provides access control technique to the transmission medium, analyzes the incoming data strings and identifies the *flags* and other characteristic *bit patterns* in them, provides the outgoing data strings with flags, error checking, and other bit patterns, and provides the reliable transfer of data along the communication link. The layer is generally responsible for transfer of data over the channel by providing the synchronization and safety aids for error-free transfer of data to the addressed destination. Examples of standards, related to this, are HDLC, SDLC, and some similar protocols.

The network layer is responsible for establishing the transmission paths and direct communication links within the network by specification of individual connections along the path connecting two terminal devices that intend to exchange the data via the network. It sets up the *routes* for data packets under transfer, builds the data packets for sequential transport, and reassembles them at the receiving end (this in order to restore the completed data sent). The layer also provides internetworking services and multiplexing procedures. The best known network layer standard is the X.25 protocol.

The transport layer provides the interface between the data communication network and the upper three layers of the model. It determines a reliable, cost-effective, and transparent data transfer between any two network participants, optimizes the service of the network layer, increases their reliability by end-to-end error detection and recovery, monitors the service quality, and interprets the session layer messages. The layer specifies *how* the logic transport connections between the corresponding terminal devices should be established, supervised, released, etc. Examples of transport layer protocols are the ITP of XEROX Network System and the IP/TCP (Internet Protocol/Transmission Control Protocol) of DOD, USA.

The session layer provides the mechanism for control of dialogue with the transport layer, for code and character translation services, modification of data frames, and for syntax selection. This enables text compression, encryption, and conversion between specific terminal characteristic and a *virtual* or *generic* model of application programs. The layer opens, structures, controls, and terminates a *session* and defines the transport connections to be used in the session. It further relates the addresses to the names, controls the dialogues, determines the art of dialogues (*half-duplex or full duplex*), and duration of transmission. Finally, it provides a check point mechanism so that the failure, occurring between the points, can be recovered.

The presentation layer provides the syntax of data presentation in the network. It does not check the content of data, nor interprets their meaning. It rather unconditionally accepts the data types generated by the application layer and uses the syntax selected for their presentation. In this way the layer provides the independence of user data from their presentation form that makes the data transfer independent of user's device. Presentation layer generally adapts the user application to the entire network by uniquely presenting each information to be transferred, mainly by translating it into a neutral language, appropriate for the communication system. The translation includes the formats, code, line length, and other specifications of the data.

The application layer, ultimately, supports the specific application requiring the communication system. The layer is in charge of *semantics* of generated data. It is equipped with the necessary management functions and manipulating mechanisms for supporting the *distributed applications*. Examples of protocols at this level are *job management, file transfer, electronic mail, business data exchange,* etc. Within the layer, two basic application service elements are present, the:

- Common application service elements (CASE), such as password check, log-in, commitment, concurrence, recovery, pear association setups etc., and the
- Specific application service elements (SASE), such as message handling document transfer, files and job transfer, file manipulation, database access and data transfer, directory service, system management, etc.

Generally speaking, the layering concept of computer network architecture, in which the network functions and protocols are hierarchically ordered, provides for:

- Logical, easy understandable decomposition of complex communication systems
- Standard hardware and/or software interfaces as boundaries between network functions
- Symmetry of functions in the adjacent layers
- Use of a standard language for system description, unique for network designer, vendor, and user
- Easy integration of user-devices and application programs, originating from different vendors

In the following, the individual network layers will be described in more details.

Physical Layer

This is the fundamental layer of a network, placed on the physical boundaries of the network. The layer implements *physical interface* between the devices of the network, and converts the bits, passed between them. The layer defines the:

- Mechanical specifications, by defining the pluggable connectors to be used
- Electrical specifications, by defining the voltage levels, timing of voltage transitions, the pulse rates etc.
- Functional specifications, by defining the operations performed through assigning the meaning and/or functionality to the signals used and to the relevant pins of the connector
- Procedural specifications, by defining the required sequence of events needed for transmitting data

The specifications listed above should enable implementation of the basic services at the physical level of the network, such as:

- Physical connections of individual terminal devices via a common transmission medium
- Bit sequencing in the sense of *bit delivery* at the destination point in the order they have been submitted at the source point
- Data circuit identification, consisting in identification of the physical communication path required or determined by higher layers
- Service quality specification by selecting the quality parameters of the transmitting path
- Fault detection of bit strings received from the data link layer

The services enable implementation of operational functions:

- Activation and deactivation of physical connections in physical links
- Activities management within the physical layer
- Synchronous and asynchronous bits transmission

The protocols used by the physical layer are transmission medium independent, so that the terminal device can easily be attached to coax, twisted pair or optical fiber line. Also the terminal devices themselves can be any equipment from passive terminals up to the intelligent computer systems. This is activated by specifying the:

- Data terminal equipment (DTE) and
- Data circuit-terminating equipment (DCE)

The transmission medium, the physical path between the transmitters and the receivers of a data transfer system, conveys the messages between terminal stations of the system. The media, dominantly used within LAN, are twisted-pairs, coax cables, and optic fibers.

Twisted pairs are wires made of copper or of steel coated with copper, 0.015 to 0.056 inch thick, usually several numbers of which are wrapped with a single outer sheath (telephone lines). Without a single repeater, the lines can be used in point-to-point connections for digital data transfer at a distance of up to 2 Km. Taped twisted-pairs have a lower transmission domain, but they still are a suitable communication medium of LANs.

Coax cables, consisting of a central conductor around which is a conducting shield separated from the conductor itself and the surrounding of the cable by isolating layers, are widely used within the LANs, preferably as:

- 50 ohm coax cable, used for baseband data transmission with *Manchester encoding technique* for a transfer distance of up to a few kilometers, and with the transfer rate up to 10 Mbps.
- 75 ohm coax cable, mostly used for broadband data transmission, for transfer distance of up to 1 km, and with transfer rate of 50 Mbps (or more).

The use of coax cable is preferred because their noise immunity is superior to that of twisted-pairs, especially in the higher frequency range. However, the coax cables are considerably more costly than the twisted-wire pairs.

Optical fiber, [11–12], an extremely thin, flexible glass or plastic fiber for data transfer with light as information carrier, is used as a transmission medium in communication systems where a transfer rate of 50 Mbps or more is needed over a distance of several kilometers, without signal repeaters. The fiber has already been used for data transfer since 1970. Still, recent developments in this field, supported by the industry, have gained broad interest by the users of data communication systems. Being an optical transmission medium, the fiber is nonsensitive to the electromagnetic interference. This is its main advantage, but it is still a more costly transmission medium than the twisted-pairs or coax cables. Optical lines are particularly suitable for explosion-proof areas. Furthermore, they do not radiate an electromagnetic field, have a relatively low signal damping ratio (line repeaters are required every 15 Km or more only), and they have a relatively low error rate (1:1000 as compared to metallic transfer media). However, their main advantage lies in their relatively high *channel capacity*, this due to the relatively broad transmission bandwidth of 500 MHz or more [13].

Fiber optic communication, which has developed as a true alternative to copper-based communication, is characterized by:

- Electromagnetic noise immunity
- Low losses (< 1.5 dB/km) in signal transmission

- Low weight
- Small fiber diameter
- Wide communication band
- Cross-talk absence between adjacent signal paths
- No electromagnetic radiation

However, the main drawbacks of fiber optic communication are:

- The remote elements on the fiber optic networks cannot be powered via the fiber as a data transmission medium
- Coupling and tapping within the network is difficult

In the local area networks with the bus topology, both *baseband* and *broadband* data transmission techniques are used. Baseband transmission technique, in which the messages are transmitted as a series of direct current pulses (by which the information content of the message is encoded), enables the transfer of only one message at a time along the transmission line. Signals are represented by two voltage levels (out of three possible levels : *low*, *high* and *zero*), one of them representing the 0-bit and the other one the 1-bit of the binary signal. The transmission medium of the network is only *time-shared* among the participants, attached to the network. For this reason, even in the case that the transmission medium is permanently used for exchange of messages, the capacity of the network as a communication channel is relatively low.

For the baseband data transmission 50 ohm coax cables are used, enabling a maximum transfer rate of 10 Mbps. When using twisted pairs, only a transfer rate of 1 Mbps is achievable that appears to be satisfactory for low cost networks. In both cases one needs the signal repeaters at each 1 km of distance.

The main advantages of the baseband system are:

- Low investment and installation cost
- Easy installation, maintenance, and reconfiguration(extension).

The disadvantages are:

Single transmission channels

- Lower channel capacity
- Shorter transmission distances
- Grounding difficulties

Broadband transmission technique, in which the messages are transmitted as a series of modulated pulses, enables a higher network transfer capacity. By the use of this transmission technique, many transmissions can take place simultaneously, each transmission using its own *carrier frequency*, i.e., its *own transmission channel*. The number of possible channels basically depends on the frequency characteristics of the transmission medium used, but nevertheless, the capacity of a broadband communication link is higher than the capacity of the baseband link.

For presentation of signals to be transmitted, *amplitude, frequency,* and *phase modulation* issued. At the receiver end they have to be demodulated for extraction of DC component of the transmitted pulses. At both ends of the transmission path modems are used for preparation of sending signals and the use of transmitted signals. The codes used for signal transmission are (see Fig. 10):

Nonreturn to Zero (NRZ). 0 and 1 are represented by pulses of equal width and equal amplitude but of different polarity (low and high). The technique has two main drawbacks:

* Identification of starting and ending points of individual bits in a string of bits of the same polarity is difficult, creating some synchronization problems
* Due to the unbalanced number of different polarities there is always a DC component present within the signal

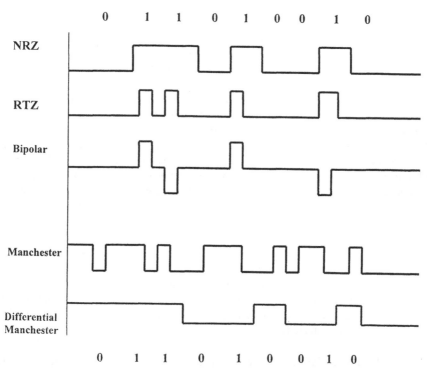

Figure 10 Encoding transmission signals.

These drawbacks are the main reasons for avoiding the use of NRZ-encoding technique for data transfer within the practical communication systems.

Return to Zero (RTZ). A pulse of half-symbol width is used for representing 1 and no pulse for 0. The technique has the property that clocking and synchronization is possible when strings of 1's are transmitted, but not when strings of 0's are transmitted. The transferred signal is not also dc free.

Bipolar. Positive and negative pulses of the same size are used alternatively for 1's, and no pulse for 0's; no dc component and a relatively low ac component for the case that 1's and 0's occur with the same probability. There is a clocking with 1, but no clocking with zero.

Manchester Code. The individual bits are split-up into two halves: in the first half the direct is transmitted and in the second half the complementary value, e.g. a 1 is represented by a *High* followed by a *Low*, and a 0 by a *Low* followed by a *High*. The code has a suppressed dc component, and a relatively insignificant low frequency component. It also has self-clocking properties, i.e. the sending and the receiving devices are synchronized during each bit transferred.

Differential Manchester Code. A zero is represented by a transition at the beginning of the bit, and a one by no transition at this point (self clocking).

The *synchronization* of data transfer is a serious reliability problem of communication networks. A terminal device, wanting to communicate with another device first has to identify the communication capability and the readiness of the receiving device and then to initiate the communication process. The first step is done by sending a signal to the receiving device, informing it that its cooperation is required. This is in fact the initial step in the communication protocol with the objective to synchronize the receiving device *before* the data is actually transferred, and later on, to keep it synchronized during the entire data transmission process. Earlier, the *asynchronous transmission* was preferred as the "best" synchronization because it transfers data one character (or one byte) at a time. Each character is marked by its *start* and *stop* code, usually preceded by a zero bit and followed (closed) by one or two *one-value* bits (Fig. 10), so that the receiver recognizes the one-to-zero transition as a start of a new character. Of course, the receiver is also instructed about the length of the characters (in bits). Because the characters are sent independently from each other, the transmission speed can be time-variable. Due to the required 2 or 3 bits for "characterize synchronization" as overhead, the asynchronous transmission is also less efficient.

For efficient and reliable data transfer along the network, the packages of data are built, based on *framing principle*, each frame containing the (in addition to the data to be transferred):

- Preamble, defined by the *sync byte* and other field *before* the user data field

- Postamble, defined by the fields *behind* the user data field ending with the final sync byte

so that the frame length includes the preamble, data and postamble length

Data Link Layer

The layer mainly provides functional and procedural means for establishing, maintaining, and releasing the data-link connections. It implements the following functions:

- Establishing, splitting, releasing of data-link connections
- Delimiting and synchronization
- Sequence and flow control
- Error detection and recovery
- Control of physical interconnections
- Data-link layer management

Unlike the physical layer, which provides only a raw bit stream service, the Data Link Layer is expected to guarantee a reliable physical link and to manage it during the completed process of data transfer by flow control and error recovery.

Flow Control. Regulates the flow of sequential messages between the communicating terminal devices by adapting their speed to each other, by preventing the overload of the receiving device, and by minimizing the idle state of the network. For this purpose the *end-to-end protocols* and some *congestion control mechanisms* are available within the layer. For efficient flow control traffic routing algorithms are used. In *connection-oriented protocols,* in simplest cases, the *stop and wait* or *hand shaking* approach is used.

Error Recovery. Helps implement error-free data transfer and includes:

- Error detection
- Error handling

Error detection is carried out using *redundancy* in the data stream provided by the sender and removed after error check at the receiver. Depending on the degree of redundancy used, various degrees of reliability of transferred data can be achieved in an *overload-free* data transmission along the established transmission path. The simplest approach is to use the *parity bits* in order to identify the correct data when having the selected *even* or *odd parity* over a predetermined data length, for instance over a byte or character. The approach can be extended by sending the data as a sequence of blocks, each block containing a number of *check bits* heading or following the information bits. This is known as the *block code* approach.

However, the most common error detection form, relying on the *binary cyclic code,* is the *cyclic redundancy check.* Once the error in transmission is detected, the recovery of error can be achieved by:

- Reconstruction of received data and correction of errors
- Repeating the transmission of the erroneous blocks

Logic link control (LLC), as a sublayer of the data link layer, copes with particular problems in data transmission such as:

- Data encapsulation and decapsulation
- Framing
- Addressing the data blocks

Data encapsulation consists in forming the blocks to be transmitted by combining the data and the control information, which could be:

- Address of the sending and/or receiving device involved in data transfer *error detection field*
- Protocol control field

Data decapsulation is the process in which the user data is extracted from the information block received. It is, as a rule, followed by *data reassembling*, a process in which the user data fields of sequentially transferred data blocks are chained to give the original data string encapsulated for transmission purposes. In this way meaningful messages are generated.

Framing is the operation of finding the boundaries between the successive frames, i.e. the location of the position where the previous frame stops and the following frame starts. The framing process also includes the idle fill between two frames, such as the intermitted synchronous bit pipe, which should be separated.

Three main framing techniques are in use in the computer networks:

- Character-based framing, relying on special communication control characters for idle fill and for delimiting the frames by indicating their beginning and their end
- Bit-oriented framing, based on flags as special strings of bits for each idle fill and each frame
- Length fields, based on frame length stored in the header

Of course, framing becomes a problem when the transmission error occurs and the receiving device cannot identify the end of a frame. Here, under circumstance, the identification of the starting flag of the next frame could, in connection with the CRC procedure, be of use.

Addressing is an operation similar to *framing* in which the incoming data blocks are analyzed with the objective to identify, at each node, *how* to forward the received information to the destination device. The problem can efficiently be solved using in the header of each block the address of the sending and receiving device, along with the *identification number* of the session to which the data transfer process belongs. The last item is particularly essential when different blocks of the same session are transmitted over different paths in the network.

Physical layer and data link layer, described up to now, are the most essential layers for implementation of industrial LANs, particularly for those used in distributed computers control systems, in which no complex network structures are used.

Network Layer

The network layer provides the means for establishing, maintaining, and terminating the network connections. It also supports the exchange of network services of data units between the transport entities. The layer also provides the transport entities, disregarding the functional details of networks or subnetworks used or the routing and relaying functions needed for data transmission.

The services provided by the network layer are:

- Connections of network entities (routing and relaying)
- Identification of service access points
- Bookkeeping of addresses used for unique identification of transport entities
- Exchange of service data units
- Selection of quality of service parameters
- Errors notification
- Sequencing for ordered delivering
- Flow control
- Release of network connections
- Reset of logical connections
- Segmenting and block building
- Management of layer services and functions

Network address is required for identification of the destination entity. Based on address of the calling and the called network service access point, the Network Layer determines which network entities participate in the data transfer. The layer selects the route to be established within the network, whereby the *network service access* point address is a fictive term defining the point where the network layer services to be used for data transfer are located. Such addresses are only the *points* within the system, representing the end points of data transfer path through the network layer defined by *called, calling,* and the *responding* address.

Routing and *relaying* are alternative networking functions, carried out either in a *fixed* manner, using some fixed tables, or in an *adaptive* manner, based on the monitored traffic flow within the route. *Fixed routing,* or *static routing* cannot achieve a high throughput. It is thus used in relatively small-size networks or in the networks where the transmission efficiency is not essential. The *dynamic routing* or *adaptive routing* adjusts the routes to time-variable traffic conditions, optimizing the resources available within the network. It also makes use of flow control to avoid the congestion when the traffic becomes very high. In order to provide the network connection between the transfer entities the Network Layer takes the logical connections within the network, such as virtual circuits.

As network complexity grows and the user applications increase in number and quality, the need arises for accessing various resources and/or data files located in various networks. This is nearly the regular case in distributed computer systems for industrial plant automation [14] and computer integrated manufacturing [15]. The way to support the user in such access requires building the larger networks out of subnetworks by their *interconnecting* or *internetworking.*

Transport Layer

The transport layer belongs, along with the session layer, presentation layer, and the application layer, to the group of *higher network layers,* whereby the physical layer, data link layer, and network layer belong to the group of *lower layers.*

The principal objective of this layer is to provide reliable *user-to-user transport services*, rather than the *device-to-device services*. For this purpose the layer directly cooperates with the neighboring layers, the network and session layers.

The transport layer provides all functions and protocols required by the session layer for optimal network throughput, minimal delay time for establishment of connections and transfer of data, minimal error rate and transport costs, requested security and priority, etc. The layer can generally help when establishing the alternative network connections, implementing multiplexing the communication paths and the like.

According to the ISO standard, the following functions have to be available in the transport layer:

- Translation of transport addresses into networking addresses
- Establishing and releasing the transport connections
- Sequence control on entire individual connections
- Monitoring the quality of services
- End-to-end error detection and recovery from error
- Flow control on entire individual connections
- Segmenting, blocking, and concatenation of data strings
- Transport paths multiplexing
- Transport supervision

Five classes of connection-oriented transport protocol are defined in ISO 7498 Standard:

- Class 0: simple protocol class
- Class 1: error recovery class
- Class 2: multiplexing
- Class 3: error recovering and multiplexing
- Class 4: error detection and recovery class

In industrial LANs usually the TCP/IP (transmission control protocol/internet protocol) is implemented. It is a general transport protocol for data transfer over long distances, through potentially noisy communication channels. It routes the packets and delivers them to any specific host computer within the Internet using a 32-bit addressing identifier. The routing between the source and the destination device within the same subnetwork of the Internet is said to be within the same hop. Otherwise, when the destination device is within another subnet, the routing of a packet is distributed, making use of a number of hops.

The IP header includes the grade of service, data unit length, identifier, flags, lifetime, user protocol, header checksum, source and destination address, options, etc. The services of the protocol include the data request, data indication, and errors, all of them relying on source and destination network, source and destination host, user protocol, options, grade of service, and data specification.

Although the TCP/IP protocol is a multi-layer communication protocol very similar to the OSI model, it still uses three additional internetwork protocol layers:

- Transport control protocol (TCP), a high-reliable connection oriented protocol
- User datagram protocol (UDP), a less reliable datagram
- IP datagrams layer, underlying the network protocol of internet virtual network

Both TCP and IP are transmission technology independent protocols and thus network and user independent. The 32-bits long IP address is in fact an identifier related to the corresponding subnetwork identifier and the host identifier. The host address specified by the host identifier must be unique and sufficient to withstand any future network modifications. It should also supports the routing strategies used within the network.

Session Layer

The purpose of the layer is to manage the activities related to the orderly data exchange by establishing, monitoring, and terminating the data transport processes called *sessions*. It enables the exchange of data in a reliable dialogue form by means of:

- Session-to-transport connection mapping
- Session connection flow control
- Session connection relies
- Session connection recovery
- Session layer management.

In addition, the layer performs the functions like:

- Communications dialogue structuring that determines the sequences and the duration of individual speakers, along with the dialogue type simplex, half-duplex, or duplex
- Establishing the data synchronization checkpoints, required for error recovery by retransmission of data in case of transmission error discovery between any two checkpoints

In structuring and conducting dialogs the session layer uses the token principle, with the following tokens specified:

- *Data token*, needed for control of half-duplex connection
- *Synchronize-minor token*, required for setting the minor synchronous points
- *Major and activity token*, the key token for setting the major synchronization points and managing the activity structure
- *Release token*, in charge of release of connections

The basic session layer part is its *kernel* that integrates the majority of functions enabling the connect, transfer, and disconnect of layer services.

Presentation Layer

The presentation layer has a number of functions and services for *exact* implementation of a *specific* mode of information transfer. For example, the layer is competent for a reliable *data encryption, data compression, file transfer* and the like. Under its arbitration the selection of the syntax to be used also for data transfer and related additional activities are to be carried out.

Data encryption transforms the original data into an abstract, non-intelligible format in order to increase the degree of data content protection and guarantee a highly-protected data exchange. Besides, the encrypted data are immune against their continual modification or utilization. The main objective of data compression is to compress the total volume of data, or the total number of bits to be transferred. This saves the data storage costs and the communication time required for their transfer.

The presentation layer is of negligible interest for implementation of industrial LANs.

Application Layer

The application layer is a direct *interface layer* between the OSI styled communication system and the user terminal devices. It is *a top layer* of OSI model and has only one neighboring layer, depending on the character of the communication system, the presentation, session, transport or network layer. In the distributed computer systems and the related local area networks the application layer, in the absence of higher communication layers, also directly communicates with the data link layer.

The principal objective of the layer is to provide services to the user application processes outside of OSI layers, that include:

- Data transfer
- Identification and status check of the intended communication partner
- Communication cost allocation
- Selection of quality of services
- Synchronization of communication dialogue
- Initiation and release of dialogue conventions
- Localization of error recovery and identification of responsibilities
- Selection of control strategies for ensuring of data integrity

The *applications* are the collections of information processing functions defined by the *user*, rather then by the user terminal device. In modern automation systems the applications are even distributed and multi-user defined.

The protocols of the application layer supports:

- Supports the transfer of files, exchange of documents, sending and receiving the messages etc.
- Ensures that the agreed semantic is used by the cooperating partner
- Help the user access the resources adequate for the given application

In the production engineering and manufacturing for the MAP/TOP standard, file transfer, access and management, virtual terminals, document and graphic interchange, and manufacturing message service are available for applications.

III. FIELD BUS SYSTEM

Instrumentation elements installed in the field, such as sensors, actuators, transmitters, drivers, PLCs, etc., are increasingly becoming smarter due to the microelectronics integrated. This has enabled gradual placement of some data acquisition and preprocessing functions there [16]. In the sequence, the problem of intercommunication among the elements and data transfer to higher-level devices are becoming important and have to be solved using modern communication technology. The earliest proposal to use *field multiplexers* and *remote terminal units*

(RTUs) for this purpose has not been seen as the appropriate one. More attention was focused at development of an open, high-performance communication link, based on some international standards, enabling a better interconnection of final instrumentation elements. Therefore, the activities of IEEE in the field of local area networks have been welcomed by the industrial automation society as a promising way to create a communication link at the field instrumentation level. The competent professional organizations (IEC, ISA, and the IEEE itself) have been motivated to start the work on standardization of a special communication system, the field bus, appropriate for direct transfer of field data [16].

To be standardized the field bus was defined as a serial, digital communication network capable of supporting field instrumentation devices on a shared data transfer medium. Its application is envisaged to be the automation of process and manufacturing plants. For this purpose the bus should be provided by enhanced service functions for intelligent devices that should help minimize user intervention in setting up and operation of industrial automation systems and maintain safety, timelines, and efficient bus utilization, even at low bit rates.

A major advantage of the field bus should be the integrity of transferred data in severe noisy environments. When applied in manufacturing automation systems, the bus should:

- Minimize the manual efforts required by staff
- Achieve optimal response times for "just in time" manufacture
- Implement direct links between the production level and the manufacturing level

To achieve this, the bus standard was supposed to meet at least the following general requirements:

- *Multiple drop* and *redundant topology*, of total length of 1.5 km or more for data transmission over *twisted pair, coax cable*, and *optical fiber*
- *Single-master* and *multiple-master* bus arbitration in *multicast* and *broadcast* transmission mode
- *Access time* of 5-20 sec or a *scan rate* of 100 samples per second
- *High-reliability* in data transfer with *error detection capability*
- *Galvanical* and *electrical* (>250V) *isolation*
- *Mutual independency* of bus participants
- *Electromagnetic compatibility*

The requirements have been worked out by IEC TC 65C, ISA SP 50, and IEEE P 1118. However, for a long time, no agreement has been achieved on final standard document because four bus candidates have been proposed for standardization:

- BITBUS of Intel

- FIP of AFNOR
- MIL-STD-1553 of ANSI
- PROFIBUS of DIN.

BITBUS, that can interconnect up to 250 secondary stations, is based on RS 485 specifications and uses the screened twisted pairs for transfer of NRZI encoded data. The transmission is based on IBM SDLC protocol, that enables the data transfer rate of 64 Kbps to 2.4 Mbps. For transmission error detection the 16 bit CRC technique used.

Factory instrumentation protocol (FIP), a bus system jointly developed by French industries and academic institutions, enables a data transfer rate of 3 Mbps. For encoding the data to be transferred Manchester code is used, supported by parity check and CRC error detection technique.

MIL-STD-1553, a bus system originally developed for Air Force and Navy requirements, also uses the screened twisted pair for interconnection of up to 31 bus participants. The transfer rate of data, encoded in Manchester II bi-phase, is 1Mbps. For error detection, 1 parity bit and 3 Manchester code violation bits are assigned to each 16 bits of data.

Process field bus (PROFIBUS) is a German field bus proposal, also based on RS 485 specifications. Up to 127 bus participants can be interconnected by the bus over a screened twisted pair of 1.500 m total length. For transfer of NRZ encoded data the synchronous mode is used, enabling the transfer rate of up to 500 Kbps. Error detection is based on one parity bit per 8 data bits.

The search for the best field bus standard proposal has taken much time. Apart from the above proposals, a number of additional bus implementations that, at least in their specific application areas have been seen as *de facto* standards, have been launched such as CiA, FAIS,

IEC/ISA, Interbus-S, ISU-BUS, LON, Merker, P-Net, SERCOS, Signalbus, TTP, etc. Typical specifications of such proposals are the following:

- RS 485, twisted pairs, coax and/or fiber used as a data transmission medium
- Master/slave, token, and time-division multiple access used as common medium access control protocols
- 20-256 participants to be attached to the bus
- 30 Kbps to 10 Mbps transfer rate to be implementable
- 100 m to 2 km of total bus length to be achieved
- HDLC as link control.

Although the work on definition of a field bus international standard is still going on within the IEC-TC65, some of the initial field bus proposals are in use. For instance, the PROFIBUS is already available on the market. Its implementation includes the OSI layers typical for the field bus: the physical layer, logical link layer, and the application layer with the possibility of their interconnection to the network layer. No use is made of transport layer, session layer, and the presentation layer.

In the meantime, various PROFIBUS implementation alternatives are available to optimally fit into specific applications, such as:

- PROFIBUS-FMS, the actual bus version representing the DIN 19245 standard, that is a RS 485-based fiber optic bus of up 1.5 Mbps data transfer rate at the total distance of 1.2 km, along which up to 126 participants can be attached
- PROFIBUS-DP, a low-cost, high speed version for up to 12 Mbps transfer rate
- PROFIBUS-ISP, an explosion proof version for 1-19 mA signals data transmission with the maximal rate of 31 Kbps

In Japan, the interest of users is concentrated on FAIS (factory automation interconnection system project), that is expected to solve the problem of a time-critical communication architecture, particularly important for the production engineering. The final objective of the bus standardization work is to support the commercial process instrumentation with the built in field bus interface. However, also here, finding a unique standard proposal is extremely difficult.

The field bus concept is certainly the best answer to the increasing cabling complexity at the field instrumentation level in processing industries and manufacturing. Using the field bus all field devices are interfaced in a unique way, and easily internetworked to the computer bus systems and local area networks at higher levels of integrated automation systems. The major benefits resulting from this fact are:

- Enormous decrease of cabling and installation costs
- Straightforward adaptation to any future sensors and actuator technology
- Easy configuration and reconfiguration of plant instrumentation
- Automatic detection of transmission errors and cable faults
- Facilitated implementation of *hot back up* by the communication software
- Resolved problem of common mode rejection, galvanic isolation, and of noise and crosstalks due to digitalization of analog values to be transmitted.

The field bus is viewed as an integral part of a factory automation system at its low hierarchical level, enabling a high integrity communication among sensors, actuators, and programmable and other local controllers in a process or manufacturing plant (Fig. 11). Its key features should be:

- Time critical communication possibility
- Intrinsic safety

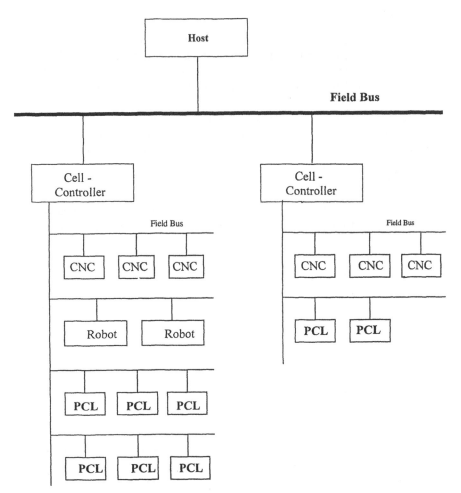

Figure 11 Field bus example.

- Robust operability in severe electromagnetic environment
- High availability against common failures in transfer medium or master station.

The field bus standard, proposed by IEC SC 65C, is based on the OSI model of ISO and has a layered structure consisting of:

- Physical layer
- Data link layer and

- Application layer

whereby the set of most important functions from the network and transport layers is incorporated into the data link layer and the functions from session and presentation layers into application layer. In addition, a *system and network management agent* is incorporated into the model, acting on all layers. The structure of the field bus standard is shown in Fig. 12.

The physical layer of the field bus enables *transparent* data transfer in *synchronous* mode using the serial, *half- duplex* mode of transfer. Without using a repeater, the implementable transmission rate is alternatively 31.25 Kbps, 1 Mbps, or 2.5 Mbps over a distance of 1.900 m, 750 m, and 500 m respectively. The recommended media are:

- Shielded twisted pair in various bit rate, power, and coupling variants
- Single 100/140 optical fiber with passive reflective star couplers or active star couplers
- Twin 62.5/125 optical fiber with passive transmitive star couplers or active star couplers
- Wireless

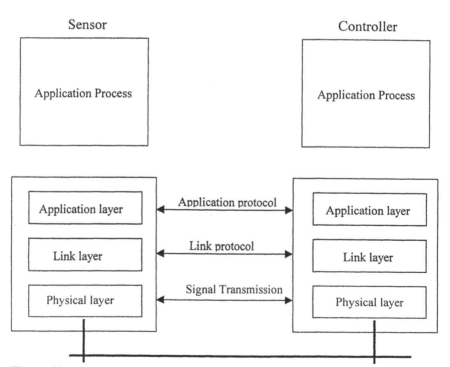

Figure 12 Layered structure of field bus.

The signals, transferred along the bus, are self-locked and Manchester bi-phase encoded. The synchronous transmission used saves the start and stop bits, not required in such transmission. Only for initial synchronization of the receiver and the transmitter is sending of a preamble is required Also for marking the initiation and the termination of messages the start and end delimiters are required. The resulting frame is shown in Fig. 13 [17].

The physical layer receives data units from data link layer and generates the physical layer protocol by adding to them the preamble and the frame delimiters.

The *data link layer* of the field bus includes the OSI:

- Medium access control sublayer
- Data link control sublayer

subdivided into three virtual levels with the following assignment:

- Path access and scheduling
- Bridge operation
- Connections and connections data transfer, bridge coordination, and data link service

When transmitting data, the *path access and scheduling level* forms the protocol data unit from user data and control information, computes and appends the pertinent check sequence and forwards the completed frame to the Physical Layer Interface. When receiving the data, the level carries out the error check, based on a 16 bit FCS generator polynomial

$$G(x) = X^{16} + X^{12} + X^{11} + X^{10} + X^8 + X^7 + X^6 + X^3 + X^2 + X + 1$$

Bridge-operation level is required when the field bus is bridged to other LANs.

Data transfer, bridge coordination, and data link service levels generally manage the interactions with the upper layers and arbitrate the sequencing of active connection end points and of connectionless transactions.

The *access control sublayer* of the field bus uses a combination of the token passing and poling methods. The devices attached to the bus are considered as *master stations* or as *slave stations.* Poling enables the device having the token, i.e. the master, to poll the servers.

Preamble	Start Delimiter	Data From Physical Layer	End Delimiter

Figure 13 Field bus physical layer frame.

The *data link control sublayer,* in the absence of higher layers, directly supports the application layer. The data link layer protocol of the field bus provides a set of communication details that help information exchange between the incompatible systems. It also supports the timely transfer of data and control information.

Fig. 14 shows the outline structure of the data link protocol data unit.

The *application layer* of the field bus enables user programs to access the bus communication environment. It facilitates solving the distributed application problem and extends the ISO application layer functionality by supporting the conveyance of time-critical application requests among industrial automation devices.

Application layer services are provided by application entities within the application process, that represent those elements of a real system that, by information processing, implement an application. Application processes are components of a distributed system. They communicate and cooperate in order to render services required for distributed applications. The communication can follow the:

- Client-server mode or in
- Publisher-subscriber mode

Client-server mode of interaction between the application processes is used for acyclic, bidirectional data transfer between the client, sending a request to a server, and a server processing the request and sending an affirmative or disapproving response to the client.

Publisher-subscriber mode of interaction between the application processes is used for cyclic data transfer between a single publisher emitting the information and a specific or a group of subscribers receiving it. The transfer can be implemented in *push mode* or in *pull mode.*

To represent the application layer of the field bus, *object oriented design* is used, objects being the entities with a well-defined behavior, usually the real-world entities. From the standardization point of view, devices are not real hardware implementations as they are, but rather network nodes with some parameters assigned to them. They are seen as *virtual field devices*

The *system and network management agent,* incorporated into the bus model, renders the following services:

Frame Control	Destination Address	Second Source Address	Source Address	Parameters	Application Layer Protocol Data Unit	FCS

Figure 14 Field bus data link protocol.

- Management of communication system, monitoring the attachment and removal of bus devices, and device identification
- Devices instantiation, initiation and braking the communications, downloading and uploading of device configuration
- Management of application processes
- Fault detection and management for maintenance support and system integrity
- Management of communication performance parameters to reach the time critical communication requirements

The system management model is based on *manager* and *agent* function. The functions locate, create, delete, connect, start, and stop the resources used to perform individual functions, monitor the system, and look for possible changes in the system configuration. The following device classes are standardized:

- SM Type 1: fixed application process, fixed object dictionary, preconfigured devices, with no user configuration permitted
- SM Type 2: fixed application process, fixed object dictionary, user configurable within prescribed limits
- SM Type 3: fixed application process, object dictionary loadable, modifiable, and extendable
- SM Type 4: programmable application process with loadable functionality and free defined function library
- SM Type 5: any service and optional function.

Devices to be attached to the network require a tag, an up to 16 characters string assigned by the user, and a unique address assigned by the *system and network management kernel.*

The configurable applications are implementable using the block modeling approach. By linking the selected block, previously uniquely assigned the tag by the user, the user can implement the desired applications. The resulting configuration, implemented as software, carries out the modeled application, for assigned parameter values of the blocks, by executing the algorithms being the constituent parts of the blocks selected and produces the output parameters. The output parameter values, after the block algorithm has been executed, are broadcasted on the network to be used by the block requiring them as their input parameters.

In the field bus standard three types of blocks are defined, the:

- *Function blocks*, implementing input, output, control, and calculation functions
- *Transducer blocks*, implementing input, output, and display functions
- *Physical block*, implementing the functions, related to the alarm, event, trend, and display of objects

To each block a number of different types of parameters are assigned, depending on the block type and its application.

IV. MAP/TOP: MANUFACTURING AUTOMATION PROTOCOL/TECHNICAL OFFICE PROTOCOL

The LAN standardization documents, proposed within the IEEE Project 802, although being accepted by the ISO as draft proposals for corresponding international standards, did not cover some essential *real-time application* aspects [18], such as the application in industrial automation, where hundreds or thousands of programmable controllers, sensors, actuators, and similar instrumentation elements have to be interconnected. Point-to-point interconnection of such elements, as originally practiced within the factory floor, enormously increased interfacing costs and called for a more elegant solution [19]. In order to reduce the cabling costs, while improving the quality of plant control, General Motors Corporation (GMC) Task Group was formed with the objective to specify an *industrial communication standard* for data exchange between equipment originating from different manufacturers, and to indicate the way for a reliable interconnection of numerous programmable controllers and industrial robots, distributed over a large number of *automation islands* within the company. This was expected to reduce the total wiring and cable lying costs, that have reached 50% of total instrumentation investments. In addition, from the perspective of annual installation of up to 500 new programmable controllers, the communication problem was running so out of control so that a prompt solution was unavoidable [20].

The Task Force of GMC has started the work by investigating the OSI reference model of ISO as the standardization basis. In cooperation with DEC, HP, and IBM in 1982 the group formulated the first industrial communication protocol launched as manufacturing automation protocol (MAP), that was not a new standard but rather a list of recommended, already existing international LAN standards for implementation of individual OSI layers. Soon, serious efforts were made by Boeing Computer Services Company to standardize the communication between the computers, peripheral devices, and work stations in office automation systems, where the technical office protocol (TOP) was standardized in a similar way, i.e. also as a list of recommended existing LAN standards enabling the realization required services, such as *electronic mail, word processing, file transfer, database management, graphic and business analysis,* etc.

The general structure of MAP/TOP implementation standards has been formulated in the MAP Version 2.1 and TOP Version 1.0 as follows:

* Application layer: ISO FTAM (DP) 8571 with:
    MAP: MMFS and CASE
    TOP: File transfer, limited file management
* Presentation layer: ASCII and binary encoding

- Session layer: ISO session IS 8327 (session - full duplex)
- Transport layer: ISO transport IS 8073 class 4
- Network layer: ISO Internet DIS 8473, connection and for
- Logic link layer: ISO LLC DIS 8802/3, type 1, class 1
      MAP: ISO TPB DIS 8802/4, TPB MAC
      TOP: ISO CMA/CD DIS 8802/3
- Physical layer:
      MAP: 5 MB carrier band, 10 MB broad band
      TOP: 10 MB CSMA/CD, 4 MB token ring

Later on, for small enterprises also a MINI-MAP version has been specified (Fig. 15).

The MAP/TOP standard has been proposed for implementation of backbone bus oriented internetworked LANs (Fig. 16), [21]. The internetworking was supposed to be implemented using network bridges—NB, network gateways, and routes—R [22].

OSI MODEL	MAP	MINI-MAP
Application	ISO CASE/MMFS,FTAM	MMFS
Presentation	Null	Null
Session	ISO Session Kernel	Null
Transport	ISO/NBS Class IV Transport	Null
Network	ISO Internet Connectionless	Null
Data Link	IEEE 802.2	IEEE 802.2
Physical	IEEE 802.4	IEEE 802.4

Figure 15 MAP/MINI-MAP structure.

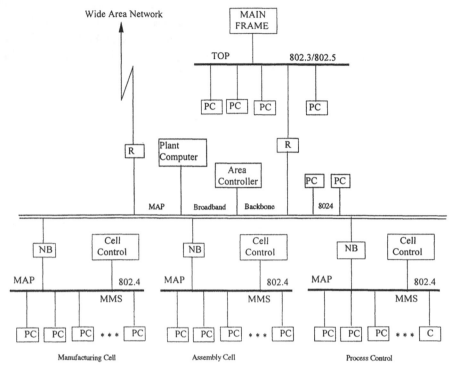

Figure 16 Broadband backbone network.

A *backbone* is basically a bus oriented local area network, interconnecting a number of *different* local area networks. Because of increased communication traffic on them, the backbones operate in broadband mode that enables a high effective transfer rate. For instance, many Ethernet systems can be attached to the MAP, used as a backbone in factory automation.

General specification of the application layer, as given in the OSI model of ISO, was not of direct use in applications, such as in process, production, or quality control. Its layer specification has met rather the communication requirements in *commercial* application fields, such as financial and administrative sectors. With the advances in integrated plant automation, in which the *process, production,* and *management* data have to be merged and multiple access to data bases enabled, the communication demands within the enterprises have considerably increased. This has been a decisive reason for starting the work on definition of *specific protocols* for automation in the industry, like the work on definition of manufacturing automation protocol and technical office protocol.

The rationale behind the international standardization of a communication system such as MAP was that the systems planning engineers, as a rule, try to implement their systems using the *best, low-price* subsystems (devices) available on the marketplace. The individual subsystems, however, because they are selected from different vendors, are not expected to be mutually compatible. That increases their intercommunication difficulties and requires a costly hardware and software interfacing. In 1980, as the work on MAP started, General Motors reported that over 20,000 programmable controllers and some 2,000 robots of different origin have been installed in the manufacturing workshops of the enterprise that had to intercommunicate via a complex communication network, incorporating various network topologies. All this could be simplified and the interfacing, programming, and cabling costs reduced by replacing the complex internetworking by a simple, say communication bus-based, internationally standardized system. This will also simplify the attachment of any device to the system, as long as the device interface obeys the proposed standard.

Boeing, DuPont, Eastman Kodak, Ford, McDonnell-Douglas, General Electric, and a large number of process instrumentation and computer vendors have had similar problems. They have all welcomed the General Motors initiative to establish a task force that should formulate the specifications for a data communication standard under the name manufacturing automation protocol. The work was completed in 1982 and a MAP document prepared that was revised in 1984 and 1985.

The work on MAP protocol completely relied on the results of IEEE Project 802 and the OSI model of ISO, so that for the implementation of version 2.1 of MAP the following was used:

- ISO DIS 8802/4 (ISO token passing ring) for the physical layer
- ISO DIS 8802/2 (ISO logic link control) for data link layer
- ISO DIS 8473 (ISO Internet) for the network layer was recommended, and for implementation of its application layer.
- Manufacturing message format standard (MMFS)
- File transfer access and management (FTAM)
- Common application service element (CASE)

Manufacturing message protocol (MMP), the most relevant application layer protocol in factory automation, defining the common language within the network, was incorporated into MAP standard version 2.0 and 2.1 as an Appendix. It, however, became the central application protocol in the MAP/TOP 3.0 version (Fig. 17). The protocol includes the functions such as:

- Reading and writing of process variables
- Event-driven triggering of actions
- Alarming and alarm handling
- Supervision of equipment status
- Data archive

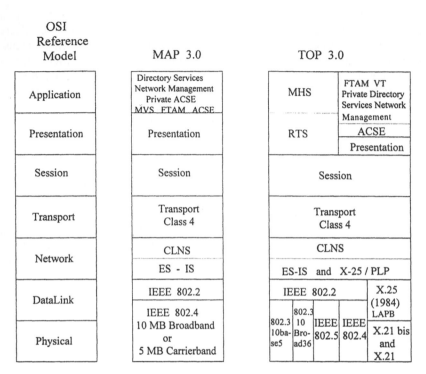

Figure 17   OSI reference model of MAP and TOP 3.0.

The MMS services are organized on *client-server* principle [23–25]; the client device applies, using a message, to the server device for services. The server executes the service and sends the execution results to the client. A server is viewed as a *virtual manufacturing device* (VMD) lending the services to the clients.

The MMS uses *abstract objects* to describe the MMS devices and the MMS services. The assignment of services to the objects and vice versa is fixed in the standard document, so that to implement the standard a mapping of objects on the real data is required because the objects are presented through the data structures.

The objects, available within the MMS, are:

- Transaction
- Domain
- Operator station
- Journal entry
- Event condition
- Event action

- Event enrollment
- Semaphore entry
- Semaphore
- Named type
- Scattered access
- Named variable list
- Named variable
- Unnamed variable
- Program invocation
- Domain

Specific service groups are assigned to the objects. The call for service includes the reference to the group name and the related parameters, such as:

- Domain management
- Virtual manufacturing device support
- Environment and general management
- File management
- Operator communication
- Journal management
- Event management
- Semaphore management
- Variable access
- Program invocation management

File transfer, access, and management (FTAM) basically controls various items of intelligent equipment at the cell level, such as programmable logic controllers, numerical controllers, robot controllers, etc. At factory level, FTAM supports the cell controller to exchange messages with various higher level automated equipment, containing the part or component information relevant to the cell controller. FTAM creates, cancels, reads, and writes individual files. File access and management services include the protection and use of file attributes like file name, type of access structure, name of presentation contest, file size, access request, structure of the current access and its name, future file size, and the date and time of creation of its data.

Common application service elements (CASE) specifies the:

- Connection building between two application devices
- Connection termination between two application devices
- Permission to a premature termination of a connection
- Permission to a lower layer to terminate a connection earlier
- Transfer of data between two application devices

The *client-server concept,* used within the MAP and the FIELD BUS, was been born with the advent of personal computers (PCs) and their internetworking using PC LANs. Numerous computer and communication vendors have made considerable effort in paving new ways in a general solution of PC internetworking. Among the most eminent ones are: AT&T, Datapoint, IBM, Intel, Nestar, Novell, Wang, Xeros, and 3COM. For instance, various alternatives to Ethernet have been proposed from different vendors for internetworking of personal computers. As a typical example the EtherSeries of 3COM is to be mentioned, with valuable complements such as Etherlink, EtherPrint, Ethermail, etc. Presently, a multitude of powerful PC networks has been launched to the marketplace, the well-known being he G-Net of Gateway, HiNet of Digital Micro, Locallnet2O of Sytec, Multilink of Davong, Net Board of PC Office, Netware/S of Novell, Omninet of Corvus, PC-Net of Santa Clara, PC Network of IBM, StarLAN of AT&T and Intel, and the Token Ring of IBM.

The IBM token ring, a paragon for the creation of the corresponding IEEE proposal 802.5, has still retained its popularity as a baseband 4 Mbps LAN using twisted pairs as physical medium. The ring has a physical header that includes the delimiter, physical control, and the source and destination address, along with the physical trailer containing the frame check sequence, ending delimiter, and, again, the physical control. besides that, the ring supports the use of IBM's system network architecture (SNA), as well as the network basic input/output system (Netbios).

The individual computers in distributed computer systems have not only to face the intercommunication problem, but also the problem of access to the common resources. Both problems have been solved in many different ways. The use of the client-server principle however has proven to be the most preferable one. The multicomputer communication systems, organized according to this principle, have the advantage of being capable to transfer the information, on request, from the client concerned to its server, that executes the client requests and returns to it the execution results. For this, two *send* and two *receive* commands are required, as usual in message-passing operations-based systems, whereby here the commands involved are sent at language level as remote procedures calls using the request-reply message exchange protocol in which, for example, the following primitives are used:

- DoOperation - used by a client for remote operations call
- GetRequest - used by a server acquiring a client reply
- SendReply - used by a server to send the reply message

Still, due to transmission failures, in addition to the usual request-reply protocol also the request-reply-acknowledge reply is required.

In modern factory automation technology, fiber optic communication systems are in long-time use as bus or ring oriented local area networks, known as fiber optic LANs [26–27]. The most typical representatives are the *fiber optic*

MAP for factory automation and the *fiber optic field bus* for industrial plant auto-mation. In addition, much work has been done to standardize the FDDI, a high-speed real-time fiber optic communication system capable of transferring the data along a *dual ring* with the rate of 100 Mbps, connecting up to 500 nodes at a length of 100 km, with the distance between the nodes of 2 km. Two rings—the *primary ring* and the *secondary ring*—are used for data transmission, the secon-dary ring serving as a *back up ring*, whereby the individual stations can be *single-connected* or *dual-connected*, i.e. attached to one or to both rings. Recovery from failure is supported both by secondary ring, as well as by *optical bypass switches*, enabling the inative node to pass the traveling light directly, i.e. without active power, from one to the other adjacent neighbor.

The basic technical parameters of FDDI are:

- 100 Mbps transfer rate
- 125 MB group encoding
- 10 Mbps sustained data rate
- 4500 B maximal frames
- 2nd window transmission
- 6 fiber sizes for cable plant
- *Timed token rotated protocol*
- Restricted token mode
- Synchronous and asynchronous frames
- Dual-ring connection
- Optical bypass switch
- Up to 50500 participating stations
- Up to 100 km cabled fiber
- Up to 2 km links

As a backbone network for medium performance LANs, the FDDI can be used to replace the IEEE 802.3, 802.4 and/or 802.5 standards for processor-to-processor interfacing or for building distributed processor networks [28,29]. In the mean time, also the low-cost, copper-based FDDI implementations have been in-ternationally standardized [30].

V. INTERNETWORKING

In industrial automation systems, different types of data transfer networks are re-quired to cover the communication and interfacing demands in process and pro-duction control, production planing and management, etc. Generally, at various automation levels various communication links are required: from the field bus at the field level up to the public network at the enterprise level. Moreover, the net-work participants (i.e. the computers and other intelligent terminals), wherever placed within a heterogeneous network structure, should be able to exchange the data with other participants and to access the data files placed anywhere within the network. For this purpose the *internetting* or *internetworking* approach is used, [28,31,32], where the subnetworks are individual networks within the complex network structure. Depending on the type of subnetworks, involved in a network, two basic network types are identifiable, the:

- Homogeneous networks
- Nonhomogeneous or hybrid networks

Internetworking homogeneous networks are equivalent to the physical extension of the basic subnetwork that is *repeated* in the total network structure. For this, *repeaters* are required, the interconnection facilities operating at the physical link layer level. The resulting network virtually becomes a single network.

Network repeaters reside at the physical level of OSI model and are used for internetworking homogeneous subnetworks. They can be viewed as *network extenders* because they simply amplify and transmit the received signals, in including their possible contentions, enlarging in this way the area of coverage of the subnetworks. They virtually make a multinetwork behave like a single network. The simplest application of a repeater is its use as a *bus extender* (Fig. 18), where the repeater is functionally involved in physical levels of both the bus and the bus extension. For complex network topologies the use of a repeater is a rather difficult task. Also the application of repeaters in fiber optic LANs is relatively complex task requiring the electro-optical and opto-electrical converters.

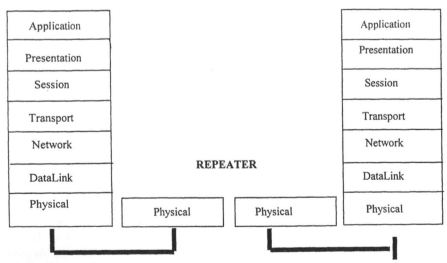

Figure 18   OSI repeater.

For internetworking the non-homogeneous subnetworks, depending on the type of individual subnetworks to be internetworked, the following facilities could be required:

- *Network Bridge*, operating at data link layer
- *Network Router*, operating at network layer
- *Network Gateway*, operating at higher layers

Network Bridges, residing at the data link layer, establish interconnections at the medium access control level. They basically take packets from one subnetwork and put them on another subnetwork, the physical media of interconnected subnetworks being electrically independent. Bridges are used both in homogeneous and heterogeneous networks with subnets having the same link control protocols, but different medium access control protocols. When used for internetworking, bridges take into consideration both the physical and the data link layer of subnetworks to be bridged (Fig. 19). This is required for reading the frames received from one station, say A, and for sending them, when designated, to the next, B station. The role of the bridge is to filter and forward traffic between the networks it bridges. The forwarding direction can be bidirectional, so that the bridge also must be capable of working bidirectionally. The bridge listens to the transfer activities an the subnetworks attached to it and checks the destination address of every packet received in order to determine the forward direction based on a list of addresses stored. However, although the bridges recognize the signals they receive and identify the packets, they analyze neither the structure of the packets nor their contents.

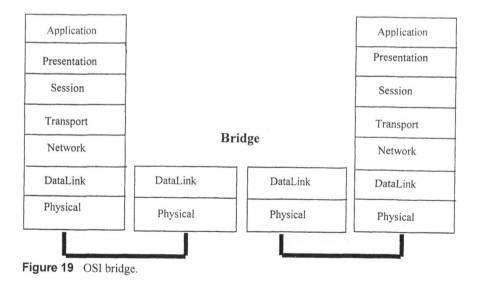

Figure 19 OSI bridge.

Two *routing algorithms* are most frequently used in network bridges:

- Transport bridge algorithm or spanning tree topology and
- Source routing bridge algorithm

In transport bridges the process of *forwarding* and of *learning* takes place. Both processes are required for proper functioning of the routing algorithm, based on source and/or destination address. In the source routing bridge the algorithm also takes account of the route that the frame should follow.

Network bridges, being important features of multinetwork systems, should at least meet some minimal requirements concerning the:

- Reliability
- Availability
- Security
- Error rate
- Transit time
- Frame life time
- Convenience and
- Performance

If the bridge interconnects two local area networks, it is called *local bridge*. Internetworking with the public area networks requires *remote bridges*.

Network routers (Fig. 20), operating at the network layer, provide the packets routing in nonhomogeneous networks having the same higher-level protocols. They are high-efficient interconnection means, that basically pass the packets from one network to another, taking care that the packets reach their correct destinations. Unlike the bridges, routers not only recognize the packets, but also analyze their structure to figure out the packet destination. For this purpose they are provided with the *routing tables* related to the structural description of the internetworked system, to be used for determination of appropriate routing by considering the transport layers of both networks.

The routing tables of the router can be *static* or *dynamic*. In both cases a number of possible, alternative routes should be indicated. The minimal requirements to be met by a router are the same as ones given above for the network bridge.

Analyzing the packets received from one subnetwork the routers decide to forward only the messages relevant to the other subnetwork. For instance, broadcast messages of one subnetwork will not be transferred to other subnetwork. This, and the like, reduces the traffic load on both subnetworks.

Network brouters, a symbiosis of network bridges and network routers, combine the best features of both of them. They simply route the packets they understand and bridge those they do not understand.

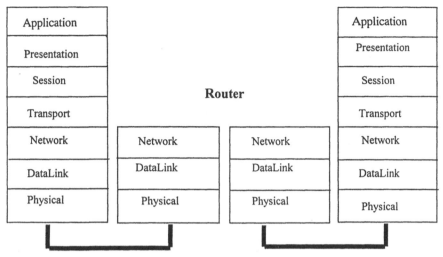

Figure 20   OSI router.

Network gateway, is an internetworking alternative, very similar to the router, the difference between them being that a gateway has more complex problems to solve than the router. This requires all higher levels of both networks to be considered, when implementing a network gateway. A gateway accepts messages from one (the sending) network, translates them (in the hierarchical order of protocol) into messages understandable for the other (the receiving) network, and forward them to this network. The gateway thus serves as a *host node* within the internetwork.

Network gateways basically interconnect dissimilar subnetworks by protocols translation at all hierarchical layers, including the application layer, using complex communication software. The most appropriate application is the internetworking between LANs and public communication networks.

REFERENCES

1. D Popovic. Concept in Data Transmission in Distributed Digital Control: In International Seminar on Distributed Control, New Delhi, 1995.
2. RH Grenier, GS Metes. *Enterprise Networking*. New York: Prentice Hall,1992.
3. B Jabbari et al. Network issues for wireless communications. IEEE Communications magazine, 33(1) 88–98, 1995.
4. NJ Muller. *Intelligent Hubs*. Artech House Publ., 1994.
5. A Santamaria, FJ Lopez-Fernandez, eds. *Wireless LAN Systems*. Artech House Publ. 1994.
6. D Popovic. Information Technology Pertaining to Process Control: Challenges and Strategies for Standardization , ISI Bulletin, 39:113-117, 1987.
7. W Stalling. *Handbook of Computer Communication Standards*. Indianapolis: Howard W. Sams & Company, Ind., 1987.

8. S Mirchandani, R Khanna, eds. *FDDI Technology and Applications.* New York: Wiley, 1993.
9. G Dickson, A Lloyd. *Open Systems Interconnection.* New York: Prentice-Hall, 1992.
10. MR Tolhurst, ed. *Open System Interconnections.* London: Macmillan Education Ltd, 1988.
11. DG Bake. *Local-Area Networks with Fiber-Optic Applications.* Englewood Cliffs, NJ: Prentice-Hall, 1986.
12. GE Kaiser. *Local Area Networks.* New York: McGraw-Hill, 1989.
13. N Chinone and M Maeda. Recent trends in fiber-optic transmission technology for information and communication networks. Hitachi Review 43 (2):41-46, 1994.
14. D Popovic, VP Bhatkar. *Distributed Computer Control for Industrial Automation.* New York: Marcel Dekker, Inc., 1990.
15. UD Block. *Data Communication and Distributed Network.* Englewood Cliffs, NJ: Prentice-Hall, 1987.
16. P Pleinevaux, J.-D. Decotigme. Time critical communication networks-field buses. IEEE Network 2 (3):55-63, 1988.
17. T Ozkul. *Data Acquisition and Process Control Using Personal Computers.* New York: Marcel-Dekker Inc., 1996.
18. A Ray. Distributed data communication networks for real-time process control. Chem. Eng. Communications 65 (3):139 -154, 1988.
19. GJ Blickley. Connecting the plant floor to management mainframes. Control Engineering 35(5):23-31, 1988.
20. B Tinham. Developing map for the process industries. C & I, 119-125,. 1987
21. MH Moris. Map 3.0 debuts at the enterprise networking event. Control Engineering 35(4):83-84, Ozkul,1988.
22. A Ray. Networking for computer-integrated manufacturing. IEEE Network 2(3): 40-47. 1988.
23. B Bates. *Client/Server Internetworking.* New York: McGraw-Hill, 1996.
24. A Berson. *Client/Server Architecture, 2nd ed.* New York: McGraw-Hill, 1996.
25. J Martin. *Client Server Application Development.* Prentice Hall International, 1993.
26. G Kaiser. *Optical Fiber Communications.* McGraw-Hill, New York, 1986.
27. D Popovic.Fiber-Optic Communication System in Industrial Automation. In: D. Popovic, ed., *Analysis and Control in Industrial Automation.* Vieweg, 32-49,1990.
28. J McConnel. *Internetworking Computer Networks—Interconnecting Network and Systems.* Englewood Cliffs, NJ: Prentice-Hall, 1988.
29. M Takahashi., et al. High-speed optical fibre network system for distributed computer control. Hitachi Review 35(4):183 -188, 1986.
30. H Westphal, D Popovic. The Role of FDDI in CAD/CAM Systems, 10th ISPE/IFAC International Conference on CAD/CAM, Robotics and Factories of Future, Aug. 21-24, Ottawa, Canada, 1994.
31. RP Davidson, NJ Muller. *Internetworking LANs: Operation and Design,* 1992.
32. DE Taylor Jr., ed. *The McGraw-Hill Internetworking Handbook, 2nd ed.* New York: McGraw-Hill, 1997.

6
Conceptual Design

Dobrivoje Popovic
University of Bremen, Bremen, Germany

A search for the most appropriate automation structure in production engineering and manufacturing has converged to distributed, hierarchically organized systems. Mechatronic systems have directly profited from the general advances in evolution of plant automation concepts, based on distributed processing and decision making possibilities. Mechatronic process in engineering design and product development has therefore benefited from these developments. This chapter introduces the reader into the conceptual design of distributed structure of production systems while the issues related to control and optimization of mechatronic processes are described in Chapter 13 of this book.

I. DISTRIBUTED SYSTEMS STRUCTURE

The evolution of *plant automation concepts*, from very primitive up to the most modern ones has followed closely the progress in instrumentation and computer technology [1,2]. Technological progress has, again and again, allowed computer vendors the possibility of updating their system concepts in order to meet the users' growing plant automation requirements. Emerging computer and communication technology [3] have by the 1980s encouraged the users to enlarge automation objectives in process control and to imbed them into the broad objectives of production and enterprise management. As a result, the *integrated automation concept* was created that encompassed the automation functions of the entire company [4]; plant and production control, production planning and scheduling, administrative, financial, and enterprise management functions, as shown in Fig 1. This was viewed as an opportunity to optimally solve some related problems, such as efficient resources utilization, production profitability, product quality, human safety, and environmental demands.

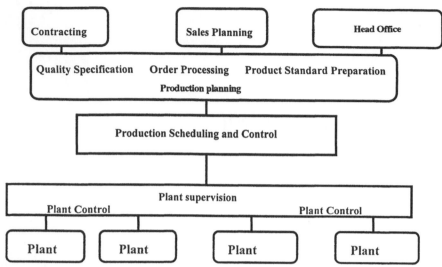

Figure 1   Integrated automation concept.

Contemporary industrial plants are inherently complex, large-scale systems imposing complex, mutually conflicting automation objectives to be simultaneously met. Effective control of such systems can only be feasible when for this purpose the adequately organized, complex, large-scale automation systems, like distributed computer control systems, are used [5,6], in which separate computers are assigned to individual automation functions (Fig. 2). This has since been recognized in steel production plants [7], where 10 million tons of steel are produced per annum, based on the operation of numerous work zones and the associated sub-systems such as:

- *Iron zone,* with coke oven, blast furnace, and palletizing and sintering plant
- *Steel zone,* with basic oxygen and electric arc furnace, direct reduction and continuous casting plant, etc.
- *Mill zone,* with the hot and cold strip mills, plate bore, and wire and wire rod mill, etc.

Here, the integrated automation concept includes, in addition to the above zones, the automation of:

- Laboratory services, such as
- Test field
- Quality control
- Analysis laboratory

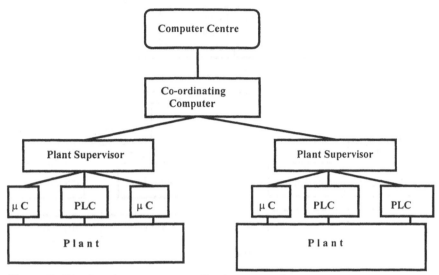

Figure 2 Distributed computer system for plant automation.

- Energy management center
- Maintenance and repair department
- Control and computer center
- *Plant management departments*, where all the required production and administrative data processing are carried out, statistical reviews prepared, and market prognostics data generated
- *Enterprise management departments*, such as finance, account, personnel, training department, fire and health services, etc.

The difficulties of satisfactory control and management of complex steel production plants are additionally augmented by adaptation needs to the randomly changing process and production parameters, this for example due to the quality variations in the raw materials or due to the possible fluctuations in the production line. This is to be understood in the following way: although the individual subsystems are specific batch processing plants, they are firmly incorporated into the downstream and upstream processes of the entire plant, which involves the coordinated schedule and control of the entire production process.

Computer-based manufacturing systems are also distributed systems (Fig. 3) in which separate computers execute individual automation functions [8]. The systems include most various functional units, delegated to different enterprise departments and distributed throughout the enterprise campus, such as:

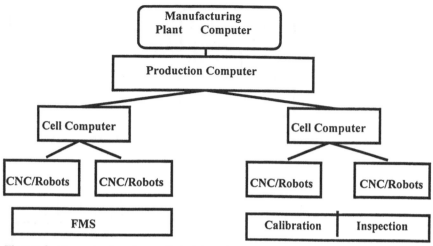

Figure 3 Computer-based manufacturing system.

- *Manufacturing process,* which includes the most various material processing operations, such as drilling, grinding, forging, machining, cold
  extrusion, etc.
- *Material handling,* responsible for automatic, computer-controlled moving of material parts in various stages of processing throughout the manufacturing plant
- *Assembly and finishing,* a sequence of operations carried out for integration of individually manufactured parts into a final product
- *Quality control and assurance,* in charge of planning, organization, and
  execution of observation and measurement steps required for individual
  quality inspection of manufactured parts and products

Also in computer-based manufacturing systems a number of auxiliary departments and services have to be integrated by the automation concept as follows:

- Product development
- Production planning
- Production scheduling and control
- Computer-aided design
- Inventory control

Contemporary automation concepts in manufacturing are mainly based on
the automation of:
- Flexible manufacturing systems (FMS)

- Computer-aided design (CAD)
- Computer-aided manufacturing (CAM)
- Business data processing (BDP)

Flexible manufacturing systems are themselves distributed production systems [9,10], consisting of a number of complementary and supplementary machines, the majority of them being of the CNC type, working under integrated computer control. The principal system units are (Fig. 4):

- *Machinery units*, containing the manufacturing equipment to which the numerically controlled machines, assembly and test facilities, etc. belong, organized in independent manufacturing cells as "islands of automation"
- *Workpiece handling units*, containing the facilities—such as robots, conveyors, computer-guided vehicles and other special transport utilities—that supply in a timely way unmachined workpieces from the storage to the machine systems units and transport the machined workpieces from the machine system units to their next destination
- *Tool handling units*, that take care that the right tools, particularly the cutting tools, are available in the right position and in the right time available, whereby the main function is to monitor tool wear during cutting by measuring the cutting force and power, tool temperature, vibrations, etc.
- *Transport and control units,* where the NC, CNC, PLC, and PC facilities and the production internal vehicles are placed

In flexible manufacturing systems robots, as flexible material handling tools for transport of work-pieces from one workplace to another, build cardinal links feeding the numerically controlled machines and cooperating with them. Besides, they are handling mechanisms for assembling and other operations, such as:

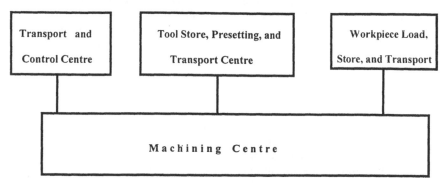

Figure 4 Flexible manufacturing system.

welding, painting, etc., so that the flexible manufacturing system is built out of a collection of machine tools fed by robots cooperating with other material handling facilities and coordinated and controlled by a distributed computer system. Consequently, special care has to be taken in planning, programming, and control of robot activities, particularly in collision-free trajectory planning and in synchronization of robot movements. Here, problems arise because there are various resources available out of which the appropriate ones have to be selected at the appropriate moment in order to guarantee the optimal use of all available facilities. In addition, each robot, each machine, each assembly station has its specific performance parameters, some typical capabilities and capacity limits, execution and cycles time, etc. Thus, the manufacturing system requires a computer support to manage all this.

The computer system required for the above manufacturing activities also has to coordinate control activities at the lowest, i.e. at machine level of the manufacturing plant, where the planning and control activities are related to the computer-based numerical control. Here, the programming and control tools are concentrated on direct control of movements of machine components, such as turning spindles, and on moving workpieces and tools along a programmed path, on changing the tools, on control of cutting fluid, etc. In contemporary Computer-based Numerical Control systems, off-line and on-line programmable microcomputers and microcontrollers are integrated parts of the systems, that increase the flexibility of control systems, increase their accuracy through higher sampling rate and adaptivity of control algorithms.

A computer-aided design unit comprises a computer group with the design software for development of product models, drawing, and other graphic forms, usually in an interactive mode with the designer. The software is a powerful tool for geometrical modeling of mechanical products and components using finite-element analysis, of strain, stress, and thermal distribution in structured objects. To the CAD unit belongs numerous electronic input/output devices, drawing and printing facilities, and interfaces to the production system for files exchange.

Computer-aided manufacturing, again, contains the computers and the associated devices required for supporting the manufacturing process in all its phases, from the process and production planning, production scheduling, up to the machining and quality control. The unit is seen as a component of a compact CAD/CAM unit, used for:

- NC, CNC, and robot control
- Geometric modeling
- Design analysis and optimization
- Design review and evaluation
- Documentation and drafting
- Simulation of manufacturing processes and systems
- Process planning and scheduling
- Manufacturing cell design

- Design of complex operations of metalworking
- Design of tools
- Product inspection for quality control

Business data processing includes activities, such as:

- Material resources planning
- Inventory management
- Stock control and forecasting
- Plant and machine maintenance
- Market analysis
- Manufacturing costs estimation
- Billing and shipping of products

The FMS, BDP, CAD, and CAM are viewed as four major "islands of automation," dispersed throughout the factory, needed to be mutually interconnected for data and program exchange and to be interfaced to the plant or factory computers in charge of coordinated supervision and overall control of process and production flow.

The distributed nature of a production system in which numerous machines and robots, control and communication devices, monitors and other intelligent man-machine interfaces distributed within the factory building and the enterprise campus, require for their harmonic cooperation an overall automation concept such as the one underlying the computer-integrated factory or enterprise automation.

Modern, computer-based systems for process and manufacturing plant automation have to be based on distributed structural concepts because of the distributed nature of industrial plants. In the plants, not only the control instrumentation is widely space-distributed, but also the departments to be integrated into the automation concept, so that the collection and processing of process, production, and management data requires distributed intelligence and a well-conceived hierarchical multicomputer architecture [3].

Multicomputer systems required for automation of distributed industrial plants have been in use in the last two decades. During this time, their architectural concepts and their elements have been steadily improved. The systems have also been better and better adapted to the problems they had to solve. This was mainly possible because of tremendous progress in hardware and software technology required for systems implementation, and because of increased awareness of the users that the new concepts and the emerging technologies can better solve their automation problems. Presently, a large number of such distributed computer control systems, launched by different vendors, are available on the market, including the powerful software that supports their enterprise-wide application.

II. HIERARCHICAL SYSTEMS ORGANIZATION

The remarkable development of automation concepts over the last decades was not solely a consequence of outstanding technical innovations, but to a considerable extent also a consequence of steadily increasing *market demands* for *high-quality products*. The manufacturers of such products, in order to compete successfully on the market, had to increase their efforts in achieving a possibly efficient production and product quality control. This, again, could only be guaranteed by a wide use of computer-based automation systems. Consequently, there was a general interest in development of such systems and of new, more advanced methods and tools for their implementation and practical use.

On the other side, the demands on more advanced automation technology has also motivated the instrumentation, control, computer, and communication engineers to work on new approaches for efficient solving the current automation problems. This has given a remarkable impetus to application oriented engineering disciplines, such as:

- Signal processing [11]
- Systems analysis [12,13]
- Model building and systems simulation [14,15]
- Adaptive and self-tuning control [16–18]
- Intelligent control [19–21]

In addition, an arsenal of hardware and software tools has been developed, comprising:

- Mainframe and microcomputers
- Personal computers and work-stations
- Parallel and distributed computer systems [22,23]
- Intelligent sensors, transducers and actuators
- Modular and object oriented software [24]
- Expert, fuzzy, and neuro-software [25]

All this has contributed to develop advanced automation systems, in which the most recent computer and communication architectures and the related methods of automation are operationally integrated.

Nevertheless, progress made in automation technology has adequately promoted progress in plant representation concepts. When considering progress made in automation technology one has also to consider simultaneous progress in automation concepts. This particularly holds for identification and precise definition of functions and activities within the plant to be automated in order to meet the required production objectives. The related functions and the activities have been identified, ordered, and classified according to the task they are executing. As a result, a general *hierarchical concept* of *functional levels* has been worked out, into which the automation tasks from the plant field level up to the enterprise management level have been included [7], [26], as shown in Fig. 5.

- *Direct process control level*, with process data collection and preprocessing, plant monitoring and data logging, open-loop and closed-loop control of process variables
- *Plant supervisory control level*, at which the plant performance monitoring, optimal, adaptive and coordinated control is placed
- *Production management level*, dedicated to production dispatching, inventory control, production supervision and rescheduling, and production reporting
- *Plant(s) and enterprise management level*, which tops all the activities within the enterprise, such as market and customers demand analysis, sales statistics, order dispatching, monitoring and processing, production planning and supervision

Although it would be possible to distinguish a greater number of functional levels, the experts have still decided that the specification of the above levels would be sufficient enough for a description of structural and functional features of any system in the processing industry, envisaged for integrated automation [27,28].

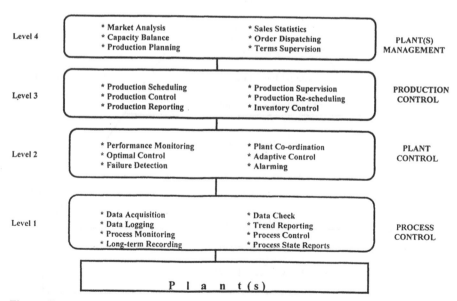

Figure 5   Hierarchical functional levels of processing plant.

In the manufacturing industry, slightly different functional levels have been identified, defining the following general hierarchical structure, shown in Fig. 6.

- *Machinery* or *control level*, i.e. instrumentation level, at which the machine parameter values are acquired through sensor elements and the actuating elements are energized
- *Cell* or *group level*, i.e. machine control level, at which the operation of a group of machines and processes within a manufacturing cell are mutually coordinated
- *Shop floor level*, i.e. process control level, at which the supervision and coordination of several manufacturing cells are executed
- *Plant level*, i.e. production monitoring level, at which the internal plant monitoring and control are accomplished, along with production scheduling and control
- *Enterprise level*, i.e. production planning level, at which the business data processing and the decision making activities are placed

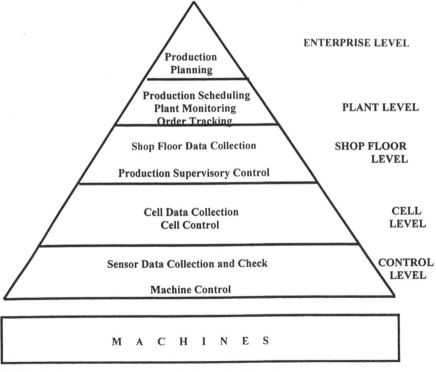

Figure 6 Hierarchical function levels of manufacturing plant.

The total distributed control system of Honeywell's TDC 2000 system (Fig. 7), which was a pioneering distributed computer control system was not hierarchically structured, but its control loops were distributed throughout the plant via some intelligent process input and output modules, designed for free programmable 8-loop controllers. The modules have been interfaced with the computer and with a data highway.

The successor of the TDC 2000, the TDC 3000 system, was already a real *distributed, hierarchically organized system* in which *all* hierarchical levels of plant automation have been built, mainly based on the following subsystems (Fig. 8):

- Various distributed process control facilities, such as
- Basic controllers (BC)
- Single strategy controllers (SSC)
- Multi-function controllers (MC)
- Programmable controllers (PC)
- Universal and enhanced operator stations (US and EOS)
- History, application, and computing modules (HM, AM, and CM)
- Data highways as communication links
- Local gateways as internetworking elements

Since that time a large number of distributed computer systems have been launched on the marketplace [7].

For implementation of distributed computer control systems, besides the computers and their peripherals, various interfaces and communication links are required for exchanging data between the individual system parts. Moreover, due

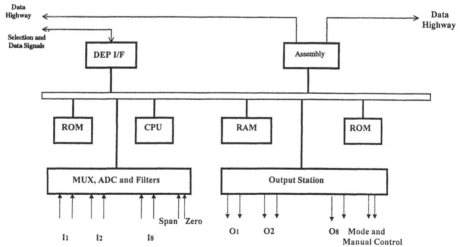

Figure 7 TDC 2000 system of Honeywell.

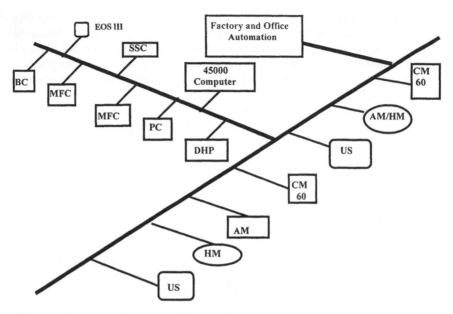

Figure 8 TDC 3000 system of Honeywell.

to the hierarchically organized automation functions of different operational complexity and the complexity of data to be exchanged at different transfer rates, *different types* of data communication networks are needed at different hierarchical levels, for instance (Fig. 9):

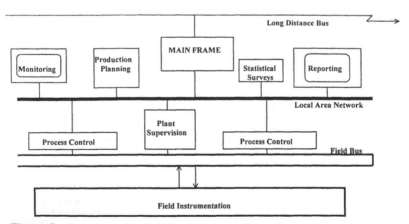

Figure 9 Communication links in plant automation systems.

- *Field level* requires—for collection of sensor data and for distribution of actuator values—an appropriate communication link, such as a field bus, capable to manage hundreds and thousands of end devices
- *Process* and *production control level* requires—for interfacing programmable controllers, supervisory computers, and the relevant monitoring and command facilities—a high-performance bus system, such as a token bus
- *Production* and *enterprise management* levels requires—for data and program exchange with the computers, workstations, and other intelligent terminals distributed within the factory building and enterprise campus—a high-performance local area network as a system interface, and for data exchange with the remote terminals a long-distance communication link

In computer-aided manufacturing systems, [29–31], apart from internationally standardized communication networks, the MAP and TOP standards take an important position, specifying the communication links required for *factory integrated automation systems*. Although it is difficult to present a general architecture of a MAP/TOP-based distributed, hierarchically organized multicomputer system, the system configuration, shown in Fig. 10, seems to be the most typical one [7]. Also here, like in the distributed computer systems for process plant automation, each functional level requires an adequate communication network. For instance:

- *Machinery* or *control level* requires—for interconnection of sensing, actuating, and display elements to their controllers, as well as for controllers parameter tuning and set points provision—a real-time bus such as field bus
- *Cell* or *group level* requires—for sequencing batch jobs of similar work pieces of products, for supervising various material handling, test, and calibration devices—a local area network or a MINI-MAP
- *Shop floor level* requires—for interconnection of device controllers and for data exchange between them—a MAP bus system as a *backbone bus* to which various devices and bus systems are attached
- *Plant level* requires—for production scheduling, monitoring and control, as well as for test, calibration, and inspection services—interface to the MAP *backbone bus*
- *Enterprise level* requires—for exchange of business and production planning data the interface to a TOP bus system and for data exchange with the customers, deliverers, and corporate extensions—a wide area network or an interface to public network

Presently, the majority of commercially available distributed computer systems for plant automation use at all communication levels the most internationally standardized networks. This simplifies the problem of mutual compatibility of devices delivered by different computer and instrumentation manufacturers, and

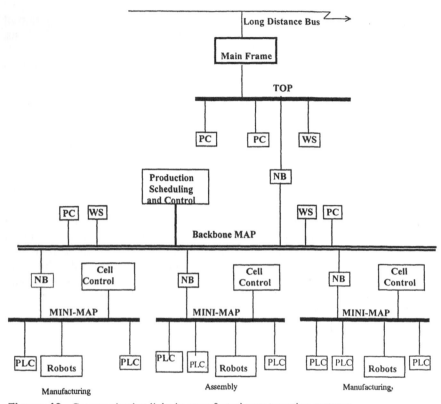

Figure 10 Communication links in manufacturing automation systems.

facilitates the design of powerful, low-cost multicomputer systems by integrating the subsystem with highest performance-to-price ratio.

Although there are a vast number of different communication standards used in design of commercially available distributed computer control systems, a comparative analysis shows that:

- *Small-scale* plant automation systems, managing only the field and the process control level, basically use the bus-based communication links (Fig. 11), like a field bus, and for higher level automation purposes a suitable link to the mainframe
- *Medium-scale* plant automation systems, additionally managing the production planning and control level, are local area network oriented and can include, for data exchange with the host computer, a long-distance bus or a bus coupler (Fig. 12)

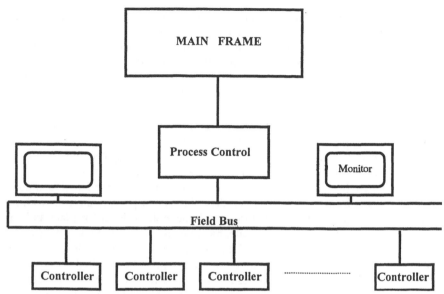

Figure 11   Small-scale plant automation system.

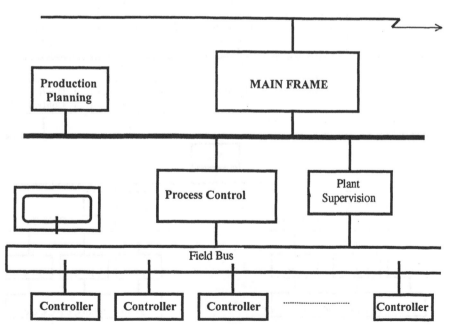

Figure 12   Medium-scale plant automation system.

• *Large-scale* plant automation systems designed for *integrated plant automation* require various communication facilities, such as local area networks, and a number of bus couplers, network bridges, network routers, and network gateways. When applied in manufacturing plant automation, the systems could even incorporate different MAP/TOP-based *backbone buses* and local area networks (Fig. 13).

The standardization of the OSI model of ISO has facilitated the design of mutually compatible system interfaces, so that the hardware from different computer and instrumentation manufactures can easily be integrated into a system. This enables the designer to optimally select the hardware subsystems.

Although the computer manufacturers design their distributed computer control systems for a wide application, they still cannot provide the user with *all* facilities and *all* functions and required at *all* hierarchical levels of plant and enterprise automation. The user, as a rule, has to plan his distribution system in the way required for the specific application, taking into account possible later system extensions. Thus, the final configuration of the system to be installed is essentially defined by the user, whereby the user has, above all, to clearly formulate the premises under which the system has to be built and the functions that should be implemented. This can be done [13] when knowing that the system should:

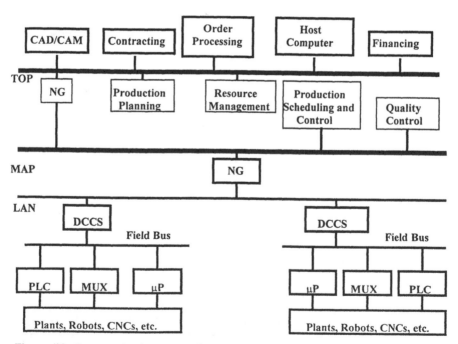

Figure 13 Large-scale plant automation system.

- Cover all functions of process monitoring and control and enable the plant operator an optimal interaction with the plant via sophisticated man-machine interfaces
- Offer a transparent view into the plant performances and the production status
- Provide the plant management with the extensive up-to-date production reports, as well as with the statistical and historical reviews of collected data
- Improve plant performances by minimizing the learning cycle and start-up and set-up trials
- Enable easier replacement of raw materials currently used by new, less costly ones
- Permit faster adaptation to the market demands
- Implement the basic objectives of plant automation: increase the production volume and product quality, decrease the production cost, and improvement of work conditions

Based on the above premises, the distributed computer control system should include a rich spectrum of *hardware features,* such as:

- At control level:
    attachment possibility for most types of sensors, transducers and actuators
    reliable and explosion-proof installation
    hard-duty and fail-safe version of control units
    on-line system reconfiguration possibility with a high degree of system expandability
    guaranteed further development of control hardware in the future by the same vendor extensive provision of on-line diagnostic and preventive maintenance features
- At supervisory and production level:
    wide choice of interactive monitoring options, designed to meet the required industrial standards
    multiple computer interfaces to integrate different kind of servers and workstations using internationally standardized bus systems and local area networks
    interfacing possibilities for various external data storage media (floppy discs, streamer, etc.)
- At management level:
    wide integration possibilities of local and remote terminals and workstations.

In addition, the system should include a rich spectrum of *special software packages*, particularly:

- At control level:

    a full set of pre-processing, control, alarming, and calculation algorithms for measured process variables that is applicable to a wide repertoire of sensing and actuating elements

    a versatile display concept with a large number of operator friendly facilities and screen mimics

- At supervisory level:

    wide alarm survey and tracing possibilities, instantaneous, trend, and short historical reporting features that include the process and plant files management

    special software packages and block oriented languages for continuous and batch process control and for configuration of plant mimic diagrams, model building and parameter estimation options, etc.

- At production level:

    efficient software for on-line production scheduling and re-scheduling, for performance monitoring and quality control, for recipe handling, and for transparent and exhaustive production data collection and structured reporting

- At management level:

    professional software for production planning and supervision, order dispatch and terms check, order and sales surveys and financial balancing, market analysis and customers statistics etc.

It is extremely difficult to completely list all operational functions and services important for powerful use of multicomputer systems in plant automation. Thus, the aspects summarized above present only the most important ones. They will be subject of a detailed discussion in the following paragraph.

III. FUNCTIONAL LEVELS

The design concept of hardware and software of a specific distributed computer control systems for industrial plant automation has to follow strongly the specification of automation problems the system has to solve [32,33]. The specification should be the result of a previous extensive *plant analysis*, carried out by the users' experts in automation in cooperation with the plant operators and plant managers. Should the design be aimed at integrated plant and enterprise automation, cooperation with the staff of the commercial departments to be integrated into the automation system is also required. Furthermore, in order to properly select the hardware and software required for implementation of automation functions at all hierarchical levels, a detailed specification of such functions and of related services should be prepared [26]. In the following, a review of the most essential functions to be implemented at individual plant automation levels is given (see Fig. 5).

At the field instrumentation level [1,34], where the field mounted devices and the related interfaces are placed for plant data collections, plant monitoring and control the details should be specified for:

- Sensors, actuators, and field controllers to be connected to the system, their type, accuracy, grouping etc.
- Acoustic and optical alarm annunciators
- Back-up concept to be used
- Digital displays and binary indicators to be installed in the field
- Completed plant mimic diagrams required
- Keyboards and local displays, handpads, etc.
- Analog and digital signal conditioners
- Physical transducers for process parameters
- Signal and noise removing filters
- On/off drivers for blowers, power supplies, pumps etc.
- Status indicators such as closures
- Pulse generators and counters
- Multiplexers
- Local indicating and display devices

At process control level the details should be specified concerning the:

- Individual control loops to be configured, including their parameters, sampling and calculation time intervals, reports and surveys to be prepared, fault and limit values of measured process variables, etc.
- Structured content of individual logs, trend records, alarm reports, statistical reviews etc.
- Detailed outlay of mimic diagrams to be displayed
- Actions to be effected by the operator
- Type of interfacing to the next higher priority level exceptional control algorithms to be implemented

At this level, the *data acquisition functions* should be specified that include the operations needed for sensor data collection. They usually appear as initial blocks in an open-loop or closed-loop control chains, and represent a kind of interface between the system hardware and software. In the earlier process control computer systems, the functions have been known as *input device drivers* and have usually been a constituent part of the *operating system*. To the functions belong:

- Analog data collection
- Thermocouple data collection
- Digital data collection
- Binary/alarm data collection

- Counter/register data collection
- Pulse data collection

As parameters, usually the input channel numbers, amplification factors, compensation voltages, conversion factors, etc. are to be specified. The functions can be executed *cyclically* (i.e., *program controlled*) or in *event-driven* (i.e., *interrupt controlled*) mode.

Next to the data acquisition functions the *algorithms* should be specified, required for collection and processing of sensor data and for calculation of control loops set point values and other command values for the devices and elements situated in the plant field. Here belong the algorithms for:

- Input signals conditioning
- Signals test and check
- Basic and advanced control algorithms
- Dynamic compensations
- Output signals conditioning

Input signal conditioning algorithms are mainly used for preparation of acquired plant data, so that the data can—after being checked and tested—directly used in computational algorithms. Due to the fact that the measured data have to be extracted from a noisy environment, the algorithms must include features such as separation of signal from noise, determination of physical values of measured process variable, decoding of digital values, etc.

Typical signal conditioning algorithms are:

- Local linearization
- Polynomial approximation
- Digital filtering
- Smoothing
- Bounce suppression of binary values
- Root extraction for flow sensor values
- Engineering unit conversion
- Encoding, decoding and code version

Test and check functions are compulsory for faultless application of available calculating algorithms which always have to operate on true values of process variables. Any error in sensing elements, in data transfer lines, or in input signal circuits delivers a false measured value which, for instance applied to a control algorithm, can lead to a false, even to a catastrophic control situation. On the other hand, all critical process variables have to be continuously monitored, e.g. checked against their *limit values* (or *alarm values),* whose crossing certainly indicates the *emergency status* of the plant. This should immediately be communicated to the plant operator via a display or hard copy log.

For the above reasons the *test and check algorithms* play an essential role in each distributed computer control system. They usually include:

- Plausibility
- Sensor/transmitter
- Tolerance range
- Higher/lower limit
- Higher/lower alarm
- Slope/gradient
- Average value

As parameters, the higher/lower values for the above tests have to be specified in engineering units. During plant monitoring, they are subject to possible changes via the operator console. As a rule, most of the anomalies detected by the described functions are, for control and statistical purposes, automatically stored in the system, along with the instant of time they have occurred.

Dynamic compensation algorithms are needed for specified implementation of control algorithms. Typical compensation algorithms are:

- Lead/lag compensation
- Dead time compensation
- Differentiation
- Integration
- Moving average building
- First-order digital filtering
- Sample-and-hold function
- Velocity limitation

The constants, present in the individual functions, have to be specified as related parameter values after the control or monitoring loops have been configured.

Basic control algorithms mainly include different versions of PID algorithm such as:

- PID, PD, PI
- PID-ratio
- PID-cascade
- PID-gap
- PID-auto-bias
- PID-error squared

As parameters, the values of proportional gains, integral resets, derivative rates, sampling and control intervals, etc. of individual control loops have to be specified after the loop configurations.

Advanced control algorithms, required for improved control of individual process parameters, include:

- Feedforward
- Predictive
- Deadbeat
- State-feedback
- Adaptive
- Self-tuning

Output signal condition consists in adapting the calculated output values (i.e. set point, actuating, and other values) to be transmitted to the final elements in the field. The adaptation includes:

- Calculation of full, incremental or percentage values of output signals
- Calculation of pulse width, rate or number of pulses for outputting
- Bookkeeping of calculated signals, lower than the resolution of final elements
- Monitoring of end-values and speed saturation of mechanical, pneumatic and hydraulic actuators

The output values could be analog, digital, and pulse values (i.e. the pulse width, pulse rate, and/or the number of pulses).

At the plant supervisory level the functions are to be defined, related to the supervisory plant control that includes:

- Plant status monitoring and plant alarm management
- Optimal and adaptive process control
- Batch process control
- Recipe handling
- Plant energy management

Plant monitoring and *plant alarm management functions* belong to the operational facilities. The plant operator monitors are provided for easy plant surveillance and for alarm detection, localization, and removal by manual interventions. Here, the decision should be made concerning the:

- Monitor screen appearance by defining the areas to be present (e.g. the message, overview, main, and command area)
- Display modules, to which the standard modules (such as plant, area, group, loop overview, etc.) belong as well as the user defined modules (such as plant, area, and group mimic diagrams)

Optimal process control, the advanced control strategy [35], based on:

* Mathematical process model
* Optimization method and the performance index selected
* Set of controllable process variables is a basic approach for optimal product quality achieved by optimal use of energy and raw materials. In engineering practice
* Throughput optimal control
* Energy optimal control
* Cost optimal control
* Optimal quality control is favored, related to the *least squares error* as *performance index* to be minimized

Adaptive process control is used for implementation of optimal control under severe internal and/or external conditions, mainly for automatic compensation of unavoidable drift of process parameters and for unpredictable environmental changes. For adaptive control, the trace is steadily kept on dynamic process behavior by *process parameter estimation,* and by subsequent compensation of corresponding controller parameters with the objective of keeping the parameters of optimally tuned control loops constant.

In modern control theory, the term *self-tuning control* is used for adaptive control in a noise poisoned environment. In a self-tuning system control parameters are based on measurements of system input and output, automatically tuned to result into a sustained optimal control. Parameter tuning itself can be affected by the use of measurement results to estimate actual values of system parameters and, in the sequence, to calculate the corresponding optimal values of control parameters.

Batch process control is basically the sequential, well-timed *stepwise control* for achievement of optimal operating conditions of batch processes. The control generally includes some *binary state indicators,* the status of which is taken at each *control step*—in addition to a preprogrammed time interval—as a decision support for the next control step to be made. The functional modules required for configuration of batch control software are:

* *Timers,* to be preset to required time intervals or to the real-time instants
* *Time delay modules,* time-driven or event-driven, for delimiting the control time intervals
* *Programmable up-count* and *down-count timers* as time indicators for triggering the pre-programmed operational steps
* *Compactors* as decision support in initiation of new control sequences
* *Relational blocks* as internal message elements of control status
* *Decision tables,* defining for specified input conditions the corresponding *Output conditions* to be considered

In a similar way *recipe handling* is carried out. It is also a batch-process control action, based on stored recipes to be downloaded from a mass storage facility containing the completed *recipes library file*. The handling process is under the competence of a *recipe manager*, a batch-process control program.

Recipe handling basically requires a strong operator's interaction, accomplished via a monitoring and command facility, needed for defining the raw material and final product tanks, reactor vessels to be used, dosing strategy to be applied, the total product quality required etc.

Energy management software takes care that all available kinds of energy (electrical, fuel, steam, exothermic heat, etc.) are optimally used, and that the short term (daily) and long term energy demands are predicted. It continuously monitors the generated and consumed energy, calculates the efficiency index, and prepares the relevant *costs* re*ports*. In optimal energy management strategies and methods are used that are familiar in optimal control of stationary processes.

For implementation and execution of functions at all hierarchical levels listed above, the configuration of overall *computer software* is substantial [24,36]. In this respect computer vendors have equipped their distributed computer control systems with a large number of different *software packages* [37] classified as:

- *System software*, i.e. computer oriented software as a set of tools for development, generation, test, run, and maintenance of programs to be developed by the user
- *Application software*, to which the monitoring and control loop configuration, and communication software belong

System software is a large aggregation of different *compilers* and *utility programs*, serving as systems development tools. They are used for implementing functions, not implementable by any combination of program modules stored in the library of functions. When developed and stored in the library, the application programs extend their content and allow more complex control loops to be configured. Although it is, at least in the principle, possible to develop new programmed functional modules using any programming language available, in process control systems the use of *high-level programming languages* such as *real-time languages* and *process oriented languages* are preferred.

Real-time programming languages are favored as support tools for implementation of control software because they provide the programmer with the necessary features for sensors data collection, actuators data distribution, interrupt handling, and for programmed real-time and difference-time triggering of actions. Real-time FORTRAN is an example of this kind of high-level programming language.

Process oriented programming languages, go one step further in supporting even the planning, design, generation, and execution of application programs (i.e., their tasks). The major characteristic of such languages is *multitasking*.

Process oriented programming languages are higher-level languages with *multitasking* capability, to enable the programs, implemented in such languages, to be simultaneously executed in an interlocked mode so that a number of real-time tasks are executed synchronously, both in time or event driven mode. The two best known process oriented languages are:

- *Ada*, capable to support implementation of complex, comprehensive system automation software in which for instance the individual software packages, generated by the members of a programming team, are integrated in a cooperative, harmonic way
- *PEARL* (process and experiment automation real-time language), particularly designed for laboratory and industrial plant automation, where the acquisition and real-time processing of various sensor data are to be carried out in a multitasking mode

In both languages, a large number of different kind of data can be processed, and a large-scale plant can be controlled by decomposing global plant control problem into a series of small, well-defined *control tasks* to run *concurrently*, whereby the start, suspension, resumption, repetition, and stop of individual tasks can be preprogrammed, i.e. planned.

At the production management level a vast quantity of information should be provided not familiar to the control engineer, such as information concerning:

- Production scheduling
- Production monitoring and control
- Production reporting
- Inventory control

Much of this is to be specified in a structured, alphanumeric or graphic form, this because—apart from the data to be collected—each operational function to be implemented needs some *data entries* from the lower neighboring layer, in order to deliver some *output data* to the higher neighboring layer or vice versa. The data themselves, irrespective of their origin, have to be well-structured and organized in *data files* for better management and easier access. This holds for data at all hierarchical levels, so that in the system at least the following data files are to be built :

- *Plant data files*, containing the parameter values related to the plant
- *Instrumentation data files*, where the technical data are stored related to the individual final control elements and the equipment placed in the field
- *Control data files*, mainly comprising the configuration and parameterization data, along with the nominal and limit values of the process variable to be controlled

- *Supervisory data files* required for plant performance monitoring and optimal control, for plant modeling and parameter estimation, as well as production monitoring data
- *Production data files* for accumulation of data relevant to raw material supplies, energy and products stock, production capacity and actual product priorities, for specification of product quality classes, lot sizes and restrictions, stores and transport facilities etc.
- *Management data files*, for keeping track of customer orders and their current status, and for storing the data concerning the sales planning, raw material and energy resources status and demands, statistical data and archived long-term surveys, product price calculation factors etc.

Consequently, through process analysis a large number of plant, production, and management relevant data should be collected, a large number of appropriate algorithms and strategies selected, and a considerable amount of specific knowledge by interviewing various experts elucidated, before the required volume and structure of the distributed computer system can be finalized. In addition, experience shows that good system design implies good cooperation between the user and the computer system vendor, because at this stage of the project the user is not quite familiar with the vendor's system, and because the vendor should on user's request implement some particular algorithms, functions, or strategies, not available in the standard system version. This can even be more complicated if the user has not definitely decided what distributed system he would prefer, so that the cooperation with different vendors is needed.

It is a substantial final work step, after finishing system analysis, to entirely *document* the results achieved. This is particularly important because of the complexity of plants to be automated and because of hierarchical distribution of functions to be implemented. For this purpose, detailed instrumentation and installation plans should be worked out using standardized symbols and labels. This should be completed with the *list of control and display flow charts* required. The programmed functions to be used for *configuration* and *parameterization* purposes should be summarized in a tabular or matrix form, using the *fill-in-the blank* or *fill-in-the-form* technique, *ladder diagrams*, *graphical function charts*, or in special *system description languages*. This will certainly help the system designer to better tailor the hardware and the system programmer to better style the software of the future system.

At the enterprise management level the central computer system is situated. It consists of a number of interconnected computers and computer-based terminals executing specific automation functions distributed within the plant. Typical functions located at this level are:

- Market and customer analysis
- Order and sales statistics
- Order dispatching

- Delivery terms supervision
- Order dispatching and sales promotion strategies
- Production planning
- Production capacity and order balance
- Price calculation guidelines
- Productivity and turnover control
- Financial survey

For manufacturing plants, planning the distributed computer control systems should encompass the automation functions summarized below.

At the *machinery* or *control level*, where mainly the sensor data acquisition and actuating values distribution are located, the automation functions required are identical with those required at the field instrumentation level of plant automation.

At *cell* or *group level*, where, in addition to the programmable logic controllers, the workstations and supervisory facilities also are concentrated. They are responsible for sequencing batch jobs, material handling and calibration, various functions that are required for:

- Optimal task decomposition
- Optimal resource selection
- Job progress and system status reporting
- Dynamic batch routing decision making
- On-line rescheduling
- Task and process monitoring

The design of modern manufacturing cells relies on the concept of *group technology*, dealing with the problem of simultaneous processing of some families of parts on a group of NC machines, this in order to minimize the intercellular material handling cost. The individual NC machines of the cells are equipped with some part buffers, tool and pallet changers, and are supported by robots or by automated guided vehicles. Thus, robot control is of vital importance at this automation level. For this purpose the following should be specified [38]:

- *Robot path generation* and *robot motion control* needed for servicing the NC machines and for material handling
- *Robots task scheduling* in the multirobot environment, particularly the dynamic job scheduling for achieving the manufacturing objectives despite the robot or machine failures
- *Robots synchronization* by just-in-time use of relevant robot actions planned for the actual task execution and prediction of possible mutual collision of robots in multirobot environment
- Robots state monitoring, failure detection, and troubleshooting, required for robot tasks rescheduling

At the *shop floor level* the automation functions are concentrated that coordinate the production and support jobs on the shop floor. They are responsible for allocation of resources to jobs, for achieving the right production time of product quantities of specified quality, notwithstanding the random events such as the machine failures, machine tool damages, breakdown of equipment, shortage of parts and supplies, etc. For this purpose, the functions are administrated by:

- Task manager
- Resource manager

The support functions to be incorporated into this hierarchical level are shop-floor data based and are able to balance production planning data and data collected from shop-floor sensors, status reporting signals, and alarm messages. The collected data are also used for:

- Scheduling of job orders on the work groups
- Machine loading and job sequencing
- Preparing the order status reports

In manufacturing practice the number of jobs generally exceeds the number of processing groups, so that the arriving jobs have to queue before being scheduled. For the processing of jobs queue the well-known selection rules are used: first-input, first-output (FIFO), earliest due date (EDD), shortest processing time (SPT), critical time ratio (CTR), etc.

At *plant level* the functions are located that support:

- Factory automation
- Manufacturing planning and control
- Product and process management

Factory automation functions encompass the activities needed for integration of manufacturing functions within individual areas of shop floor level. These activities are as follows:

- Material processing
- Assembly
- Material handling
- Inspection and test

Manufacturing planning and control functions primarily enclose the shop floor related material and facility planning and job scheduling. The functions also control the manufacturing and report on its status.

Product and process management functions generally help design, simulation, analysis, and documentation of product and processes.

At the *enterprise level* the majority of enterprise management functions are accommodated related to:

- Marketing
- Strategic planning
- Manufacturing and resources management
- Financing

Manufacturing systems that include all functional levels discussed above, are known as *computer integrated manufacturing systems*, or CIM systems. The idea of introducing such system was to integrate the available resources, technologies, and business processes with the objective to meet the specific customer demands and general market requirements. The integration was a result of market compulsion to manage thousands of different activities and to process thousands of data files needed for mass customization of products. It was urgently required for improving the overall system performances by coordinating and harmonizing individual activities with overall enterprise goals. This should support the global production optimization by integrated decision making and control actions, as well as by integration of information flow and product and process data.

Computer integrated manufacturing is strongly dependent on databases available at individual functional levels of the manufacturing factory that are generally clustered into:

- Machines related databases
- Tools related databases
- Parts related databases
- Material handling system related databases

Machines related databases include the details concerning the status of NC and other machines and the accompanying equipment, machine parameters, command and control routines, trouble shooting, diagnostics, and maintenance, etc.

Tools related databases substantially contain the databases on status and life of all classes of tools available.

Parts related databases incorporate the information concerning the status, priority, quality, etc. of the parts being in the manufacturing process.

Material handling system related databases, finally, hold mostly the relevant status, commands, and diagnostics data of the material-handling equipment.

Recently, neural networks have been proposed for implementation of *intelligent manufacturing* [39,40].

IV. INTERLEVEL COORDINATION

Generally speaking, in a distributed computer control system the automation functions are predominantly stored there where they are needed, i.e. the functions needed at a hierarchical level, are also stored at that level. However, to activate and run the functions at a level at a certain time, relevant *initial data* are needed on which the function should operate. Such data are as a rule provided by the neighboring levels, from their databases. In this way the system functions and the relevant data, distributed across the automation levels, are mutually interconnected [41]. This represent the basic interlevel coordination of automation activities. In a similar way the data, generated by the functions at one functional level and required there, are stored at that level. For instance, data required for direct control and plant supervision are allocated at the field level, i.e. next to the plant instrumentation, whereby the data required for production control are allocated at a higher level, close to the plant manager. The stored data are, however, used for interlevel coordination in the following way: a *selected data set* of every level is transferred to its neighboring levels, needed there for initiation and run of automation functions.

Of course, the organization of data within a hierarchically structured system requires—for efficient generation, access, updating, protection, and transfer of data between different files and different hierarchical levels—a flexible *data management system*. This is of fundamental importance because data is basically organized in files, assigned to some relevant databases distributed within the system. Thus, the concept of database structure, the local and/or global relevance of each database, and the procedures of their programmed access is of foremost importance.

In distributed computer control systems, because the data is hierarchically organized, the concept of *data exchange* between the individual hierarchical levels is also essential. This exchange takes place in both directions, i.e. "up" and "down" the hierarchy. However, the intensity of the data flow in the "upward" direction decreases, and in the opposite direction increases because, for instance, the sensor data are less important for higher hierarchical levels and the customer data less important for the lower hierarchical levels. On the other hand, the data exchange rate between the "lower" hierarchical levels is higher, and the response time shorter than between the "higher" hierarchical levels because the automation functions of lower levels execute their *real-time tasks*, whereas those of the higher levels execute some planning and scheduling tasks.

The content of individual database units (Fig.14) is basically determined by their position within the hierarchical system. For instance, the database unit at the process automation level (Fig. 15), situated at the process control level, contains the data necessary for data acquisition, preprocessing, and check, for plant state monitoring and alarming, for open and closed-loop control, and for reporting and logging. Beside this, the database unit contains, as long term data, the specifications concerning the loop configuration and the parameters of individual functional

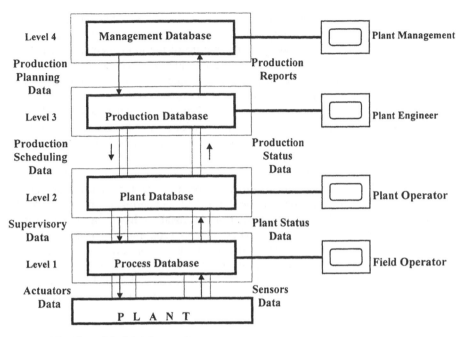

Figure 14 Hierarchical database units.

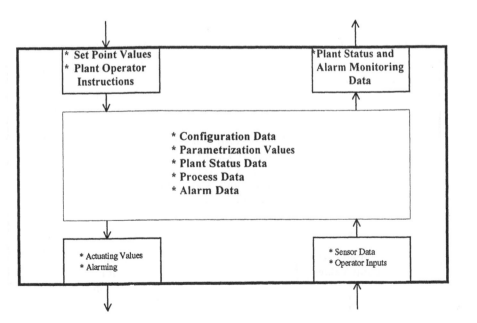

Figure 15 Database unit of process automation level.

blocks used. As short-term data it contains the measured actual values of process variables, the set-point values, calculated output values, and the received plant status messages, etc. Depending on the nature of the implemented functions, the origin of collected data, as well as the destination of generated data, the database unit at this level has to be efficient under real-time conditions, in order to handle a large number of short-life data, having a very fast access.

To the next "higher" hierarchical level only some actual process values and plant status messages are forwarded, along with a short history of some selected process variables. In the reverse direction, calculated optimal set-point values for controllers are to be respectively transferred.

In order to meet the required real time data access conditions, at this hierarchical level a semi-conductor memory is preferable as the database storage medium. Usually, temporary data is stored in RAM, and permanent data in ROM. For the purpose of a safe system restart, an initial database has to be stored on ROM and loaded into RAM after restart, where it can be continuously be updated.

The database unit, situated at the plant automation level, contains data concerning the plant status, the basis of which the monitoring, supervision, and operation of plant is carried out (Fig. 16). As long-term data, the database unit contains the specifications concerning the available standard and user-made displays, as well as

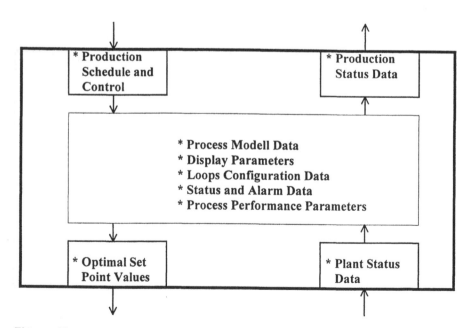

Figure 16 Database unit at plant automation level

data concerning the mathematical model of the plant. As short-term data the database contains the actual status and alarm messages, calculated values of process variables, process parameters, and optimal set-point values for controllers. At this hierarchical level, a large number of data is stored whose access time should be with in a few seconds. Here, some calculated data has to be stored for a longer time (historical, statistical, and alarm data), so that for this purpose hard disks (or at least floppy-disks) are used as back-up storage. To the "higher" hierarchical level, only selected data is transferred for production scheduling and directives are received.

The database unit at the production automation level (Fig. 17), contains data concerning the products and raw material stocks, production schedules, production goals and priorities, lot sizes and restrictions, quality control as well as the store and transport facilities. As long-term data the archived statistical and plant alarm reports are stored on bulk memories. The data access time is here in no way critical. To the "higher" hierarchical level, the status of the production and order processing, as well as of available facilities necessary for production re-planning is sent, and in the reverse direction the target production data.

The database unit at the enterprise automation level finally, situated at the corporate or enterprise management levels (Fig. 18), contains data concerning the customer orders, sales planning, product stocks and production status, raw material and energy resources and demands, status of store and transport facilities, etc. Data stored here is long-term, with the access needed each few minutes up to many

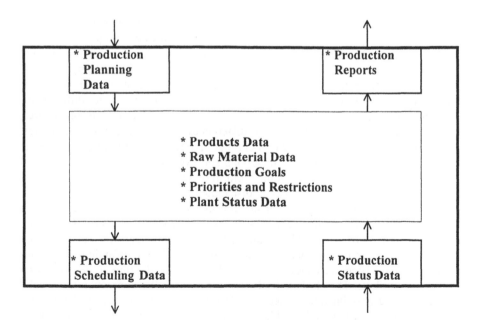

Figure 17 Database unit at production automation level.

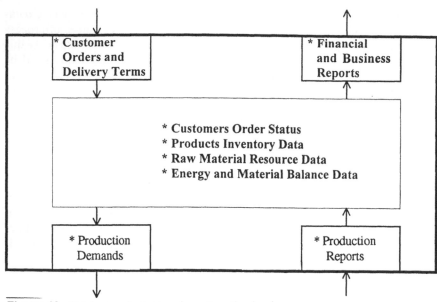

Figure 18 Database unit at enterprise automation level.

weeks. For this reason, a part of database can be stored on magnetic tapes, where it can be deposited for many years for statistical or administrative purposes.

The fact, that different databases are created, placed at different hierarchical levels and stored where they can be deposited for years in different computers, and that they are administrated by different database management systems or by different operating systems, makes the access of any hierarchical level difficult. Inherent problems here are: the problems of formats, log output procedures, concurrency control and other logical differences concerning the data structures, data management languages, label incompatibilities, etc. In the meantime, some approaches have been suggested for solving some of the problems, but there is still much creative work to be done in this field in order to implement a flexible, level-independent access of any database in distributed computer system.

Another problem, typical for all time-related databases like, the real-time and production management databases, is the representation of *time related data*. Such data has to be integrated into the context of time, a capability that the conventional database management systems do not have. In the meantime, numerous proposals have been made along this line which include the time to be stored as a universal attribute. The attribute itself can, for instance, be transaction time, valid time, or any user-defined time. Recently, four types of time related databases have been defined according to their ability to support the time concepts and to process temporal information:

- *Snapshot databases*, i.e. databases that give an instance or a status of the data stored concerning the system (plant, enterprise) at a certain instant of time, but not necessarily corresponding to the current status of the system. By insertion, deletion, replacement and similar data manipulation a new snapshot database can be prepared, reflecting a new instance or state of the system, whereby the old one is definitely lost.
- *Rollback databases*, i.e. a series of snapshot databases, simultaneously stored and indexed by transaction time, that corresponds to the instant of time the data has been stored into the database. The process of selecting a snapshot out of a rollback database is called *rollback*. Also here, by insertion of new and deletion of old data (e.g., of individual snapshots) the rollback databases can be updated.
- *Historical databases*, in fact the snapshot databases in valid time, i.e. in the time that was valid for the system as the databases have been built. The content of the historical database is steadily updated by deletion of invalid data, and insertion of actual data inquiries. Thus, the databases always reflect the reality of the system they are related to. No data belonging to the past is kept within the database.
- *Temporal databases* are a sort of combination of rollback and historical databases, related both to the transition time and the valid time.

Building and management of databanks at all hierarchical levels of distributed computer systems is facilitated by the fact that such systems are in the majority of cases delivered by a single manufacturer. Consequently, the design and implementation of the system underlie a unique concept. Besides, the system software provided by the vendor includes—apart from the process monitoring and control functions—also the functions needed at higher hierarchical levels. For instance:

- The *Contronic S* system of MANESMAN/Hartmann and Braun contains a number of engineering and production supervisory functions
- The *INFI 90* system of Bailey includes in its basic configuration an engineering workstation, equipped with a variety of computer aided design programs
- The *PLS 80E* system of ECKARDT accommodates at its enterprise level a powerful management information system, relevant for production and plant management
- The *PMS* system of Ferranti contains an information management system and an PMS supervisor that can support distributed process control systems as well as manufacturing systems in which islands of automation have to be linked together
- The * *RS-3* system of ROSEMOUNT incorporates batch orchestration, batch recipe building, system resources management and other high-level functions

In manufacturing systems, particularly in computer integrated manufacturing systems [42–45], database building and management is much more difficult for the following reasons:

- Manufacturing systems are structurally much more complex than process plant control systems
- Manufacturing automation systems are heterogeneous systems as far as their hardware and software concerns

The complexity of manufacturing systems consists of a huge number of machines, tools, operation, products, and accompanying activities that requires a huge number of automation functions to be implemented and a huge number of databases to be built. This was highlighted in Section III.

The heterogeneity of manufacturing systems results from the fact that no computer systems vendor can supply the user with the *full* collection of functions and databases required. The situation in manufacturing practice is rather so that the planned multicomputer system has to incorporate a number of different existing computers and be complemented by some new computers. Hardware and software compatibility problems arise that have to be solved. Although hardware compatibility is solvable relatively simply using international communication standards (field bus, local area networks, MAP/TOP systems, etc.), the problem of software compatibility is much more difficult to solve. For instance, databases of such systems are frequently different in their *logical* and *physical* structures because they have been generated and are to be administrated by different data management systems.

REFERENCES

1. L Kane, ed. *Advanced Process Control Systems and Instrumentation.* Houston: Gulf Publ. Co., 1987.
2. EF Schagrin. Evolution of distributed computer systems, Chemical Engineering Progress 76(6): 72-75, 1980.
3. A Ray. Distributed data communication networks for real-time process control. Chem. Engg. Communications 65(3):139-154, 1988.
4. JE Rijnsdorp. *Integrated Process Control and Automation.* Amsterdam: Elsevier, 1991.
5. G Coulouris, J Dollimore, T Kindberg. *Distributed Systems - Concepts and Design,* New York: ISA Internat. Conf., 2nd ed.,1994.
6. MP Lukas. *Distributed Control Systems: The Evaluation and Design.* New York: Van Nostrand, 1986.
7. D Popovic, VP Bhatkar. *Distributed Computer Control for Industrial Automation,* New York: Marcel Dekker, Inc., 1990.
8. FLM Amirouche. *Computer Aided Design and Manufacturing.* Englewood Cliffs, New Jersey: Prentice-Hall, 1993.
9. WM Luggen. *Flexible Manufacturing Cells and Systems.* Englewood Cliffs, New Jersey: Prentice-Hall, 1991.

10. RA Maleki. *Flexible Manufacturing Systems: The Technology and Managements.* Englewood Cliffs, New Jersey: Prentice-Hall.1991.

11. LH Sibul, ed.*Adaptive Signal Processing.* New York: IEEE Press, 1987.

12. PH Laplante. *Real-time Systems Design and Analysis,* IEEE Press, New York, 1993.

13. D Popovic, ed. *Analysis and Control of Industrial Processes.* Vieweg-Verlag, Braunschweig, Germany, 1991.

14. N Viswanadham, Y Narahari.*Performance Modeling of Automated Manufacturing Systems.* Englewood Cliffs, New Jersey: Prentice-Hall, 1992.

15. LE Widman, KA Loparo, NR Nielsen. *Artificial Intelligence, Simulation, and Modeling.* John Wiley, New York, 1989.

16. KJ Astrom, B Wittenmark. *Adaptive Control.* Addison Wesley, Reading, MA, 1989.

17. PE Wellstead, MB Zarrop. *Self-Tuning Systems.* New York: Wiley, 1991.

18. PJ Gawthrop. Self-tuning PID controllers - algorithms and implementations. IEEE Trans. on Automatic Control 31(3): 201-209, 1986.

19. J McGhee,MJ Grandle, P Mowforth, eds. *Knowledge-Based Systems for Industrial Contro.,* London: Peter Peregrinus Ltd., 1990.

20. D Popovic and VP Bhatkar. *Methods and Tools for Applied Artificial Intelligence.* Marcel Dekker, Inc.: New York, 1994.

21. DA White, DA Sofge, eds. *Handbook of Intelligent Control-Neural, Fuzzy, and Adpative Approaches,* New York: Van Nostrand Reinhold, 1992.

22. H El-Rewini, T Lewis, B Shnver. Parallel and distributed systems – from theory to practice. IEEE Parallel Distributed Technology 1(3): 7-11, 1993.

23. A Umar. *Distributed Compüuting: A Practical Synthesis.* Englewood Cliffs, New Jersey: Prentice-Hall, 1993.

24. RL Gloss. *Real-Time Software.* Englewood Cliffs, New Jersey: Prentice-Hall, 1983.

25. CH Chen, ed.1996. *Fuzzy Logic and Neural Network Handbook.* IEEE Press, New York.

26. TJ William. *Analysis and Design of Hierarchical Control Systems.* Amsterdam: Elsevier, 1985.

27. D Popovic. State-of-the-art and future trends in industrial process automation, Electronics Information and Planning 13: 686-691, 1986.

28. JD Shoftler..Distributed computer systems for industrial process control. *IEEE Computer* 17: 11-18, 1984.

29. DC Chang, RA Wysk, HP Wang. *Computer-Aider Manufacturing.* Englewood Cliffs, New Jersey: Prentice-Hall, 1991.

30. A Kusiak. *Intelligent Manufacturing System.* Englewood Cliffs, New Jersey: Prentice-Hall, 1990.

31. PN Rao, NK Tewari, TK Kundra. *Computer-Aided Manufacturing,* New Delhi: Tata McGraw-Hill, 1993.

32. L Sha, SS Sathaye. A systematic approach to designing distributed real-time systems. IEEE Computer 26(9): 68-78, 1993.

33. KD Shere, RA Carlson. A methodology for design, test, and evaluation of real-time systems, IEEE Computer 27(2): 34-48, 1994.

34. JT Fink. Advanced engineering tools for process automation, Control Engineering 35(9): 23-26, 1988.

35. AE Niesenfeld. Control of complex industrial process - a survey. Digital Computer Application, Vienna, 31-34, 1985.

36. EJ Kompass. Reviewing PC-based software for control engineering. Control Eng. 35(11): 57-60, 1988.

37. MS Shatz, JP Wang. Introduction to distributed software engineering. IEEE Computer 20(10): 23-31, 1987.
38. CR Asfahl. *Robotics and Manufacturing Automation.* John Wiley, New York. 1992.
39. HD Cihan. *Artificial Neural Nerworks for Intelligent Manufacturing.* New York: Chapman & Hall, 1994.
40. CH Dagel, ed. *Artificial Neural Networks for Intelligent Manufacturing.* London: Chapman & Hall, 1994.
41. MT Ozsu, P. Valduriez. Distributed database systems – where are we now? IEEE Computer 24(8): 68-78, 1991.
42. DD Bedworth MR Henderson, PM Wolfe. *Computer-Integrated Design and Manufacturing.* New York: McGraw-Hill, 1991.
43. PG Ranky.*Computer Integrated Manufacturing: An Introduction with Case Studies.* Prentice Hall, Englewood Cliffs: New Jersey, 1986.
44. N Sing.*Systems Approach to Computer-Integrated Design and Manufacturing.* New York: Wiley, 1996.
45. DN Chorafas. *Manufacturing Databases and Computer Integrated Systems.* Ann Arbor, Michigan: CRC Press, 1993.

7
Computer Aided Design of Automotive Control Systems Using MATLAB®, Simulink®, and Stateflow™

Richard J. Gran
Mathematical Analysis Company, Northborough, Massachusetts

I. COMPUTER AIDED DESIGN OF AUTOMOTIVE CONTROL SYSTEMS USING MATLAB, SIMULINK, AND STATEFLOW

This chapter shows how an automotive engine controller can be developed using computer aided design tools. The chapter is divided into the following four sections:

1. The development of a model and a control system for an automotive engine
2. The development of the control system logic
3. The prototyping of the control systems
4. A look at the future of computer aided control for mechatronic systems and for systems in general

The assumption is made that the reader is familiar with standard control system design techniques. We also assume that the reader is familiar with the workings of an internal combustion engine. In particular we assume that the reader is familiar with the way the engine combines air and fuel to provide a combustible mixture that will power the engine, and the fact that the engine consists of four strokes.

An overview of the computer aided design tools MATLAB®, the Control System Toolbox™, and Simulink™ is presented in the Appendix to this chapter. These tools are used extensively in the development of the engine controller design in this chapter. Before readers continue with this chapter, they are advised to refer to this appendix first.

II. CONTROL SYSTEM DESIGN FOR AN AUTOMOBILE ENGINE

This section first discusses the engine model and how it is implemented in Simulink, followed by a description of the control system design process using MATLAB, Simulink and the Control System Toolbox. This example is based on a paper by Cook and Crosley [1] and a detailed simulation that was done by the MathWorks staff [2]. Once the control system is designed, Simulink is again used to provide performance analysis of the resulting design. In particular, the nonlinear control design block set is used to develop the optimal control for the entire nonlinear system.

MATLAB, Simulink and Stateflow™ are computer aided system design products created by The MathWorks in Natick, Massachusetts (the product names are registered trademarks of The MathWorks). MATLAB is a high level computing language that permits rapid analysis, design and implementation of complex systems. The MATLAB language was first introduced in the early 1970s as a teaching tool and, in the intervening years, its extreme flexibility and its user-friendly syntax have been significantly augmented. Additional tools have been added that allow block diagrams and state charts to be used for simulating dynamic systems (Simulink) and discrete event driven systems (Stateflow).

The tools from The MathWorks that allow this iterative design process to be used to develop a controller for an automobile engine (or for any complex system for that matter) are shown in Figure 1.

At the top level of this process (the modeling stage) the designer is confronted with the problem of modeling the system using first principles and/or the data obtained from testing. If the test data is noisy, or if it is not obvious how the data is related to the test inputs, the engineer will develop a model using signal processing and system identification procedures. The resulting model is then simulated (in the simulation stage) using Simulink and Stateflow. For the automobile engine controller we will develop here, both physics and measured data are used to develop the model.

After the model is complete, the system is linearized to perform a linear control design analysis (the Design or Synthesis stage). This analysis verifies that the controller is stable and performs adequately in the presence of modeling uncertainty, nonlinearities, and other engine dynamics. This process consists of alternately analyzing the system and simulating it in the nonlinear model, in the process perhaps adding more complexity to the nonlinear simulation. Once the design is verified through this iterative linear control design process, the nonlinear system

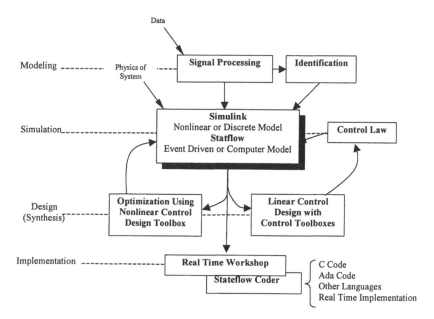

Figure 1 The design process that is described in this chapter.

can be optimized using the Nonlinear Control Toolbox. This allows the parameters in the controller to be "tuned" to insure that the system performs optimally for a wide variety of command inputs.

The last stage (Implementation) in the design process is the automatic generation of the embedded code for the system. This stage uses Real Time Workshop and the Stateflow coder to create the code for the controller so that the system may be tested and ultimately to generate the required code for operating the engine in the production vehicles.

The following sections describe each of these steps.

A. The Engine Model

An engine has five subsystems that are required to achieve the desired operation. These are:

1. The Throttle: specifies the air–fuel mixture input to the engine based on the deflection angle of the throttle. This is a nonlinear function since the airflow depends on atmospheric and manifold pressure and the throttle angle.

2. The Intake Manifold: models the mass flow rates of the input and the output based on the ideal gas law and the temperature and pressure in the manifold.
3. Mass Flow Rate: this models the air–fuel mixture mass that is pumped into the cylinders from the manifold. This model is empirically determined.
4. Intake to Power Stroke Delay: this models the fact that a four-stroke engine fires once in every two complete revolutions of the engine crankshaft.
5. Torque Function and Engine Speed: this is a model of the dynamics of the engine. The model has a derivation of the torque developed as a consequence of the air–fuel mixture being burned and the torque from the load (the automobile) based on the operating environment of the auto.

Additional components such as exhaust gas recirculation, etc. can be added to this model, but in order to make the exposition as clear as possible, the model has been significantly simplified.

The Throttle

The throttle causes the air–fuel mixture to change when acceleration and deceleration are required. The flow through the throttle can be either subsonic or supersonic. Since the mass flow rate through the throttle body changes with the speed of flow, the nonlinear properties of the flow are modeled in two different blocks. The block that is used is determined by the speed of the airflow in the system. Thus the model is as shown in Figure 2.

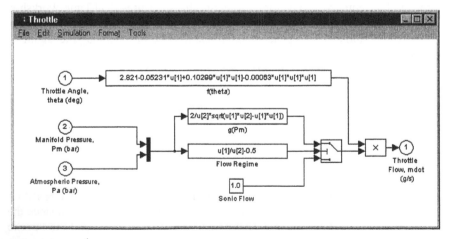

Figure 2 Throttle model.

The input to the throttle subsystem is the throttle angle (in degrees). The output of the subsystem is the mass flow rate in grams per second. When the output flow rate is subsonic, the output is the product of two functions. These functions are empirically determined and are modeled in Simulink using a function fit to the data. Thus the output is given by:

$$\frac{dm_{ai}}{dt} = g(Pm)(2.821 - 0.0523\theta + 0.103\theta^2 - 0.000630\theta^3)$$

Where θ = Throttle Angle in degrees, Pm = Manifold Pressure in bar, and m_{ai} is the mass of the air flowing into the intake.

The nonlinear function g(Pm) is given by:

$$g(Pm) = \begin{cases} 1; & Pm \le P_{amb}/2 \\ \dfrac{2}{P_{amb}}\sqrt{PmP_{amb} - P_m^2} \end{cases}$$

The switch in the value of g models the sonic and subsonic flow regime. Notice that because g = 1 when the pressure is below half the ambient pressure, the model causes the mass flow rate to be a function of the throttle angle only. The nonlinear function g is plotted in Figure 3.

The Intake Manifold

The intake manifold pressure will change when the air mass changes. To evaluate this change, the ideal gas law is used. From the gas law, the volume, pressure and temperature of the gas in the manifold is given by:

$$P_m V_m = nRT$$

Where P_m is the gas pressure, V_m is the volume of the manifold, R is the specific gas constant and T is the temperature.

The mass of the gas is related to the molecular weight of the air. Since we are interested in the rate of change of pressure, we can assume that the volume and temperature of the manifold stay constant and we can solve the ideal gas law equation and differentiate to give the pressure derivative as:

$$\frac{dP_m}{dt} = \frac{RT}{V_m}\left(\frac{dm_{ai}}{dt} - \frac{dm_{ai}}{dt}\right)$$

Where m_{ai} and m_{ao} are the masses of the air at the inlet and outlet of the manifold.

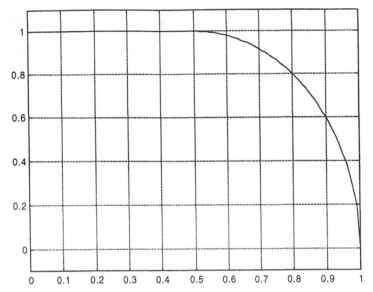

Figure 3 Plot of the function g used in the throttle block.

Mass Flow Rate Into the Engine

The amount of air pumped into the engine is a nonlinear function of the engine speed and the manifold pressure. From measurements of these date, an empirical model can be developed. For example, a typical set of engine data was fit with a 2-dimensional quadratic least squares function to give:

$$\frac{dm_{ao}}{dt} = -0.366 + 0.0898NP_m - 0.0337NP_m^2 + 0.0001N^2P_m$$

Simulink models this equation using a block called a function block. This block has an input that is denoted by u. If the input has two or more values, u is a vector "**u**," and the components of u are denoted by u[1], u[2],.... The block is shown in Figure 4.

Since the fuel-air mixture flows into the engine once in every two revolutions, the output of these blocks must be sampled and held. The fuel–air charge that the engine burns is sampled in the Simulink model using the integrator block. Looking at this block in Fig. 5, the integrator block has two inputs and two outputs. These input and output ports are set by a dialog box that is opened by double clicking the integrator block. The input is the mass flow rate into the engine. The trigger causes the integrator to sample the integrated input whenever the engine cycles through two complete revolutions.

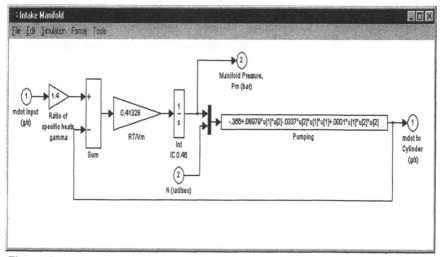

Figure 4 Modeling the intake manifold.

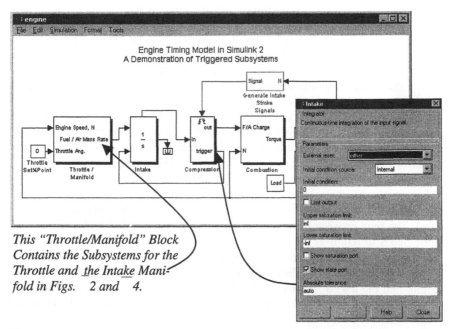

Figure 5 Top level Simulink block diagram of the engine.

The Intake to the Power Stroke

In a four-stroke engine, the complete pattern of intake, compression, firing (expansion), and exhaust takes two complete revolutions. Thus the engine delivers power only once per 4π (in radians) revolutions of the crankshaft. To model this, the integrator block of Simulink is used. The complete top-level block diagram of the engine is shown in Figure 5. The bottom half of this figure shows the dialog box that appears when the integrator block is opened by double clicking on the block. Simulink integrators can be set up so that:

- They can be reset with external signals (triggered based on a rising, falling or either slope). If no trigger is used, the external reset box is specified to be none (the default).
- They can have an initial condition based on an external signal (in which case an additional input port is added to the integrator icon). When the initial condition is internal, the "Initial condition:" box is used to specify the initial condition.
- The output can be limited; in which case the upper and lower saturation limits can be specified in the dialog box.
- The saturation port can be shown on the integrator icon so that the onset of saturation can be used to trigger other events.
- The integrator state port can be explicitly shown.

In the implementation for the engine, the block is reset based on an external signal that changes in either direction. There is no saturation on the integral, and the "show state port" check box is on. In the block diagram of the engine, the integrator has two input ports and two output ports. These are the reset (on the bottom left side of the integrator) and the input to the integrator on the top left side.

The output from the integrator is the integral of the input, but because the engine cycle requires that the integration be reset based on the output of the integrator, the integrator has an algebraic loop around it. For this reason, the state port in the integrator is used. This port is explicitly included in the integrator to allow for this sort of reset. If the integrator output were used, Simulink would generate an error message that an algebraic loop was detected and the simulation might be incorrect. This is only one of the ways that Simulink can handle algebraic loops. There is an explicit algebraic loop solver block that can also be used.

The other attribute of this block is the way the compression stroke is modeled. The time between the intake stroke and the power stroke is the time the engine needs to return to top dead center after the intake stroke is complete. This time is ¼ of the total engine cycle, or is π/N, where N is the engine speed in radians per second. This is implemented in the block diagram by the reset on the integrator and the trigger directly from the engine rpm. If we look inside the blocks labeled "compression:" and "generate intake stroke," we see that the block sets in Figure 6.

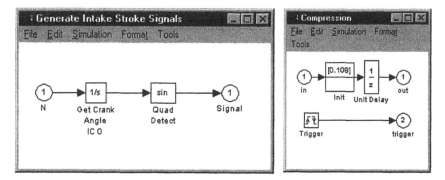

Figure 6 The blocks for the signal stroke signal and the compression.

The initial condition block label "init" in the compression block allows the discrete delay represented by 1/z to be set to 0.108 when the simulation starts.

Torque Function and the Engine Speed—The Vehicle Dynamics

The torque developed by an engine depends upon the energy released in the combustion stroke, and the efficiency of the energy transfer into the crankshaft—the mechanical system that converts the stroke into rotating motion. Obviously this is a function of the mass of fuel–air mixture in the cylinders, the air to fuel ratio (how lean or rich the mixture is), and the spark advance (the closer the spark timing is to the point where the engine is at top dead center, the smaller the amount of torque that will be created because of the time it takes for the combustion wave to propagate through the fuel–air mixture). An empirical model for this function of three variables is used to establish the engine torque. Thus, we model the torque as:

$$\text{Torque} = a + bm_a + c\left(\frac{A}{F}\right) + d\left(\frac{A}{F}\right)^2 + e\sigma + f\sigma^2$$

Where m_a is the mass of the fuel-air mixture in the cylinders; A/F is the air to fuel ratio (we will model the case for stochiometric mixtures only so that A/F = 14.6); σ is the spark advance angle in degrees before top dead center; and the constants in the curve fit are: a = -181.3, b = 379.4, c = 21.91, d = -85, e = 0.26 and f = -0.0028.

Notice in the entire derivation only one cylinder is acting. In a real engine, four, six or eight cylinders produce the torque, and the actual torque from each cylinder is summed at the crankshaft. This example models a four-cylinder engine. The integrator triggering accomplishes the development of the required torque from each of the cylinders by multiplexing the torque developed. The Simulink block that models the torque equation is shown in Fig. 7 below.

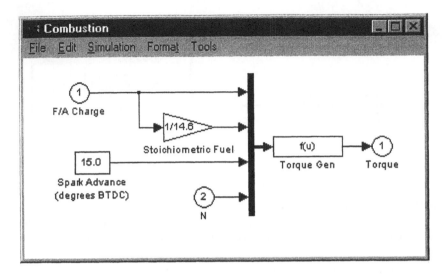

Figure 7 The model of the engine torque block in the Simulink engine model block labeled "combustion."

The equation in the function block labeled "Torque Gen" (Figure 7) is the equation above. After the load torque is subtracted from the engine torque, the resulting acceleration is integrated in the "vehicle dynamics" subsystem block (Figure 8). In the final block diagram in Figure 5 there is a block that generates the trigger command for the intake stroke. This block simply integrates the engine speed to get the total accumulated crankshaft angle (the initial condition on the

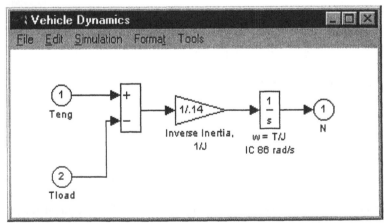

Figure 8 The vehicle dynamics block where the engine speed N is computed from the total torque.

integral is zero), and then takes the sine of the angle to create an output that is positive during the intake stroke and negative when the compression stroke takes place. This command is then passed to the trigger block where a one-sample delay is used. In this case, the 1/z block inherits its sample time from the input, so the fuel charge is used in the combustion block at the proper time for the combustion stroke.

B. The Engine Response

The Simulink block diagram is now complete.

The response of the engine to a load change is simulated. The block labeled load contains a step function set up to cause a load torque profile that introduces a 17 NM torque at t = 0. The engine responds to this torque change. The open loop response of the engine with this torque profile is shown in Figure 9.

As can be seen, the engine speed starts at 800 rpm and very quickly drops to 600 rpm because of the load. The response then oscillates at a period of about 2.25 seconds. There also is a large overshoot indicating that the system is under-damped. This speed response is undesirable, and leads to the need for a control system for the engine. A control system is clearly required to improve the engine's response to a load.

The typical speed controller on an engine is a simple proportional plus integral control loop, so the control system design will be a relatively straightforward PI design.

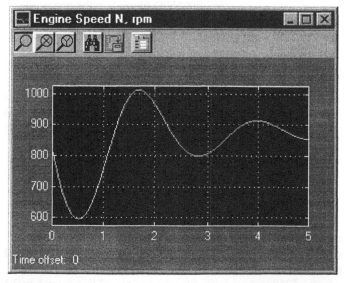

Figure 9 Open loop response of the engine model response to a step in load torque of 17 NM applied at t = 0.

For most modern engines, the exact timing of the fuel injection and the spark is determined as a function of the various parameters in the engine, and is then loaded as a table look-up into the control computer. Since this example is relatively simple, and no attempt has been made to model some of the more subtle characteristics of the engine, the control design will mainly illustrate the use of the tools rather than an explicit control system.

C. The Control Design

The control system design can be accomplished using any of the linear control techniques and any of the toolboxes that the designer would like to apply. However, since this example is so nonlinear, it is instructive to illustrate a nonlinear design tool in MATLAB called the Nonlinear Control Design toolbox, or "NCD." NCD provides a Simulink interface that allows the user, through a graphical user interface, the ability to specify a time response constraint. The NCD toolbox develops control system parameters (gains, time constants, filter characteristics, etc.) that force the response of the nonlinear system (through the Simulink simulation) to satisfy the constraints (if possible).

The first step in using the NCD toolbox is to develop a structure for the controller. As was indicated above, the engine can be controlled using a simple PI (proportional plus integral) control. Thus, we modify the engine Simulink model to include a PI control as shown in Figure 10.

This diagram is essentially the same as the block diagram of the engine itself, except that the control system has been added.

The NCD block is added to the block diagram in Figure 10 by opening the NCD blockset from Simulink and dragging the NCD block into the engine simulation block diagram. After this is done, a set of initial values for the gains K_p and K_i must be set. These are set in the MATLAB Command window. We have selected initial values of 0.05 and 0.1, respectively, for these gains. The next step is to open the NCD window by double clicking on the NCD block. The window will open with a time axis equal to the simulation time, and a set of constraints, in red, that will specify where the time response must fall. The typical set of constraints that is specified when the window is first opened is for a step response. Since the engine controller simulation specifies that the engine speed is to be maintained at 2000 rpm and also since the simulation starts at an engine speed of 750 rpm, the optimization constraints need to be moved from their normal values to provide the desired specification on the response. In Figure 11, the NCD window is shown along with the constraints that have been imposed. Loosely speaking, these constraints are:

1. A rise time of 1 second or less, specified by the vertical line at 1 second
2. An overshoot of 100 rpm, specified by the horizontal line from 0 to 2 seconds at 2100 rpm;
3. An undershoot of 250 rpm, specified by the horizontal line from 1 to 2.5 seconds at 1750 rpm;

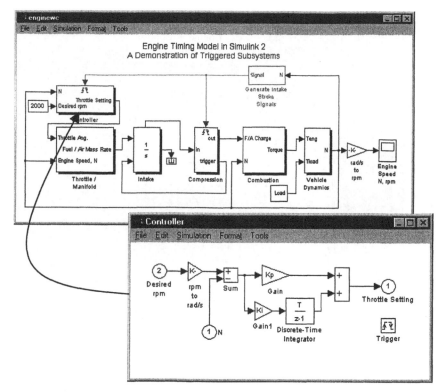

Figure 10   Engine simulation with a PI controller added.

4. An undershoot of 250 rpm, specified by the horizontal line from 1 to 2.5 seconds at 1750 rpm;
5. A settling time of 2.5 seconds, specified by the two horizontal bars above and below the 2000 rpm steady state response (these are at 2050 and 1995 rpm respectively; and
6. In addition, to the settling time of 2.5 seconds, the constraints in step 4 also provide a steady state accuracy specification.

When the NCD simulation is started (using the start button on the display), the optimization begins by first plotting the time response for the initial gains (K_p = 0.05 and K_i = 0.1). After this initial response is plotted, the NCD toolbox perturbs the gains and numerically estimates gradients and Hessians for the output. A modified form of a Newton iteration determines the search for the best values of the gains. There is a complete description of the algorithms and, as always is the case for The MathWorks products, the m-files that perform the optimization are open and accessible so that the user may modify and add to the procedures as seen fit.

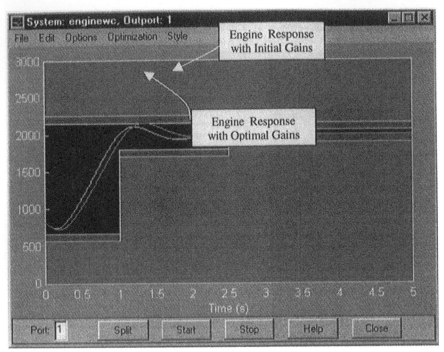

Figure 11   Using the nonlinear control design (NCD) toolbox to optimize the control gains for the engine.

The final control gains for the optimal control are $K_p = 0.0363$ and $K_i = 0.0747$. The final (optimal) response is plotted on the NCD graph in the window for comparison with the original response.

In general, the proper way to use the NCD toolbox is to have a reasonable design for the control first, and then use this design appoach to improve the response. The nominal controller here is a PI controller and as such its transfer function is given by:

$$H(z) = K_p + K_i \frac{T}{1 - z^{-1}} = K\left[\frac{1 - \alpha z^{-1}}{1 - z^{-1}}\right]$$

Where the transfer function has a discrete zero at $\alpha = K_p/(K_p + K_I T)$ and a pole at 1.

When using the NCD toolbox, the optimization parameters do not have to be gains; they can be any parameter in the simulation. Thus, if a noise filter is needed to improve the time response, the filter time constant can be used as one of the optimization paramters, and the noise response will be optimized along with the transient and steady-state characteristics.

Before we look at how code is generated for this control system, let's assume that the Simulink model for the control system is the simple two input single output block shown in Figure 10 (the block labeled "controller"). The first step in developing code for this control system is to isolate the control block. In Simulink, open a new model through the File menu under New and Model. Once this window is open, copy the control block into the subsystem and add input and output ports. Note the enable subsystem icon is kept in this block so that the entire controller will be triggered. When the C code is generated from this block, the sample time will be computed using the integration time. When the block is actually implemented, the user only needs to modify the code to read a clock at the times that the block is executed to get the sample time. Remember that the sample time is used in the discrete integrator to perform the integration. Figure 12 shows the blocks created by this process.

The blocks in Figure 12 will be augmented with the code needed to do fault detection and analytic redundancy management that will be developed in the next section. The combined models will then be used with RTW to create the embedded code for the control system. The new MathWorks computer aided design tool that allows modeling of event driven systems is called Stateflow.

II. STATEFLOW FOR THE MODELING EVENT DRIVEN SYSTEMS

The block diagram representation of a system is ideal for modeling systems that are continuous time or that have recursive (sampled) time structure. When a system is driven by an event such as the closing of a switch, the failure of a sensor, the need to turn on a subsystem based on a state crossing, or any number of other complex events, then other modeling approaches are more useful. The MathWorks has introduced a new product called Stateflow [3] that will model an event driven system using a diagram where the basic object is a state that is either active or

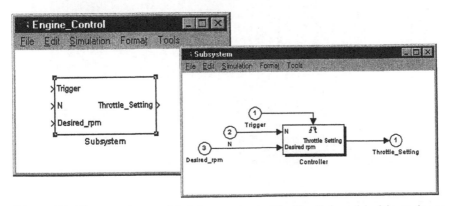

Figure 12 The control system block is carved out of the Simulink model of the engine as a subsystem for translation (using RTW) into C code.

inactive. A directed arrow that is annotated with the action that will cause the transition to take place indicates transitions among the states in the diagram. The easiest way to understand the notation used to create a Stateflow diagram is to look at a typical diagram.

A. A Typical Simulation Model

Figure 13 shows the Stateflow model for a simple sensor failure in an automobile engine. In this example, the sensor measures the engine speed, one of the variables that we need to control the engine. We assume that the sensor is either failed or functioning normally.

Stateflow uses a simple syntax to represent the state of a system. In Figure 13, for example, there are three states represented by the rounded rectangles. The sensor failures are denoted by the two states labeled speed_norm and speed_fail in the center of the larger state labeled Speed_Sensor_Mode. The outer rectangle that encloses the two sensor states provides a hierarchy in the diagram. The meaning of the hierarchy is that whenever either of the sensor states is active (failed or operating) the state Speed_Sensor_Mode is also active. Later, when we assemble a complete state model for the engine failure monitor, this state will be one of four states in parallel that will represent the possible failures in the engine (the other three states will be oxygen sensor, pressure sensor and throttle sensor failures). A directed arrow denotes the transitions between states with a label attached. To

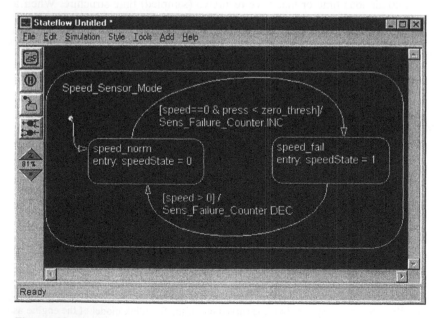

Figure 13 Stateflow model for the speed sensor failure.

draw these diagrams, Stateflow uses the same click and drag approach as Simu-link. The user only needs to position the cursor anywhere on the border of the state and click and drag (except at the curved areas where the cursor will change to a two sided arrow that allows the user to increase or decrease the size of the state rectangle). An arrow will be created that the user can connect to any other state in the diagram (including the state from which the arrow starts).

After the arrow is placed, there will be a small question mark alongside. By clicking on this question mark, the cursor changes to an insert bar that will allow the user to type in a command that represents the condition for the transition to take place and information about what action will be caused by the transition.

In the example we have constructed, the command for moving from the normal speed sensor state (speed_norm) to the failed state will occur when the measured speed drops to zero and the pressure is less than a quantity called "zero_thresh." Note that the condition is placed inside of square brackets. The second part of the transition is an action that is executed when the transition takes place. In the transition to the failed state, this action is the setting of the state "Sens_Failure_Counter.INC." When we construct more of the Stateflow diagram, this action will be set by each of the four failure states when a failure occurs, and we will see that this action will be used to count the number of failures.

When the sensor failure goes away (detected by the speed becoming greater than zero) a transition to the "speed_norm" state occurs, and the state "Sen_Failure_Counter.DEC" will be set (this structure is a subset of the state "Sen_Failure_Counter" where it is used to decrement the number of failures counter).

There are two additional features in the Stateflow model of Fig. 13 that need to be discussed. First, there is a transition that starts at a small dot going into the state speed_norm. This transition defines the initial condition for the state diagram. Thus, when the system starts, the sensor is assumed to be operating, and we will be in the speed_norm state. Secondly, each state has a notation that begins "entry:" and the expression that follows the colon in this syntax will be executed whenever a transition causes this state to be entered. Thus, at start up, the initial transition into the state "speed_norm" will set a flag called "speedState" to 0. Similarly, the transition to the state "speed_fail" will set the same flag to 1.

As we have developed the Stateflow model, it should be clear that there is an implicit hierarchy of states. In essence each state is a structure. Thus at the top level of the hierarchy the structure is the name of the state. Thus Speed_Sensor_Mode is the top state in this diagram and its variable name in the code that will be generated by the diagram is simply Speed_Sensor_Mode. The state "speed_norm" is a substate of this state and its name is therefore Speed_Sensor_Mode.speed_norm. This convention carries through to all states in the diagram. The property of a state is inherited from its children. Thus, whenever either speed_norm or speed_fail is active then so is Speed_Sensor_Mode.

In building the state diagram, the icons at the left part of the screen are used. The topmost icon is the state, and the third icon is the initial condition transition.

The second icon of the set is called "History." This icon, when placed inside a state that contains multiple children, causes transitions that come back to the top state to transition to the state that was active when a transition last left the top level state. Figure 14 illustrates the use of History and also illustrates the use of the fourth icon, the connective.

The additional states in Figure 14 follow the same syntax that was described above. Since this diagram is a little more complicated than the single failure state diagram, let's walk through this picture step by step. We will start at the top level of the hierarchy. This state is called "Fueling_Mode" and since its border is a dashed line, we know that it is a parallel state in a much larger set of states (this state will actually be active at the same time as the states that monitor the four sensors for failures). Inside this state are two states called "Running" and "Fuel_Disabled." The Running state is further divided into two states that are called "Low_Emmisions" and "Rich_Mixture" respectively. Lastly, inside the

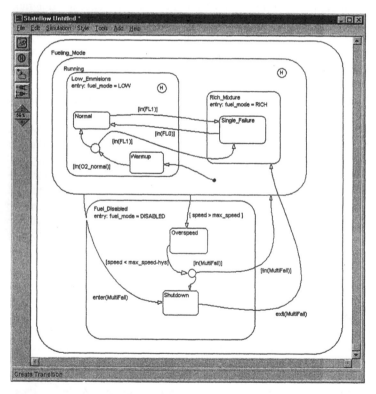

Figure 14 The state flow diagram for the various fueling modes. When any sensor fails, the fueling mode switches to "RICH" causing the engine to run rich through a command that is sent out from Stateflow to Simulink.

Low_Emmisions state are two states that represent the normal operation of the engine and a state that the engine is in during warm up. The normal sequence of events that are modeled in this diagram are as follows:

1. On entry, the engine is in the "Running" state because we enter the state "Warmup" that is a child of the Low_Emmisions state that is in turn a child of the Running state.
2. When O2_normal becomes set, the transition between the Warmup state and the Normal state will take place through a conditional branch that is denoted by the connective circle. This connective allows IF-THEN-ELSE statements and loops to be created in Stateflow.
3. The connective will cause a transition to the Rich_Mixture state if the event in (FL1) is true (i.e. if the machine is *in* the state FL1).
4. If the flag FL1 is not set, the transition with no action on it (the arrow to the Normal state) is executed as the "ELSE" in the IF-THEN-ELSE statement.
5. The engine is in the normal state and is metering fuel based on the Low_Emmisions criteria. If any failure occurs, the engine transitions along the arrow with the [in(FL1)] flag test to the state labeled "Rich_Mixture."
6. The engine will stay in the rich mixture fueling state as long as only a single failure has occurred.
7. If the single failure corrects itself, the transition will return to the Low_Emmsions state along the arrow labeled [in(FL0)].
8. At the top level, the engine will be disabled in the "Fuel_Disabled" state. This state will set a flag entry_fuel_mode = DISABLED on entry into the state.
9. The way we enter the state is through the MultiFail state. This state is set by the failure counter whenever the total number of failures is two or greater.
10. The return from the multifail state will go to the state that was last active in the Running state because of the history symbol in this state.

The complete model of an engine controller in Stateflow is integrated into Simulink so that external events that depend upon the dynamic behavior of the system may be used to stimulate the transitions and the transitions can be used to cause changes in the operation of the system. In the example we have started, the sensors are modeled along with methods for deducing a single failed sensor's information from the remaining three operating sensors. The next section describes the Simulink part of the model.

B. The Engine Simulation with Failure Monitoring and Control

The complete Simulink block diagram for the engine controller that was developed in Section I has been modified to include the Stateflow model of the control computer's supervisory functions.

The block diagram of the supervisory controller is shown in Figure 15. The Stateflow block is integrated into Simulink and provides outputs that are used in the model to alter the dynamic behavior of the system based on the logic etc. that takes place in the block. The Stateflow block is the rounded rectangle at the center of the diagram. As can be seen, inputs to the block are also being passed from the Simulink model. Thus, the failure states of the sensors can be forced by flipping the switches in the diagram. Each of these switches are animated, so that by double clicking on the switch its state transitions from that shown to the alternative state. Note, these are essentially what are known as double pole single throw switches. In all cases, the switch state change will model the loss of sensor information by setting the value to zero. As we saw when we were developing the sensor failure model in Section II.A, the sensor failure is detected when the value of the sensed state is zero.

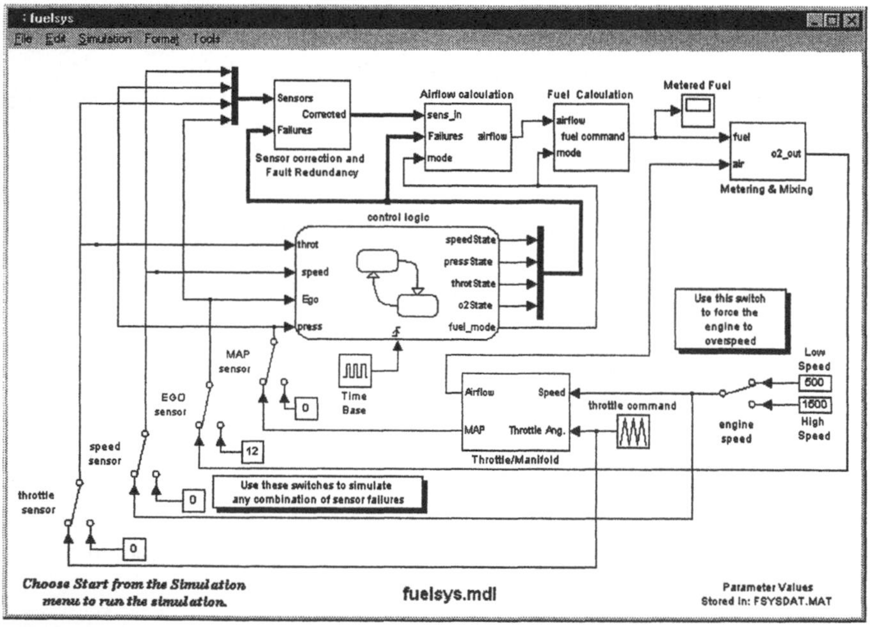

Figure 15 Simulink model for the fuel system with the supervisory controller modeled in Stateflow.

Another feature of the Simulink model is the way the computer operations have been synchronized with the timing in the simulation (and implicitly with an external clock). This is done through a trigger that is added to the Stateflow block. The trigger is connected to a time base that is modeled as a square wave (at the bottom of the Stateflow block).

Most of the blocks in this diagram are the same as we developed in Section I. However to accommodate the different modes of operation, some of the blocks have been modified. For example, the air-fuel mixture was assumed to be stochiometric in Section I. To allow for a rich mixture when any single failure occurs, the air-fuel ratio is set to 80% of stochiometric (a rich mixture) and the total fuel is set to 0 when the engine is off. Figure 16 shows the block that performs this calculation. In this block, the switch that controls the fuel ratio is changed by the input from Stateflow called "fuel_mode."

Stateflow changes the state of "fuel_mode" based on the number of failures. Figure 14 shows the way that this is changed. In each of the states in this figure, there is an entry condition. An entry condition means that when the state is activated (entered) the entry condition is executed. Thus, for example, when the Low_Emmisions state is activated (at start up for example), the entry: condition causes fuel_mode to be set to "LOW." When a transition causes the state Rich_Mixture to be activated, the entry: condition causes fuel_mode to be set to "RICH." The third possible value for fuel_mode is "DISABLED," which occurs when the Fuel_Disabled state is entered.

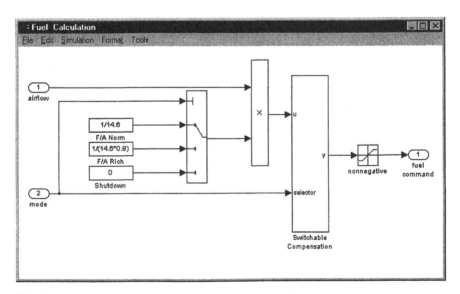

Figure 16 The fuel calculation block and how it is modified based on the Stateflow output "mode."

The switch block in Simulink requires numeric values for the various switch positions. Thus in Figure 16, the variable "mode" must be 1, 2 or 3 so that the switch will cycle through the three positions. The conversion from the names LOW, RICH and DISABLED to the numeric values 1, 2, and 3 is accomplished by setting the values of LOW, RICH, and DISABLED to 1, 2 and 3, respectively, in MATLAB. Stateflow keeps a running dictionary of all of the data in the state diagram (both internal and those used to communicate with the outside world). Using this dictionary and Stateflow's debugger, the system designer can track all of the variables used and also insure that the code that will ultimately be generated is accurate and provides the required functionality.

As an additional aid in debugging and verifying the Stateflow code, when Simulink is executed and there is a Stateflow model in the block diagram, the state transitions and the flow of data in the Stateflow model are all animated. Transitions that take place will highlight the state, and the path taken will also be highlighted. This feature makes it very easy to follow the flow of both the signal and the logic.

To complete the discussion on this model, the remaining Stateflow pieces will be described in the following sections. There are also Simulink blocks that are used to provide the analytic redundancy and interact with the Stateflow model. These blocks are also described.

The Remaining Failure Modes

Four possible sensor failures need to be monitored. We have already seen the Stateflow model for the speed sensor failure. The remaining three sensor failure

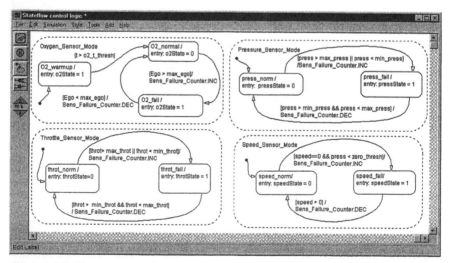

Figure 17   The remaining failure states in the Stateflow model.

states are very similar. The remaining sensors that are monitored are the pressure, oxygen and throttle sensors. The Stateflow model for monitoring all of the sensors is shown in Figure 17. In this figure, all four-engine sensor states are enclosed with rectangles with dashed lines. This indicates that the states are in parallel and each of them can be active at the same time. The states "Pressure_Sensor_Mode'"" and "Throttle_Sensor_Mode" are essentially the same as the state for the speed sensor that was already discussed. The only difference is the condition for the transitions into the failed state and back to the normal state. The "Oxy-gen_Sensor_Mode" state is slightly different in that an additional state is provided to indicate that the sensor is warming up. The warm up is based on time, and the transition from the warm-up state to the O2_normal state is based on the time t (which is set by Simulink to zero when the simulation starts) being greater than O2_t_thresh, a parameter that is defined in the Stateflow model.

The Failure Counter

The Stateflow model of the counter that keeps track of the number of failures is shown in Figure 18. The counter uses a pair of variables called INC and DEC that are generated in Stateflow by transitions from the normal state to the failure state for any one of the four sensors. Thus, the counter will be incremented by INC and decremented by DEC, and the states corresponding to no failures (FL0), and one, two, three or four failures (FL1, FL2, FL3 and FL4 respectively) are entered whenever the appropriate number of failures have been counted. Notice, the failure counter states FL2 through FL4 are included in a super-state denoted Multi-Fail. This state is used in the Stateflow model to denote that not enough information about the engine is available to properly run the engine, and the engine is shut down. The logic for this operation is shown in the Stateflow model of Figure 14.

The Analytic Redundancy Blocks in Simulink

The block in Figure 15 that is labeled Sensor correction and Fault Redundancy is a block that contains models of all four of the sensors on the engine. This block is

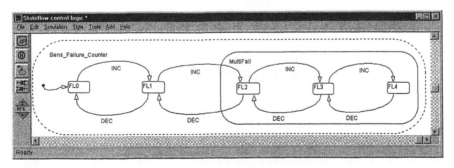

Figure 18 The Stateflow model for failure counting.

part of the control system; it provides a pseudo-measurement through the analytic redundancy contained in the sensed data. As an example, the block that generates an estimate of the engine speed when the speed sensor fails is shown in Figure 19. The figure also illustrates the Simulink block that selects a particular vector component from a vector and the block that performs a table look-up. The block is an enabled subsystem that is enabled by the speed sensor failure indicator.

The input to the block is a vector called "Sensors" that contains the measured data from the four sensors. The throttle is the third component of the vector and the Manifold pressure is the second component. As can be seen, the selector block has a switch like icon that indicates which of the vector components is being selected. The table look-up function uses data stored in the MATLAB work space. Since the table is 2-dimensional, the data is stored as a matrix and the vectors corresponding to the x and y axis of the table are stored as vectors. The table look-up block has a dialog window that allows the user to specify the name of the table and the names of the variables that specify the axes. Figure 20 shows a 2-dimensional plot of the look-up table that is used. As can be seen, the table essentially reconstructs the speed implied by the pressure and throttle position. The speed (in rpm) is highest when the throttle angle is at its maximum (90 degrees) and the manifold pressure is at its lowest (0 bar). The speed indicated by the table in this case is 8213 rpm. At the other extreme, the speed when the throttle angle is 0 and the pressure is close to maximum (1 bar) is 290 rpm. The actual engine will not necessarily have the tabulated speed when it is operating with the speed sensor failed, but it will be close enough that the engine may still be controlled. As we saw in the Stateflow model of the control logic, the engine is placed into a rich operating mode when this condition occurs. Any further failures will shut down the engine since calculating the speed will then be impossible.

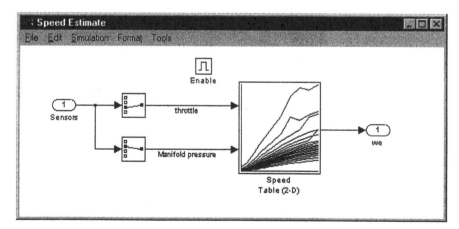

Figure 19   The analytic redundancy calculation that provides an estimated speed measurement based on throttle position and manifold pressure.

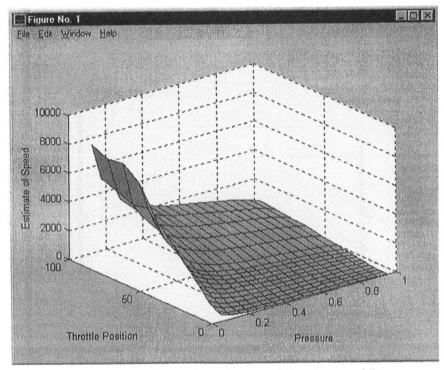

Figure 20 A plot of the 2D data used to estimate speed when the sensor fails.

An analytic redundancy block exists for every possible single failure, and these are all contained in the Sensor correction and Fault Redundancy block. The details of this block are shown in Figure 21.

There are many more details in this engine model simulation. The model is contained in the demonstration (demos) file that is shipped with Stateflow. This model may be exercised and the various components of the system can be investigated readily using MATLAB, Simulink and Stateflow. In the next section we will develop embedded code directly from the Simulink and Stateflow model using Real Time Workshop.

III. REAL TIME WORKSHOP FOR GENERATING CODE FOR THE CONTROL SYSTEM

The PI controller for the engine was carved out of the engine simulation so that the computer code for the control code could be generated using real time workshop (RTW). As we saw in Section II, this piece of the control system is actually only a

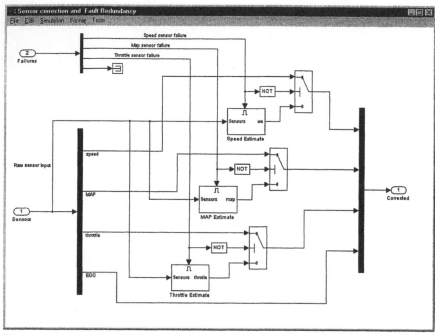

Figure 21   The sensor correction and fault redundancy block provides estimated measurements when a sensor failure occurs.

small part of the overall controller, since the analytic redundancy blocks and the control logic defined by the Stateflow model must also be included. Thus we must add to the PI controller model, shown in Figure 12, the blocks that are required for the remaining part of the controller, shown in Figure 15.

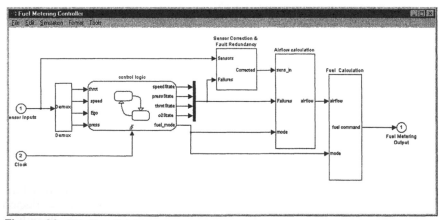

Figure 22   The Simulink block for the final control system.

The complete model of the controller was "carved" out of the Simulink blocks to create the controller shown in Figure 22. Several features of this model should be described. First, each input from an external source needs to come into the computer via an input port. Since there are four sensors, we have grouped them into a vector. Second, the output from the computer will be the fuel metering. This needs to be sent out via a port.

After the code for this model is generated, the way the data enters and leaves the computer must be coded. Data may be continuously loaded into a memory location through direct memory access (DMA), in which case the computer simply uses the data in the appropriate memory location as if it were always the correct value. Alternatively, the data might be developed asynchronously causing a computer interrupt whenever the data is ready. In either case, code is required to properly process the data and insure its integrity. This is not included in the nominal RTW code processing (except when the code is generated for the dSPACE system or for the VXWorks system). The input and output ports placed in the Simulink model act as a placeholder for these I/O ports in the C-code.

A. The Real Time Workshop Code Generation Process

The process for generating the code from the model is extremely easy. In the Simulink menu is a section called Tools. Pulling this menu down gives access to the options of setting the RTW Parameters, or RTW Build. The parameters menu provides the options for selecting the solver used; in this case the selection is a fixed step with no continuous states, since the controller is totally discrete. When the code is executed and the options selected, the status of the process is provided in the MATLAB command window. An example of this output is shown in Figure 23. A section of the C-code generated is shown in Figure 24. This section of the code is a table look-up. The table look-up is a built in Simulink block. The code for this block comes out of a set of meta-code descriptions that The MathWorks calls a target language compiler description, or a TLC file. Each of the blocks in Simulink has a corresponding TLC file. These files may be edited and changed in any way the user might require.

The MathWorks has created a very clever process to automate the conversion of the Simulink and Stateflow models to code. The Simulink model file is really an m-file, with a .mdl file type. These .mdl files contain the complete description of the Simulink model, including description of the blocks used, their size, location, color, and other characteristics when drawn on the screen, and their connections. The code conversion first strips out the description of all of the drawing specific information. The resulting file is stored with a type of .rtw (real time workshop). The next step is to go to the .tlc file for each of the blocks and use its description of the code structure for the block (in the target language). An example of a .tlc file is shown in Figure 25. The figure shows the "Gain" block. The .tlc file is a meta-code description of the gain block. The file describes the code to be generated and the variables that need to be instantiated for the particular application.

```
MATLAB Command Window                                                    
File  Edit  Window  Help

StateFlow Generating code for "Fuelsys", targeting "rtw" ...

Options Chosen:
Inline event broadcast optimization option.
Flow-graph optimization option.

    name: 'FUELSYS_RTW.C'
    date: '25-Jun-1997 12:18:12'
    bytes: 85583
    isdir: 0

Code generation successful for machine: Fuelsys.
### Invoking Target Language Compiler on Fuelsys.rtw
### Loading THW TLC Function libraries
### Initial pass through model to cache user defined code
### Creating source file Fuelsys.c
### Creating part 1 of registration file Fuelsys.reg
### Creating parameter file Fuelsys.prm
### Creating model header file Fuelsys.h
### Creating part 2 of registration file Fuelsys.reg
### TLC code generation complete.
### Successful completion of RTW build procedure for model: Fuelsys
```

Figure 23 Diagnostics generated by real time workshop.

Thus if the gain block's name is changed to "control_gain," the coding process will cause the code generated to use this name when the block is converted to C-code. Each of the variable names in the block will be converted into code using the names that were placed on the signal lines in the block diagram, making it easy to understand and modify the resulting code.

To see these features in more detail, we convert a simple gain block to code. Figure 26 shows all of the pieces of this conversion. The code that results from the .tlc file is a full gain block multiplication since the gain was not equal to zero or plus or minus one. The name assigned to the output of the block is rtB.Control_Gain indicating that the output variable in the code is a structure with the top level name of rtB, and the next level set to the name used in the Simulink block (Control_Gain in this instance). The inputs to the gain block are always called rtU at the top level of the structure, and are given the name of the block output where it is connected (Input in this instance), as a substructure (the variable name for the input is rtU.Input). The general name for an output is always rtY, and this is also a structure, and the sub-structure is the name of the block that the gain block is connected to (Output in this case). The gain itself is also a structure, with a name rtP and a sub-structure name also taken from the name of the block in the Simulink diagram (rtP.Control_Gain in this case).

The complete description of the way that the conversion process takes place is shown in Figure 27. As the figure illustrates, the code generation process has a structure that is based on the Simulink and Stateflow model. Simulink allows the user to specify a custom block using a file structure called an s-function. These custom blocks may be compiled along with the Simulink model, but since s-functions are already C-code, they don't have to be created, only compiled and linked. In the process of creating a model, a special purpose .tlc file can be used to describe how the s-function custom block should be converted. Once the model is

```
Sys_s10_Speed_Es_Output(int_T tid)
{
/* Lookup2D Block: <S14>/Look-Up Table (2-D) */
    {
    int_T xIdx=0,yIdx=0;
    real_T xlo=0.0,xhi=0.0,ylo=0.0,yhi=0.0;
/* Compute the column index corresponding to the scalar input: rtB.s7_Switch
*/
    if (rtB.s7_Switch == 0.0) {
        yIdx = 0;
        } else {
        yIdx=rt_GetLookupIndex(&rtP.s14_Look_Up_Table__2_D_.ColumnIndex
[0],19,rtB.s7_Switch);
        }
    ylo = rtP.s14_Look_Up_Table__2_D_.ColumnIndex[yIdx];
    yhi = rtP.s14_Look_Up_Table__2_D_.ColumnIndex[yIdx+1];
/* Compute the row index corresponding to the scalar input: rtB.s24_Switch
*/
    if (rtB.s24_Switch != 0.0) {
    xIdx=rt_GetLookupIndex(&rtP.s14_Look_Up_Table__2_D_.RowIndex[0]
,17,rtB.s24_Switch);
    xlo = rtP.s14_Look_Up_Table__2_D_.RowIndex[xIdx];
    xhi = rtP.s14_Look_Up_Table__2_D_.RowIndex[xIdx+1];
        }
/* Compute the block outputs */
    if (rtB.s24_Switch == 0.0) {
    rtB.s14_Look_Up_Table__2_D_ = INTERP(rtB.s7_Switch,ylo,yhi,
    rtP.s14_Look_Up_Table__2_D_.OutputAtRowZero[yIdx],
    rtP.s14_Look_Up_Table__2_D_.OutputAtRowZero[yIdx+1]);
        } else {
    real_T Zx0ylo,Zx0yhi;
    Zx0ylo = INTERP(rtB.s24_Switch,xlo,xhi,
    rtP.s14_Look_Up_Table__2_D_.OutputValues[xIdx][yIdx],
    rtP.s14_Look_Up_Table__2_D_.OutputValues[xIdx+1][yIdx]);
    Zx0yhi = INTERP(rtB.s24_Switch,xlo,xhi,
    rtP.s14_Look_Up_Table__2_D_.OutputValues[xIdx][yIdx+1],
    rtP.s14_Look_Up_Table__2_D_.OutputValues[xIdx+1][yIdx+1]);
        rtB.s14_Look_Up_Table__2_D_INTERP(rtB.s7_Switch,ylo,yhi,Zx0ylo,
Zx0yhi);
        }
    }
/* Outport Block: <S13>/we */
    rtB.s10_Speed_Es = rtB.s14_Look_Up_Table__2_D_;
```

Figure 24 C-code generated for the engine control 2D table look up block.

converted to C-code, the code and the header file are combined in the "Make" step using the model make file with type .mk. At this step, any special purpose files needed for real time operation are brought together with the support files. These files are the C-code for the MathWorks programs and blocks (such as the blocks for the 2-deimensional table look-up above, and the numerical integration routines or solvers, like ode45.c, that might be needed). After the C-code is compiled and linked with all of the support files, the target processor code may be run. All of these steps are completed automatically.

A major feature of The MathWorks products is their open architecture. All of the files, the model (.mdl), the header (.h), the C-code (.c), the target language code (.tlc), and the make (.mk) files are ASCII files that can be edited and modified by the user in any way. This "open" feature of The MathWorks files is an

```
%% $RCSfile: gain.tlc,v $
%% Copyright (c) 1996-97 by The MathWorks, Inc.
%%                      Function:              FcnEliminateUnnecessaryParams
================================================
%% Abstract:
%%     Elimate unecessary multiplications for following gain cases when
%%     inlining parameters:
%%     Zero: memset in registration routine zeroes output
%%     Positive One: assign output equal to input
%%     Negative One: assign output equal to unary minus of input
%function FcnEliminateUnnecessaryParams(y,u,k) Output
  %if LibIsEqual(k, 0.0)
   %if ShowEliminatedStatements == 1
    /* %<y> = %<u> * %<k>; */
   %endif
  %elseif LibIsEqual(k, 1.0)
   %<y> = %<u>;
  %elseif LibIsEqual(k, -1.0)
   %<y> = -%<u>;
  %else
   %<y> = %<u> * %<k>;
  %endif
%endfunction
%%                              Function:                          Outputs
================================================
%% Abstract:
%%     Y = U * K
%%
%function Outputs(block, system) Output
  /* %<Type> Block: %<Name> */
  %assign rollVars = ["U", "Y", "P"]
  %roll sigIdx = RollRegions, lcv = RollThreshold, block, "Roller", rollVars
   %assign y = LibBlockOutputSignal(0, "", lcv, sigIdx)
   %assign u = LibBlockInputSignal(0, "", lcv, sigIdx)
   %assign k = LibBlockParameter(Gain, "", lcv, sigIdx)
   %if InlineParameters == 1
    %<FcnEliminateUnnecessaryParams(y, u, k)>\
   %else
    %<y> = %<u> * %<k>;
   %endif
  %endroll
%endfunction
%% [EOF] gain.tlc
```

Figure 25 An example of a .tlc file—gain.tlc.

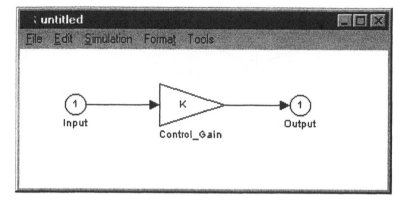

Figure 26 RTW code for the gain block.

important part of the overall process for converting models into code, since it allows the user to customize the code generation process for any desired target processor (or multi-processors for that matter).

B. Using Simulink for Code Generation with dSPACE and VxWorks Hardware

There are several ways that the code generated by RTW can be implemented in a real time environment. The MathWorks ships RTW with a DOS environment that allows code to be generated that will run on a stand alone DOS machine. In addition, there is a generic real time code that is not specific to any hardware. This code can be modified to incorporate into any operating system that the user would like.

The most robust of the code development environments are dSPACE and VxWorks. This section describes these hardware implementations.

dSPACE

The code that was generated in the previous section is not targeted to any specific machine, and as a consequence the user must make the code modifications needed

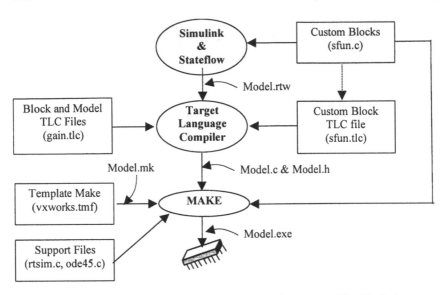

Figure 27 Real time work shop (RTW) code generation process. The blocks in gray are customizable by the user. At each step in the code generation process a file, "Model.type," is generated where "type" is as shown in the figure. In addition, the support files that are created along the way are indicated in each block.

to insure that the system operates in real-time, as desired. While this process is not difficult, it is time consuming and is not necessarily productive if the code is only going to be used in a single processor. For this reason, several manufacturers of DSP and computer boards have adopted Simulink and Stateflow as the coding paradigm for their proprietary operating systems.

The company dSPACE in Germany has created a complete hardware implementation that connects in a seamless way to MATLAB, Simulink and Stateflow. A set of DSP boards are available for the embedded system, along with an operating system that provides the required real time operation. The processor boards available are the DS1002, a floating point processor board, that is built around the Texas Instrument TMS 320C30 floating point digital signal processing chip. This card is used in a standard PC environment. The DS1003 processor is a parallel DSP Board that uses the TMS320C40 processor, and the DS1201 Multiprocessor Board, which is used in conjunction with the DS1003 Processor Board, and also uses the TMS320C40 chip. The hardware also consists of a complete line of I/O boards that provide analog to digital and digital to analog conversion, encoder inputs, pulse width modulation (PWM), and input to the CAN or VAN bus. There is also a prototyping board that will support the development of custom interfaces.

All of the boards in the dSPACE system can be placed in a specially designed enclosure that allows rapid connections to the host computer and also contains the required power supplies. The "AutoBox" enclosure runs off a DC input and allows the user to work in vehicle applications (for automotive and aerospace applications).

The various boards are interconnected with a high speed PHS bus that provides much faster communications rates. This means that the generated code can run at faster sample rates because the PHS bus has a lower computational overhead.

dSPACE provides software in the form of device drivers, an interrupt service routine, communications between the host processor and the DSP boards, data acquisition and monitoring, and interactive control panels that allow the user to control the process in a visual way. A custom template make file is provided that allows Real Time Workshop to automatically build the required code. The interface between Simulink, Stateflow, and RTW and dSPACE is through a Real-Time Interface (RTI), supplied by dSPACE, that includes support files that extend the Simulink block library by including a complete library of blocks for the dSPACE I/O device drivers. These additional support blocks are connected in the Simulink diagram using the same procedure that was used to build the Simulink diagram. The ability to add the drivers to the diagram makes it easy to control the way the actual control system is operating in real time. RTW and the dSPACE hardware are optimized for speed. After RTW generates the code the make file invokes a post processor that revises the code for speed, inserting as it does the appropriate device drivers and debugging tools. The drivers are coded "in-line" to enhance execution speed (approximately 6 microseconds per function call is saved this way).

Communication between Simulink and the dSPACE hardware is also part of the RTI. This means that any parameter in the Simulink diagram may be changed while the dSPACE boards are controlling the process, and the change will migrate from the Simulink model and cause an automatic update of the changed parameters that reside in the real time hardware.

TRACE is a software module that is provided by dSPACE to acquire data and monitor it as the process runs on the DSP boards. The TRACE browser allows selection of the data within the system based on the names that were assigned to the blocks in Simulink. TRACE is a virtual oscilloscope. It allows the user to record and graphically display all of the signals and parameters that are inside the DSP boards. Just like an oscilloscope, the display can be free running or triggered by a crossing of a specified level. The data that is collected can be typed as floating or integer and the distinction between static and dynamic variables can be made in the data acquisition. Data may be plotted in any time interval selected, in the form of strip charts, and in multiple plot windows. In addition, data may be saved in the MATLAB .mat format so the data may be post processed using MATLAB.

COCKPIT is another tool provided by dSPACE. This allows the user to create a virtual instrument panel that has the appearance of meters, indicator lights, push buttons, sliders, and digital displays. The developer can operate the system from the COCKPIT display if he chooses.

VxWorks

VxWorks is a real time operating system that has become an industry standard. The system was introduced in 1987. The operating system contains a fast, efficient, and highly scalable object oriented kernel that has over 1000 application programming interfaces that range from real-time operation to network utilities. The system contains a shell, a browser and a debugger that scales down as the target processor memory requirements get smaller.

VxWorks is a product from the WindRiver systems. The development environment provided by WindRiver includes StethoScope, a real time data visualization, profiling and debugging tool, a tool called WindView that detects system level problems, Wind C++ Gateway for the Object Center C++ development environment, and Tornado.

The Mathworks, Inc. supports the VxWorks operating system. The conversion of a Simulink model into embedded code is almost identical to the process used for dSPACE code.

IV. FUTURE DIRECTIONS FOR COMPUTER AIDED CONTROL OF MECHATRONIC SYSTEMS

This chapter has attempted to show the process whereby a designer develops a device that will control a mechatronic system. To put the process in perspective, Figure 1 (in Section I) showed schematically the steps that we went through in the development of the engine controller in this chapter.

These design steps are:

1. The modeling of the system using real data and physics. This combines judicious modeling of devices from first principles (as we did for the pressure in the manifold using the ideal gas law) with the numerical analysis of data obtained from testing components in the system (as we did for the airflow through the throttle). The two approaches allow the user to build a model that will be adequate for his design. The Simulink model that results can then be exercised to determine the characteristics of the system, and the developer can then hone in on the requirements he would like to satisfy. This step is often called a requirements flow down.

2. The requirements for the system performance are parceled out across the various subsystems, and ultimately are formulated as design specifications for the control system alone. This was done in the example by seeing how the engine performed without a control. A PI controller was then developed to improve the performance, and the values of the gains were specified to insure stability and adequate performance. The next step was to set error bounds on the time response of the engine and optimize the

control system (we used NCD for this, but there are many other ways of doing the same thing). This step is the control design and optimization step.

3. The next part of the design is the development of the control modes and the performance measurement and monitoring software. We used Stateflow for this step. The logic and the computer flow are tested in the combined Stateflow and Simulink simulation. This step is the performance verification and validation step. The designer insures that the system will meet the requirements and that the system works properly in the presence of failures and other problems.

4. The design is coded into the target processor and tested. This is the final step in the design process. The completed design, including the embedded software, is then ready for production implementation.

The design cycle is complete when the production system is in service. Even here, though, the engineer will need to maintain the ability to run the code and test new design concepts. The need to improve production devices and include new operating modes, etc. makes it inevitable that the production code will be changed as the system runs through its life cycle. The ability to make rapid changes with the assurance that the changes are not interfering with or negating a previously correct operation are important reasons for the automation of the design process.

This chapter has shown the current state of the art for automatic code generation, analysis and synthesis of systems that use computers for their control. The future directions in this process are toward more and more automatic coding and integrated development. In the future, a process is required that insures the developed code is tested in every possible path, and that no single point failure can cause the system to fail.

The code that is generated needs to be documented. One of the great features of a block diagram representation of a system is that it is self-documenting. The flow is illustrated in a graphical way, and as such it makes it easy for a future developer that was not involved in the early design process to pick up the "code" (the block diagram) and follow the design. A last attribute of the automatic process is the ease with which it can be adapted to changes in both computer hardware and system hardware. Software can also change (C replaced with C++ for example). The ability to modify a few simple files and thereby adapt to all of the external changes is a very important feature of this design approach.

The design software developed by The MathWorks, Inc. is making life easier for designers of complex systems. This process is just now being accepted. In the future, this process will be the norm. Also, more and more of the features required to automate the code generation process will be included in future releases of MATLAB, Simulink, Stateflow and their related toolboxes.

ACKNOWLEDGMENTS

The names dSPACE, TRACE, COCKPIT are trade marks of the company dSPACE, GmbH. VxWorks, Tornado, and StethoScope are trademarks of WindRiver Systems Inc. In addition to MATLAB, Simulink, and Stateflow, the names Real Time Workshop and RTW are trademarks of The MathWorks, Inc. Any other trademarks and registered trademarks referred to in this chapter are the property of the trademark owner.

The author would like to acknowledge the following contributors to this section. The engine timing model and support for the development of the Simulink model for the engine was provided by Ken Butts and Jeff Cooke of Ford Motor Company. The Ford Motor Company is acknowledged for permission to use this model in this chapter. The engine timing model is based on a paper by J.A. Cooke and P.R. Crossley [5]. Bill Aldrich developed the Stateflow model for the fuel system, Stan Quinn has added significant improvements and changes to the fuel system model and he reviewed this manuscript for the author. Paul Barnard developed pieces of the control model. Bill, Stan and Paul are members of the staff of The MathWorks, Inc.

APPENDIX   OVERVIEW OF MATLAB, THE CONTROL SYSTEM TOOLBOX AND SIMULINK

A.1   THE MATLAB LANGUAGE STRUCTURE

MATLAB [4] is a language that has its roots in Algol and related high level languages. Its main feature is that every object in the language is automatically treated as an array. In terms of structured programming languages, all objects in the language are structures that contain all of the information needed to identify the object. Thus, in MATLAB typing the following line will create a 3x3 matrix:

```
a =
1  2  3
4  5  6
7  8  9
```

The matrix object that was created by this instruction consists of the following information (this information is available in MATLAB by typing "whos"):

```
Name     Size    Bytes      Class
a        3x3      72        Double array
Grand total is 9 elements using 72 bytes
```

The objects in MATLAB may be double precision arrays of any dimension (not just matrices), sparse matrices, string variables, complex variables, structures and cells. In addition, the language allows object-oriented programming and operators that are overloaded. The structures, cells, and object oriented programming features are all new to MATLAB 5 that was released in January of 1997.

Since MATLAB uses a structure to define the objects in manipulates, the language permits matrix operations to be performed in a single instruction. For example, if b were a 3x3 matrix, then typing a*b will automatically perform the requisite matrix product. Similarly, the operators / and \ have been defined to be equivalent to inverting the matrix and multiplying (from either the right or the left). Thus a\b is $a^{-1}b$ and a/b is ab^{-1}.

Not only are these constructs a basic part of the language, but every operation in the language recognizes the structure and acts accordingly. For example, if a is a 3x3x3 array and the instruction b = cos(a) is executed, every element in b will be the cosine of the corresponding element in the array a. This feature means that loops (in this case three loops) do not have to be written to cause operations on arrays.

The other major feature of MATLAB is its built-in matrix operations. Eigenvalues, eigenvectors, determinants, matrix inverses, and many other matrix operations are executed with a single command. This capability is where the name MATLAB originates. In the 1970s, Cleve Moler, then at the University of New Mexico, decided to build a teaching tool for linear algebra based on the subroutine packages LINPACK and EISPACK that had just been developed at Los Alamos National Laboratory. MATLAB, or Matrix Laboratory was the resulting tool. From this starting point, where all of the numerical methods were state of the art, MATLAB has maintained a complete set of built-in matrix algorithms that are robust and numerically as accurate as possible. All computations are done with double precision using the IEEE floating point standard (including inf for infinity and NaN for Not a Number.

MATLAB also has an extensive graphics capability that is illustrated by all of the figures in this chapter. There also is a graphical user interface (GUI) building tool, called "guide™," that builds user interfaces rapidly and easily using a WYSWYG tool with click and drag capability.

MATLAB commands can be edited into programs called m-files. These programs may be scripts or functions. The difference between a script and a function is in the first instruction that appears in the file. If the instruction is of the form "function [a b c d...] = name (x, y, z,...)," where "name" is the name of the function, then a, b, c, d, ...are the outputs returned by the function and x,y,z,... are the inputs. The m-file is automatically a function that will be executed whenever the command line parser sees the instruction [a b c d...] = name (x, y, z,...). Any variables generated internally by the function do not appear in the MATLAB workspace, so variables generated within functions do not clutter the work space. If the m-file does not start with function, the instructions are executed as a "main" program or a script.

Figure A.1 shows a list of the built in operators that are available in MATLAB. These permit operations on matrices, arrays, structures, and classes that are both numeric and logical. The operation is denoted by the symbol on the right, and the name of the operation is used to overload the operator for a particular object.

In addition to these basic commands, MATLAB comes with a wide variety of language constructs that permit programs to alter their execution sequence (such as for loops, if-then-else, case, etc.). There are m-files that provide matrix,

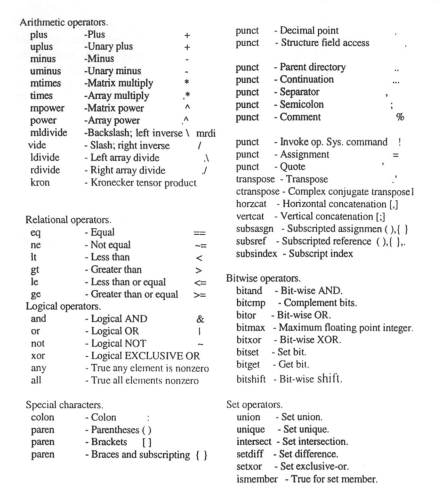

Arithmetic operators.

plus	-Plus	+
uplus	-Unary plus	+
minus	-Minus	-
uminus	-Unary minus	-
mtimes	-Matrix multiply	*
times	-Array multiply	.*
mpower	-Matrix power	^
power	-Array power	.^
mldivide	-Backslash; left inverse \	mrdi
vide	- Slash; right inverse	/
ldivide	- Left array divide	.\
rdivide	- Right array divide	./
kron	- Kronecker tensor product	

Relational operators.

eq	- Equal	==
ne	- Not equal	~=
lt	- Less than	<
gt	- Greater than	>
le	- Less than or equal	<=
ge	- Greater than or equal	>=

Logical operators.

and	- Logical AND	&
or	- Logical OR		
not	- Logical NOT	~
xor	- Logical EXCLUSIVE OR	
any	- True any element is nonzero	
all	- True all elements nonzero	

Special characters.

colon	- Colon	:
paren	- Parentheses ( )	
paren	- Brackets [ ]	
paren	- Braces and subscripting { }	

punct	- Decimal point	.
punct	- Structure field access	.
punct	- Parent directory	..
punct	- Continuation	...
punct	- Separator	,
punct	- Semicolon	;
punct	- Comment	%
punct	- Invoke op. Sys. command	!
punct	- Assignment	=
punct	- Quote	'
transpose	- Transpose	.'
ctranspose	- Complex conjugate transpose	'
horzcat	- Horizontal concatenation [,]	
vertcat	- Vertical concatenation [;]	
subsasgn	- Subscripted assignmen ( ),{ }	
subsref	- Subscripted reference ( ),{ },.	
subsindex	- Subscript index	

Bitwise operators.

bitand	- Bit-wise AND.
bitcmp	- Complement bits.
bitor	- Bit-wise OR.
bitmax	- Maximum floating point integer.
bitxor	- Bit-wise XOR.
bitset	- Set bit.
bitget	- Get bit.
bitshift	- Bit-wise shift.

Set operators.

union	- Set union.
unique	- Set unique.
intersect	- Set intersection.
setdiff	- Set difference.
setxor	- Set exclusive-or.
ismember	- True for set member.

Figure A.1 MATLAB operators—definition of symbols used to manipulate arrays in the language.

vector and scalar functions such as Bessel functions, matrix exponentials, and a wide range of elementary functions that provide the user with sines, cosines, tangents, and the related hyperbolic functions for both real and complex arithmetic. The feature of MATLAB that it automatically reverts to complex arithmetic when needed is very powerful. For example, if the angle of a complex variable is required, the MATLAB command imag(log(z)) gives the correct answer (since the complex variable z is given by $Re^{i\theta}$, taking the log gives logR + iθ so the imaginary part of this expression is the desired angle θ).

The extensive command structure of MATLAB is augmented by many "toolboxes." A toolbox is a collection of m-files and graphical user interfaces that perform specific design functions. For example, the Control System toolbox performs simple control system analyses (Bode, root locus, Nichols, Nyquist plots).

There also is a toolbox that will design control systems that are insensitive (robust) to variations in the system dynamics, and there is a tool box that will selectively optimize a parameter in a system by forcing the response to have some desired characteristic. A list of the toolboxes available is shown in Figure A.2.

The next sections show some of MATLAB's features through some simple examples. Since the main thrust of this chapter is on mechatronic systems, the examples that are shown illustrate both mechanical and electrical systems. The examples shown use MATLAB. In Section I we used these examples to illustrate how the same computations can be achieved with Simulink. For a detailed description of MATLAB, and how it is used, see the MATLAB Users Manual [4].

A.2 MATLAB LANGUAGE EXAMPLES—THE CONTROL SYSTEM TOOLBOX

The MATLAB language allows many different approaches for developing systems. Starting with MATLAB 5, the Control System Toolbox uses MATLAB's object oriented programming feature to define system objects that are called LTI objects (Linear Time Invariant Objects) [5]. The objects can be created using a constructor (in MATLAB, constructors reside in a subdirectory that starts with the symbol @). LTI objects can be built using transfer function, state space model or zero/pole/gain forms. The constructors for each of these are called tf, ss, and zpk respectively. Constructors create objects that may be manipulated using arithmetic rules whose meanings have been altered. This "overloading" of the operators is a very powerful feature of MATLAB, and the user may modify and/or create new rules for these operators if needed. For example, to create an LTI transfer function object the convention is that the numerator and denominator are vectors with the powers of s descending. Thus to create a transfer function with the numerator s and denominator $s^2 + 7s + 12$, the vector for the numerator is [1 0] and the denominator is [1 7 12], and the constructor tf is used as follows:

» sys1 = tf([1 0], [1 7 12])

Transfer function:

```
        s
-------------------
s^2 + 7 s + 12
```

The same system can be converted from the transfer function form to the zero/pole/gain form using the constructor zpk as follows:

Tool Box	Description
Communications	*A set of blocks that simulate communications systems*
Symbolic Math	*An object oriented MATLAB front end to the MAPLE kernel*
Mapping	*Map tools, projection & data interpretation in terms of maps*
Wavelet	*A set of tools and a gui for analyzing data with Wavelets*
Partial Differential	*A MATLAB 2D pde solver using finite elements*
Financial	*A set of tools for the financial community*
LMI Control	*Linear Matrix Inequality toolbox for Lyapunov control design*
QFT Control	*Quantitative Feedback Theory design based on Horowitz*
Fuzzy Logic	*Fuzzy logic design tool with a gui for building fuzzy blocks*
MPC Tools	*Model Predictive Control design tools*
Frequency Domain ID	*System ID in the frequency domain*
Higher Order Spectral	*Spectra for nonlinear random systems*
Statistics	*A set of tools for statistical analysis and inference*
Image Processing	*A set of tools for image processing and analysis*
Neural Networks	*Create control systems and logic based on neural networks*
Mu-Analysis	*Control system design using modern robust control*
Signal Processing	*A set of tools and a gui for signal processing*
Splines	*A set of m-files for creating splines of any order*
Optimization	*Minimization, iterative solutions to nonlinear equations*
Robust Control	*Robust control tools: sigma plots, loop transfer recovery, ...*
System Identification	*System ID using auto-regressive moving average models*
Control System	*The general control system tool set*
Related Products	
Simulink	*Block diagram manipulation tool that allows rapid simulation*
Stateflow	*Modeling of Discrete and Event Driven Systems*
Fixed Point Block Set	*Does fixed point calculations in Simulink*
DSP Block Set	*Blocks for building block diagrams of DSP processes*
Nonlinear Control	*Optimization of a system using its Simulink diagram*

Figure A.2 MATLAB tool boxes and related products.

```
» zpk(sys1)

Zero/pole/gain:

       s
---------------
(s+4) (s+3)
```

The last constructor is the state space constructor ss. Operating on sys1 with ss causes the state space model to be generated along with names for the states, the inputs and the outputs. The default names for the states are x1, x2, ... The default names for the inputs are u1, u2 etc. and for the output are y1, y2, etc. Thus the lti object has information in it other than the dynamics. In addition to the variable names, all lti objects can be continuous or discrete (for continuous systems, they have a sample time of zero), they can have a time delay associated with the input or output, and they can have a string of data such as a description of the system. To illustrate these, the state space model is formed using ss as follows:

» ss(sys1)

a =

	x1	x2
x1	-7.00000	-3.00000
x2	4.00000	0

b =

	u1
x1	1.00000
x2	0

c =

	x1	x2
y1	1.00000	0

d =

	u1
y1	0

Continuous-time system.

The conversion of this system from a continuous to a discrete system uses a MATLAB function that is built into the control toolbox called "c2d". Thus, if the sampling time is one second and the conversion is to be with a zero order hold on the input, the syntax for the conversion is:

» sys2 = c2d (sys1,1,'zoh')

Transfer function:

$$\frac{0.03147\ z - 0.03147}{z^2 - 0.0681\ z + 0.0009119}$$

Sampling time: 1

The options for the conversion are: zero order hold ("zoh"), first order hold ("foh"), Tustin ("tustin"), pre-warped Tustin ("prewarp"), and a match of the poles and zeros of the transfer function for single input single output systems ("matched"). For the pre-warped Tustin, the match in the transfer functions (the critical frequency) is specified as an additional argument in c2d.

The powerful feature of the lti object is the overloading of the operators. In the process of modeling a complex system there are three subsystems (sys1, sys2, and sys3) that are to be connected so that sys1 and sys2 are in parallel and the resulting system is in series with sys3. The MATLAB command for causing this concatenation is:

sys = (sys1+sys2)*sys3

When the resulting system (sys) is created, it will be an lti object and its form will be inherited from the objects that appear on the right hand side. Thus, if any one of the objects is a state space form, the object sys will be also. Note that the objects only need to be lti objects, sys2 could be a transfer function, and sys3 could be a zero/pole/gain object and this instruction would be correct. MATLAB commands that perform these operations are (commands follow the » symbol):

» sys1 = tf([1 0],[1 2 1])

Transfer function:

```
        s
  ---------------
  s^2 + 2 s + 1
```

» sys2 = zpk([-1 -2],[-2 -3 -4],10)

Zero/pole/gain:

```
    10 (s+1) (s+2)
  ---------------------
  (s+2) (s+3) (s+4)
```

» sys3 = tf([1 2],[1 1])

Transfer function:

```
  s + 2
  ------
  s + 1
```

» sys = (sys1+sys2)*sys3

Zero/pole/gain:

$$\frac{11\ (s+2)^\wedge 2\ (s+0.3195)\ (s^\wedge 2 + 3.044s + 2.846)}{(s+1)^\wedge 3\ (s+2)\ (s+3)\ (s+4)}$$

During the preliminary design phase of a control system, the ability to manipulate systems as a single object can be extremely useful. MATLAB allows the object to be used as the input for all of the control system functions. Thus the syntax extends beyond the concatenation of systems. All of the analysis tools for Bode plots, root locus, Nyquist plots, and Nichols charts are available by typing the name of the function with the argument the lti object "sys." Thus the command bode(sys) generates the bode plot shown in Figure A.3.

The usefulness of this syntax for creating control systems is obvious. During the preliminary design phase of a mechatronic system there are hundreds of decisions that need to be made that all interact. The ability to take a slice of the puzzle

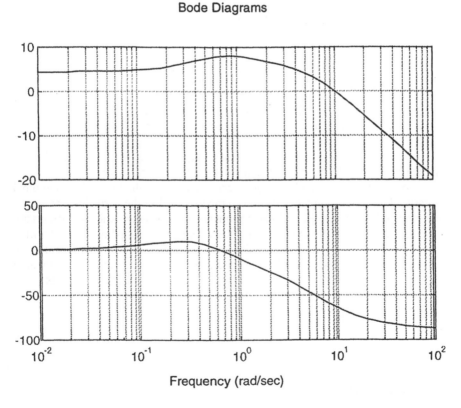

Figure A.3 Bode plot for lti object sys 1.

and analyze it by itself in a simple and easy to use programming environment is very powerful. MATLAB is the computational engine that allows these applications. The next section illustrates Simulink, a graphical programming environment that allows simulation models to be built, control systems to be structured, analyzed, and synthesized, and code to be developed. Simulink is illustrated with an automotive engine controller. The control system is developed using MATLAB"s NCD toolbox in addition to Simulink. Simulink also allows code to be generated directly from the diagram. Also, a tool for creating state charts called Stateflow is available. Stateflow allows event and logic driven systems to be drawn in a simple graphical format and also allows code to be created directly from the diagram.

A.3 SIMULINK MODELS

MATLAB and the augmentations provided by the toolboxes provide wonderful intuitive tools for analyzing and synthesizing systems. By themselves these tools give a knowledgeable analyst most of what is needed for control systems design and analysis. However, today the business environment pays a significant premium for accuracy and speed in design. Thus, the ability to simulate a system and design components for that system in a rapid and accurate format is very important. MATLAB provides part of the answer to this need, but it is the visual aspect of a block diagram that really makes it possible to quickly design complex systems and at the same to verify that they are accurately portrayed. Simulink [6] achieves these goals.

Simulink is a block library that has a complete set of extensible blocks for linear and nonlinear continuous or discrete time systems. Figure A.4 shows the blocks that are available.

To build a simulation model, all that is required is a click and drag. The icon in the library is clicked and dragged into a separate window that is the model.

As the blocks are added, they inherit the name of the block from the block library. The name of the block can be changed to reflect the variables that are being modeled. These names will also be inherited by the automatic code generation system that is part of Simulink called "Real Time Workshop" or RTW. To illustrate how a simple system can be built, let's simulate Galileo's experiment at the Leaning Tower of Pisa. A weight is dropped from the tower, and we want to know its position and velocity as functions of time. The equation for this system is:

$$\frac{d^2 z}{dt^2} = -32.2$$

with

z(0) = 150 ft.
dz/dt(0) = 0.

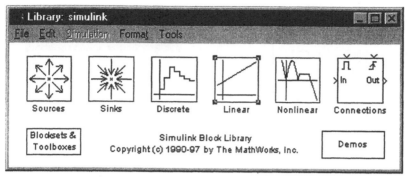

Figure A.4 Simulink block library.

The simulation for this system requires two integrators and a constant. Integrators are in the linear system block set and the constant is in the "sources" library.

Clicking and dragging the integrators and the constant block into the Simulink model window (untitled in Figure A.5) creates the model. This gives the start of the model as shown in the figure. To change the constant value in the Constant block to 32.2 (it is set to 1 in the block library), the user simply double clicks on the block and a dialog window opens that allows the user to specify the value of the constant. The last step is to connect the blocks together. This is accomplished by clicking on the arrow port at the block output and with the mouse button held down, dragging the line that results until the cursor changes to a cross at the input port where the connection is to be made. When this is complete, the block diagram, and the simulation is complete. If the user would like, he can specify the names of the variables on the lines in the block diagram. Simply double clicking on any line will place a question mark on the line that the user can then type a name. For this example, we will change the names of the variables to position and velocity and we will also rename the integrator blocks to position integrator and velocity integrator. The final block diagram appears in Figure A.6.

Notice that the fact that the mass hits the ground is used to stop the simulation through the block labeled "stop." The figure also shows a unique Simulink block called a "Scope." This allows the user to look at a graph of any of the variables generated in the simulation. The last step is to simulate the system using the "Start" command in the Simulation pull-down menu. The result is the plot of the position as a function of time shown in the Scope window in Figure A.6.

The user has access to many different commands within Simulink that allows selection of the type of integration, the length of the simulation, the variables, if any, that will be sent to MATLAB"s command window, and the ability to generate code directly from the diagram. Unlimited numbers of subsystems may be included in a Simulink block diagram. A subsystem lives in its own window. When this window is closed the subsystem appears as a block with as many inputs

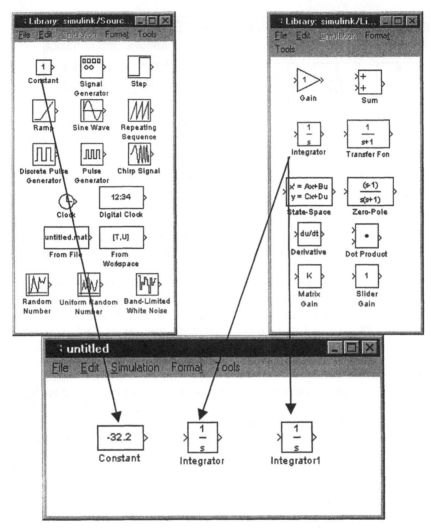

Figure A.5 Creating a Simulink model for the "Leaning Tower of Pisa" is as simple as clicking and dragging a block from the library.

and outputs as required. Subsystems may be nested in any order desired. Subsystem blocks help to keep the signal flow in the system clear and concise.

A subsystem block may also be covered over with a "mask" to allow a generic block to be built. For example, if you often have a need for a block that takes the absolute value of the input and then raises the result to some power, you could build a masked block by first building the absolute value and then raising that to a

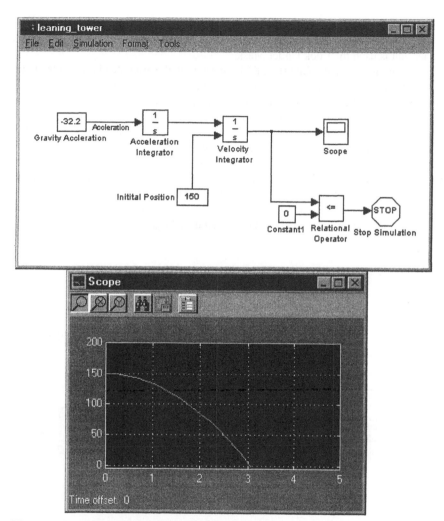

Figure A.6 "The Leaning Tower of Pisa" example using Simulink—final block diagram and the output of the simulation on the scope.

power using nonlinear blocks. The resulting subsystem block is masked using the "Mask Subsystem" command in Simulink"s edit pull down menu. The subsystem will become a block that can be used over and over again with different values for the parameter power. With a masked subsystem, clicking on the block opens a dialog window that is created when the system is masked. The dialog window in this case contains the value of the Power (sent into the block as the variable "power"). Figure A.7 illustrates these steps. The masked subsystem is shown first.

The dialog box to the right appears when the user double clicks on the subsystem block, and the block diagram under the mask is seen when the pull down menu under edit is used to "Look Under Mask."

Among the many features of Simulink that the user needs to understand, perhaps none is more powerful than the "Enabled" and "Triggered" subsystem feature. Within the Simulink connections block set are two blocks that really are not blocks at all, but are behavior modifiers. The first of these, the Enabled block, causes a subsystem to be turned on or off based on a signal. When the enable block is placed in a subsystem, an enable input appears on the subsystem block whenever the value of the input at this port is greater than zero. The block itself can be used to specify what happens to the internal subsystem states when the enable signal goes away (i.e. are the states held at their last value or are they reset?). A triggered subsystem will be executed once when a trigger signal appears at the input port. The trigger may be set to be rising, falling, or both. Triggered subsys-

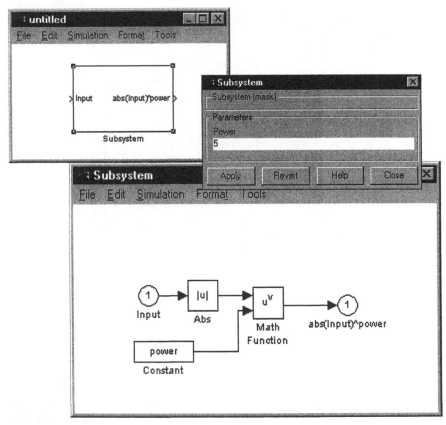

Figure A.7   Using the mask command to create a masked subsystem.

tems must always be discrete. The subsystems that are not enabled are not executed when the Simulink simulation is started. Since they are only executed when the enable signal is greater than zero, there is no penalty in the simulation speed.

REFERENCES

1. JA Cooke and PR Crosley, IEE Conference, Control 91, Conference Publication 332, Edinburgh, U.K, March 25-28,1991, vol. 2, 921-925.
2. *Simulink Automotive Examples Brochure*. Natick, MA: The Mathworks, Inc., 1997.
3. *Stateflow Users Manual*. Natick, MA: The Mathworks, Inc., 1997.
4. *MATLAB Users Manual*. Natick, MA: The Mathworks, Inc., 1997.
5. *Simulink Users Manual*. Natick, MA: The Mathworks, Inc., 1997.
6. *Control Toolbox Users Manual*. Natick, MA: The Mathworks, Inc., 1997.

8

Rapid Prototyping of Mechanical and Electronic Subsystems of Mechatronic Products

Adam Postula
University of Queensland, Brisbane, Australia

Periklis Christodoulou
CSIRO Manufacturing Science and Technology, Brisbane, Australia

The objective of this chapter is to provide the reader with an overview of methods, tools and the latest technologies used in rapid prototyping of mechatronic systems. The tools and methods for system level development are covered in other parts of this book; here we focus on rapid prototyping of mechanical and electronic hardware. The mechanical hardware provides the crest of the system and without its prototype it is very difficult if not impossible to fully evaluate system functionality, customer response, performance, reliability, etc. The electronic hardware provides the kernel for the intelligent control of the mechatronic system. In the case where high speed or volume production is needed the controller must usually be completely implemented in hardware as an Application Specific Integrated Circuit (*ASIC*). A prototype of a such controller should be available as early as possible to allow performance evaluation of the complete system.

Figure 1 shows the generalized process of mechatronic product development. The idea of the product must be converted into a technical description that is more precise and allows first-proof analysis. For this purpose a simplified model of the system and its working environment should be created. This may involve development of a CAD solid model of the object, definition of physical properties and/or modelling of the required behavior. Simulation

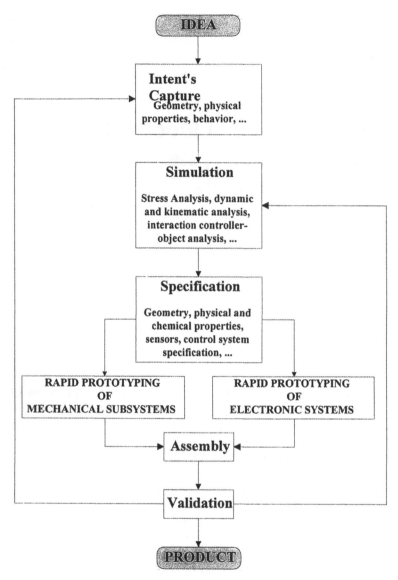

Figure 1   Generalized diagram of mechatronic product development.

of the mechatronic system allows the designer to experiment with various design options and leads to a more precise specification of the product.

   Complete *simulation* is usually not practical since the detailed model of a complex system can be too difficult to develop and too slow to execute on a computer. A simplified model is adequate only to a limited extent and for this reason *simulation* cannot fully replace the experiments with actual hardware.

Rapid prototyping of mechanical and electronic subsystems provides a means to produce an accurate physical model of the product and allows its evaluation in conditions very close to real situations. Since a complex mechatronic product usually needs to be fully tested to prove its validity before financial commitment to production, rapid prototyping is of great importance.

Rapid prototyping of mechanical parts is a complex process and consists of many preparation steps before the actual prototyping technique is applied to produce a physical object. Section I provides an overview of the whole process and different techniques for rapidly prototyping the mechanical parts of a mechatronic system.

Rapid prototyping of electronic hardware is closely related to the electronic design process spanning the simulation, synthesis and reprogrammable circuit technology for hardware implementation. Section II provides an introduction and overview of the concepts, methods, techniques and reprogrammable circuit technology applied in the electronic circuit prototyping process.

I. RAPID PROTOTYPING OF MECHANICAL SUBSYSTEMS

The term, "rapid prototyping" in manufacturing refers to the ability to make prototypes of almost any shape quickly and accurately, including internal cavities. The term, "free-form fabrication" refers to the fact that no special tooling, with regard to the shape, is required. This and developments in information technology have revolutionized the product development process.

Rapid prototyping may be quite involved and might entail *simulation* of the process that will be used to make the prototype as shown in Figure 2. This exemplifies the process of manufacturing of a prototype cast tool that may be used to manufacture prototype parts. The upper shaded area, in Figure 2, represents the part of the process described in Figure 1, while the rest of the diagram is specific to the rapid prototyping route. In the loop "design briefing-solid modelling" the idea of the design intent is captured. Then performance of the product is simulated from a stress or dynamic point of view. This may result in design changes. When the *simulation* results are satisfactory, the solid model of the product (specification) is an input for the casting design process. The product is reviewed from the casting process point of view and accordingly changed in an iterative process, "design-casting process *simulation*," until both the design intent and the casting requirements are satisfied. Finally, a prototype tool is manufactured using rapid prototyping techniques.

The ability to make prototypes cheaply within a few days allows many design iterations without actually incurring the cost of the normally required production tooling. Prototypes are built for visual inspection, testing the design and market assessment. It is now possible to manufacture very complex internal cooling, heating, lubricating or any other form of ducting channels within the bulk of the material that conforms to the shape of the tool or part.

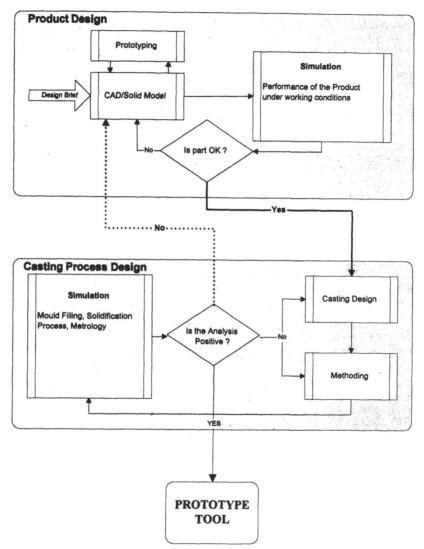

Figure 2   An example, cast prototype tool development.

Prototype parts are no longer the ultimate aim, now that prototype tool development has become a reality. In many cases, instead of providing prototype parts it is more appropriate to manufacture prototype tools to produce parts that will have the same properties as the future production parts. This makes the required testing of the parts possible. *Rapid Prototyping and Tooling (RP&T)* is about manufacturing of fully functional parts through the use of prototype tools.

Because the existing prototyping techniques can use a limited number of materials, several conversion techniques have been developed. Basically, *rapid prototyping* provides the shape and the conversion technology transforms it into the part made of the material of choice.

There are three basic limits to any *rapid prototyping* (*RP*) technique: material, size of the prototype that can be manufactured and geometrical accuracy. The conversion technologies are mainly limited by the geometrical accuracy of the part. While the cost of prototyping using *RP* techniques is still quite high, it is however far cheaper than conventional prototyping in many cases. It is reasonable to manufacture 1 to 5 prototypes using any of the *RP* techniques when material is not a problem. When the required number of parts is higher, tooling techniques may be more appropriate. Depending on the number of parts several possibilities exist (see Figure 3). Some of them are described in section I.B.

A. Rapid Prototyping Techniques

All *RP* techniques described below use the same idea of manufacturing a three dimensional object by building it layer by layer. Using 3-dimensional computer aided design (3D CAD) software an electronic solid model of the object is first developed and then divided into a set of slices to be sequentially built by a relevant *rapid prototyping* process. In each case the slicing process is adjusted to the *rapid prototyping* technique to be used as the thickness of the slices and the direction of slicing affect the feature's definition and the manufacturing cost/time of the prototype. Thinner slices give better accuracy; however the time required for building the prototype is greater and this affects the overall cost. In general, the process itself and the materials used define the minimum thickness of the layer and the direction of slicing and for these reasons each

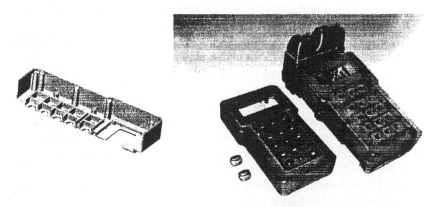

Figure 3 Prototype ATM assembly. (Courtesy of G. Tow, Keycorp. Limited, Sydney, Australia.)

of the *rapid prototyping* developers has developed its own proprietary slicing software. Usually STL files are used for further processing. More information can be found elsewhere in the literature ([1], [2], [3], and [4]).

It must be noted that files created by X-ray tomography, optical scanning systems (e.g., laser digitisers) or ultrasonic scanning that carry the information about existing 3-dimensional objects can be used to reproduce the objects by most *RP* techniques. The most widely used *RP* process for these purposes is stereolithography.

Once the slicing process is finished, the parts are built and then post processed. Post processing involves removal of any supporting structure, polishing, post-curing and inspection. The post processing is usually labor intensive and time consuming.

Due to the nature of this book and limited space many *rapid prototyping* techniques will not be described. Among them are solid ground curing (SGC), precision stratiform machining (PSM) and high speed machining. Initial information on these techniques can be found in *Prototyping Technology International '97*, published by UK & International Press (ISSN 1367-2436). There are also a number of additional technologies derived from the ones described below.

Stereolithography

The stereolithography (SL) *rapid prototyping* technique uses liquid photopolymers that cure under ultraviolet light. Basic steps involve deposition of a resin layer of a required thickness and then curing the part of the resin layer that will form the prototype. Figure 4 shows the schematic diagram of the equipment. A layer of the resin is deposited on the movable platform and the first slice of the prototype is built by curing the required parts of the layer using an ultraviolet laser beam, which is controlled by a computer program. This process is repeated until the prototype is built.

The deposition of the new layer is known as the recoating process. Each layer of resin is laid using a curtain coating system [5] in which a liquid curtain is formed by pumping the resin to a pouring head. The coating head deposits a layer through a narrow gap by moving the liquid curtain over the substrate. The coating head is wider than the vat and the resin falling outside the vat flows back to the resin reservoir.

The required support for overhangs or undercuts is built in the form of a sparsely hatched structure that is easy to remove. The model resides in the liquid polymer during the build. When the building process is finished, the platform rises and liquid resin drains back into the vat. The part is removed and the excess resin wiped away. Then the parts are placed in a curing chamber lined with UV light bulbs for a short period of time. Finally, any support structure is removed and the part is polished.

There are a number of manufacturers of stereolithography equipment. 3D Systems Inc. has dominated the market in the U.S. The machines manufac-

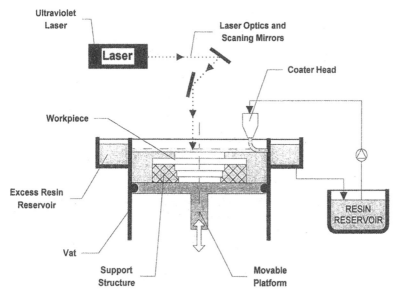

Figure 4 Schematic diagram of the stereolithography process.

tured are categorized by the working envelope (building volume). 3D Systems manufacture three types of SL machines, SLA-250, SLA-350 and SLA-500. The acronym SLA stands for stereolithography apparatus and the numbers indicate the working envelope in millimeters, e.g., 250 × 250 × 250. EOS builds two types of stereolithography machines, STEREOS DESKTOP with working envelope 250 × 250 × 250 mm and STEREOS MAX with working envelope 600 × 600 × 400 mm.

Photo-polymer resins are available from Ciba-Geigy, AlliedSignal, Du-Pont and UBC Chemicals. Ciba-Geigy has developed a series of resins for SLA machines, SL5170 for SLA-250, SL5180 for SLA-500, SL5190 for SLA-350. The Exactomer is from Allied Signal and Uvecryl from UBC Chemicals. Some selected properties of these resins are shown in Table 1.

In general, SL resins are hygroscopic and have a low heat deflection temperature; thus an SL prototype part's accuracy diminishes over time. Under humid conditions or at elevated temperatures SL parts will degrade, swell, warp and distort. However, AlliedSignal announced [8] a new resin Exactomer HTG 35X which is water-resistant, retains high modulus at temperatures up to 100°C and has a glass transition temperature of ~130°C.

Best applications for SL are visual and conceptual models, parts where detail and accuracy are important, master patterns for secondary processes, and snap feature applications. The overall accuracy of SL parts is ±100 μm and generally the relative error is of the order of 0.25 to 0.5%.

Table 1   Selected Material Properties of Some Commonly Used SL Resins[a]

Property	SL5170	SL5180	Exactomer 5201 AR	Uvecryl ST-100
Tensile strength, MPa	60	43	48	39
Tensile modulus, MPa	3960	2700	1379	2100
Tensile elongation at break, %	7–19	6–16	6–7	4
Felxural modulus, MPa	2977	2527	—	1950
Flexural strength, MPa	108	88	—	63
Impact strength, J/m	33	38	33	22
Hardness, Shore D	85	84	84	75
Glass transition temperature, °C	65–90	60–85	78–90	53
Thermal linear expansion coefficient	—	—	—	$7.9\ 10^{-6}$

[a]From Refs. 6 and 7.

Laminated Object Manufacturing

The Laminated Object Manufacturing (LOM) *RP* technique utilizes sheet material to build layer-wise, 3-dimensional objects. Sheet material is placed on the working platform (Figure 5) and is then laminated by a heated roller travelling across the top surface in a single reciprocal motion. The adhesive under the top sheet of the material is activated and adheres to the bottom sheet. The corresponding cross-section is cut using a laser beam. This laser also crosshatches the part of the sheet that is to be removed. The power of the laser beam can be adjusted to the thickness of each layer and type of material. Postfinishing is manual and involves removal of the crosshatched material, sanding and then protective coating of the part. The post processing may vary with different materials.

Any sheet material can be used: paper, plastic, polymer composite prepregs, metals, ceramics and ceramic matrix composies [9]. Paper sheet is the most popular; the finished part possesses properties similar to those of wood and is fairly inexpensive. The choice of material depends on application and its availability in the form of sheets. Commercial versions of LOM machines are using paper (LOMPaper), plastic, and recently introduced fiber-reinforced glass-ceramic sheet materials [10]. Selected properties of LOMPapers are provided in Table 2.

Several ceramic materials like SiC thin sheets, ceramic matrix composites (SiC/SiC ceramic type layer/ceramic fiber layer) and polymer matrix composites (glass/epoxy resins) have been successfully tried [3] with the LOM process. New materials require different post-processing which often involves additional post-curing or other heat treatment process.

Applications of LOM include sand casting patterns, dispensable hydroforming tools, patterns for investment casting, prototype tools for blow moulding and injection moulding. For visualization and form fitting, paper can be

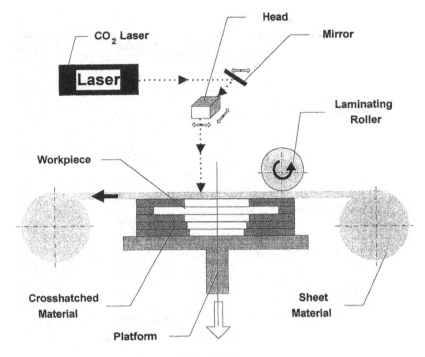

Figure 5 Schematic diagram of the LOM process.

used. For snap-fit and limited functional testing, plastic materials are recommended. The relative dimensional accuracy of the LOM process depends on overall size and is of order of ±1 to ±2% for small parts (up to 30 mm) and ±0.25 to ±0.5% for bigger parts. Surface finish is not as good as SL.

Helisys Inc. offers two LOM machine models, LOM-1015 and LOM-2030H with working envelopes 245 × 367 × 367 mm and 813 × 559 × 508 mm respectively. Both machines are controlled by the proprietary software, LOMSlice™, operating on Microsoft Windows NT platforms [9].

Selective Laser Sintering

Selective Laser Sintering (SLS) begins with spreading a thin layer of powder on the working platform using a roller mechanism. Powder is transferred from the powder supply container in which the movable platform maintains the powder surface at the required level (see Figure 6). Next the CO_2 laser beam draws a cross-section of the corresponding layer of the part by bonding the particles and fusing the adjacent layers. When bonding (sintering) is finished, the platform is moved down by the thickness of the next corresponding layer and the process is repeated until the build is completed.

There is no need for support structures as the surrounding powder supports overhanging elements of the part being built. When the build is finished, the working platform is lifted, the part is removed and excess powder is

Table 2 Selected Material Properties for LOMPapers[a]

Property	LPS038	LPH042
Tensile strength, Mpa		
Transverse		1.4
In-plane	66	26
Tensile modulus, Mpa		
In-plane	6679	2524
Comprehensive modulus, Mpa		
Transverse	814	407
In-plane	9310	2193
Comprehensive failure strain, %		
Transverse	51	40.4
In-plane	1	1.01
Thermal conductivity, $Wm^{-1}K^{-1}$		
Transverse	0.07	—
In-plane	0.23	—
Thermal linear expansion coefficient, $°C^{-1}$		
Transverse	149×10^{-6}	185×10^{-6}
In-plane	13.1×10^{-6}	4×10^{-6}
Glass transition temperature	90	
Deflection temperature, °C	138	77
Density, kg m^{-3}, sealed (unsealed)	893 (1449)	900 (–)
Ash content, %	2.7–3.1	—

[a]From Ref. 10.

brushed away. Then a glass-bead blaster is used to remove any excess powder. Finally, if required, the part is lightly sanded to improve surface finish.

SLS offers a wide range of materials for prototyping. Nylon, ProtoForm™ Composite, Polycarbonate, TrueForm™ PM Polymer and RapidSteel™ Metal which are commercially available from the SLS machine vendor DTM Corporation. Many powder materials can be used to build a prototype and a sand-binder system for making ceramic shells for casting is among them. According to vendor information [11], the tolerance in the x-y plane (plane perpendicular to the build direction) may range from 0.125 mm to 0.5 mm with a typical tolerance of ±0.25 mm.

Applications of SLS include conceptual models, patterns for casting, snap fits and living hinges, prototypes of blow-moulded parts, functional prototypes, patterns for investment casting and functional prototype tools.

The opportunity to make parts and tools using the variety of materials, that SLS presents, is very appealing. Even more appealing is the ability to make parts with varying material composition in a controlled manner. SLS applications are being researched in the following three areas [12]:

1. Combination of SLS with other powder oriented technologies like Hot Isostatic Processing (HIP).

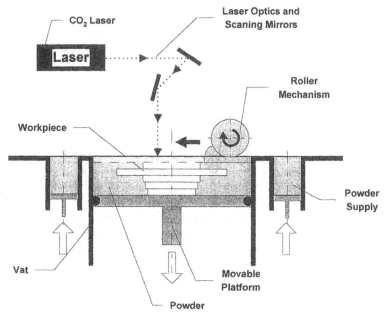

Figure 6 Schematic diagram of the SLS process.

2. Direct SLS processing that will not require any post-processing steps. Here, development of materials and process parameters will lead to manufacturing functional parts and tools without any post processing.
3. Multi-Material SLS (M2SLS) techniques have the capacity to manufacture parts with complex geometry and controlled composition. The ability to control the chemical composition at arbitrary points of the part will allow the development of unique components with tailored properties.

Table 3 Selected Properties of Materials for SLS[a]

Property	Nylon	TrueForm	ProtoForm	Polycarbonate
Tensile strength, MPa	36	10	49	23
Tensile modulus, MPa	1400	1.1	2828	1220
Tensile elongation at break, T	32	1.2	6	5
Flexural modulus, MPa	870	NA	4330	1050
Impact strength, J/m Notched Izod	70	8.1	68	53
Glass transition temperature, °C	186[b]	69	193[a]	150

[a]From Refs. 6 and 11.
[b]Melting point.

DMT Corporation offers both the Sinterstation 2000 and Sinterstation 2500 with flexible architecture for a wide variety of SLS applications. The working envelope for the Sinterstation 2000 is defined by a 304 mm diameter cylinder which is 381 mm high.

Fused Deposition Modeling

Fused Deposition Modeling (FDM) is similar to the previously discussed *RP* technologies in that it builds a 3-dimensional object layer by layer. Layers are formed by extruding a thermoplastic material through a temperature-controlled nozzle and depositing it on the working surface. Material is fed in the form of a 0.178 mm diameter wire, which is heated to a semi-solid state before being deposited. Movements of the head are controlled in the x, y and z directions. A two-head delivery system is implemented. One head is used for building the part and the second for support building if required. To ensure that the part is built on a level surface and is easily removed from the build platform upon completion, the support material is used to build the first layer. Each successive layer is built at a 90° angle to the previous layer. Upon completion, the part is removed from the foam platform and the supports are easily removed. Surface finishing usually is required (see Figure 7).

A spool-based filament material feeding system facilitates quick changes of the modelling material, when required, without significant waste. ABS, MABS, E20 Elastomer and wax are the commonly used materials for FDM modelling. MABS is a grade of ABS for medical applications. Both plastic materials are available in multiple colors. Selected properties of these materials are provided in Table 4 [13].

Wax is used for investment casting applications while ABS materials are ideal for prototype parts that normally would be manufactured by injection moulding. Best applications are for small functional parts, thick-walled parts and scaled-down concept models.

Stratasys Inc. offers three types of FDM machines: FDM8000, FDM2000 and FDM 1650 with working envelopes 457 × 457 × 609 mm for FDM8000 and 254 × 254 × 254 mm for the latter two. The proprietary software Quick-Slice 4.1 along with SupportWorks for automatic creation of support structures provides a fully automated layering process. Layer thickness can range between 0.178 to 0.356 mm. The system can operate on Hewlett Packard, Silicon Graphics, Sun Microsystems and NT workstations [13].

Three Dimensional Printing

Three Dimensional Printing (3DP) was developed at Massachusetts Institute of Technology and is similar to SLS. A thin layer of powder material is spread on the surface of the working platform and then this powder is selectively fused by ink-jet printing of a binder material. The process is repeated until comple-

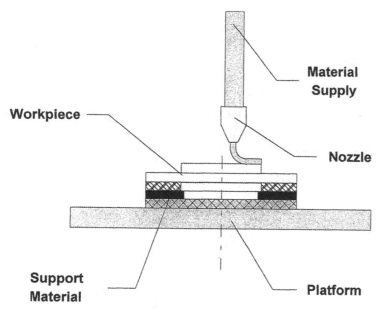

Figure 7 Schematic of the FDM process.

tion of the build. Unbound powder supports unconnected parts and undercuts and allows creation of internal volumes.

3DP can create almost any geometry. The feature size is limited by the droplet diameter, which is typically 100 μm. Any powdered material can be used including ceramics, metals, cermets and polymers. A multiple nozzle system can be used to control chemical composition of the printed part at ~100 μm resolution. The multiple nozzle system can also be used to scale both size and build rate of the process [14] (see Figure 8).

Table 4 Selected Material Properties for FDM Process[a]

Property	P400 ABS	P500 MABS	ICW05 Wax
Tensile strength, MPa	35	F37	3.5
Tensile modulus, MPa	2486	1968	276
Elongation, %	50	NA	>10
Flexural strength, MPa	66	59	4.3
Flexural modulus, MPa	2624	1775	276
Notched impact, J/m	108	178	17
Hardness, Rockwell (R), Shore (D)	R105	NA	33D
Softening-Vicat or Point R&B, °C	104(V)	NA	81(R&B)
Specific gravity, kg/m3	$1.05 \cdot 10^3$	NA	$1 \cdot 10^3$

[a]From Ref. 13.

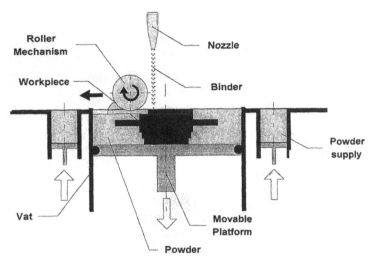

Figure 8   Schematic diagram of the 3DP process.

3DP can be used to manufacture concept models and functional parts. Combined with other technologies it can be successfully used to develop injection moulding tools [14] or investment castings (Direct Shell Production Casting [15] - Soligen Technlogies Inc. USA).

There are other similar processes which concentrate on concept models and literaly provide 3D printing. They print 3-dimensional CAD models. The drop-on-demand jet technology is offered by Sanders Prototype Inc. (SPI ModelMaker II system). Basically, it employs a plotter with dual drop-on-demand jet systems traversing over a precision elevator table on which the layered model is constructed. It also incorporates a horizontal milling cutter that travels on the same plotter carriage as the dual plot assembly and maintains the required layer thickness. The dual jet system allows the use of two materials —thermoplastic material ProtoBuild™ for the part and the second, a fatty-ester-wax ProtoSupport™ for support structure, if required.

The system is claimed to have an exceptionally good accuracy of ±26 μm, layer thickness can be controlled in the range of 13-76 μm, has a resolution of 100 μm in X-Y plane, and working envelope size of 304.8 × 52.4 × 228.6 mm [16], [17].

3D Systems developed the 3D printer Actua 2100, implementing a similar technique called Multi Jet Modelling (MJM) that employs 96 jets. Parts are built from thermopolymer material and are limited to a working envelope of 250 × 200 × 200 mm. Layer thickness is ~33 μm, x-y resolution of 85 μm and droplet placement of ±100 μm [18].

Finally, Stratasys Inc. USA offers another 3-dimensional printing system for concept modeling which is the Genesis 3D Printer. This technique does

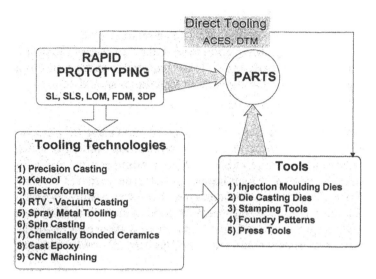

Figure 9   Rapid prototyping routes.

not use jet technology but employs an extrusion process. A high strength polyester compound is extruded through a 0.333 mm orifice at a controlled rate. Layers are laid on a metallic substrate that rests on a table. Supports are built from the same polyester and are perforated at joints with the model for easy removal. The accuracy is claimed to be ~0.356 mm and working envelope is 203 × 203 × 203 mm [13,19].

B. Conversion Technologies

Ultimately, it is the product that needs to be tested prior to being released to the market. It is best to carry out tests on a prototype that was manufactured in the same way as the product. SLS and 3DP have the potential to make functional parts in the material of choice. Even then the prototype will not exhibit the same properties as the future production part due to differences in manufacturing. Also due to manufacturing cost, *RP* techniques may be used to make a limited number of parts, up to 5 generally. This leads to the need to manufacture prototype tools rather than prototype parts in order to address both cost and properties of the prototype parts. Figure 9 shows schematically the connections between *RP* techniques, conversion technologies and tooling for prototype production purposes.

All mass production technologies, such as injection moulding, die casting and forging, use dies. Rapid prototyping can be used to make prototype dies for direct use on production machines (direct tooling or bridge tooling). If this is not a viable solution, a master pattern may be manufactured by *RP*. This master pattern can be used to make a substitute tool (soft tooling) or may be transformed into a prototype production tool with all required properties (hard tooling). Finally, when an *RP* technique warrants that properties of the prototype are satisfactory, it can be used to provide the required prototype parts but in very limited numbers. Otherwise conversion techniques should be used.

Direct Tooling

Direct tooling or bridge tooling allows production of parts using *RP* parts directly as die inserts. This is possible in the case of injection moulding, which does not involve extreme working conditions. A good example is the use of SL inserts for injection moulding of high impact polystyrene parts for a switch activator component [20]. Two hundred parts have been successfully produced within acceptable tolerances. The inserts can be solid or shell-like, filled with backing material to improve the tool's performance and allow easy introduction of a cooling/heating system when required. Basically all *rapid prototyping* techniques can be used for this purpose.

Vacuum casting

The combination of *RP* techniques with one of the conversion technologies that will produce a prototype tool that has very limited life time is called soft tooling, since this usually creates a tool that is not as hard as a normal production tool. The most common example is the use of the SL technique to manufacture a room temperature vulcanization (RTV) silicon rubber tool that in turn is used to make a number of parts from RTV polyurethane.

In this procedure, an SL pattern, or any other *RP* pattern, is used to make a silicon rubber tool. The pattern is put into a container which is then filled with the liquid silicon rubber material. When the silicon rubber is cured the mould with the *RP* pattern is removed from the container and is then cut through so the pattern can be removed (silicon rubber material is transparent and the pattern and parting line, marked prior to this process, is visible). Next, the silicon rubber tool is placed in a vacuum chamber in which the polyurethane material is prepared and poured into the mould. When casting is finished, the mould is placed in the oven to facilitate the curing process. Preparation and casting of the polyurethane under vacuum conditions results in good quality parts (no air bubbles and good filling).

There is a wide range of polyurethane materials that can be used. The material properties can be significantly changed by adding metal powder. Also the color and translucency of the material can be controlled. This technique faithfully reproduces every detail of the pattern—even finger prints left accidentally on the pattern surface. Up to 50 parts can be manufactured from one mould, which, due to poor thermal stability, quickly deteriorates. This technique is widely used for making parts that normally would be made using injection moulding or for making wax patterns for investment casting.

Spin casting

The spin-casting technique can also be classified as soft tooling. It requires an *RP* pattern and spin casting equipment. Spin-casting employs both RTV (liquid) silicone or heat-cured (vulcanized) silicone moulds. In this process, centrifugal force is used to inject a molten zinc alloy (or other low temperature melting alloy) or liquid thermoset plastic into a cylindrical multi-cavity mould, assuring good and rapid filling of the cavities.

An *RP* pattern is used to make either RTV moulds or heat-cured silicone moulds. If the *RP* pattern cannot be exposed to elevated temperatures and increased pressure without distortion, an RTV mould is made first, which in turn is used to make several new patterns by spin-casting in either low- melting-temperature alloy or a high-temperature-resistant, quick-setting, liquid, thermoset plastic. The new patterns are then used to make a multi-cavity heat-cured silicone mould, which is more heat-resistant, more durable and has better dimensional stability. The advantages of spin-casting are as follows [21]:

1. A large number of parts can be manufactured very quickly.
2. The process is relatively cheap.
3. Zinc Spin-Cast parts are comparable in strength and physical properties to die cast zinc and aluminium parts and are accepted for testing purposes. Also plastic parts have virtually identical properties to general purpose plastic injected parts.
4. Surface finish and detail reproduction are excellent.

A more detailed description of the process and its applications can be found in work by Mosemiller and Schaer [21].

Metal Spray Tooling

Thermal metal spraying coating and *RP* are an excellent combination for making prototype or even production tools (hard tooling). A wide variety of materials can be readily used to address requirements pertaining to such phenomena as friction, adhesion, abrasion, oxidation/corrosion, self-lubrication and reflectiveness.

Thermal metal spraying is a process where metal is first molten and then atomized and propelled by compressed air onto a substrate surface. Material to be sprayed is fed into the gun in the form of wire or powder. There are several types of metal spray process: arc spray, flame spray, plasma-arc spray, high-velocity oxygen fuel (HVOF) spray and the detonation-gun process (D-gun process). More information on thermal metal spraying can be found on the relevant World Wide Web sites [22].

The procedure is fairly straight forward. An *RP* model of two halves of a tool or part is manufactured and coated with release agent. Then the pattern is coated with a layer of thermal spray material (zinc, copper, steel, ...). The layer formed is then reinforced using a backing layer (filled epoxy resin or ceramic material). Finally, the pattern is separated from the coating giving a single mould half.

Electroforming

This is another hard tooling technique that combines *RP* with electroplating to manufacture prototype, or even production tools or functional parts. Electroforming involves electro-deposition of a coating on a mandrel (master pattern) which is subsequently removed. Typically, an *RP* pattern is electrolytically coated (electroplating) with Ni or Cu. Backing material is used to strengthen the tool before separating it from the pattern. This technique is used to make injection moulding die inserts or EDM (electro-discharge machining) tools for making or finishing dies.

Keltool

The 3D Keltool™ process involves the use of SL master patterns, RTV intermediate silicone moulds, 3D Keltool fused powder metal inserts and master unit die frames. SL master patterns are used to make intermediate RTV silicone moulds, which in turn are used to obtain the tool. A patented mix of metal powder and binder material is poured into an RTV silicone mould and allowed to cure. The cured green tool is demoulded from the RTV mould and fired at high temperature to fuse the metal particles and remove the binder. Finally, the fused part is infiltrated with copper giving a fully dense mould [20].

Keltool inserts have an outstanding tool life; for neat plastics 10 million parts have been produced from a single insert. For glass-filled resins, lifetimes of 0.5 to 1 million parts have been achieved. The size of Keltool inserts is limited to a 15 cm envelope due to loss of accuracy if the part is larger.

Precision Casting

Frequently, casting is the preferred technology for the conversion process. Investment casting, precision sand casting and the Shaw process are the competing casting techniques. Several elements determine the selection of a particular casting technology. Among them are the number of prototypes, the required accuracy and the shape complexity.

Investment Casting. Investment casting is probably the most accurate but most risky as it uses a sacrificial pattern. It is suitable for complex and shell like castings, and for one to three prototypes. If more parts are required the process may become too expensive, depending on *RP* pattern cost. This process involves the use of an *RP* pattern, which in turn is coated with several layers of ceramic material. The pattern is then burnt out in a flash-fire furnace producing a ceramic mould for subsequent casting. There is only one chance to get it right or a new pattern will have to be made. Mainly for this reason, *simulation* of the casting process needs to be done in order to validate the casting process as shown in Figure 2. Basically, the design of the part is assessed from the casting point of view and, if necessary, changed accordingly. Then the required feeding and gating system is designed and validated by a *simulation* process. When the results of the analysis are positive, the relevant sacrificial pattern is produced and an investment casting shell made.

QuickCast™ combines the stereolithography prototyping technique with investment casting. An SL QuickCast pattern differs from a normal SL pattern. These are built in a honeycomb-like fashion with a strong external skin to reproduce the required shape. If the pattern was solid, the ceramic shell would be cracked in the burning-out process due to large differences in the thermal expansion coefficients between ceramic and SL materials. Stresses developed in the QuickCast™ patterns during the burning out process cause collapsing of the pattern as the internal walls are very weak so the crack-free shell can be made [23].

Other *RP* techniques can be used to make investment castings. The SLS technique uses the specially developed TrueFormTM material; FDM's wax patterns are readily suitable for this process and so are LOM patterns. More about the casting accuracy using these processes can be found in the work of Riek et al. [24,25].

The Direct Shell Production Casting™ process eliminates the need for patterns and core boxes by producing ceramic casting moulds directly from CAD files using a modified 3-dimensional printing technique [15].

Precision Sand Casting.   This process uses reusable patterns, is fairly accurate and cheap, and is characterized by quick turn around. A limited shape-complexity can be achieved. Precision sand casting requires patterns with parting lines and core boxes for reproducing the internal cavities of the casting. Any of the *RP* techniques mentioned can be used to provide the required patterns. Classical sand casting technology is used, except that casting moulds are made using a mixture of fine zirconia sand, ceramic flour, and organic binder.

Shaw Process.   This process produces castings of an accuracy comparable with investment castings and complex shapes are permissible. The mould is developed by pouring ceramic slurry over the pattern. When the pattern is removed the mould is fired to achieve its final strength. The ceramic slurry is a mixture of ethyl silicate and refractory. It is a fairly expensive process and still requires cores and dividable patterns in order to make the cavities. As in precision sand casting, any of the *RP* techniques can provide the required patterns and cores.

Chemically Bonded Ceramics

Chemically bonded ceramics (CBC) developed by CEMCOM Corp., COMTEC 66, use a backing material that has extremely low shrinkage (0.02%) and a thermal expansion coefficient of the order of $13.5.10^{-6}/°C$ which is similar to that of steel and nickel. It has good compressive strength (50,000psi) that makes this material especially well suited for backing purposes [26]. The bonding is achieved at room temperature and thus any heat distortion issue is completely avoided.

CBC reproduces any feature accurately and gives a high surface finish. Quite often CBC can be used as the tool material on its own. However, it is mostly used in combination with other techniques as a backing material as in the case of Nickel Ceramic Composite (NCC) tooling. The NCC tooling employs SL patterns, nickel electroforming and CBC material to make small and large tools. It is said that NCC tools are capable of production of ten thousand parts [26].

C. An Example

Keycorp Limited, New South Wales, Australia is involved in design and manufacture of equipment for "point of sale" and automatic transaction machine (ATM) terminals. Four prototype zinc castings assemblies of a vandal-proof ATM were produced within three weeks. The castings consisted of front and rear housings, keys (or buttons) for data entry and separate housings for a magnetic card reader. The pictures (Figure 9) show the front housing which accommodates the buttons. In total, the assembly consisted of 22 separate castings.

The work included: interaction with client regarding minor modifications to their 3D CAD files to aid the casting process; building of stereolithography patterns from these CAD files; using these patterns as "loose patterns" for creating precision sand moulds; and final casting of the required zinc alloy.

II. RAPID PROTOTYPING OF ELECTRONIC SUBSYSTEMS

The controller of a mechatronic system is usually composed of a digital hardware module and control software. The boundary between software and hardware is well defined in the case of microprocessor based designs considered in Chapter 3 of this book. This makes the prototype of a microprocessor based control readily available and flexible since the software part is easy to modify. If a very high performance controller is required the microprocessor based design is usually not suitable and an Application Specific Integrated Circuit (*ASIC*) must be developed. In such an implementation the hardware takes over the functions performed by the microprocessor software and the overall performance is greatly improved. The new complicated hardware requires much more design effort and makes additional demands regarding the prototype since an *ASIC* circuit is not easy and cheap to modify and the production lead time is long. This makes the issue of prototyping the high performance electronic hardware even more crucial for the success of the whole mechatronic product.

The following sections provide an overview of available design methods, tools and implementation technologies for *rapid prototyping* of *ASIC* based controllers.

A. The Development Process and Implementation Technology

The design process and development environment for rapid prototyping of electronic hardware must provide the designer with methods, procedures, tools and implementation technologies that assure a very short time between the idea and the finished design.

The essential components in rapid prototyping of electronic hardware are *simulation, synthesis* and *reprogrammable circuits.* The *simulation* allows efficient experimentation with the functionality of the hardware even before it has been completely designed and can also provide accurate performance estimation. This reduces the potential for very costly mistakes at the top level of the design hierarchy. The *synthesis* provides a very quick and efficient way of converting the simulated behavior into the hardware, freeing the designer from the error prone and tedious tasks of producing detailed designs of electronic circuits. The *reprogrammable circuits* technology allows quick implementation of the design and easy changes to the hardware during the development.

In an ideal rapid prototyping environment the designer should be able to experiment with hardware as quickly as software programmers can do with software by using compilers for high level programming languages. The latest developments in hardware *synthesis* bring us closer to this ideal since they demonstrate efficient automatic compilation of behavioral specification into digital hardware.

The design flow with *simulation* and *synthesis* is shown in Figure 10. The designer works mainly in the *simulation* environment which should support development of hardware models on different levels of abstraction. The design work starts from the purely behavioral-program like descriptions that are simulated, verified, and refined by adding more information about functionality and timing. The *simulation* models can be developed by the designer and/or imported from other environments that can greatly reduce the development time for larger designs. Once the functionality of the design is verified, the designer can proceed with *synthesis* which can be performed manually and/or by the computer tools. The design modules—"cores," *synthesis*ed or manually designed elsewhere and provided with *simulation* models can also be mixed with the design. The *synthesis* produces a completely structural description of the design that is functionally correct by construction but should be simulated in the same environment to check the timing characteristics of the final circuit.

This requires back annotation of the *simulation* model with the delays computed after synthesis. The tightly coupled *simulation-synthesis* development loop assures that the *synthesis*ed design can always be checked functionally against the verified model of a higher abstraction level.

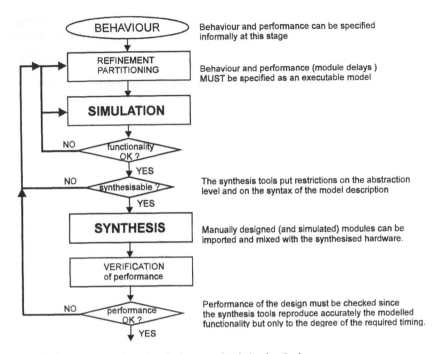

Figure 10 Interaction simulation–synthesis in the design process.

Field Programmable Circuit technology is fundamental for quick and efficient conversion of the *synthesis*ed design into a working prototype ready for tests in the field. This technology provides programmability of both the logic on the chips and the interconnections between the chips. The logic resources on the chips are approaching hundreds of thousands of gates and the time for reprogramming is milliseconds. The programmable interconnections chips allow about 900 pins to be interconnected. Both the logic and interconnections can be fully reprogrammable allowing complete re-use of the same chips in different designs. Even more, the logic can be reprogrammed on the fly while the chips are performing their function; this really blurs the boundary between hardware and software.

Flexibility of various implementation technologies is compared in Table 5. General purpose processors and microcontrollers offer the highest flexibility but at the cost of relatively low performance. Programmable technology rivals the microcontrollers in flexibility and exceeds them in performance. Both microcontrollers and programmable circuits are equally suitable for small production quantities. When the performance is essential, prototyping with programmable technology offers another advantage—a relatively easy migration to Application Specific Integrated Circuit solutions for larger production quantities.

Table 5 Flexibility of Different Integrated Circuit Technologies

Technology	Performance	Size (no of gates)	Development cost	Production volume	Flexibility
General purpose processors	Low	Limited by size of memory	Low	Low to medium	10
Embedded micro-controllers	Low to medium	Limited by size of memory	Low	Low to medium	8
RAM Field Programmable Gate Arrays	High	Up to 100k equivalent gates	Medium	Low to medium	7
Fuse Field Programmable Logic	High	Up to 16k equivalent gates	Medium	Low to medium	4
Gate Arrays	Very high	Up to 1000k equivalent gates	High	High	2
Full Custom	Very high	Millions of transistors	Very high	Very high	1

B. The Design Methodology

Simulation

Principles of Digital Simulation. Simulation is a process of emulating and observing the behavior of an electronic circuit in response to the changes of input signals. Figure 11 shows the components of this process. The simulated circuit is placed in the *simulation* test bed where the input stimuli can be produced, the functional model of the circuit executed, and the resulting output signals observed. The input stimuli is a set of combinations of signals chosen by the designer to verify the functionality of the circuit. The stimuli can be generated within the test bed as a part of the model or changed interactively by the designer The designer can decide about proper operation of the circuit by comparing the timing waveforms of the circuit outputs to the expected values. The timing waveforms of the inputs and outputs are included in the *simulation* report. The comparison of the expected and simulated values can be built in the *simulation* process and the designer informed only in case of disparity. This is the preferred method in *simulation* of larger designs.

Simulation of digital circuits can be based on a very simple timing model assuming a unit time delay. All the delays in the circuit can then be expressed as multiples of that delay. The *simulator* executes the model every time unit to sample the inputs and compute the outputs as shown in Figure 12a. This mechanism is ineffective since low speed signals are unnecessarily oversam-

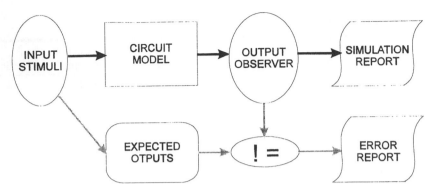

Figure 11 Components of the simulation process.

pled. The resolution of the *simulator* is also limited causing problems with *simulation*s of asynchronous sequential circuits.

Modern digital simulators are based on the principle of event driven *simulation* shown in Figure 12b. The *simulation* model is executed only if there was a change in any of the signals or some predefined time period lapsed. This allows efficient and accurate *simulation* of large designs.

A typical *simulation* environment is presented in Figure 13. The designer uses the text and schematic editors to prepare the *simulation* model. The graph

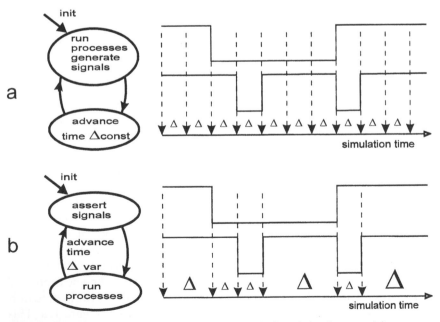

Figure 12 Principles of digital simulation: a. unit time delay, b. event driven.

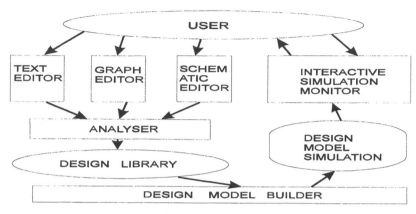

Figure 13 Typical simulation environment.

editor can be used to specify complicated sequences in the form of state diagrams of extended Finite State Machines. The analyzer checks the syntax and semantics of the combined circuit description and builds an internal representation of the design. The model builder converts this representation to the executable code that can be run on a host computer. The progress of the *simulation* can be controlled and observed by the designer in the interactive *simulation* monitor which also allows a step debug mode similar to software debuggers.

VHDL as a Simulation Language. To apply the top down development strategy in the *simulation-synthesis* environment the designer must be able to describe a model of the design on different abstraction levels. As the work progresses the design model description changes from behavioral to structural with all the possible combinations of both on the way. The *simulation* language should combine good features of a programming language with specific requirements of hardware description such as timing, parallelism, modularity, and explicit structure.

A hardware description language for *simulation* must support not only the development process but should also enforce compatibility between different commercial simulators to ensure that a design developed in one system can be replicated in another. In a commercial environment this cannot be achieved without a standard. The standardization effort undertaken by IEEE resulted in the wide acceptance of two hardware description languages: *VHDL* and *Verilog*.

VHDL stands for Very high speed integrated circuits Hardware Description Language and has been accepted as IEEE 1076 standard in 1987. This standard was updated in 1993 [27] and is still evolving with the development efforts focused on different aspects of the design process supported by the standardized packages for, e.g., *synthesis*, mathematical functions, testing [28]. Another concurrent development is Analog *VHDL* which aims at the *simulation* of analog and mixed analog-digital circuits [29].

The *VHDL* language is rich and versatile and can be efficiently used in *simulation*s from the system to the circuit level [30–32]. *VHDL* supports structured, top down design methodology at the same time allowing the designer much freedom in experimentation with different design solutions.

A sample of *VHDL* code describing the same circuit in three different styles is shown in Figure 14; of course this simple example does not show the real power of the language.

When needed, the *behavioral, data flow*, and *structural* descriptions can be freely mixed even within one design module. This allows the designer to simulate the already designed hardware modules together with purely behavioral specifications of other, not finished parts of the design.

As is shown in Figure 15 the functional testing of the design model can also be described in *VHDL*. A much more sophisticated test pattern generation than the one shown is possible if processes are used to describe sequences of signal assignments. This example also shows that a designer can use assertions to efficiently check the design for malfunctions, spikes, and other conditions.

```
entity and_or is  -- this is the design module
   port (a, b, c, d : in bit ; z : out bit ) ;
end and_or ;

--this is the behavioural description        --this is the data_flow description
architecture  behaviour of and_or is         --of the and_or design
begin                                        architecture  data_flow of and_or is
   process (a, b, c, d)                          signal s1, s2 : bit ; -- signal declarations
      variable t1 , t2 : bit ;               begin
   begin                                        s1 <=  a and b after 2ns ;
      t1 :=  a and b ;                          s2 <=  c and d after 2ns ;
      t2 :=  c and d ;                          e <=  s1 or s2 after 3 ns ;
      e <=  t1 or t2 after 5 ns ;            end data_flow ;
   end process ;
end behaviour ;


--this is the structural description of the and_or gate design
use work.all  ;  -- use library work where the components are designed
architecture structure of and_or is
   -- component declarations can be moved to a package
   component and1
      port(i1, i2 : in bit ; o: out bit )
   end component ;
   component or1
      port(i1, i2 : in bit ; o: out bit )
   end component ;

   signal s1, s2 : bit ;  --signal declarations

   -- configuration can be moved to a separate unit
   for all : and1 use entity and1(fast_architecture) ;
   for all : or1   use entity  or1(slow_architecture) ;
begin
      C1: and1 port map(a,b,s1) ;
      C2: and1 port map(c,d,s2) ;
      C3:  or1 port map(s1,s2,e) ;
end structure ;
```

Figure 14   A simple example of VHDL design.

```
entity test is -- this is the test bed unit
end entity ;

-- this is the implementation of the test bed unit
architecture and_or of test is

     signal ta, tb, tc, td, te : bit ;
     component and_or1
          port (a, b, c, d: in bit ; e: out bit ) ;
     end component ;
     for all : and_or1 use entity and_or(data_flow) ;

begin
     -- this is the component under test
     TestC : and_or1 port map(ta, tb, tc, td, te) ;

     -- these are test patterns for the pins of the and_or circuit
     ta <= '0', '1' after 10ns, '0' after 20 ns, '1' after 50 ns ;
     tb <= '0', '1' after 15ns, '0' after 35 ns, '1' after 55 ns ;
     tc <= '1', '0' after 10ns, '1' after 30 ns, '0' after 40 ns ;
     td <= '0', '1' after 20ns, '0' after 40 ns, '1' after 60 ns ;

     --this statement checks if the gate produced correct result
     --and reports error if this did not happen
     assert ( (ta and tb) or (tc and td) ) = te
          report " Mafunction of gate TestC " severity warning ;

end and_or ;
```

Figure 15 Functional testing of a design in VHDL.

Figure 16 shows relations between the most important elements of *VHDL*. The *VHDL entity* defines the interface of the design module with *ports* defining *signal* (pin) names and *generic* holding the user defined parameters such as delay, area, power, etc. The *architecture* defines the functionality of the design module. There can be many architectures corresponding to the same entity. This way the same module can be described on different abstraction levels and with different styles for various purposes, e.g., *simulation*, *synthesis*, testing. The entity and architecture are elements of component which conceptually corresponds to a design module such as integrated circuit, gate, flip-flop, subsystem. Components are connected with signals. The *configuration* allows the designer to specify which entities and architectures are used for particular components.

The essential element of *VHDL* is a *process* that is executed by the simulator concurrently with all other processes although execution of the statements within the process is sequential, as in a procedural programming language. Complex processes can be easily structured by use of *functions and procedures*. Process can be seen as a *VHDL* programming construct allowing the designer to describe behavior of the hardware in a purely abstract way, not related to the implementation. Signals are assigned new values within processes with a *signal assignment* statement. The signal can be delayed with the

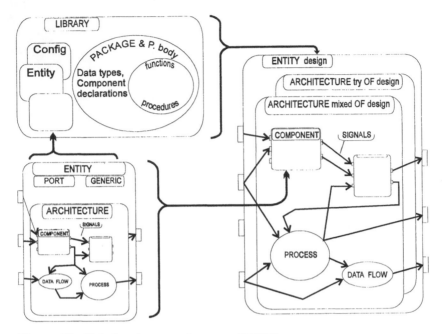

Figure 16   Relations between elements of VHDL.

after clause specifying the delay in time units. A process can be halted and made to wait for change of signals or lapse of time with the *wait* statement.

    VHDL data flow statements are short hand notation for a process assigning a value to a single signal. All *data flow* statements and components are translated to equivalent processes, this way the whole design can be simulated as a set of concurrent communicating processes.

    VHDL defines an *event driven simulation* mechanism shown in Figure 12b. The execution of the *simulator* starts with the elaboration and initialization phase when all the signals are assigned the values and all the processes are run for the first time. This is *simulation* time 0. From now the *simulation* proceeds with execution of all the processes that were scheduled for execution at that time and/or whose input signals have changed. Execution of the processes results in new signal values produced. These values are to be available to other processes only after some delays as prescribed in the processes that generated them. Execution of a process can also result in halting it until the prescribed time when it should be again revoked. After all the processes are executed, the *simulator* advances the *simulation* time to the earliest point when any of the signals is to be asserted, or when the process execution was scheduled. All the signals are updated at this point of time and the cycle repeats.

    The described mechanism minimizes the computational effort by executing processes only when there is a chance that they can produce new values of signals. It allows the *simulator* to handle with equal efficiency delays ranging from picoseconds to hours. The disadvantage is that this kind of *simulator* is much more complex in implementation than a unit delay *simulator* since elaborate scheduling of events must be handled.

One can also see *VHDL* as a programming language tailored for hardware descriptions. *VHDL* has a rich set of predefined data types. Hardware related types such as *time, bit, boolean, bit_vector* aid the designer in describing hardware on lower abstraction levels. The *integer, real, record, array,* types can be used for modeling on higher abstraction levels. New data types and subsets of types can be created by the designer. *VHDL* is a strong typed language and requires declaration of all elements, e.g., signals, variables, components, before they are referred in the code. Typical programming language control constructs, such as conditional statements, loops and assertions, are available in the behavioral view and with the exception of loops have their counterparts in the *data flow. Libraries* and *packages* provide means of efficiently structuring larger designs by encapsulating commonly used functions, procedures and data types and components.

Verilog [33,34] originated in 1983 as a proprietary language and gained popularity as the first widely available, simple and efficient, *simulation* language available from Cadence [35]. It has grown in features and capabilities over the years, has been released to the public domain, and standardized [36]. The *Verilog* language philosophy is similar to *VHDL* and it supports the same design methodology—structured, top down design. It is closer to hardware and more restrictive than *VHDL*, which is a universal language with a somewhat wordy syntax. *Verilog* being smaller and more concise has some advantages over *VHDL* not only in code writing but also in the process of building *simulation* tools.

Most of the Electronic Computer Aided Design vendors provide *VHDL* and/or *Verilog* simulators usually interfaced or tightly integrated with the *synthesis* systems, [37–40].

Simulation is especially time consuming for gate level designs since a large amount of circuit details must be processed. *Simulation*s of large designs can run for days and that precludes any possibility of interactive debugging. The problem can be alleviated by use of specialized hardware accelerators to speed up gate level *simulation*s.

The principle is to "implement" the design on the *FPGA* (Field Programmable Gate Array - Section II.B) based accelerator by mapping the design gates to the real *FPGA* cells in the accelerator (in general the coprocessor does not need to be *FPGA* based). The coprocessor emulates the design which runs many times faster than the *simulation* on a general purpose computer. This emulation also efficiently supports debugging since appropriate circuitry is automatically added to the *FPGA*. The interface software run on a host computer can read the state of *FPGA* cells and annotate it back to the gate level schematic or the source code. The designer can single step, set breakpoints, query signals and read back internal nodes. It is also possible to connect real

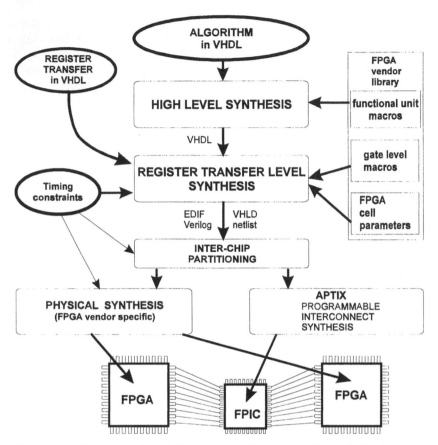

Figure 17   The synthesis path from an algorithm to reconfigurable hardware.

hardware modules (already designed and implemented or standard circuits such as microprocessors) to the emulator. The software mapping of the design onto the emulator uses principles of hardware synthesis described in the next section. Capabilities of the *simulation* accelerators range from 250,000 to 16 million gates depending on the configuration [41–43]. The time speed up range from 20 to 120 depending on the type of a design. Considering the time and effort the accelerators save they are worth their high price.

Synthesis

From Behavioral Model to Field Programmable Circuit.   *Synthesis* of digital circuits is a complex process and is usually divided into hierarchy levels shown in Figure 17. Each of the stages produces results that can be simulated to validate the result of that stage. The tools in different stages accept different abstraction levels of hardware descriptions and concentrate on different transformations and optimizations.

The *High Level Synthesis* starts with a purely behavioral description of functionality in *VHDL*, *Verilog* or in any other language (can also be graphical). The synthesis process in this stage transforms the program-like hardware description to a *register transfer level* where the data path units, the controller and their interaction is defined. The *synthesis* tools have access to a vendor specific library of components in order to make good optimization of the data path. The output format is usually *VHDL* or *Verilog*.

The *Register Transfer Level (RTL) synthesis* concentrates on optimization of the controller and data path units. Testability is usually also handled on this level. The result is a highly optimized gate level design presented as a netlist of connections in EDIF, *Verilog*, *VHDL* or proprietary netlist formats.

Large designs usually require partitioning into different reprogrammable chips. This is performed by tools that try to minimize the number of interconnections between the partitioned modules. As the partitioning process is computationally very difficult the designer is usually provided with an option of guiding the tools. The output of this stage is in a format accepted by the next stage tools, usually EDIF (Electronic Design Interchange Format) or proprietary *FPGA* formats such as XNF.

The task of *physical synthesis* is to map the connection of gates (a netlist) onto the resources of the *FPGA*. It involves converting and mapping the gates onto the cells of the chip in such a way that the interconnections and delays are minimized. Finally a pattern for programming the chip is produced. This pattern is usually in a JEDEC format for use by a programming instrument or in a proprietary format if sent directly to the RAM based chips.

In the rapid prototyping environment there is a need for reprogramming the logic and the connections between the chips in larger designs. This is achieved by programming the *FPICs*—Field Programmable Interconnect Circuits. The programming pattern is prepared by a proprietary system and directly sent to the *FPIC* chips.

High Level or Behavioral Synthesis. The role of *high level synthesis* is explained in more detail in Figure 18, which shows the basic elements of the design process on this level.

The *behavioral synthesis* accepts program like descriptions of hardware functionality and the constraints on the design. The constraints can be global timing requirements: request to perform the whole function within a specified time; local timing requirements: request to perform an operation within a specified time frame; and cost requirements: request to minimize the chip area or number of gates. The library of components in the target technology can also be seen as a structural constraint forcing the *synthesis* to use specific types of components.

The *synthesis* system produces a design that fulfills the constraints and is optimized for given criteria such as performance, cost, or both. The designer is in the position of a software programmer working with a compiler but here the compiler produces hardware.

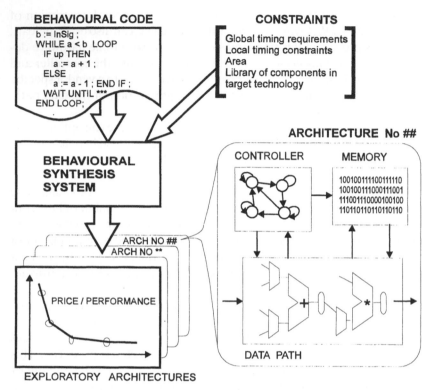

Figure 18   Exploration of design options with high level synthesis.

The task of *synthesizing* hardware is much more complex than software compilation because of the multitude of optimization options. Users of software compilers know that they do not produce results as optimal as programmers working in assembler. This also applies to behavioral hardware compilation. The results are not as optimal as hand made designs but are produced in hours instead of weeks. This allows very quick experimentation with different architectures and design options.

A good *high level synthesis* system provides the designer with a graphical interface allowing easy experimentation, intervention in the design decisions, and observation of the results. This not only helps to find bugs and improve the *synthesis* results but is also an important factor in acceptance of the tools in the hardware designers' community.

There is a widespread belief that designers cannot delegate crucial architectural decisions to automatic tools with no talent. In fact the same argument was raised many times in the past regarding the physical and logic *synthesis* tasks that has become an almost exclusive domain of the *synthesis* tools. *High level synthesis* now plays a role mainly in the architectural exploration of a design but with progress in research and development in this area we can expect *behavioral synthesis* tools to compete with human designers.

The typical process and structure of a behavioral *synthesis* system is shown in Figure 19.

The behavioral code is translated to an internal graph based representation called a *Control and Data Flow Graph* (*CDFG*) that represents the *data flow* and control dependencies in the code. This graph is usually extended with elaborate structures of place holders for the structural data supplied by consecutive optimization steps. All the optimization stages read and write to this data structure. This way the internal representation becomes the focal point of the process and holds all the information about the design.

The library of *generic components* provides information about components available for the design and their parameters in different technologies. The cost and timing of the components is taken into account by scheduling which optimizes the reuse of resources. Since reusing the same component, e.g., an adder, can incur the cost of a multiplexer not smaller than the adder itself the scheduler must use information relevant to the specific technology

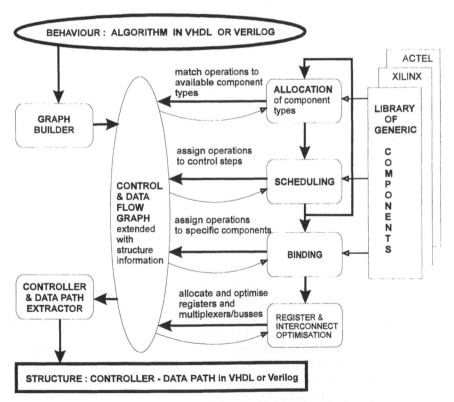

Figure 19 Typical structure and modules of a high level synthesis system.

to make appropriate decisions. The generic library can be complemented with components specific to a particular design and the rest of the system can be made aware of their functionality. Another option is to generate specialized components with the help of logic *synthesis* tools. Both approaches make the generic library itself a complex system.

Allocation and scheduling are interdependent and performed in some systems iteratively or as one optimization step. The allocation decides what types of components should be used in the design. The first trial is usually made by simple mapping of operators of different widths to appropriate component types in the generic library. The allocation is improved later after the scheduling supplies information that indicates reuse of components. The allocation can make, e.g., a decision that instead of two types of adders, 16 bit and 12 bit, only one type is used—a 16 bit adder. Allocation does not assign a specific component to an operation nor does it decide about the number of components.

Scheduling is a complex optimization process which given component types, data and control dependencies in the code, and constraints on the cost and timing, assigns operations to control steps in such a way that the number of components and control steps is minimized. Since the scheduling decides whether to reuse components, it greatly influences the cost and performance of the design. It should be stressed that many decisions made by the scheduling process are based only on the estimates of the final cost and performance. Factors such as post placement timing and influence of component reuse on routing are very difficult to estimate accurately. Estimation tools can be used to provide more accurate information but they are small *synthesis* systems in themselves making the whole process very computationally intensive. Another more practical solution is to use statistical data for estimation.

The binding finally assigns operations to be performed on the specific components. It tries to do this in such a way that the interconnection and placement cost is minimized.

The register optimization reduces the number of registers holding data values between control steps. The interconnect optimization decides whether to use multiplexers and busses and tries to minimize their number.

The final result of the *synthesis*: the controller and data path structure can be derived from the *CDFG* which has been filled with structural information about the data path by the above described optimization steps. The data path is usually described as a netlist of generic library components. The controller states and signals can be inferred from the scheduling information stored in the *CDFG*. The controller is usually described as a Finite State Machine (*FSM*). Both the data path and the controller can be further optimized by the *RTL synthesis* tools.

Most of the commercial and research systems [44–49] apply the *synthesis* strategy described above called vertical *synthesis*. There are also systems that work with the so-called horizontal approach [50]. The controller and data path are derived directly from the behavioral code and represented as extended

Petri nets. Local transformations applied to these two units provide incremental improvements of the whole design.

High level synthesis systems usually restrict the style and constructs that can be used as hardware descriptions. This leads to establishing a description policy and a subset of the *VHDL* or *Verilog* which is synthesizable by a particular system. The *synthesis* results are very much dependent on the description styles and the designer is usually presented with a set of detailed directives and examples of what constructs result in what synthesized hardware. This helps not only to achieve better optimizations but also to ease in relating the results to the source code for the debugging.

High Level Synthesis remains a very interesting research field with many open key issues such as memory *synthesis* and optimization, source code style independence, reusing old designs, accurate estimation of cost and performance, incorporating design for test, optimization across processes, and partitioning.

   c. Register Transfer Level synthesis. *Register transfer level synthesis* [51] is usually considered as controller and logic *synthesis* with test synthesis and retiming included as optional stages. The main difference between behavioral (high level) and *RTL synthesis* is that the latter deals with the designs where the operations are already scheduled, assigned to the control steps by the designer or higher level *synthesis* tools. Figure 20 shows optimization flow and the main tools in the *RTL synthesis*.

Finite State Machine (FSM) controller *synthesis* starts with a consistency check of the *FSM* specification, which is especially needed in the case of manual *RTL* designs. Consistency of transition conditions, reachability of the states, and completeness of the specification is checked and reported to the designer.

After the controller passes the consistency check the *synthesis* attempts to minimize the number of states by finding equivalent states, those which on the same conditions generate the same or nonconflicting output signals and transit to the same or equivalent states. This optimization reduces the number of states and leads to smaller circuits.

The next step is the state assignment that assigns unique codes to the states in such a way that the next state and output logic of an *FSM* can be efficiently minimized. The state assignment greatly influences not only the cost but also the performance of the design. It is one of the most researched optimization problems in *synthesis* and the automatic tools produce good results. The designers can still compete with the tools but at the expense of an enormous amount of time spent on this optimization. There are different strategies of state assignment for different *FPGA* technologies, depending mainly on the type of cells available on the chip. The designer can direct the tools towards an appropriate strategy for a chosen technology.

The combinational logic optimization and technology mapping ultimately makes decisions about the cost and performance of the controller and the data path units. Boolean minimization is only the first step leading to a more efficient logic. The minimal number of gates resulting from the Boolean representation is not always the best solution for the overall performance of the circuit. Boolean expressions must be efficiently mapped to the gates available in the particular implementation technology and the timing and interconnections between the gates should also be minimized. *Logic synthesis* tools excel in optimization of irregular combinational logic and the designers still have an edge with *synthesis* of regular logic such as arithmetic circuits. On the other hand there are specialized tools called module generators that can provide the designer with highly optimized arithmetic logic. Module generators do not perform logic *synthesis* in a strict sense but rather produce an instance of a highly optimized, parameterized module.

Retiming optimizes the overall timing of the design and relies on balancing the delays of the combinational modules in between the registers. Since the maximum clock frequency is limited by the longest delay in a block that cannot be reduced further by logic optimization methods, the idea is to split that combinational block. A clocked register can be inserted in between the split parts and this will reduce the longest delay. In general the split parts are combined with other combinational blocks to balance the delays and this requires moving the registers according to a complex set of rules preserving the circuit functionality. Retiming is usually performed as a separate optimization step although in Figure 20 it is included in general timing optimization.

Register transfer level design can be produced by a human designer or higher level *synthesis* tools. In the former case the *RTL* description usually contains statements with operators that are mapped to the data path units such as adders and subtractors. The *RTL* tools analyze data dependencies between those statements and attempt to minimize the number of data path units by reusing the same unit for performing operations of the same type from many statements. Resource sharing is very powerful since it gives the designer more freedom in the *RTL* description. The designer is able to control the tools regarding which operations and to what extent they should be shared.

Test *synthesis* is one of the most important and useful modules in the *RTL* tools suite. It analyzes the design and inserts the necessary circuitry to make the circuits testable. The most popular method for providing testability is the scan path design. It relies on circuit flip flops to form a long shift register allowing writing and reading the state of the whole design in the test mode. The test *synthesis* also produces the test vectors for the circuitry; that relieves the designer from a very time consuming task and usually provides a better test coverage than manual test generation.

RTL synthesis puts specific requirements on the description style, syntax, modeling and design methods for *synthesis*. The *synthesi*zable *VHDL* or *Verilog* subsets are sufficiently large to allow much freedom for the designer. Usually a list of the *synthesis*able constructs does not describe the subset sufficiently

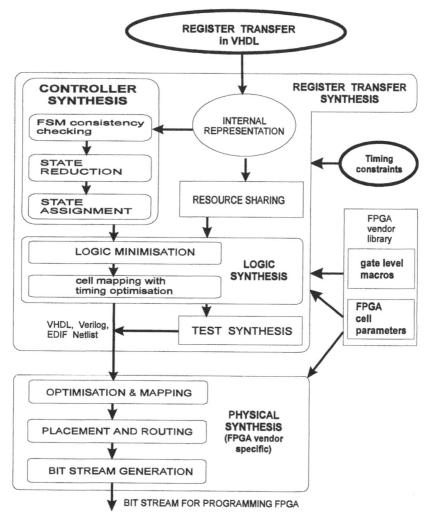

Figure 20 Register transfer and physical synthesis for FPGAs.

because also the context in which the constructs are used may determine the amount and type of *synthesis*ed hardware. The designer is required to adopt the design style that is understood by the tools in order to achieve the best *synthesis* results.

RTL synthesis is now well established as the industrial practice supported by a wide range of sophisticated design tools. All major vendors of electronic CAD provide RTL *synthesis* from *VHDL* and/or *Verilog* as parts of their design systems. There are also a number of smaller systems specializing mainly in *synthesis* for *FPGA*s or *Programmable Logic Devices* [52–55].

Many of the design systems support graphical input for the hierarchical and parallel Finite State Machines combined with truth tables, Boolean expressions and schematics. The tools usually provide complete controller *synthesis* with consistency check, state minimization, assignment and test. The output is *VHDL* or *Verilog* or proprietary formats of Programmable Logic tools. The graphical tools not only ease the task of specification but also provide the designer with a very good *simulation* interface [56–59].

d. Physical synthesis. Physical synthesis [60] as the closest to the hardware is vendor specific since the results of optimizations depend very much on intrinsic knowledge of the device architecture and features.

The optimization and mapping stage converts the gate design supplied by the logic *synthesis* tools and maps it onto the type of resources available on the *FPGA* chip. This mapping is still logical since the specific physical cells are not yet assigned. *FPGA* architectures differ greatly from the simple gate model and usually the mapping is not straightforward. Several simple gates can be packed into one *FPGA* cell and choosing which gates should be placed together is a difficult and time consuming task. If a design block such as an adder-accumulator is to be mapped onto the *FPGA* the gates of one bit full adder should be placed in the same cell as the corresponding flip flop of the register to minimize routing and delays. Keeping the constraints as above for a whole design is difficult and quite often the mapping is less efficient than a human designer would produce.

One way to avoid problems with mapping of larger regular blocks such as adders, ALUs, registers, counters, etc., is to use so-called hard macros in the *synthesis*. Hard macros are pre-placed units that will not be changed by the mapping process, in this way preserving not only functionality but also the layout and timing of this design block.

Placement and routing are the stages where the design already mapped to the cells is finally placed on the chip and the cells are interconnected. The cells that are to be connected should be placed close to each other to reduce the cost of routing. Of course this objective can generate contradictory requirements from many densely connected cells. Physical characteristics of the *FPGA* cells such as placement of inputs and outputs and number of connections to the routing mesh must also be taken into account. On top of this, the timing aspects must be considered and if there are any interconnections (nets) marked by the designer as time critical they should take priority. The placement and routing process is very computationally intensive and usually requires iterative improvements especially when many time critical nets are specified [51]. It is also one of the most important stages regarding the overall performance of the *FPGA* design.

Finally the bit stream specifying the configurations of all the programmable cells and interconnections of the chip must be produced. This bit stream is sent to the *FPGA* device which in the programming mode configures its logic and interconnections according to the downloaded bit stream.

C. Programmable Hardware Technology

Field programmable logic circuits form an almost ideal basis for implementation of a digital circuit prototype. Since their programming is easy and quick the prototype development time is reduced to design and debugging time only. In contrast, when a design is to be implemented on a mask programmable gate array, the lead time is at least four to six weeks to the first samples. Another extremely important aspect is the cost of correcting a design mistake which in the case of mask gate arrays usually tens of thousands of dollars and requires weeks at the manufacturers plant. Reconfiguring a field programmable logic circuit does not incur any cost and is performed in a fraction of a second on a designer's desk.

This chapter uses the term *FPGA* to cover different families of field programmable logic circuits, since this is the popular use of this acronym. The term Field Programmable Gate Array (*FPGA*) originated in the XILINX company as the name of their product—the first RAM based programmable circuits with the *Logic Cell Array* architecture.

FPGA Architectures

Field programmable logic circuits differ widely in their internal structure and the principle of programming [61]. Figure 21 shows in a schematic way the basic structures of logic and interconnections of *FPGA*s built according to different architectural concepts:

1. Logic matrix: The logic resources form an AND-OR programmable matrix. The outputs of the matrix are connected to the configurable input/output cells that have at least one flip-flop. The number and width of AND and OR gates determines flexibility of the circuit. This is the most popular structure for smaller circuits called *Programmable Logic Devices* (PLDs) or *Programmable Array Logic* (PAL) produced by a number of makers. The rationale behind this architecture is that in applications such as finite state machine based controllers the amount of combinational logic is more important that the number of flip-flops.
2. Hierarchical Logic Blocks: A natural evolution of logic matrix structures. A small number of relatively large logic blocks corresponding in functionality to PLDs are interconnected by a bus like structure. This is the structure of popular Altera devices [62]. The programming is usually based on EPROM cells (require exposure of the chip to ultraviolet light for erasing) or EEPROM devices (electrical erasing and programming). The rationale behind this architecture is that a digital design consists of modules that are easy to implement as PLDs. Within the module the interconnections between the resources are

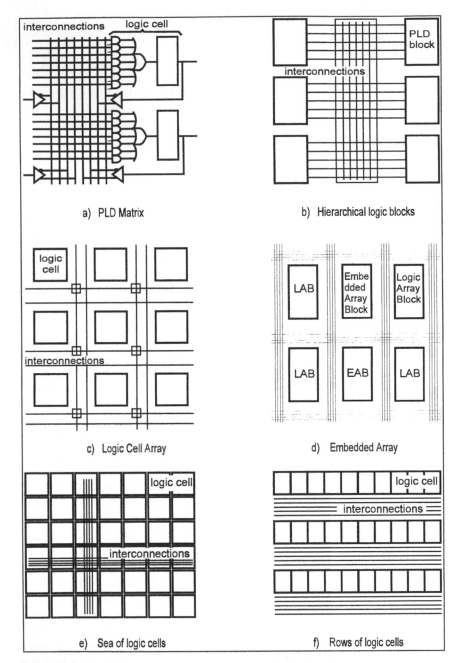

Figure 21 Architectures of field programmable logic circuits.

dense but only a few signals are usually taken to other modules; hence the bus like interconnection structure provides the best trade-off.

3. *Logic Cell Array*: A large number of small size logic cells composed of a look up table and flip-flops are evenly placed on the chip and interleaved with routing resources. Input/output cells differ in structure from logic blocks and are placed around the chip close to the pins. This is a typical structure of Xilinx XC4000 [63], ORCA [64] and other devices. The programming is based on RAM cells storing the configuration information both for the logic and routing resources. The rationale for this architecture is that combinational logic circuits are usually interleaved with flip-flops in any digital design and this should be reflected in the *FPGA* structure for efficient mapping of the design to the chip.

4. Embedded Array: Evolution of the hierarchical block structure with blocks of dedicated functionality. In the Altera Flex10k family of *FPGA*s the dedicated blocks are memories but in general the blocks can also be specialized logic resources such as Arithmetic and Logic Units.

5. Sea of Logic Cells: The name relates this structure to the sea of gates architecture of mask gate arrays. Small function units consisting of a look up table and a flip-flop are evenly distributed over the chip. The routing resources are abundant and hierarchical, allowing dense connections within the small area and gradually less lengthy lines to connect to more remote blocks. This is the architecture of Xilinx XC6200 and Atmel [65] devices. The rationale is that abundant and hierarchical routing resources make timing delays smaller and it is easier to fully utilize the chip's vast logical resources.

6. Rows of Logic Cells: This is the architecture closest to the ordinary masked gate arrays. The small logic cells are arranged in rows while the routing resources are placed in channels in between the rows. The logic cells are small and multiplexer based and there are no flip-flops. If needed they must be built from the gates and this makes it more desirable to use latches than clocked flip-flops. This is the structure of Actel [66] devices. The advantage of this architecture is that it is easier to adapt design tools previously developed for gate arrays and to produce a masked gate array from a prototype.

FPGA Programming Techologies

Field programmable logic circuits differ not only in their structures but also in the way they store the configuration information. Figure 22 shows principles of different programming techologies:

1. Fuse: The programmable logic devices, PLD and PALs, made in bipolar technology use fuses for reconfiguration. The programmable device is manufactured with all the possible connections made with the fuses and the programming is performed by blowing the fuses. In

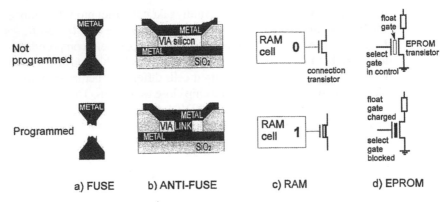

a) FUSE   b) ANTI-FUSE   c) RAM   d) EPROM

Figure 22   Principles of different programming technologies.

the programming process the fuse must be addressed and an appropriately high current applied to burn it out. The high current drivers take much space on the chip and their area is usually as large as the area of the AND_OR logic matrix. Still the fuses are the best solutions for the quickest *Programmable Logic Devices* [67].

2. Anti-Fuse: Programming is performed by applying relatively high voltage (about 10 V) and high current (5 mA) to melt the insulator layer of the anti-fuse. An unprogrammed anti-fuse exhibits a gigaohm resistance, that changes to about 80 ohms in the programmed state. The device is very small but as in the case of fuses the high voltage and current drivers are needed. Also some extra steps in the fabrication are required. This technology is used by Actel and QuickLogic [68]. The advantage is that after programming, the device retains its configuration even without supply voltage. The disadvantage is that the anti-fuses as well as fuses are not reprogrammable and in case of design mistake or modification the whole chip must be changed.

3. EPROM and EEPROM: Programming is performed by activating transistor switches made of special transistors with an additional floating gate. The control of the select gate over the transistor is blocked if the floating gate is charged by applying high programming voltage. The charge remains in the gate for a very long time (tens of years) since it is fully insulated. The charge can be erased by exposing the chip to the ultra-violet light (EPROM) or reverse voltage (EEPROM). The clear advantage over the fuses is that the configuration is erasable. EEPROM transitors are twice as big as EPROMs but they allow reprogramming without removing the circuit from the board (in-circuit reprogramming).

4. SRAM and FLASH: Programming is made as in the SRAM or FLASH memories by storing the information in the static latches made of

transistors. In the case of SRAM the configuration must be reloaded every time the circuit is powered up. This is done from additional serial EPROM. SRAM based devices are very popular due to easy reprogramming in the prototyping environment. This capability is even more enhanced in the new Xilinx 6200 devices that look like a standard memory to a microprocessor which can reprogram parts of the chip on the fly. The FLASH memory holds the information permanently [69] which makes the chips fully in-circuit reprogrammable but also non-volatile—a clear advantage for end products.

Representative RAM based Programmable Gate Arrays

We focus this section on the representative *FPGA* architectures that are RAM based. The choice of the products presented here was guided mainly by how well they represent popular *FPGA* architectures and development trends. There are about 15 makers of programmable logic circuits and we cannot even try to briefly characterize their products in the limited space of this book. The interested reader is referred to the web site [70] containing an updated list of *FPGA* related links where the web addresses of the manufacturers can be found. The data sheets for new and old products are easily available on the World Wide Web.

Comparing Oranges with Apples. Before we analyze capabilities and features of different *FPGAs* the reader should be made aware of the difficulties in comparisons due to architectural differences.

The size of the *FPGA* device is usually expressed in the number of equivalent gates: two input NAND circuits. This is the measure borrowed from masked gate arrays and not quite relevant to *FPGAs* since their granularity (logic resources per cell) is usually much larger than a mask gate. Additional difficulty arises when comparing sizes of the devices with and without dedicated blocks such as memories. Xilinx proposed a measure based on the number of logic cells defined as a combination of a 4 input logic block and a dedicated register. Altera uses an equivalent NAND gate measure weighted with the statistical utilization of the device in different designs. The differences between these estimates can be surprisingly high. Devices that have about the same capacity in the logic cell measure differ more than 70% in the equivalent gate count.

If the size estimations differ so much between *FPGA* makers how can we compare the equivalent gate count of *FPGAs* to the mask programmable gate arrays? A designer's way to get a fair comparison would be to implement a number of designs in both technologies and check the number of gates. This method is also supported by the PREP benchmark suite available from Altera. An interesting research report [71] suggests that dependent on a design type (e.g., ratio of flip-flops to combinatorial logic) the gate estimation of *FPGAs* is exaggerated from 1.5 to about 4 times.

The speed of different *FPGA* chips is characterized by the delay of combinational blocks and toggle rate of the flip-flops. This is a good performance

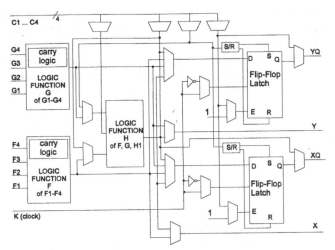

Figure 23 XC 4000E series configuration logic block CLB.

measure for PLDs and PALs but can be very misleading if taken as a measure of attainable speed of the *Logic Cell Array* based design. The delays of routing can exceed the delays of the combinatorial logic and cannot be easily predicted until the design is placed and routed. A more accurate comparison of timing can be achieved by checking the performance of hard macros supplied by the producer. Since the hard macros—counters, adders, multiplexers, arithmetic blocks, etc.—are pre-placed and routed their timing is predictable and can be used for performance comparison of different *FPGA*s.

Xilinx XC4000. This family of *FPGA* is representative for *Logic Cell Array* and includes chips from about 3k to 100k equivalent gates (100 to 2300 logic cells).

The logic cell called *CLB*, as shown in Figure 23, consists of two 4 input look up tables, two flip-flops and the configuration circuitry. The cell is flexible and has a number of configuration options including 5 input logic block, 32 × 1 or 16 × 2bit memory, and 16 × 1 dual port memory. The high speed carry logic for arithmetic circuits is also available. Other resources on the chip include tri-state buffers allowing efficient bus oriented designs and wide decoders making address decoding easier.

The input/output cell also has a flexible structure. The input and output latches can be used to ease the interface problems to asynchronous signals. The skew rate, polarization of clocking, and polarization of output signals can also be controlled. An additional feature of this chip I/O system is the circuitry supporting boundary scan JTAG standard.

The *CLB*s can be interconnected by the abundant wiring resources shown schematically in Figure 24. Single-length lines connect the closest switch matrices. They are used to connect neighbor *CLB*s or to lead non-critical signals

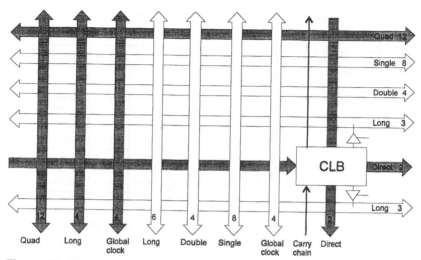

Figure 24 Routing resources of XC4000 series.

via switch matrices to more distant modules. The double length lines connect every second switch matrix and the long lines cross the chip and are used for carrying low-skew signals. The new versions of XC4000 have increased routing capacity (shaded lines) and a new feature is quad lines that run past four *CLB*s before entering the switch matrix.

a. Xilinx XC6200. This is a new family of *FPGA*s representing the sea of cells architecture. It includes chips with size ranging from about 10k to 60k equivalent gates (assuming 4–6 gates per function cell).

As can be seen in Figure 25 the chip is highly regular. The cells are arranged in 4-by-4 blocks, which form 16-by-16 tiles, which finally are arranged as the 64-by-64 central array. The cells are much simpler than in the *Logic Cell Array* type *FPGA*s and each cell contains both the programmable logic and routing resources. The cell is built of a flip-flop and a combinatorial logic block capable of implementing any two input functions or any type of 2-to-1 multiplexer. Also any cell can be converted to two bytes of RAM or ROM memory. The internal delay through the logic is about 2 ns, the register setup time about 3 ns.

The interconnect structure of XC6200 is also hierarchical. As is shown in Figure 25 all the cells are directly connected to the nearest neighbors. Within the 4-by-4 block there are horizontal and vertical Length4 FastLANEs spanning the block and within the tile there are Length16 lines; the length of the lines increases with the hierarchy of blocks. Special switches are provided on the 4-by-4 block level to interconnect between the length4 and longer lines. It is easier to visualize the interconnections in three dimensions with each hierarchy level above the lower one. The interconnection delays vary from 1 ns for the nearest neighbor to 3 ns for the chip-length FastLANE wires.

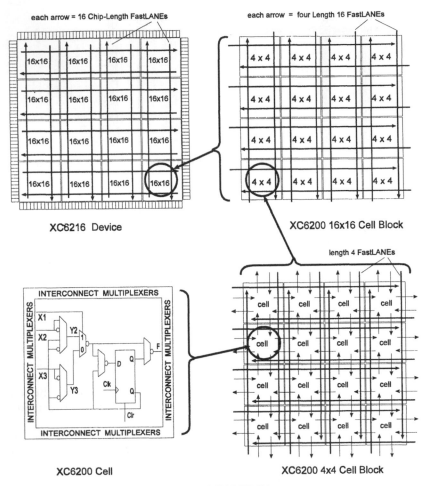

Figure 25   Architecture of Xilinx XC6200 FPGAs.

Since the architecture is so symmetrical regarding both the logic and routing resources it supports position independent design blocks that can be easily relocated to another area within the chip.

One of the most innovative features of this architecture is its dynamic partial reconfiguration capability and the ultra fast configuration (about 2 ms for a whole chip). The dynamic partial reconfiguration means that parts of the chip can be reconfigured while the rest of the chip is working. It makes it possible to implement hardware in a more software oriented way with function or procedures implemented in hardware but loaded only when their execution is needed. Loading can be performed very quickly from low cost EPROM memories or even from a host microprocessor system.

XC6200 chips have a dedicated host processor interface called FastMAP which can be configured for 32, 16 or 8-bit busses. The FastMAP provides high-speed access to all internal registers by mapping them into the host processor memory address space. Due to this very efficient data interchange between the processor and the chip the XC6200 *FPGA* family is very suitable for building specialized coprocessors for control or computation applications. This is the very fast growing area of Custom Computing Machines.

Altera Flex 10k. This family of *FPGAs* represents the embedded array architecture. It includes chips with size ranging from about 10k to 100k equivalent gates including RAM cells (according to the manufacturers measure in equivalent gates).

The Flex 10k *FPGA* contains Embedded Array Blocks (EABs) each of which can be used to create RAM, ROM, FIFO or dual-port RAM with 2k bits. The word width can be set to 1,2,4,8 bits. Access time for EAB is as low as 7 ns. EABs can also be used to implement complicated logic functions in the form of large look-up tables. For example a 4×4 bit multiplier can be implemented in just one EAB.

A Logic Array Block (LAB) is the main logic resource of the chip and contains 8 Logic Elements (LEs) made of a 4 input look-up table, flip-flop, carry and cascade logic, and about 30 local interconnect channels. The delay through the look-up table is as low as 1.5 ns. Each LE can operate in a normal, arithmetic, up-down or clearable counter mode. These predefined modes make very efficient and optimized use of the resources.

Signal interconnections are provided by FastTrack Interconnect—row and column channels running the entire length of the device. The number of FastTrack channels increases with the size of the device providing ample resources for efficient interconnections.

The chip can be configured from the configuration EPROM, serial input, parallel inputs, or a special JTAG controller.

Embedded Versus Distributed Memory. Nobody questions the fact that memory resources should be provided on *FPGA* chips but there is a debate on what is the most suitable memory architecture [72].

Memory in logic call array architectures such as XC 4000 or ORCA [ ] is distributed evenly throughout the chip since it is embedded in the logic cells, which when needed are configured as small memory slices. This makes it possible to fine tune the amount of memory to the requirements of the design since both the depth and width of memories can be customized in small increments. The *memory blocks* can be placed close to the modules using the data and also their geometrical shape can match other blocks. On the other hand the address decoding circuitry must be created from logic cells and is usually slower than the address decoder of an embedded array.

Embedded memory arrays have a predefined interconnection structure which limits configurability. The data width and the depth can be changed only

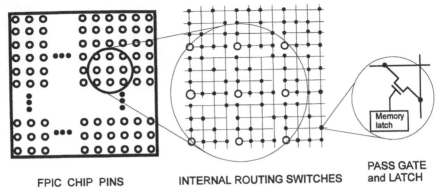

FPIC CHIP PINS   INTERNAL ROUTING SWITCHES   PASS GATE
                                                             and LATCH

Figure 26   Architecture of aptix field programmable interconnection circuit.

in power of two increments. They are usually faster than distributed memories, e.g., access time of Flex 10k EAB is about 7 ns. The memory modules are dedicated but can also be used to implement complex logic functions, hence these resources are not completely lost when memory is not needed. Routing resources of the memory and logic are shared and that can also limit the use of logic resources. Some new research results [74] point to the need of dedicated memory to memory routing resources.

It seems that both architectures find good applications and the distributed memory is better in a general purpose logic design while the embedded arrays have clear advantages in more structured computationally oriented designs.

Programmable Interconnection Devices

Programmable Interconnection devices are available from only a couple of manufacturers [75,76]. The APTIX-Field Programmable Interconnection Circuit (*FPIC*) is a device that has over 1000 leads of which 936 can be used for interconnections. It also includes debugging interface allowing the user to observe internal signals. The *FPIC* structure is shown in Figure 26.

The chip uses a sophisticated routing architecture that creates a large switch matrix between all pins. All interconnects are programmable bypass gates connected to static RAM memory elements. The propagation delay from pin to pin is about 5 ns. This quickly deteriorates with a larger fan-out and is about 20 ns when a signal is propagated from one pin to six other pins.

New generation *FPIC*s provide a number of buffers to improve performance of high fan-out nets.

APTIX also produces prototype boards with *FPIC*s and general purpose pin-hole areas where *FPGA*s of different types or other devices such as memories, microcontrollers, or dedicated chips can be inserted. The board provides downloading of the configuration data to both *FPGA*s and *FPIC*s.

Core Program - Design Reusability

*FPGA*s have reached logic densities that allow whole electronic subsystems to be implemented on one *FPGA* chip. The design of such complex circuits is very time consuming and is usually based on reusing design modules made in house for other designs, supplied by the *FPGA* manufacturer, or made by a third party vendor specializing in *FPGA* design. Those modules are called cores and are becoming a booming market in the field of *FPGA* design.

The library of simple design modules provided by the *FPGA* vendor usually includes simple gates, flip-flops, and a set of hard and soft macros. The hard macros are pre-placed which makes their timing predictable. The soft macros are not pre-placed and their timing can very much depend on the final placement and routing. The functionality of the macros usually cover fairly simple blocks such as multiplexers, comparators, counters, registers, adders, and small ALUs.

In the design of digital subsystems there is a need for far more complicated modules such as UART communication devices, PCI bus interface circuits, Digital Circuit Processing circuits such as digital filters, correlators, and FFT modules. The *cores* with this functionality are available from the *FPGA* vendors or third party developers. An example of this initiative is the library of LogiCORE modules from Xilinx and their AllianceCore programme.

The LogiCores are optimized for target *FPGA*s and usually provided in a netlist format together with the functional *simulation* model and a test bed. In case of timing critical cores appropriate constraint files for the placement and routing are also supplied. It is possible to purchase complete *synthesis* models in *VHDL* or *Verilog* which allow the designer to change their functionality when needed. The vendors provide extensive documentation and technical support. which is essential in the case of complex cores. There are about 15 AllianceCore partners supplying a wide variety of cores.

Migration from Prototype to Product

The prototype is usually only an intermediate step to a product. There is usually expectation that the conversion from the prototype design to the final one will be an easy step. It is not always the case and complexity of the conversion task depends very much on how different the prototype and the target technologies are.

A prototype *FPGA* design is usually slower than a corresponding masked gate array to be used in a product. If the designed chip communicates with other chips (and most probably it does) there is a problem of scaling the performance and matching the timing of the prototype to the product. In synchronous designs it can be solved by scaling up the system clock and preserving timing relations of the prototype in the final design. Usually limited redesign is needed anyway. In case of asynchronous circuits a major redesign would be required.

If the design has been made with the *simulation* and *synthesis* tools any redesign is easier since the *synthesis* tools support re-targeting the same designs to different technologies. Of course when the design is to be optimized for larger volume production more work must be done to achieve the best performance and area tradeoffs. Also *simulations* must be repeated to test the timing properties of the new design.

If performance of the prototype is sufficient for the final product there is an easy and straightforward way to undertake conversion to mask gate arrays. The *FPGA* makers provide families of masked gate arrays that have one to one correspondence to their *FPGAs*. These chips have the same packages, pin-outs, and I/O pins electrical characteristics. Usually they also include scan-path logic for easier testing. The power consumption is reduced, as is the cost, in volume production.

Converting a design is done at the *FPGA* vendor site from the programming patterns of the prototype. The designer should provide the vendor with the functional test patterns. The first samples of mask gate array are usually delivered in four to six weeks. An example of this approach is Xilinx HardWire mask gate arrays and the DesignLock conversion process that guarantees a turnkey, automatic conversion from the *FPGA* preserving all the characteristics of the original *FPGA* design.

Future Trends

The development of efficient RAM based *FPGAs* was a milestone in the evolution of digital hardware. Up until that time hardware was hard (hard to change) and software was soft (easy to change). Nowadays with XC6200 the hardware is treated as the RAM memory of a microprocessor and can be reconfigured on the fly. This really blurs the boundary between software and hardware.

One natural trend in the evolution of *FPGAs* is to enhance the reconfiguration capabilities even further. Presently the dynamically reconfigurable chips need to load the data from the off chip memory and this is a relatively slow process, since it is sequential. If it was possible to just swap each bit of memory with a context memory on the chip the reconfiguration could be done in just one clock cycle [77,78].

Of course the main limitation in the above approach is the amount of memory elements on a chip but this increases rapidly with advances in technology. At present the Xilinx chip XC4085 fabricated in 0.35 micron CMOS technology uses 16 million transistors. Xilinx plans to move before the turn of the century to 0.18 micron technology with about 60 million transistors on a chip. This amount of transistors makes the idea of configuration swapping viable.

There is also a possibility of using external storage sources and still achieving rapid reconfiguration by reloading the *FPGA* optically. Some research on this idea has been reported in [79].

FPGAs have already achieved such densities that they can be used as co-processors for control and computational applications. This requires efficient interface to standard microprocessors. This is a clear trend and representative examples of commercial chips are XC6200 and ORCA OR3000 families. We can also expect that the embedded array architectures will include not only dedicated *memory blocks* but slices of ALUs or even whole mini–risk processors. This will greatly enhance the capabilities of these *FPGAs* in coprocessor applications.

REFERENCES

1. PF Jacobs. *Stereolithography and other RP&M Technologies*. New York: SME and ASME Press, 1996.
2. PF Jacobs. *Rapid Prototyping & Manufacturing*. New York: SME, 1992.
3. RP Chartoff, JS Ullett, DA Klosterman. Advanced Materials Processing. Prototyping Technology International'97 the International Review of *Simulation*-based Design, Rapid Prototyping & Manufacturing, published by UK & International Press, 1997.
4. RD Stoddart. Drop-on-demand yields precision tooling patterns. Prototyping Technology International'97 the International Review of *Simulation*-based Design, Rapid Prototyping & Manufacturing, published by UK & International Press, 1997.
5. I Lauwers, J Meyvaert, P Kruth. Layer deposition in SLA and SLS. Prototyping Technology International'97 the International Review of *simulation*-based Design, Rapid Prototyping & Manufacturing, published by UK & International Press, 1997.
6. TA Grimm. SLS and SLA different technologies for different applications. Prototyping Technology International'97 the International Review of *Simulation*-based Design, Rapid Prototyping & Manufacturing, published by UK & International Press, 1997.
7. MA Johnson. Acrylated resins for SLA applications. Prototyping Technology International'97 the International Review of *Simulation*-based Design, Rapid Prototyping & Manufacturing, published by UK & International Press, 1997.
8. Rapid News TCT 2(2), March 1997: 49.
9. SS Pak, DA Klosterman, B Priore, RP Chartoff. Tooling and low volume manufacture through laminated object manufacturing. Prototyping Technology International'97 the International Review of *Simulation*-based Design, Rapid Prototyping & Manufacturing, published by UK & International Press, 1997.
10. http://helisys.com.
11. http://dtm-corp.com.
12. LL Jepson, JJ Beaman Jr. Functional Parts from Advances in SLS. Prototyping Technology International'97 the International Review of *Simulation*-based Design, Rapid Prototyping & Manufacturing, published by UK & International Press, 1997.
13. http://www.stratasys.com.
14. E Sachs, H Guo, E Wylonis, J Serdy, D Branchio, M Rynerson, M Sima, S Allen. Injection moulding tooling. Prototyping Technology International'97 the Interna-

tional Review of *Simulation*-based Design, Rapid Prototyping & Manufacturing, published by UK & International Press, 1997.

15. Y Uziel. Seamless CAD to metal parts. Prototyping Technology International'97 the International Review of *Simulation*-based Design, Rapid Prototyping & Manufacturing, published by UK & International Press, 1997.

16. RD Stoddart. Drop-on-demand yields precision tooling patterns. Prototyping Technology International'97 the International Review of *Simulation*-based Design, Rapid Prototyping & Manufacturing, published by UK & International Press, 1997.

17. http://www.sanders-prototype.com.

18. http://www.3dsystems.com.

19. M Stanley. Fused Deposition Modelling for product development. Prototyping Technology International'97 the International Review of *Simulation*-based Design, Rapid Prototyping & Manufacturing, published by UK & International Press, 1997.

20. PF Jacobs. Recent Advances in Rapid Tooling form Stereolithography. 3D Systems publication P/N 70270/10-15-96.

21. Mosemiller, L Schaer. Combining *RP* and Spin-Casting. Prototyping Technology International'97 the International Review of *Simulation*-based Design, Rapid Prototyping & Manufacturing, published by UK & International Press, 1997.

22. http://www.corrosion.com, http://luv.postech.ac.kr/~temlab/research/tsc.html, http://www.sulzermetco.com, http://www.metalcoatings. com.

23. CW Hull, PF Jacobs. *Stereolithography and QuickCast: Moving Towards Rapid Tooling*. 3D Systems Valencia, California, June 1995.

24. T Riek, P Christodoulou. Rapid Product Development Using Rapid Prototyping and Investment Casting. 4th Asian Foundry Congress Proceedings, Broadbeach, Queensland, Australia, 27th to 31st October, 1996: 229-240.

25. T Riek, P Christodoulou. Comparison of Dimensional Accuracy of Castings Using Rapid Prototyping Techniques. WORLD CONGRESS, Manufacturing Technology Towards 2000, 3rd Conference on Rapid Product Development Proceedings, 15th to 17th September 1997, Cairns, Australia: 473-482.

26. S Wise. Net Shape Nickel Ceramic Composite Tooling from *RP* Models. Rapid Prototyping, 3(1), 1997, published by the Rapid Prototyping Association of the Society of Manufacturing Engineers.

27. JM Berge, A. Fonkoua, S Maginot, J Rouillard. *VHDL'92*. Kluwer Academic Publishers, 1993.

28. IEEE web: http://standards.ieee.org/catalog/press/*VHDL*MODS/TUTORIAL/HTML/HOMEPG.HTM.

29. A- *VHDL* : http://server. *VHDL*.org/analog/wwwpages/tutorial.html.

30. JM Berge, A Fonkoua, S Maginot, J Rouillard, *VHDL Designer's Reference*. Kluwer Academic Publishers, 1992.

31. J Mermet. *VHDL for Simulation, Synthesis and Formal Proofs of Hardware*. Kluwer Academic Publishers, 1992.

32. RE Harr, AG Stanculescu. *Applications of VHDL to Circuit Design*. Kluwer Academic Publishers, 1991.

33. DE Thomas, P Moorby. *The Verilog Hardware Description Language*. Kluwer Academic Publishers, 1991.

34. E Sternheim, R Singh, Y Trivedi. *Digital Design with Verilog HDL*. Automata Publishing Company, Cupertino, Ca 95014, 1990.

35. Cadence Website: http://www.cadence.com.

36. *Verilog* IEEE Standard 1364-1995 reference manual, Website: http://standards.ieee.org/catalog/it.html.

37. Mentor Graphics, Website: http//www.mentor.com.

38. ViewLogic Website: http//www.viewlogic.com.

39. Synopsys Website: http//www.synopsys.com.

40. Model technology Website: http://www.model.com.

41. Paradigm XP - *Simulation* Supercomputing, Zycad, Website: http://www.zycad.com.

42. HDL-ICE ASIC Emulation, QUICKTURN Design Systems, Website: http://www.qckturn.com/prod/hdlice/hdlice2.htm.

43. H Verheyen, G Lara. Rapid Prototyping Using System Emulation Technology for DSP Design Validation, APTIX, Website: http://www.aptix.com.

44. D Gajski, N Dutt, A Wu, S Lin, *High Level Synthesis - Introduction to Chip and System Design*, Kluwer Academic Publishers, 1992.

45. RA Walker, R Camposano, A survey of High-Level *Synthesis* Systems, Kluwer Academic Publishers, 1991.

46. P Michel, U Lauther, P Duzy. *The Synthesis Approach to Digital System Design.* Kluwer Academic Publishers, 1992.

47. Behavioral compiler, SYNOPSYS Website : http://www.synopsys.com.

48. YC Hsu, KF Tsai, JT Liu, ES Lin. *VHDL Modelling for Digital Design Synthesis.* Kluwer Academic Publishers, 1995.

49. A Hemani, A Postula, B Karlsson, M Fredriksson, K Nordquvist, B Fjellborg, Using High-Level *Synthesis* System SYNT in an Industrial Project, Asian Pacific Conference on HDLs - APCHDLS'93, Brisbane 1993.

50. Z Peng, A formal methodology for Automated *Synthesis* of VLSI systems, Dept. of Computer and information Science, Linkoping University, Sweden.

51. G De Micheli et al. *Design Systems for VLSI Circuits - Logic Synthesis and Silicon Compilation.* Martinus Nijhoff Publishers, 1987.

52. Programmable IC Design, Synario, Website: http://www.synario.com.

53. HDL Analyst and Synplify, SYNPLICITY, Website: http://www.synplicity.com/

54. DATA I/O Website: http://www.data-io.com/.

55. MINC Inc Website: http://www.minc.com/Minchome.html.

56. VeriBest Graphical High Level Design, VeriBest, Website: http://www.veribest.com.

57. StateMate Magnum, i-Logix, Website: http://www.ilogix.com.

58. Peak *VHDL simulator* and Peak*FPGA synthesis*, Accolade, Website: http://www.acc-eda.com.

59. Visual Tools for *FSM* design Summit Inc. Website: http://www.summit.com.

60. JD Ullman. *Computational Aspects of VLSI*. Computer Science Press, 1984.

61. SD Brown, RJ Francis, J Rose, ZG Vranesic. *Field Programmable Gate Arrays*. Kluwer Academic Publishers, 1992.

62. Altera Corporation, FLEX 10K Embedded Programmable Logic Family, Website: http://www.altera.com/.

63. XILINX, XC4000 and XC6200 Family Field Programmable Gate Arrays, Data Books, Website: http://www.xilinx.com/.

64. Lucent Technologies, optimized Reconfigurable Cell Array (ORCA) OR3Cxxx/ OR3Txxx Series *FPGA*s, Website : http://www.lucent.com/.

65. ATMEL Website: http://www.atmel.com/.

66. Actel ACT2 and ACT3 Data Sheet, Website: http://www.actel.com.
67. AMD/Vantis Website: http://www.vantis.com/products/products.html.
68. QUICKLOGIC : Website: http://www.quicklogic.com/.
69. GateField, Zycad Website: http://www.zycad.com.
70. Len Harold, *FPGA* related WWW links, Website: http://www.mrc.uidaho.edu/ *FPGA/FPGA*.html.
71. V Tchoumatchenko, T Vasileva, R Ribas, A Guyot, *FPGA* Design Migration: Some Remarks, Lecture Notes in Computing Science 1142, Proceedings of FPL'96, pp405-409, 1996.
72. Altera Corporation, Implementing logic with the Emebedded Array in FLEX 10K Devices, Website: http://www.altera.com.
73. Lucent Technologies, optimized Architectures Using Distributed On-Chip SRAM in *FPGA*s, Website: http://www.lucent.com.
74. SJE Wilton, J Rose, ZG Vranesic, Memory/Logic interconnect Flexibility in *FPGA*s with Large Embedded Memory Arrays, IEEE Custom Integrated Circuits Conference, May 1996.
75. Aptix Website: http://www.aptix.com/.
76. ICUBE : Website: http://www.i-cube.com.
77. P Lysaght, J Dunlop, "Dynamic reconfiguration of *FPGA*s" , in MORE *FPGA*s (W.R.Moore and W.Luk, eds):.82-94, Abingdon EE&CS Books, 1994.
78. J Brown, D Chen, I Eslick, E Tau, A DeHon, "DELTA: Prototype for a first-generation dynamically programmable gate array", Transit Note 112, MIT Artificial Intelligence Laboratory, Nov. 1994, ftp transit.ai.mit.edu, transit-notes/tn112.ps.Z.
79. Vasilko, M. Ait-Boudaoud, D. Optically Reconfigurable *FPGA*s: Is this a Future Trend. Lecture Notes in Computing Science 1142, Proceedings of FPL'96: 270-279.

9

System Integration

Dobrivoje Popovic
University of Bremen, Bremen, Germany

An integration of mechatronic systems involves the interconnection of system elements of different physical natures. In order to optimally accomplish this task, a provision should be taken that the system elements are mechanically and electrically compatible and optimally tuned to each other. Besides, they should meet the international standards. Finally, during system integration, the reliability and maintainability aspects should also be considered for production and product quality purposes.

In this chapter, the design aspects of mechatronic system integration including optimality and compatibility of the system elements are presented. The guidelines to the selection and the interface of the system elements and the measurement of the resulting reliability and the robustness of the integrated system are also described.

I. SELECTION OF SYSTEM ELEMENTS

Appropriate design of computer-based systems for industrial plant automation requires, first of all, that the automation problems to be solved are technically well specified [1,2]. The specification should be based on a detailed *plant analysis* and on the knowledge elicited from plant engineers and experts of the departments that should be included into the automation system [3]. As the result, two main areas should be identified in which the system elements should be adequately selected by the mechatronic system designer [4]:

- At the field instrumentation level where, apart from sensors and actuators, the devices are situated such as field mounted instrumentation and related interfaces for open-loop and closed-loop control, plant status monitoring and alarming, etc.

- At all higher hierarchical levels where, apart from the computers, intelligent terminals and associated communication links between them are situated

In the following, a short review of the most important selection aspects of system elements and the functions to be implemented will be given.

A. Field Instrumentation Level

At this level of plant automation systems, the selection decisions should be made primarily concerning the:

- Sensors, actuators and field controllers, along with their type, accuracy, grouping, etc.
- Back-up elements and their application concept
- Local digital displays and binary indicators
- Alarm announcers: use and locations
- Plant mimic diagrams required
- Keyboards and local displays, handpads etc. available
- Field instrumentation interfaces and field communication links

Typical electronic elements to be selected at this automation level are:

- Signal transducers
- Signal filters
- Voltage-to-current and current-to-voltage converters
- Voltage-to-frequency and frequency-to-voltage converters
- Operating amplifiers
- Galvanic isolation setups
- Input/output channel multiplexers

mainly required for adaptation of terminal process elements (sensors and actuators) to the computer input/output channels. Furthermore, signal format and/or signal representation has also to be adapted by selecting the appropriate:

- Analog-to-digital and digital-to-analog converters
- Parallel-to-serial and serial-to-parallel converters

The recent development of field bus standards has considerably contributed to the standardization of *process* interface in a universal way and has facilitated the interconnections between the final field control elements and the related programmable controllers and similar digital control facilities [5].

The field bus concept is certainly the best answer to the increasing cabling complexity in production engineering and processing industries at sensors and actuators level that was difficult to manage using exclusively point-to-point links between the sensors and actuators in the field and the intelligent equipment in the *central control room*. Using the field bus concept, all sensors and actuators are interfaced to the distributed computer system in a unique way, as any external communication facility. The benefits resulting from this are multiple, such as:

- Enormous reduction of cabling and installation costs
- Easy adaptation of any future sensors and actuators technology
- Flexible configuration and re-configuration of plant instrumentation
- Automatic detection of transmission errors and cable faults data by transmission protocol
- Facilitated implementation and use of hot backup concept
- Noise and cross-talk elimination, common mode rejection, and galvanic isolation due to the digitalization of analog sensor values to be transmitted

Selection of sensors and actuators [6] as terminal control elements is a most important issue for a design engineer because the technological progress in this area is tremendous. In the past, the development of new sensors has always enabled solving control problems not solvable with the available ones. For example, development of special sensors for on-line measurement of moisture and specific weight of running paper sheet has enabled a high-precision control of the paper making process. Similar progress in the processing industry is expected with the development of new electromagnetic, semiconductor, fiber optic, nuclear, and biological sensors.

The VLSI technology has definitely been a driving agent in developing new sensors, enabling the extremely small microchips to be integrated within the sensors or the sensors to be embedded into the microchip. In this way the intelligent sensors [7] or smart transmitters have been created with data preprocessing and digital communication functions implemented in the chip. This helps increase the measurement accuracy of the sensor and its direct interfacing to the field bus.

Increasing the intelligence of the sensors has enabled that some *signal pre-processing algorithms*, originally implemented in microcomputers, such as the:

- Calibration and recalibration
- Diagnostic and troubleshooting
- Rearranging and rescaling
- Ambient temperature compensation
- Linearization
- Digital filtering and smoothing

have been implemented in the sensor itself. Such *smart sensors* are for the time being in their initial application phase. They most frequently support the existing sensing systems. It is however expected that most of the sensing and pre-processing functions will be covered by such smart implementations.

Further technical innovations are expected from emerging semiconductor, magnetic, chemical, optical, and biosensor technology. For automation of both process and manufacturing plants the advances in fiber-optic sensor technology are of particular interest.

Fiber sensors and *fiber optic devices* have since long time been one of the most promising development fields of sensor technology because the sensors developed in this field, have the following advantages:

- High noise immunity
- Insensitivity to electromagnetic interferences
- Intrinsic safety, i.e. they are explosion proof
- Galvanic isolation
- Light weight and compactness
- Ruggedness
- Low costs
- High information transfer capacity

Operationally, optical sensors are mainly based on the principle of controlled:

- Refractive index
- Absorption coefficient
- Fluorescence constant

Usually, the sensed value is extracted through light modulation of the sensing signal, which can be frequency, phase, or polarization modulation.

On the other side, according to the sensing process used for measurement of physical variables, optical sensors can be:

- *Intrinsic sensors,* in which the fiber itself carries the light and forms the intrinsic part of the sensor
- *Extrinsic sensors,* in which the fiber is only used as a transmission medium for an optical signal generated by a nonoptical sensing element

Widely in use are optomechanical sensors, mainly based on amplitude modulation of the light by:

- Cutting the light between two fibers (shutters)
- Reflecting the light emitted by one fiber and collected by another one (Y-guide probe)
- Varying the coupling between two fibers
- Changing the attenuation losses in fiber with small movements, called microbending

For example, shutter-based optical sensors are used as analog signal sensors with the gap clearance as the measured values. On the other hand, the Y-guide probes are used as flow, pressure, and vibration sensors, as well as the micro-bending based sensors that are also suitable for displacement and strain sensing.

It should, however, be pointed out that, in spite of a wealth of optical phenomena, appropriate for the use in sensing of process parameters, the elaboration of industrial versions of sensors to be installed in the instrumentation field of the plant will still be a matter of hard work over the years to come. The initial enormous enthusiasm, induced by the discovery that fiber-optical sensing, is viable, has overlooked some considerable implementation obstacles of sensors to be designed for use in the industrial environment. As a consequence, there are relatively few commercially available fiber-optic sensors applicable to the processing industries. The sensors are more in use in production, engineering, and manufacturing for dimensionality and shape tests required for product quality inspection.

The term *integrated optics* was coined analogous to *integrated circuits.* The new term was supposed to indicate that in future LSI chips the photons should replace the electrons. This, of course, was a rather ambitious idea, later to be more realistically applied to *opto-electronics*, indicating the physical merger of photonic and electronic circuits, known as hybrid, *optical integrated circuits.* Implementation of such circuits is based on thin film waveguides, deposited on the surface of a substrate or buried inside it. The waveguides could have the planer channel in coupled or branching form made of glass on amorphous or crystalline dielectric or semiconducting material as substrate. The nature of the substrate material very important: it determines the functionality of the photonic part of the circuit. For this purpose, besides semiconducting, the *electro-optic, acousto-optics*, and *non-linear* optic properties of the substrate are used also.

In the meantime there is no illusion that an entirely optical chip is possible; however, there is a series of similar chips capable of performing functions required by sensing elements.

B. Distributed Intelligent Facilities

In a complex, computer-based automation system a large variety of intelligent facilities is distributed across the factory and throughout the enterprise implementing various automation functions. The facilities, nevertheless, can generally be classified into two basic categories:

- *Computers,* where the primary or collected data and the secondary or calculated data are stored, along with the programs implementing the individual automation functions
- *Human interfaces,* supporting the interactions between the computer system and professional experts such as plant operators, plant managers, etc.

Of course, both computers and human interfaces are internetworked for data exchange, so that the selection of computers and human interfaces is strongly related to the selection of communication links for the automation system. In modern distributed computer control systems available in the marketplace, the computers, human interfaces, and the communication links are to a great extent preselected by the vendor, so that here the user-based selection consists in the selection of the distributed system itself to be purchased.

For computer selection two main criteria are important:

- Computer architecture, along with the pertinent system software
- Deliverable special application software

For selection of human interfaces as the intelligent facilities between plant and plant experts, the selection criteria are based on available software, including:

- Standard displays, such as:
  plant overview
  area overviews
  group displays
  loop displays
  alarm surveys
  trend displays
- User defined displays, such as:
  plant mimic diagram
  area mimic diagrams
  group mimic diagrams
  batch control diagrams

Human interfaces are to be distinguished from the *man–machine interfaces*: the interfaces between computers and computer experts that generate, test, integrate, operate, and maintain the computer software.

Among the human interfaces, two kind of facilities are of major importance:

- Supervisory stations
- Field control stations

Supervisory stations are placed at an intermediate level between the central computer system and the field control stations. They are designed to operate as autonomous elements of the distributed computer control system executing the following functions:

- State observation of process variables
- Calculation of optimal set-point values
- Performance evaluation of the plant unit they belong to
- Batch process control
- Production line control
- Synchronization and backup of subordinated field control stations

Belonging to the specific plant units, the supervisory stations are provided with special application software such as the software for *material tracking, energy balancing, model-based control, product quality inspection and production quality control, batch process control, recipe handling,* etc.

In some applications, the supervisory stations figure as *group stations*, being in charge of supervision of controller groups, aggregates of control units, etc. In the small-scale and middle-scale plants some functions of the central computer system also are allocated to such stations.

Process data acquisition software, available within contemporary distributed computer control systems, is *modular software*, comprising the algorithms [8] for sensor data collection and signal preprocessing, as well as for actuators data distribution [9]. The software modules implement function such as:

- *Input devices drivers*, to serve the programming of analog, digital, pulse, and alarm or interrupt inputs, both in event-drivers and cyclical mode
- *Input signal conditioning*, to preprocess the collected sensor values by applying linearization, digital filtering and smoothing, bounce separation, root extraction, engineering unit conversion, data encoding, etc.
- *Test and check operations*, required for signal plausibility test, sensor/transmitter test, high and low signal values check, signals trend check, etc.
- *Output signals conditioning* such as calculation of full and incremental output values, based on the results of the control algorithm used, or the calculation of pulse rate, pulse width, or the total number of pulses for outputting, needed for adapting the output values to the actuators driving signals
- *Output device drivers*, finally, serve the execution of calculated and conditioned output values

Process control software, also organized in modular form, is a collection of algorithms, such as:

- *Basic control algorithms* i.e. the PID algorithm and its various modifications (PI, PD, PID, ratio, cascade, etc.)
- *Advanced control algorithms* such as feed-forward, predictive, deadbeat, state feedback, adaptive, self-tuning, nonlinear, and multivariable control

Control loops configuration [10] procedure is executed in two successive steps for determination of the:

- *Structure* of individual control loops in terms of functional modules used and of their interconnections required for implementation of the desired overall characteristics of the loop under configuration, thus called the *loop's configuration step*
- *Parameter values* of functional modules involved in the configured loop, thus called the *loop's parameterization step*

Once configured, the control loops are stored for their further use with various parameter values. In some situations the parameters of the block in the loop are also stored for future use.

Generally, the functional blocks available within the field control stations—in order not to be destroyed—are stored in a ROM or in an EPROM as *firmware modules*, whereas the data, generated in the process of configuration and parameterization, are stored in a RAM, i.e. in the memory where the configured software runs.

It should be pointed out that every block required for loop configuration is stored *only once* in the ROM, to be used in *any numbers* of loops configured by simply addressing it and its parameter values. The approach actually represents a kind of *soft wiring*, stored in the RAM.

For multiple use of functional modules in the ROM, their subroutines should be written in so called *re-entrant form*, so that the start, interruption, and continuation of such subroutines with different initial data and different loop parameter values is at any time possible.

Once having the required functional blocks stored in the *library of subroutines* of the control software, and having the tool for loops configuration and parameterization, the user can program the required control loops in a ready to run form. The loops programming is here a relatively easy task because the loop configuration in this case practically means that, in order to implement the desired control loop, the required subroutine modules should be taken from the library of functions and linked together [11].

II. INTERFACE

In computer systems, an *interface*—usually called *computer interface*—is a shared boundary between the processors and the peripherals attached to it. In process automation systems the interface, usually called *process interface*, is the shared boundary between the computer and the process, i.e. between the processor and the process instrumentation. In order to operate properly, enabling the reliable data exchange between the facilities on both sides of the boundary, the interface has to be standardized in terms of its hardware and software implementation. International standardization of the interface, when correctly implemented, enables any peripheral or process control element to be connected to any computer for data exchange. It also enables the system designer to select the system elements out of the worldwide spectrum of computer and instrumentation elements available on the international market. This, for instance permits a great flexibility in interconnection and re-interconnection of equipment in the CNC machine tool industry and computer manufacturing and reduces the maintenance efforts and the related costs. That is why the computer and instrumentation industries have been since the early 60s, and particularly in the 70s after microelectronics penetrated into computer and instrumentation technology, deeply interested in international standardization of communication interfaces [12].

In integrated computer systems, beside the local and wide area networks, local buses and computer interfaces are used for data exchange between the system devices and for communication between the user and the computer. For instance, in CAD/CAM systems, a number of input/output devices are used for man-machine communication. The device technology has improved over the years, so that a rich collection of text and graphic devices (such as scanners, light pen, and image-input devices) is available on the market. On the other system end, various *soft-type output devices* (like text monitors and graphic displays) and *hard-type output devices* (like the printers, plotters, etc.) are available. For their interconnections, the local bus systems and processor interfaces are used in the majority of situations, rather than the local area networks [13].

Because the attached devices are different in speed, presentation and transfer of data, etc. different interface standards have been worked out by communication experts and most of them have also been internationally standardized. In the following, a brief review of well-known interface standards will be given, most frequently used in process and production automation systems.

A. Microprocessor Bus Systems

One of the first interface standards that has since been widely be accepted as a common computer interface was originally defined by the Electronic Industries Association (EIA). It is the RS 232C standard, defining the serial interface for interconnection of terminal devices, microcontrollers, microcomputers, CNC controllers, and the host computers within a distance of 15 meters. The achievable data transfer rate is 20kbps.

For the use in CNC applications a D-shaped, 25 pin connector is recommended, from which only 8 pins are used for the interface signals:

- Data terminal ready (DTR)
- Request to send (RTS)
- Clear to send (CTS)
- Transmit data (TD)
- Receive data (RD)

- Data set ready (DSR)
- Carrier detect (CD)
- Ring indicate (RI)

The equivalents of the RS 232 interface standards are the V.24 and V.28 interface standards of CCITT and ISO 2110 standard.

For higher data transfer rates and longer distances EIA has standardized the interfaces RS 422 (for transfer rate of 100 kbps and for distance of 1.2 km), RS 423, RS 485, and RS 491.

Later on, the International Standards Organization (ISO) and the International Electrical Commission (IEC) standardized, in cooperation with the Institute of Electrical and Electronic Engineers (IEEE), European Computers Manufacturing Association (ECMA) etc., a number of parallel interfaces for building of microprocessor-based multiprocessor systems and of process data acquisition and process control systems.

The best known bus standard for integration of microcomputer-based data processing systems is the VME bus standard, a backplane interface developed from Motorola's VERSAbus for the 16- and 32-bit MC68000 family of microprocessors. The bus standard was internationally accepted and supported by Mostec, Signetics/Philips, and Thompson CSF. It was defined as an internal communication link for interfacing the microcomputer with its peripherals, that can also include the process interface (digital and analog inputs and outputs).

Multibus I, based on the IEEE 796 standard, has been used since the mid-70s as microprocessor independent internal bus architecture. Its main advantages are the simplicity of its use and its high reliability with the MTBF of over 150,000 hours. This makes it suitable for heavy-duty applications.

Intel Corporation, after years of experience with the Multibus standards, has started working in cooperation with some of its large customers on next-generation bus structure that was called Multibus II. It is a parallel bus system, similar to the token bus, implemented as a VLSI device to be associated with a message-passing coprocessor.

A very interesting development in the area of processor independent bus interfaces represents the IEEE 896.1 standard proposal known as FUTUREBUS. It is a relatively low cost, simple architecture reliable high-performance 32-bit backplane bus for building various multiprocessor architectures with a burst data transfer rate of 100 Mbps.

Later on, in cooperation with ECMA and ANSI (American National Standards Institute), ISO standardized the SCSI (Small Computer System Interface), a general interface for building data processing systems. The standardization work was initiated because of the need for of a high-performance, flexible communication link between the processor and the peripherals in small-scale computer systems. The first version of the standard, the SCSI-1 version, was issued in 1986, to be followed by the SCSI-2 and SCSI-3 version.

In addition to the above interface standards, some additional company internal bus standards have been developed related to some specific computers or the computer families. Typical in this respect are:

- NuBus, an advanced 32-bit system bus, used as a backbone in Texas Instruments and Apple (Macintosh) computer systems
- UNIBUS and Q-Bus of DEC computer family PDP-11

With the advent of personal computers the very well known AT bus system was launched, also known as the ISA-Bus, an 8-bit bus system of 8 Mbps transfer rate. It was designed to be used in 16-bit microprocessors 80286. However, the introduction of full 32-bit processors i386 and i486 (with the 32-bits data bus and 32-bit address bus and with the high-performance operating systems Windows™ and OS/2, requiring huge work memory to be administrated) has put the extension of the 16-bit ISA bus systems on the agenda. As a result, two new bus concepts have been worked out:

- The EISA (Extended ISA) bus concept
- The IBM Microchannel concept

The EISA concept was worked out for a data transfer rate of 33 MBps, supports electrically and mechanically, as far as possible, the ISA components designed for the data transfer rate of 8 Mbps. The Microchannel of IBM, again, is electrically and mechanically entirely IBM PC/XT/AT oriented.

The advances in powerful interactive graphic systems have in the early 90s stimulated the development of special *graphic coprocessors* based on the extension of the ISA-Bus called VESA-Local-Bus or VLB-Bus, a 32-bit bus system with the transfer rate of 40 Mbps, widely used in contemporary microprocessor graphic systems. At nearly the same time the Peripheral Component Interconnect Bus, the well-known PCI Bus was launched by Intel for high-performance 64-bit Pentium microprocessors. It is an autonomous, input/output oriented communication link, operationally decoupled from the internal microprocessor bus by a PCI bridge. The bus is seen as a high-end-solution for data transfer problems in microcomputers and in workstations. The bus can reach a transfer rate of 133 Mbps for instance in the burst transfer mode within a multiprocessor system.

B. Process Interface

The first efforts in the definition of interface standards for real-time data acquisition systems have been made on the initiative of Hewlett-Packard and the IEEE and supported by a large number of computer users and computer vendors. As a result, the General Purpose Instrumentation Bus (GPIB), named IEEE-488 interface system was specified for laboratory and test field data acquisition [14]. It is a bi-directional, 8-bit parallel, byte-serial data transfer interface. Its operational concept is relatively simple: in addition to the 8 lines for data and command transfer, the bus has 8 control lines managing the signals such as:

- Attention (ATN)
- Service request (SRQ)
- Interface clear (IFC)
- Remote enable (REN)
- Not ready for data (NRFD)
- Not data accepted (NDAC)
- End-or-identity (EOI)

The main idea was to define an interface system to facilitate the integration of laboratory and test field instrumentation to be attached to a microcomputer or to a programmable device. In order to reduce training expenses for personnel that are supposed to operate the high-sophisticated electronic instruments, and in order to increase direct manpower-free communication between the instruments, a communication protocol, based on the *listener–talker principle* was implemented with centralized bus arbitration. According to the protocol:

- *Listener* is a device that, after being addressed, receives the data from the
- *Talker*, a device that, after being addressed, sends the data to the listener

In addition, the communication protocol defines the:

- *Controller* as a common device, controlling the communication on the bus

For microprocessor industrial control applications a powerful modular interface system has been developed that meets the reliability requirements of the industrial environment, the so-called STD bus or IEEE 961 standard, based on master-slave, single-master, and multimaster principles. The main advantage of the bus lies in the cost-effective modular approach of its implementation. This is particularly required in control applications where the modularity supports the design flexibility, enabling the designer to configure the entire system to optimally meet the design objectives.

In the past, the STD bus has been applied in various industries, including:

- *Process control*, for instance in the food industry for automatic control of a multiple compressor and fan system by keeping the required temperature within a refrigerating room while reducing the peak energy requirements

- *Machine control*, for instance for control of raw material feeding, mold clamping hydraulics, heating and cycling of the injection process, etc. in an injection molding machine
- *Robotics*, for positioning, rotating, raising, and lowering the head in the pick and place machine for surface-mount components, where the head is positioned over the PCB and the component is lowered in its position
- *Test field automation*, for integration of test equipment and collection of test data
- *Factory control*, for implementation of a distributed, multiworkstation data acquisition system based on the server–user principle

A particular strength of the STD bus is the variety of process interface modules available on the market, including the modules for interfacing to the field bus candidates, BITBUS and ARCNET.

In manufacturing automation systems, where a large number of programmable controllers are used, a special data transfer solution was worked out based on the INTERBUS-S system, that is an industrial automation bus system for facilitated connection of final process elements (i.e. of sensors, actuators, and controllers) to PCs. It is a real-time bus system capable of transferring, in addition to the signal values, also the accompanying messages.

III. SYSTEM RELIABILITY AND MAINTENANCE

The idea of *reliability* is relatively new in systems engineering and mathematics. It defines the *system reliability* in terms of the *probability* that the system, for some specified conditions, normally performs its operating function for a given period of time. It is a measure of *how well* and *for how long* the system will operate in view of its design objectives and its functional requirements, before it fails. The definition itself indicates that the term reliability applies to the systems that work permanently and that are subject to random failures, as the mechanical, electrical, and electronic systems do. The failure itself is not defined so that any failure within the system automatically causes the total system failure, but rather so that the failures can cause the deviations in system behavior from the specified design objectives, so that the system performance is reduced. For instance the IEEE, in its standard dictionary of electrical and electronic terms, recommends to distinguish the following types of failures:

- *Degradation failures*, that are *partial* system failures within the tolerable limits of the performance drift including some gradual failures
- *Complete failures* or *total failures* that are beyond the tolerable limits of performance drift; some are *sudden failures* and can even become *catastrophic failures*

In engineering, reliability is mathematically defined as the systems' *mean time between the failures* (MTBF), or as the systems' mean time before failure.

Our experience with real technical systems indicates that system reliability generally decreases with the time, i.e. as the older the system the less reliable it is expected to be. Consequently, the reliability function must be a decreasing, non-negative function that can take zero as its final value.

In engineering practice, the reliability function of a given system, as a basic reliability characteristic of the system, can only be estimated in an experimental way. For instance, if N similar systems are run simultaneously, and the number of systems failed after the time t is Nf(t), and of those not failed Nn(t), the system reliability at time t can be expressed as the ratio:

$$R(t) = Nn(t) / N = N n(t) / [Nf(t) + Nn(t)]$$

The complementary value:

$$F(t) = 1 - R(t)$$

is the system unreliability at time t.

Knowing the *failure rate function* $r(t) = \lambda$, the value of MTBF can be calculated as the inverse value of λ:

$$MTBF = 1 / \lambda.$$

Using the above reliability parameters, system reliability can be mathematically modeled and simulated using a given failure rate λ and the simulation results can be used for reliability system design and for system maintenance.

Mechanical and electronic systems, despite their high reliability, undergo occasional failures that are more frequent as their structure becomes more complex. The system failures occur at the very beginning of the system operation, i.e. after the system has been installed, this more frequently because of possible unreliable hardware or software parts used for system building, or because of poor system design, manufacturing flaws, or the initial misadjustments of the system. These are *early failures* of the system are said to be in the *infant mortality phase* of operation.

In the operational phase that follows, and after the above causes of failures have been removed, system failures will become rare, system reliability will become stable, and the failure rate will be constant. In this operational phase the time at which the failures really occur is not predictable, but the probability of their expected occurrence within a time period can be predicted. This is the *normal* operational phase of systems.

Finally, after a long period of active time, *wear-out* failures become more frequent because individual mechanical and electrical parts undergo failures due to wear or aging. System failures will be more frequent and the failure rate higher.

A. Systems with Repair

Once the system has failed, two situations are possible:

- The system can be repaired
- The system cannot be repaired

For systems with reparable failed components, the definition of the overall system reliability includes, besides the mean time between failures (MTBF), also the time required for components repair, i.e. the *mean-time to repair* (MTTR). The components repair effectively prolongs the total system life, i.e. it makes the system longer available. Thus, for systems with repair the *system availability* is defined as:

$$A = \frac{MTBF}{MTBF + MTTR}$$

It is a fundamental indicating factor integrating the system reliability and system maintainability, showing the portion of time, within the total time of system use, during which the system was available for execution of operations it was designed for. Computer-based systems, used for direct digital control, have to guarantee at least 99.95 percent of availability over the full year of their non-interrupted use. This figure is achieved through the selection of high-quality components, use of expedient maintenance and repair services, and, where applicable, duplication of facilities.

For *non-reparable systems*, the MTBF is the only reliability parameter essential for defining the *average lifetime* of the system.

Reliability of a computer-based automation system is determined by the reliability of systems' hardware and software components. Thus, when designing or selecting such systems from the reliability point of view, the reliability of both components should be taken into consideration.

Hardware reliability of a complex, computer-based plant automation system strongly depends on the reliability of system components (computers, communication links, terminals, final instrumentation elements, etc.) and on the redundancy and the structural interconnection of system components. Consequently, the overall system reliability can be increased by:

- *Increasing the reliability* of individual system components and by
- *Design for reliability*, i.e. by implementing a multiple, redundant system structure with a higher overall reliability

This is a general guideline for design of high-reliable distributed computer control systems and of programmable logic controllers where, for meeting the required reliability requirements of the system, one of the following alternatives is used:

- building a single-computer system out of high-reliable components
- building a multi-computer system out of less-reliable microcomputers

In the first case the resulting reliability is straightforward calculable in terms of reliability of individual components, the structure of the system, and the redundancy concepts within the structure (parallel redundancy, serial redundancy, etc). In the second case, the multicomputer system incorporates redundant computers so that the system will continue to operate even when some of the computer have failed. In this case the multicomputer system is a *fail-safe system*. When designing such systems, the design engineer has first to decide *how redundant* the multi-computer system should be to still operate even in the *worst case,* i.e. when a maximally permitted number of individual computers in it simultaneously fails. For instance, when in an *n* computer system *m* computers are needed for normal operation of the system, and when *Rc* is the reliability of individual computers in the system, the *overall system reliability* Rs, that determines the *system fail-safety,* is given by:

$$Rs = \sum_{k=0}^{n-m} \binom{n}{k} R_c^{n-k} (1-Rc)^k$$

When designing reliable systems containing *repairable components,* the decision also should be made concerning the realization of required system *availability,* using the MTBF and MTTR as two basic reliability parameters.

Present multicomputer systems have a very high MTBF value because they are built out of high-integrated, high-reliable chips, whereas their MTTR value is very low, mainly because the repairing procedure consists solely in switching over to some backup units or in exchange of the failed modules, so that the resulting system availability is nearly *absolute.*

In industrial automation systems the fail-safety implementation includes a number of support measures, such as error detection and removal, on-line system diagnosis, preventive maintenance, backup concept, automatic recovery after failure, etc. In such systems, precautions also should be taken against a total crash and possible system crash upon power failure. In both cases the output signals, representing the set-point and actuating values of control instrumentation, should be saved, i.e. they should remain frozen in their last values, that during the computer off time have to be manipulated by the back up controllers, or by manual manipulations, until the computer system has been repaired or until the power recovery.

An additional probability problem, particularly related to the failure detection of sensing elements working under severe industrial environment, is the *components failure detection.* Here, as an adequate problem solution, most frequently the *majority voting technique* is used, also known as *m from n selection approach.* It has proven to be a very efficient *on-line test method* of components operability because it detects the failure of a component by comparing its output value with the output values of other components and—by detection of deviation of the single value from the remaining values—indicating the possible failure of the specific component.

For implementation of a still higher system reliability, the majority voting technique is additionally supported by the *diversity principle*, consisting in combining the sensing elements of different manufacturers, and in connecting them to the system via *different instrumentation channels*.

The *majority voting* approach and the *diversity* principle belong to the category of *static redundancy* implementations. In the repairable systems, however, *dynamic redundancy* is preferred, based on the *backup* or *standby* concept, that attaches to each critical active element of the system a redundant element, capable to take over the function of the active element in case it fails. The standby elements can be used as:

- *Cold-standby*, that is the elements are switched off while the active element is running properly and switched on to operate when the active element fails
- *Hot-standby*, that is the elements are permanently switched on and execute in the *off-line mode* all operation of the active elements, ready to take over on-line operations from the active elements when they fail

For industrial automation systems, of outstanding importance is the reliability and the fail-safety of the system's power supply. Although the most sensitive units of the computers used are battery powered and thus reliable enough, the power failure will still disturb a considerable part of computers, peripherals, and the intelligent terminals used.

Software reliability introduces some additional features that can influence the overall systems' reliability because the possible software errors, such as conceptual, formal, and coding errors of the subroutines, as well as the nonpredictability of total execution time of the critical real-time subroutines under arbitrary operating conditions, can deteriorate the interrupt service routines and the communication protocols by not guaranteeing the *time-critical responses* required for reliable system operation. However, being intelligent enough, the software itself can perform error detection, error location, and error correction. In addition, an *off-line simulation test* of the software can well estimate the *worst-case execution time*, that would be sufficient enough for the judgement of the applicability of the simulated software to the given critical problem. For these reasons, the definition of *software reliability* is not as straightforward as the definition of the hardware reliability. Furthermore, analogous definitions of *software reliability, software availability, software failure, software security*, etc. fails here because there are some additional aspects to be taken into consideration such as:

- A software module is viewed *as reliable* if the errors and faults within the computer system do not cause its failure, even if the data to be processed or the algorithms to be executed are not correct.

- Software is also viewed as reliable, as long as the effects of its failures are tolerable.
- Correct software is, again, viewed as *unreliable* if, for deviations in input data or by any hardware fault, it generates the failures that are not tolerable.
- Highly reliable error-free, and fail-safe software is viewed as *unreliable* if—due to processor overload—cannot be initiated or terminated when needed.

At this point the fact should be underlined that preconfigured software—such as standard display panels, data acquisition, and control modules - are highly reliable software elements. They are, in order to have the *shortest execution time,* usually programmed in an assembly language and are well tested under normal operating conditions, both off-line and on-line. In some systems, the same software modules simultaneously run on two cooperative computers, one of them being operationally active and the other one serving as a hot backup.

B. Solution Concepts

High-reliability demands, placed on plant automation systems, have forced system manufacturers to conceptualize high-sophisticated reliability architectures of their systems, particularly on the system control and supervisory level. For this purpose reliability and fail-safety concepts are used that include separate, redundant digital backup units for automatic protection of failure-free, noninterrupted execution of control functions. In such cases only one backup unit is active in data acquisition, process control, and hardware diagnostics. The errors, discovered by diagnostic software, automatically trigger the security actions and shut down the active unit. The backup unit usually looks for possible errors every 100 *msec* or so, and—when discovering any systems' error—takes over the role of the shut down active unit within the parts of the second by starting some *initialization routines.* The same holds when the backup unit discovers some errors through the internal diagnostic software.

In batch process control systems very frequently a security module is provided as an additional reliability facility for monitoring the process control for abnormal execution of individual control steps, as well as for conditions to be met for the execution of the next step (normal process behavior, normal operating status of hardware and instrumentation etc.). This, for instance, is achieved by:

- *Devices status check* at the end of the step
- *Actual values check* of process variables
- *Security check of alarm signals generators*

Another reliability concept, followed in process control systems, consists in duplication of process interface and plant monitoring facilities. The concept can also be extended to higher automation levels, so that a large number of redundant structures and functional modules are available at each hierarchical level of the system that can include various functional modules, such as the watchdog and *self-test modules, self-diagnostics* and alarming functions, hardware monitoring by *on-line* test etc. In some cases it is enough to duplicate only the critical facilities at each hierarchical level, for instance by providing:

- Two processors with equal rights, capable of replacing each other
- A twin-bus to interface the processors with its instrumentation
- Duplicated the most critical analog and pulse input/output cards, etc.

C. Implementation Problems

For a long period of time, one of the major difficulties in the use of computer-based automation structures was that the safety authorization agencies refused to license such structures as *safe enough*. For the agencies, only the discrete and semiconductor-based protection circuits have been accepted as safe security provisions. However, the progress in computer and instrumentation hardware, as well as in monitoring and diagnostic software, has helped building computer-based automation structures acceptable even for applications in critical process areas, as far as they guarantee that:

- Failure of any instrumentation element at the field level, or of any programmable controller or computer at control level, does not cause a hazardous situation.
- The instrumentation element at the field level, or of any programmable controller or computer at control level, does not cause a hazardous situation.
- The entities themselves—as far as possible—initiate the adequate failure recovery procedures.

The last requirement implies that system structure at the instrumentation and control level should be relatively simple, easy to survey, decomposed into small autonomous cells rather that to be complex, consisting of multiply interconnected subsystems.

It is, however, evident that a high degree of system safety is connected with the high additional investment in redundant system elements, self-inspection and failure recovery features etc., that could only be justified by improved throughput of the production and the quality of the products due to an increased system availability.

D. Design for Reliability

Reliability aspects of distributed computer control systems imply some specific criteria to be followed in their design. This holds for the overall systems concept, as well as for the hardware elements and software modules involved. For instance, when designing the system hardware:

- Only highly reliable elements and subsystems should be selected, that have in long term testing under heavy-duty conditions shown their high reliability.
- A modular, transparent, well-structured hardware concept should be taken as a design basis.
- When applicable, the voting and diversity principle should be build in the most critical system parts, supported by error-check and diagnostic software interventions.
- A hot standby or cold standby provision should be available, along with a noninterruptible source for their powering internal communication links, particularly those at the instrumentation level.
- At the plant field level, process interface modules for analog and digital inputs and outputs should have backup facilities, powered by independent supplies.
- There should be a variety of sensor data examinations: plausibility, as well as automatic periodic interface tests using diagnostic, parity, and redundancy checks for error detection etc.
- For expedient systems maintenance and repair, extensive software provisions should be available along with a set of the most critical hardware parts.

Similar issues are related to software design:

- Modular, free configurable system software should be available with a rich library of modules that have been tested and verified for various operational conditions for a long time
- Loops and display panels configuration procedures should be simple, transparent, and easy to learn.
- A sufficient number of diagnostic, check, and test functions for on-line and off-line system monitoring, maintenance and repair should be provided.
- Special software provisions should be made for a *bumpless* switch over from *manual* to *automatic* mode to computer control.
- The software core image versions, deposited on an external nondestructive store for its later repeated loading into the work memory of the computer, should be well-protected so that their use after system recovery from power or hardware failure is reliable.
- For process and actual long-term data the technique of shared, multiple-files databases should be used.

These are, naturally, only some outstanding features to be implemented in multicomputer, industrial automation systems, the list based on experience gained from long-term applications of such systems.

E. Systems Maintenance

Maintenance, as the activity of restoring the failed system to its nominal operating conditions, can be:

- Preventive maintenance
- Emergency maintenance or maintenance on failure

Preventive maintenance is system restoring according to a regular schedule of servicing with the objective of preventing unexpected system failure. In computer aided automated systems preventive maintenance is a *predictive maintenance,* based on condition monitoring and on-line diagnostics. Thus, in the case that the system elements are subject to continuous or periodic monitoring so that their degradation is detectable early, such elements can be replaced *before* they fail. This simplifies the planning and execution of preventive maintenance.

Taguchi, Elsayed, and Hsiang (1990) have considered the effect of preventive maintenance on product characteristics and indicated that the optimal preventive maintenance schedule can minimize quality losses of products, as well as tolerances of the product characteristics. Such maintenance involves some system element checks, their possible repair, or replacements *before* their failure, which minimizes the total downtime of the production line. This, however, should be kept in balance with added maintenance costs.

Assuming that production losses are proportional to the duration of downtime of the process due to its failure, or are due to the discovered considerable deviations of product characteristics from their nominal values, the total loss of product failure per unit time, assuming there is no preventive maintenance, can be written as

$$L = C / MTBF$$

where $C = Pl \times t + c$, with Pl as loss due to the product failure and c as repair cost of the process. When using a preventive maintenance schedule each time the amount of deviation or drift reaches a value Δ, the above value of loss of product failure per unit time should be corrected by a factor depending on Δ..

The *modularity* concept in building computer hardware and software facilitates the problem of maintenance: it reduces repair time and also lowers the spare parts cost. Besides, modular systems, because being structurally more transparent, have an increased degree of *maintainability*.

The modular structure of distributed computer control systems increase their reliability and availability, while accelerating their repair time and facilitating their maintenance. The increased reliability is highly desirable because any partial or total failure of the automation system can disturb the production line, damage the plant, injure people, etc. This was, from the very beginning, the reason for seeking for highly-reliable, fail-safe systems with a high availability figure.

REFERENCES

1. G Coulouris, J Dollimore, T Kindberg. *Distributed Systems – Concepts and Design, 2nd ed.* ISA Internatl. Conf., New York,., 1994.
2. PH Laplante. *Real-time Systems Design and Analysis.* New York: IEEE Press, 1993.
3. KD Shere, and RA Carlson. A methodology for design, test, and evaluation of real-time systems, IEEE Computer.27(2):34-48, 1994.
4. D Popovic, VP Bhatkar. *Distributed Computer Control for Industrial Automation.* New York: Marcel Dekker, Inc., 1990.
5. D Johnson. *Programmable Controllers for Factory Automation.* New York: Marcel Dekker, Inc., 1987.
6. CW De Silva. *Control Sensors and Actuators.* Englewood Cliffs, NJ: Prentice-Hall, 1989.
7. MM Bob. Smart transmitters in distributed control–new performances and benefits, Control Engineering 33(1): 120-123, 1986.
8. L Sha, SS Sathaye. A systematic approach to designing distributed real-time systems, IEEE Computer 26(9): 68-78, 1993.
9. MS Shatz, JP Wang. Introduction to distributed software engineering, IEEE Computer, 20(10): 23-31. 1987.
10. D Popovic et al.Conceptual design and C-implementation of a microcomputer-based programmable multiloop controllers, J. of Microcopm. Appl. 12(2): 159-165, 1989.
11. JT Fink. Advanced engineering tools for process automation, Control Engineering 35 (9): 23-26, 1988.
12. J Di Giacomo. *Digital Bus Handbook.* New York: McGraw-Hill, 1990.
13. T Ozkul. *Data Acquisition and Process Control Using Personal Computers.* New York: Marcel Dekker, Inc., 1996.
14. AJ Caristi. *IEEE-488–General Purpose Instrumentation Bus Manual.* New York: Academic Press, 1989.
15. G Taguchi, AE Elsayed, T Hsiang. *Quality Engineering in Production Systems.* New York: McGraw-Hill, 1990.

10
Optimality of System Performance

Dobrivoje Popovic
University of Bremen, Bremen, Germany

Since the beginning of manufacturing the trend has been to group similar types of work machines together for better work organization and execution. This concept was used for production planning process, leading later on to the concept of what is known as group technology in production organization, product manufacturing, and design. The group technology concept is used in mechatronic approach of process and product design because it enables process and product development to be based on the use of modification and the use of existing development documents available within the enterprise, instead of initiating each development from the scratch.

This chapter introduces the reader to the basic ideas of production flow analysis, monitoring, and control and to the related subjects of production performance analysis and product quality control, based on product visual and acoustic inspection.

I. PRODUCTION FLOW AND PERFORMANCE ANALYSIS

A. Production Flow Analysis

Production flow analysis in manufacturing is a methodology of *group technology* that groups the production components according to the *manufacturing sequences* to be executed. The grouping is based more on the facilities to be used, rather than on operations to be executed. The analysis is to be carried out when designing new production systems, as well as when organizing existing ones [1].

Generally, production flow analysis includes the:

- Factory flow analysis

- Group analysis
- Line analysis

It helps to optimally arrange the groups of components and the required facilities to enable their maximal utility at minimal workpiece movements. To achieve this, the following has to be done:

- The available *work machines* should be *classified* and *coded* according to the job they can do and the operations they can perform.
- The *production sequences* to be carried out should be *checked* against the facilities and against the *part groups* required for carrying out the required operations.
- The *part flow analysis* should be accomplished through the identified machines, and the *total flow efficiency* estimated.
- Using identified part groups needed, *analysis of operations* to be performed should be carried out.

For solving the most difficult problem of part group analysis, graph theoretical methods, linear programming methods, similarity-coefficient based methods, as well as the methods of artificial intelligence have been used.

Graph theoretical methods consist in building graphs in which the *machines* and the *parts* represent the *nodes,* and the *processing steps* the *arcs,* connecting the nodes. The graphs are analyzed with the objective of finding the independent areas of the graphs representing the subgroups of machine parts, equivalent to the part family and machine cells. Various approaches have been proposed for graph partitioning.

Mathematical programming methods consist in minimizing the total sum of distances between the individual parts for a given number of families and the given constraints, so that no part should be member of two or more part families. *Integer programming, quadratic programming*, and *dynamic programming,* based on *heuristics*, have been proposed for the above minimization. The methods of mathematical programming are, however, relatively inefficient when used for real-time grouping purposes because they are computationally time-consuming, particularly when handling complex manufacturing problems. In such cases, much more efficient are the *matrix manipulation methods*, consisting in rearranging the rows and the columns of a matrix until the nonzero elements are brought close to the main diagonal, simplifying further mathematical operations with the matrix.

Methods of artificial intelligence, applicable to the solution of the above problem, rely on:

- Knowledge-based systems, or expert systems
- Self-organizing neural networks

Expert systems solve the machine clustering problem, essential for machine grouping, by storing in the *knowledge base* of the system the clustering strategy in terms of *production rules* or in form of *syntactic patterns*. The drawback of the approach is that the changes in manufacturing process or in machine configuration imply the *knowledge acquisition process* to be carried out anew, *before* using the expert system.

Self-organizing neural networks, based on the paradigm of *competitive learning*, can also be used as *cluster identifiers*. The networks as *pattern classifiers* are capable to detect, within the total set of input patterns, clusters of similar patterns. The identified clusters represent the part families and machine cells from the compiled *part machine incidence matrix*.

B. Production Flow Control

Another component of the methodology of group technology deals with the problem of flow control. In manufacturing, the general approach to flow control is based on production planning and scheduling. In flexible manufacturing systems, because *identical* machines are grouped into production cells, each group being capable of executing tasks on a set of various products with no time lost for setting up, the process of production planning and scheduling has to take into account the possible alternatives. Some groups could produce the same products with different capacities. In both cases, flow control also has to take into account the discrete events, such as machine failures, demand changes, required setups, etc.

Efficient flow control is based on the *dynamic model* of production scheduling under given restrictions and for given boundary conditions [2]. The model incorporates items such as the given multilevel bills of materials, flexibility of machines, availability of buffers, and profile demands to be met. On such conditions, the *flow control objectives* could, for instance, be the optimal control of *production rate* by minimizing the *buffer cost* [3].

A classical approach to the problem of production flow control was proposed [4], relying on a *four-level hierarchical system model*. In the model, the relative time-consuming events, like machine failures, are considered at the *first hierarchical level*. The production rate, directly relevant to the optimal flow control, is calculated at the *second hierarchical level* and the corresponding route splitting at the *third hierarchical level*. Short-time part dispatching, finally, is considered at the *fourth hierarchical level*. The main disadvantages of the proposed approach are, that it:

- Does not take account the possible constrains
- Ignores the problem of restricted buffer capacity
- Does not permit the consideration of the alternative machine groups capable of executing the same task

Gerschwin (1989), [5] proposed an approach to the synthesis of operating policies for manufacturing systems. He developed feedback control lows using *dynamic programming models* that account for the discrete nature of manufacturing process. The proposed approach finds, based on hierarchical structure of a production process, optimal planning and scheduling policies. At each hierarchical level, the time for controllable discrete events to occur during the production process and the related frequency targets of the events, is determined. The events, that potentially affect the production capacity, are characterized in terms of the frequency with which they occur and in terms of the degree of control the decision maker can exert over them.

For a formal definition of the proposed synthesis approach, the following terms are of major importance:

- *Resources*, the parts of the production system not being consumed or transformed during the production process, such as the production machines themselves, but not the machine tools that undergo the wear
- *Activities,* the events associated with the resources, whereby each activity has two events corresponding to the *start* and the *end* of the activity
- *Setup* or *current configuration*, a limiting factor for individual resources that determines what activities each resource can perform at a given time

Setting up machine tools is considered a special activity on the machine that can support different cutting edges at different locations. Changing a tool is an activity that removes, replaces, and calibrates the cutter. Using the terms, defined in this way, the proposed approach to the hierarchical flow control can be formulated in the following way.

Let S_{ij} (t) be the j-th *activity* of the i-th *resource*, S(t) itself being a *binary function* having the value 1 when the i-th resource is busy with the execution of j-th activity at the time t, otherwise being 0. It holds:

$$\sum_{j} s_{ij}(t) \le 1$$

because not more than one activity could be present at any i-th resource. Every resource has a *capacity* that cannot be higher than 100% , and every activity has a specific *duration* and a specific occurrence *frequency*.

Event controllability is an essential feature useful for implementation of production flow control. *Controllable events* are, any time they are required, selectable by the *decision maker*. Consequently, an activity is controllable if its initial event is controllable. It is generally supposed that the failures are *noncontrollable*. The goal of optimal production flow control is to produce a given products quantity, subject to a given product demand profile, at lowest cost. The goal should be achieved by controlling the controllable events (production operations) and by minimizing the effect of noncontrollable events (failures, repairs, setups, maintenance, etc).

A generalization of the approach, proposed by Khmelnitsky, Kogan and Maimon (1994), [6], merges the four hierarchical levels described above into one level. The approach relys on a dynamic model essential for solving the optimization problem with constraints, for instance by means of the *maximum principle* of Pontrjagin [7]. The above generalization supposes that the:

- Machines are reliable, (i.e. that there are no machine failures).
- Production process is continuous.
- Transport time between the machines, and between machines and buffers, are negligible.

The statement of the problem can be derived by considering a *flexible floor shop* that fabricates, assembles, and disassembles a set of products corresponding to a *given demand profile*. The state of a machine within the shop is characterized by the workpieces to be processed with respect to the *bill of materials*. Accordingly, identical machines within a group can be in different states, depending on the product to be produced and the production rate to be achieved. The goal of optimal flow control is to minimize the buffer effective cost by retaining the demand profile. The buffer levels $X_i(t)$, defined as the difference between the instantaneous production and production demand, is determined at any time t using the canonical form of a continuous time optimal control problem, from the equation

$$\frac{dX_i}{dt} = \sum_{k,j} W_{kj}(t) V_{ikj} - D_i \qquad X_i(t) \geq 0$$

where W_{kj} is the *control variable*, defined by actual loading for the machine group of the art k, the production capacity of which is in the state j. V_{ikj} is the *capacity* of the same art being in product i. D_i, finally, is the demand profile for the product i.

For optimal control, the buffer cost:

$$\text{Min} \int_0^T \sum_i C_i X_i(t) dt$$

should be minimized, subject to some given constraints. For minimization of buffer cost, here the maximum principle of Pontrjagin also is proposed by Khmelnitzky, Kogan, and Maimon, (1994), [6].

Hierarchical framework, developed by Pereira and Sousa (1993), [8], for *optimal flow control* in a manufacturing system, is based on a *two-level model*. At the *higher level*, optimal control problem is solved by the iterative version of the *projected gradient* type of algorithm, that determines the *optimal production rates* for the individual subsystems and for each activity as a functions of time. The *optimality conditions* for overall manufacturing system are then found using the maximum principle.

At the *lower level*, optimal strategies are further refined and the sequences of operation specified. For a given time horizon and for a specified demand profile, the algorithm determines a *suboptimal admissible schedule* for the manufacturing system considered. The algorithm used defines a list of machines, ordered by their criticality, because the more critical machines contribute, in case of production disturbances, more to the cost functional increase. Thus, the algorithm first concentrates on scheduling of the most critical machines under given *flow restrictions* and *machine constraints*. Thereafter, it allocates and sequences the machine operations in a coherent way.

A *dynamic flow control policy* for a flexible, *unreliable* manufacturing system was proposed by Long (1994), [9], whereby the system unreliability is supposed to be due to the *unreliable machines*. The proposed policy minimizes, in spite of machine failures, the difference between the number of parts *produced* and the number of parts *required*. This is achieved in a *two-stage linear programming* that first calculates the lower bound of control policy, enabling balancing the production, and then maximizes the production rate. During the manufacturing process, the control policy is dynamically adapted to the real status of the machines. This minimizes production reduction as well as possible overproduction.

Defining the state variable $\mathbf{X}$ of the manufacturing process as the difference between the *demand* $\mathbf{D}$ and the *cumulative production* $\mathbf{W(t)}$:

$$\mathbf{X}(t) = \mathbf{D} - \mathbf{W}(t),$$

and considering a manufacturing system consisting of m *machines*, manufacturing a family of n *products*, the above vectors are defined as:

$$\mathbf{X}(t) = [\, X1\,(t),\, X2\,(t),...,\, Xn\,(t)]^T$$

$$\mathbf{D} = [D1,\, D2,...,\, Dn]^T$$

$$\mathbf{W}(t) = [W1\,(t),\, W2\,(t),...,\, Wn\,(t)]^T$$

Defining, finally, the instantaneous production rate vector as:

$$\mathbf{U}(t) = [U1\,(t),\, U2\,(t),...,\, Un(t)]$$

the dynamic system equation

$$\mathbf{X'}(t) = -\,\mathbf{U}(t), \qquad \mathbf{X}(0) = \mathbf{D}$$

can be obtained.

The *machine state vector* is defined, taking into account the state of individual machines $S1(t)$, $S2\,(t)$, ..., $Sm\,(t)$, as:

$$\mathbf{S}(t) = [S1\,(t),\, S2\,(t),...,\, Sm\,(t)]^T$$

with $Si\,(t) = 1$ when the i-th machine is operational, and $Si\,(t) = 0$ when not.

The cost function to be minimized for optimal flow control is now defined as a function of the state variable $\mathbf{X}$, the machine state vector $\mathbf{S}(t)$, and the time t:

Ju = F($\mathbf{X}$, $\mathbf{S}$, t)

For minimization the *optimality principle* of Bellman was proposed [7].

The simulation results have confirmed the high efficiency of the proposed dynamic flow control policy in the sense that, in spite of machine failures, it still guarantees a higher material throughput, a better machine utilization, and a smaller production duration than any of the previous methods. Besides, the proposed policy is robust against parameter variations of the cost function. It is also not computationally intensive and can be used for real-time control within the flexible manufacturing systems.

C. Performance Analysis

Performance is the primary operating characteristic of a process or of a system, playing - together with the *product quality*—a fundamental role in manufacturing industry. A typical manufacturing system consists of interconnected, failure-prone subsystems like *NCN machines, assembly stations, robots and material handling facilities, conveyors and automatically guided vehicles* that are expected to render—through systems reconfigurability and possible operability (at least at a degraded performance level) also during the maintenance and repair time—a high flexibility and a high degree of fault-tolerance. Hence, a mathematical model of system performance has to include, apart from the normal system behavior, also production disturbances such as *failures* and *faults,* that have a direct influence on production efficiency. Both, failures and faults, can cause random *partial production stopages* in different parts of the production line or they can lead to the *total shutdown* of the production [10]. Partial production stoppages prevent the partially failed system from its total shutdown, mainly due to the buffers available at the production machines entries.

Failures by definition may occur with a certain probability at random, caused *by material wear* of system parts, for instance by the wear of machine bearings, or by *aging* of machine parts under stress. Failures need a relatively long time to be repaired. Thus, failures easily lead to production shutdowns. Fortunately, due to the relatively high reliability of system parts, their occurrence probability or occurrence frequency of failures is relatively low. For this reason also the probability of *simultaneous failure* of two or more machines is negligibly low.

Faults, on the other side, can occur at any time because they are the consequence of improper machine operation, in the sense that a *bad* part has been fed into a machine, or feeding of a proper part was *improper,* the *improper tool* for its processing was used, etc. Thus, the probability of occurrence of a fault in one machine, or the simultaneous occurrence in two or more machines, is time independent and could be relatively high. The faults, again, although having a higher occurrence frequency then the failures, do not necessarily lead to total production shut-downs. Furthermore, because being of short duration, they are successfully managed by installing the workpiece buffers on the production line.

Assuming that the *time to failure* and the *time to repair* have a continuous probability distribution, the *failures occurrence* can easily be modeled. Denoting by λ**m** the *mean throughput rate* of the production line and by τ the *average working time* between two failures, the *production line efficiency* η, as the ratio between the *average real production* within the time interval considered and the *total production possible* when no failures are present, is given by

$$\eta = \lambda m \tau$$

Similar relationships can be derived for the faults and can subsequently be used for calculation of failures and faults influence on the total *effective production time* of manufacturing line.

Another aspect of production performance is its dependency on *deadlocks,* caused not only by the failures and faults, but also by the inadequate production scheduling [11]. An FMS system can undergo a *partial flow deadlock* when the workpieces are assigned to the machines that inhibit their further flow. This is a crucial manufacturing problem that might be only partially soften by improved flow control, directing the critical workpieces to the available buffers, or better, by real-time controlled production scheduling.

For instance, designating by χ_{ijv} the *completion time* of an *operation* v, belonging to the *job* i, on the *machine* j, the deadlock-free production scheduling of a system with the total number of m machines should optimize the *performance measure:*

$$J = \sum_{i=1}^{m} x_{ijv}$$

subject to the natural set of time and physical constraints related to the manufacturing machines, material handling facilities, and the available buffers. The optimization can be carried out by any *discrete optimization method,* whereby for identification of deadlocks the *graph-theoretic approaches,* for instance the use of *Petri Nets,* are useful. Real-time production scheduling, based on optimization of the above performance measure, help also avoid deadlocks by *production rescheduling* [12].

System sensitivity on parameter deviations is another important feature, closely related to production quality. In manufacturing, system sensitivity is to be understood as an interdependency between the system efficiency and the change of *production parameters* such as the interdependency between *overall system throughput* and *machine failure rates.* Hence, sensitivity analysis helps:

* *Identify* system parameters producing considerable system errors
* *Discover* possible production bottlenecks by simulation
* *Implement* on-line system optimization

For this purpose, the development of an appropriate *sensitivity model* of the system is required, preferably based on the Markovian approach of *discrete events representation* [13]. The model represents the interdependency $p(\theta, t)$ between the *state probability* vector p and the *system parameters* such as the *failure rate* and the *repair rate* of a machine or of any utility, expressed as:

$$p'(\theta, t) = p(\theta, t) \, Q(\theta), \qquad p(\theta, 0) = p0$$

where $Q(\theta)$ represents the *infinitesimal generator matrix* of the homogeneous finite-state, continuous-time *Markov chain* of the structure-state process.

Transient analysis of the above Markov model needs numerical computations based on the approximate problem solution:

$$p(\theta, t) = \sum_{n=0}^{\infty} \pi_n (\theta) \exp$$

where $\pi_n (\theta)$ is the solution of the recursive relation:

$$\pi_n (\theta) = \pi_{n-1} (\theta) \, (I + Q(\theta) / \lambda), \qquad \pi_0 (\theta) = p0$$

The performance of a flexible manufacturing system can be considerably improved by optimal preparation of *tool scheduling* and scheduling of the consecutive *machine operations*. In this way the delayed supply of proper tools, the delayed assembly and arrangement of tool kits, of tool breakage, and of fast wear of cutting edges can be avoided, and the possible productivity reduction prevented.

Instead using the standard *trial-and-error approach* for defining the number of tools required for a planned manufacturing process, it is better to use an *event oriented tool management approach*, based on operations to be performed. Petuelli and Mueller (1996), [14], proposed to use a *setup simulation model*, based on analysis of the given NC programs, time frames, and the execution alternatives. Simulation of the model will help:

- Check the time frame
- Minimize the number of tools required
- Optimize the tool inventory

II. PRODUCTION MONITORING AND QUALITY CONTROL

A. Production Monitoring

Production monitoring is an automation task of production engineering consisting in periodic *collection* and *analysis* of parameter values, indicative for the operational efficiency and the quality of the monitored production process. Production monitoring includes the monitoring of:

- Machining processes
- Machine status
- Tool conditions

It, furthermore, accounts for *surveillance* of production malfunctions, requiring immediate *corrective actions* and/or some *fine tunings*. For instance, if the process control loops, due to the process parameter drifts, are not capable of retaining their optimal performances, the controller parameters are to be correspondingly readjusted [15].

Process monitoring, although far more direct than the process control, requires a continuous harmonic action of sensors, controllers, and actuators. Its implementation is more difficult because the interpretation of monitoring results is much more complex than the interpretation of the difference between the sensor and the set-point value within a control loop. Thus, the problem of process monitoring can be successfully solved only in relatively simple cases.

The situation has, however, been changed with the advent of modern *intelligent approaches* to the solution of monitoring problem [16.,17]. The approaches have opened new horizons to the general solution of even more complex monitoring problems, so that present computer-based automation already contains *knowledge-based software* as an intelligent tool for solving production monitoring problems, incorporated into the application software.

In manufacturing, monitoring is a substantial part of *quality assurance* program. It encompasses the detection of *malfunctions* in the manufacturing process, such as the:

- Lubricant losses
- Machine tools misleading
- Deviations caused by tool vibration or wear
- Sensor drift or failure, etc.

The capital questions in process monitoring are:

- *Which* signals have to be collected for monitoring
- *How* they should be preprocessed
- *What* signal features should be extracted
- *Which* mathematical approach should be used for the extraction

If the monitoring results are to be used *on-line* for corrective intervention in the manufacturing process, the *implementation technology* of the production monitoring system has to be adequately selected, enabling real-time signal processing by parallel computer architectures or neural networks.

Monitoring of *one-dimensional signals* could be covered by:

- Stationary signal analysis
- Local signal analysis
- Dynamic signal analysis

- Sensor fusion

Stationary signal analysis could be:

- *Statistical analysis*, based on evaluation of signal mean value and of its standard deviation
- *Trend analysis*, in which the signal drift, related to its expected value, should be detected

Local signal analysis deals with the search for areas of local signal disturbances, local *peak values* of the signal, etc., whereby the *dynamic signal monitoring* is based on analysis of *time-behavior* of the signal, carrying implicitly the status of the system, generating the signal.

In manufacturing, production monitoring is aimed at two main target objectives:

- Status monitoring of machines and tools
- Production quality monitoring and control

B.   Status Monitoring of Machines and Tools

In mechanical engineering, the analysis and diagnosis of rotating and vibrating systems was performed by processing the recorded acoustic and vibration signals using *vibration analysis* [18]. The final objective was to identify the *operational status* of the system, i.e. its *regularity* or its *failure*. This is essential for *on-line system monitoring* and for *failure diagnosis,* consisting of the following tasks:

- Eccentricity check
- Bearings damage test
- Rotor critical speed test
- Gear dynamic load test
- Resonant-whip instability test
- Rotor translateral and axial vibration
- Static and dynamic unbalance test
- Modal structure test

Vibration and sound analysis is also used for supervision of manufacturing machinery, by monitoring the:

- High-frequency bearing noise
- Backlash in gear couplings
- Rotational eccentricity
- Motor critical speed

A wide application field of vibration and sound analysis in manufacturing is on-line *tool wear monitoring,* for instance during the cutting, where monitoring results should help in order to predict its possible break [19]. For exact break prediction, in addition to the vibration and acoustic information, the continuous sensing of *cutting power, cutting forces,* and *tool temperature* is required because the power consumed during the cutting process depends on acting force, that yet depends on the status of the cutting edge. The tool wear needs an increased cutting power that influence the expected total tool life.

In robotics, a typical application of sound analysis is the slip detection between the gripper and the workpiece it is transporting.

The methods, most frequently used in vibration and sound signal analysis, are *cepstrum analysis, feature extraction,* and *pattern classification. Cepstrum analysis,* the name being derived by inverting the *spec* part of the word spectrum to *ceps,* to result in *cepstrum,* was proposed by Bogert, Healy, and Tukey in 1963, [20], for extraction of *echo signals* out of *seismic records.* The approach was soon broadly accepted in the related areas of signal analysis [21].

Originally, *power cepstrum*—referred to as the *real cepstrum* Cp(q) for a signal x(t)—was defined as the logarithm of the power spectrum:

$$C_p(q) = | \, \mathcal{F} \{ \log F_{xx}(f) \} \, |^2$$

where q represents the *quefrency* and has the dimension of time, whereby Fxx(f) is the power spectrum of the signal x(t), defined by Fourier transform

$$Fxx(f) = | \, \mathcal{F} \{ x(t) \} \, |^2$$

Later on, the definition of the power cepstrum was introduced as the inverse Fourier transform of the logarithm of the power spectrum:

$$C_p(q) = \mathcal{F}^{-1} \{ \log | F_{xx}(f) |^2 \}$$

The definition is very similar to the definition of the *autocorrelation function,* also defined as the inverse Fourier transform of signal power spectrum:

$$Rxx(\tau) = \mathcal{F}^{-1} \{ Fxx(f) \}$$

The mathematical strength of cepstrum represents its capability to detect *hidden periodicities* in frequency spectrum that, for instance, are present in echo records. Moreover, using cepstrum techniques the *harmonics* and *sidebands,* possibly present in the spectrum, can be identified because the cepstrum—as being a *logarithmic transformation*—transforms the *multiplicative* signal components (contaminated by noise or superposed by hidden signal we are looking for) into the *additive* ones, enabling their spatial separation within the frequency spectrum. This property is exploited in troubleshooting and monitoring of rotating mechanical systems because the frequency spectrum of the overall sound signal produced by the system contains the periodicities, harmonics, and/or sidebands.

A typical cepstrum application is the automated failure diagnosis of gear-boxes [22]. A direct comparison of signal frequency *spectrums* of a damaged gearbox *before* it was repaired and *after* it was repaired does not directly indicate any explicit difference. However, the comparison of *cepstrums* of the gearbox *before* and *after* its repair reveals an essential difference.

The acoustic spectral data can be analyzed using neural networks [23]. Hewitt, Skitt, and Witcomb [24], used a five-layer back-propagation network as a *self-organizing signal classifier*, trained on a set of acoustic signals generated by the engine. Similar investigations have been made by Filippetti, Franceschini, and Tassony [25], in *on-line diagnosis* of rotor faults of an induction motor, using a neural network trained on a set of experimental data belonging both to the incontestable and the faulty motor. For the feasibility study, a number of different motor state variables and motor parameters were selected as network inputs for monitoring the state of rotor bars.

Incipient fault detection plays an important role in failure detection of induction motors. Goode and Chow [26], developed a hybrid, *neuro-fuzzy* approach to the solution of the *on-line fault detection* problem (Fig. 1). A *modular* back-propagation network with two modules connected in tandem was used in the experiments. *Module 1*, a two-layer network, generates—based on motor current and rotor speed—the *membership functions* required for the fuzzy part of the system. *Module 2* of the network, also a two-layer back-propagation network, is structured so, that its first layer represents the *knowledge base* of the system containing the *fuzzy rules* required. The module is off-line trained on selected examples and with the preliminary set of fuzzy rules. During the experimental test the neuro-fuzzy system produced the results practically acceptable for *good/fair/bad* classification of motor bearing wear faults.

Neural networks are also applicable as an appropriate tool for *on-line condition monitoring* of working machines in production engineering, where the monitoring results are used for fine tuning of machine parameters or for machine maintenance or repair. Reports on this subject were published [27–29]. For example, Govekar et al. [30], reported on monitoring a *drilling machine* using a two-layer neural network for processing the acquired acoustic signals. Monostory and Nacsa [31], examined the networks application to the real-time machine condition monitoring. Wasserman [32], used neural networks to detect small cracks in the shafts of rotating machines. Javed and Littlefair [33], have compiled a comparative study of neural network paradigms for condition monitoring, and Harris, Mac-Intyre, and Smith [34], have published a general survey on neural network application in the vibration analysis.

Wear monitoring of machines and machine tools using neural networks is directly related to *on-line quality control* because by correcting the tool wear the once achieved desired product quality can be retained over a long production period. Kamarthi et al. [35], investigated the applicability of *Kohonen's feature map* to on-line identification and monitoring of flank, based on vibration records. They used *sensor data fusion* as a very reliable and robust method, giving very good results for a wide range of operating conditions. Similar results have been achieved by Elanayar and Shin [36], using radial basis function networks.

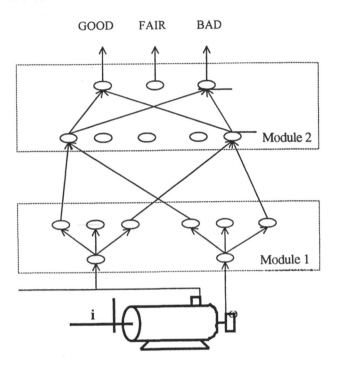

Figure 1 Neuro-fuzzy fault detection.

Liu and Anantharaman [37], examined, based on the previous results of Liu and Wu [38], in on-line detection of drill wear, the problem of *on-line sensing* of *drill wear* using neural networks. They trained a 9-14-1 structured backpropagation network to recognize the wear status of the tool using 46 thrust and torque data sets collected from a CNC milling machine. The relevant process features have been extracted from the recorded curves of the machine torque and thrust. To calculate the drill wear on-line, experiments have been carried out with the *fixed* and *variable slope* of neurons *activation function*. As a result, the *usable drill* could be distinguished from the *worn-out drill* with the absolute success rate of 100 per cent. Moreover, the network with the variable slope of activation function was capable of estimating the *degree* of the drill wear with an average accuracy of 92.27 percent.

Simultaneous monitoring and analysis of several interrelated system signals can detect irregularities in system behavior that separate analysis of individual signals would not discover. This is because of synergy in the use of multiple sensors in systems monitoring based on *signal fusion* and *multisensor integration* that incorporate—besides the signal values—also knowledge about the dynamic process behavior. In manufacturing, sensor fusion and multisensor integration are used preferably in the area of assembly and robotics.

There is a basic difference between sensor fusion and multisensor integration. Sensor fusion is a more restricted signal evaluation process. It is only a part of signal integration process. Multisensor integration, again, is a complex multistage decision process, based on the information of several sensors, some of them being previously fused within the integration scheme. Fig. 2 shows the integration of signals S1, S2,..., Sn, whereby some of the signals have been fused prior to their integration [39].

The inherent capability of signal fusion and multisensor integration to detect features of the monitored process nondetectable using individual sensors separately is interpreted as the *increased estimation accuracy* of monitored process features and as the *decreased estimation uncertainty* through *redundant information,* incorporated in individual sensor signals. The methods, appropriate for such estimation, are of a statistical nature, such as *Bayesian estimate, statistical decision theory,* and *Kalman filtering.*

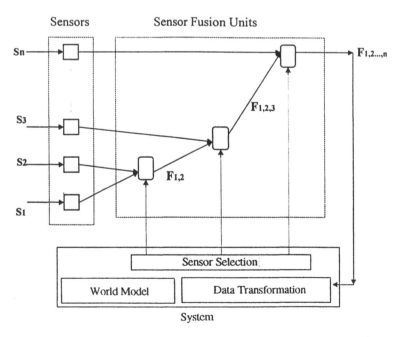

Figure 2 Signal fusion and multisensor integration.

Neural networks have been applied both to signal fusion and to multisensor integration. Rajapakse and Acharya [40], have studied a hierarchical neural architecture for enhancement of feature recognition, based on multisensor input data. Each hierarchical stage of the network contains two types of cells:

- s-cells for feature extraction
- c-cells for pattern error discrimination

At lower network layers relatively simple features can be identified and at the higher the more complex ones, as combinations of simple features, extracted at lower network layers. Feature deformation and position errors are tolerated at each layer, so that the entire network structure is insensitive to the feature transformations such as *scaling, translation,* and *rotation.*

A general approach to sensor fusion was proposed by Eggers and Khuon [41], that uses the statistically-based adaptive signal preprocessors for sensor data to be fused and a highly parallel neural network for building the final decision (Fig. 3). The approach recursively models the sensor data as a higher-order autoregressive process and identifies the segments of their quasistationarity. A neural network used for implementation of the fusion process is a multilayer perceptron having the associative memory features. It maps the feature vectors to the correct decision at its output. For tuning the network parameters, a learning algorithm is applied that minimizes a *risk function,* mathematically defined as the optimal association desired. The experimental results have shown that the combination of statistical signal processing methods and neural network techniques, results in an accelerated robust operation.

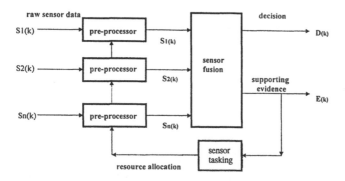

Figure 3 A general sensor fusion system

A wide application of sensor fusion and multisensor integration in robot-based material handling and assembly in manufacturing environments is primarily due to the robots' requirements for high sensory capacity. This is because the robots, used for *in-process* handling of workpieces, are *sensory controlled*. This saves higher expenses in building of more sophisticated, more expensive guidance systems. Multiple sensing is extremely important for robots having a certain mobility feature, as well as for *cooperative* and *coordinating robots* requiring intelligent systems for trajectory planning and obstacle avoidance. Sensory controlled robots have also a higher degree of autonomy and a higher degree of security than preprogrammed, sensory poor robots.

Robot manipulators need sensory information to better control their behavior, for which the position/velocity and force/torque information are basic. In addition, *tactile information* is required for *gripper control*. Integration of the above information can enhance the capabilities of a manipulator, particularly in assembly operations because the assembly is a complex area of manufacturing in which the robots are used for bringing preprocessed workpieces together for assembling the final products.

In the past, the design and implementation efforts in industrial robotics for flexible assembly systems have been enormous. In the mean time, modern approaches, relying on intelligent computer systems, have been worked out. For example, a higher degree of autonomy and intelligence has been provided to the robots by integration of different sensory signals and video data, giving the robots detailed information about the environment and the position, size, and the shape of the objects to be handled by the robot. This also facilitates the robot's trajectory planning under avoidance of obstacles.

C. Production Quality Control

The high-quality demands on manufacturing products to be marketed force manufacturers to attain a 100% quality of all produced parts and of final products. Originally, the simplest definition of *high-quality products* has only included the requirement that such products perform their functions reliably over a long period of time, without breaking down or requiring repairs. New definitions include a number of additional attributes, particularly those related to product dimensionality tolerances, to visual appearance, etc.

Advances have been made in automated visual product inspection for quality control using modern computer technology and advanced approaches in signal processing, especially in *image processing, pattern recognition,* and *artificial intelligence* [42,43]. There are numerous *vision-based* automatic inspection systems in manufacturing [44] and in various industrial sectors: electronics, textiles, wood, and glass industries, as well as in automobile and paper production.

Quality control can generally be carried out:

- *Off-line*, during the product and process design stage of the product development cycle
- *On-line*, during the process of manufacturing to help real-time process quality control by reducing manufacturing imperfections in products

Off-line quality control during the product design phase mainly includes [45]:

- *System design*, the functional *prototype design* based on new concepts and new design approaches, as well as testing of functional compatibility between distinct parts of the system
- *Parameter design*, selection of system parameters that guarantee highest product quality and lowest sensitivity to the system parameter variations
- *Tolerance design*, following the parameter design, that tightens the tolerances around the given target values of the control components in order to improve the product quality at possibly lowest extra cost

In manufacturing, *quality loss* is defined as:

$$ql(x) = q(x - xt)^2$$

where **x** is the achieved quality value, **xt** the target quality value, and ql the quality loss constant.

During product manufacturing, quality inspection is carried out whenever a workpiece has undergone a precise operation. At the same time, if needed, the critical bearings and manufacturing tools used also can be inspected. The final workpiece inspection is, as a rule, accomplished when its manufacturing has been completed.

The most frequent product quality inspection used in manufacturing is the *product dimensionallity test, product shape inspection,* and *product surface state inspection,* needed for discovering possible *roughness* or *flaws* present on the surface. Here, the visual surface inspection is considered to be the most important product inspection in manufacturing [46]. It is carried out at the end of the manufacturing processes, whereby the dimensionality inspection is carried out more frequently, after each manufacturing operation has been completed. As a result, visual surface inspection identifies the degree of deviation of the actual surface from the desired one. It belongs, generally speaking, to the category of surface *texture analysis,* to be discussed in more detail in the next paragraph.

A neural network based tool for visual inspection in manufacturing was implemented by Liatsis et al. [47]. The tool accomplish the inspection of components quality by detecting the irregularities on their surfaces. The flexibility of the tool guarantees the real-time product inspection, disregarding the *scale, position, and orientation* of the component parts. The network used for visual inspection consists of a number of *modules* (subnetworks), responsible for completion of individual tasks, such as *scaling, translation, rotation*, etc. The final module, capable of learning on-line the characteristics of the inspected species, is responsible for final recognition of *normalized objects*.

Experiments with visual inspection of different categories of screws, rivets, and blemish detection in axisymetric industrial parts have verified the high-speed, high-precision identification and classification of object quality of the tool. For instance, the different types of screws and rivets have been detected with 100% of accuracy and the blemishes in axisymetric parts with 86%.

Increased product quality awareness of the customer has created the need for a worldwide unique conformity regarding the definition and establishment of product quality,which has given a considerable impact to international standardization in this area. The ISO 9000 series of *Quality Management and Quality Assurance Standards* is the result of joint international efforts to define the related standard documents on quality systems [48]. The standard documents enable the manufacturing companies to register for a certificate related to their *quality process*. After an auditing procedure the company can get the certificate, valid for the next 6 months to 1 year, that acknowledges that the internal quality criteria and practices for quality are in accordance with the international standards.

Thus far, the following ISO 9000 standard documents have been issued:

- ISO 9001, Model for Quality Assurance in Design/Development, Production, Installation, and Servicing
- ISO 9002, Model for Quality Assurance in Production and Installation
- ISO 9003, Model for Quality Assurance in Final Inspection and Test
- ISO 9004, Quality Management and Quality System Elements: Guidelines

III. QUALITY INSPECTION AND CONTROL

Hard competition on the international market makes the economic survival of manufacturing industries highly dependent on quality. This forces manufacturing companies to take care that their products, apart from reaching a high performance reliability, also meet some specified *quality standards* at *reasonable cost*. Manufacturing of high quality products at relatively low cost are contradictory requirements that can not easily be met simultaneously. This rather calls for a well-organized *quality assurance* service, that includes efficient *product quality control*.

In production engineering, and especially in manufacturing, *quality assurance and quality control* are viewed as basic production management activities. Their implementation is based on evaluation and classification of production and product data, collected in the production field for *product quality inspection* and *production control*. The evaluation and classification of collected data consists in their comparison with some prespecified values, representing the *desired* product characteristics to be achieved. In the mean time, quality control has matured into—what is increasingly called—*quality engineering,* an engineering that integrates product metrology, inspection, and test with quality planning and management, quality cost analysis, and quality auditing [49]. The results of product inspection are needed for production process monitoring, so that by discovering some considerable production process deviations from the prescribed status values or by considerable fluctuations in time behavior of the process, alarm messages can be generated and some corrective steps undertaken to maintain product quality within tolerable limits.

Quality assurance includes actions carried out *before* the production process starts, i.e. during the production planning phase. The actions take care that the planned products will be produced as given in its manufacturing specifications. This includes the appropriate selection of work materials, tools, machines, and manufacturing operations to be used. Adequately carried out quality assurance thus facilitates quality inspection, carried out after the product has been fabricated.

Besides the specific inspection approaches used in various industrial sectors, two basic types of quality inspections are most common in production engineering, and particularly in manufacturing:

- *Visual inspection* that covers the geometric product check (mainly the dimensionality, proportionality, and symmetry check) at every production step when applicable
- *Acoustic inspection* that verifies the operability or correct functionality of final products, usually in rotating mechanical systems such as electric motors, combustion engines, compressors, pumps, etc.

Automated product inspection generally uses digital computers, equipped by associated instrumentation, to collect production field data. For subsequent processing of collected data, required for product quality control, both conventional and knowledge-based software is used [50]. This generally improves the accuracy, reliability, and the speed of product inspection. It also eliminates human factors that may lead to misinterpretation of inspection results.

Advances in computer-based product quality control are to a great extent due to progress in *image processing, pattern recognition,* and *computer vision.* The general trend to use *neurosoftware* for this purpose is of great advantage because such software provides the computer with *learning* and *recognition* capabilities, that are essential for automated quality inspection [51].

A vast number of quality control functions are implementable in a rather simple way, without using complex computational algorithms, such as the ones conventional implementations need. Using *cognitive technology,* the patterns—once learned by the network in its *training phase*—will automatically be recognized in its *operating phase.* The automatic inspection approach, based on learning and recognition, obviously resembles the approach used by human experts in manual product quality inspection.

The main advantages of using neural networks in solving specific engineering problems are based on networks:

- Learning and generalization capability
- Adaptability to the problem to be solved
- Massive parallel structure
- Fault tolerance and robustness

In the manufacturing industry, the:

- Feature extraction
- Pattern recognition
- Pattern classification

capabilities of neural networks have qualified them for solving the problem of automatic product quality inspection and control.

Successful neurosoftware application needs, however, an extensive set of experimental data to be collected in the inspection field. This means that a large number of training examples, carried out by human experts in the product inspection field, are needed for the learning phase of the simulated neural network. The selected examples, both the positive and the negative, should *be statistically representative* enough.

A. Automated Visual Inspection

The demands for application of modern methods and tools in automated product inspection, for ensuring product quality, have been increased considerably after incorporating the ISO 9000 quality standard series into *quality assurance*. For this reason, automated visual inspection of product quality has become the driving agent for cooperation of *production engineers* and *computer experts,* predominantly in the field of *image data processing* and *visual pattern recognition*, the two basic disciplines of camera-oriented product quality inspection. This is to a great extent true for manufacturing [52].

An additional aspect that has encouraged alternative ways in visual product inspection is the need for replacement of human labor involved in on-line quality control because human quality inspectors have to:

- Be skilled and trained in the field for a long time
- Work hard and repeat the same monotonous work all work-day long, often in unfavorable environments, completing a large number of inspections per work-shift

Human quality inspectors become tired. The possibility of producing erratic inspection results increases with time. Above all, experts in assessing product quality are a considerable cost factor in production because they are high-qualified and well paid. Automated product quality inspection, when introduced on the production line, not only reduces the inspection costs but also releases human labor from the heavy work load. Beside, because being faster than human expert based inspection, automated inspection also provides timely feedback and control when the production line deviates from the desired quality level.

A. Industrial Applications

A large number of examples of applied automated visual inspection is found in the *metal processing* industry for *metal surface inspection* [53,54] the and in automotive industry [55] for the individual inspection of car parts [56,57].

A number of applications in other industrial sectors have also been reported [44], particularly in:

- Paper and pulp [58,59]
- Ceramic [60,61]
- Electrical and electronic [62,63]
- Glass [64,65]
- Manufacturing [66–70]
- Photographic and printing [71,72]
- Textile [73,74]
- Wood [75]

C. Principles of Automated Visual Inspection

Figure 4 shows a typical automated visual inspection system where the object to be inspected is grabbed by a camera, the video signal digitized, stored, and pre-processed using different algorithms for *noise reduction, contrast enhancement,* and *spatial adaptation* through *rotation, translation, or zooming.*

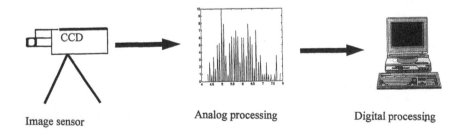

Image sensor   Analog processing   Digital processing

Figure 4 Typical layout for automated visual inspection.

The simplest image processing approach for automated visual quality inspection consists in comparison of inspected *objects* with their ideal *models* [76]. The object pattern irregularities, detected by comparison, representing the local deviations from the model, reveal the quality of the inspected objects [42], or at least enable a YES or NO decision concerning the required object quality. For a more exact quality determination, when applying the model-based approach, besides the *ideal* object model, representing the *best quality* of the object, a set of *"neighboring"* models is needed, each model representing a specified *quality class* of the object. The comparison of the acquired inspection data with the individual *quality models* stored can then be carried out by image data subtraction. The minimal difference between the object pattern and a specific model stored determines the quality class to which the inspected object belongs.

A simple application example of the described approach is given in Fig. 5. As a result of image data subtraction the inspected "circle," indicating the deviation of the inspected object from its model, has been generated at the location where the deviation is situated. This is a great advantage when inspecting large and complex objects like the PCBs or microelectronics masks.

In some cases, where only simple objects are to be inspected and the regularity of their spatial shape checked, much simpler models can be used, such as *geometrical models*, requiring only a limited number of image pixels to be checked and verified. When dealing with the determination of products tolerance deviation classes, the corresponding visual inspection is known as *dimensional verification* [77].

The actual pattern recognition process generally includes two basic steps:

- Features extraction
- Pattern classification

The final objective is to decide, based on the collected observation data, to *which* of the predefined classes of patterns the given pattern belongs and—in an ambiguous case—to estimate the probabilities that the pattern belongs to the specific, predefined quality classes. The most typical pattern in manufacturing to be classified represents the *geometrical shape* of a product, i.e. its form, dimensions, outline, etc. The extended product quality classification includes attributes such as *pattern uniformity, symmetry, spatial homogeneity,* etc.

Once defined, the pattern matching approach can be used for inspection of

- Mechanical parts
- Electronic components
- Pcb layouts
- Integrated circuits masks
- Printed images
- Casting, welding, and milling uniformity
- Texture and fabric

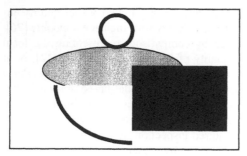

Image I

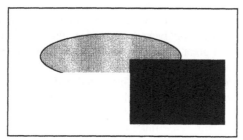

Image II

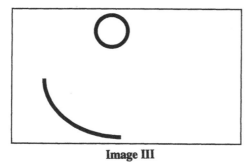

Image III

Figure 5   Image substraction.

in various industrial sectors, such as:

- Manufacturing and mechanical workshops
- Iron and steel
- Electronics
- Automotive
- Printing
- Textile, jute, and leather
- Dyeing and batiks
- Iron and steel industry

D. Image Features Extraction

From the mathematical point of view, signals are generally decomposable into frequency spectra, each frequency participating in the spectrum with a specific amplitude and phase. In case of *one-dimensional signals,* Fourier analysis is used to find the frequencies present in the signal. In case of *two-dimensional signals*, generated through image data acquisition, a set of two-dimensional frequency grids is used for signal decomposition, each grid standing for a *spatial frequency component* of the image. The individual frequency components can be calculated using *two-dimensional Fourier transform.*

Feature extraction of two-dimensional signals is the most essential step in image data processing for pattern recognition and pattern classification, where the original two-dimensional image signal $I(x,y)$, represented by, say (256×256) pixel values, is transformed into its spatial frequency equivalent using the *2D-Fast Fourier Transformation* [78–80].

Feature extraction is a process of finding, mainly by removing the *redundant* signal components, the *minimal* number of attributes (*features*), sufficient for adequate pattern description. This is equivalent to the *compression* of the original signal so, that its later *decompression* restaurates the original pattern completely. It can be shown that both the pattern data *compression* and, the pattern *feature extraction* problem can be easily solved using neurosoftware.

The lower-dimensional pattern descriptors, resulting in the feature extraction process, are referred to as *features* because they contain the *inherent properties* of the original pattern. This is essential for pattern classifiers, usually cascaded to the feature extractors, whereby—the lower the *feature vector* dimension, the simpler the implementation of the pattern classifier required for further processing of the feature vector.

Figure 6 shows a typical architecture of a neural network for solving the feature extraction problem by pattern data compression. It is a *three-layer RCE (Restricted Coulomb Energy) network* having the same number of neurons in the *input layer* and the *output layer*, but a reduced number of neurons in the *hidden layer*. For compression of input data the weights of network interconnections are tuned so that the final output values of the network correspond *one-to-one* to its

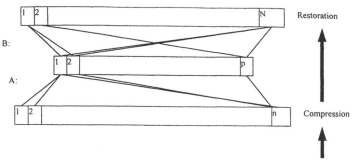

Figure 6 Data compression network.

input values (i.e., at the output of the network appears the same exact signal pattern as the one presented to its input). In this case the hidden layer of the network contains the input signal pattern in a compressed form, represented by a reduced number of neurons in this layer. The neurons of the layer, however, enable the compressed pattern to be fully restored at the network output by its *decompression*. Reducing the number of hidden-layer neurons, while maintaining the input pattern at the network output, the minimal number of hidden layer neurons can be found that still completely represents the original pattern in its compressed form. The hidden layer neurons represent the components of the *feature vector* of the original pattern at the network input.

E. Neural Signal Classifiers

The process of signal classification includes three basic operational steps:

- *Sensor data collection*, that could be—depending on the signal dimension and on its nature—acoustic or image data acquisition, required for preparing the training examples for the neural network learning phase. It is a decisive, application-dependent implementation step requiring much instrumentation field knowledge, professional skill, and field experience.
- *Feature extraction*, in the sense of simplification of signal representation by the reduced amount of collected sensor data and of building the corresponding *feature vectors* of the signal.
- *Pattern classification*, the final step that discriminates the pattern classes, presents in the extracted feature vector, and assigns the discriminated pattern to the specific pattern class, of out of a set of pre defined pattern classes, to which it belongs.

Neural networks, because of their cognitive capacity, have widely been employed both as *feature extractors* and *pattern classifiers* [81,82]. The main strength of back-propagation networks consists in the capability of *supervised learning* through training for future applications. The networks, as *nonlinear pattern discriminators*, map an **n**-dimensional input vector into an **m**-dimensional output vector by adjusting—during the learning phase—the weights of the network interconnection links. The weight adjustment is carried out by minimizing the error, at network output, defined as the difference between the *desired* output vector **yd** and the *actual* output vector **yo**. For minimization, usually the *gradient descent algorithm* is used and the *mean-squared error* is the *performance index*. The mapping itself is performed—within the network—in two steps: the input vector is first mapped on the hidden layer of the network that, in sequence, is mapped on the network output layer. The hidden layer is thus equivalent to a *data concentrator* that internally encodes the essential features of the input pattern in a compressed form [83].

F. Automated Inspection of PCBs and ICs

The use of pattern recognition technique in visual product inspection, such as automated verification of manufacturing quality of printed circuit boards and defects detection in integrated circuits is one of the most promising fields of neural network application in quality control [84]. It mainly consists in searching for irregularly shaped lines or other geometric figures that might represent the locations of product defects [85]. This is particularly essential for the inspection of integrated circuits, where the individual circuit elements have become very small, close to the optical resolution threshold, beyond which the difficulties in automatic detection of shape errors considerably increase [86].

The problem of automated visual inspection of *printed circuit boards* can be solved by splitting the entire board image into a number of smaller windows of, say 5 x 5 or 7 x 7 pixels in size and by associating each window with a list of *good* patterns collected from the field [87]. During the visual inspection, the corresponding windows of the product under test are compared with the items of the related lists to find out whether there is a match with the window patterns of the acquired image. If *no* pattern in the list matches the acquired pattern, the related part of the printed board is considered as possibly defective, and recommended for further testing.

Due to the multistep process of chip production, visual inspection of integrated circuits is broken into at least three *specific* inspection steps [63]:

- *Circuit mask inspection*, the most decisive inspection because a defective circuit mask can make subsequent production steps entirely useless. Another goal of mask inspection is to identify the error source: is it mask generation or lithographic processing
- *Process monitoring and inspection*, dedicated to timely control of produced wafers before they can be used. The main objective is to check the electrical usability of wafers and to find out local anomalies on each wafer, to locate the possible causes in the production, and to estimate the

possible consequences when using such an erroneous wafer. It should be emphasized that local wafer anomalies have to be verified by visual inspection and not by electrical measurements because there are no mathematical relationships between the electrical measurements and the possible local anomalies.

* *Final chip inspection,* or *overall chip quality control,* the final search for chip errors that have not been detected through mask inspection and electrical test. This inspection includes visual check of possible damage of the overall insulation layer, circuit interface conducting lines, contamination of chip surface, and the like.

In the *chips quality control,* the most important is the *pattern defect inspection* (PDI), widely used in *all* inspection phases. The inspection checks for local pattern defects on semiconductor wafers.

Spence (1996), [88], used the *magnetic field patterns,* produced by the current moving between the devices on the printed circuit board to inspect the operational status of the board. The possible failures of the devices on the board, change its operational status and the magnetic field pattern on it. For detection and interpretation of the pattern changes in the magnetic field, neural networks have been used, capable of classifying the possible circuit failures for repair purposes. Using a *magnetic scanner,* the magnetic field on the board has been recorded and converted into the corresponding *magnetic image* for feature extraction and classification purposes.

G. Automated Texture Analysis

Automated visual inspection plays an important role in *quality check* of produced *fabric* in the *textile industry* [73,89]. Modern inspection methods are based on automated *texture analysis* of fabric using neural networks [90,91].

The main objective of texture analysis is to determine the *class* to which the given texture belongs. The mathematical apparatus originally used includes the *two-dimensional correlation analysis, spectral theory, probability theory* and the like [92]. Later on, *texture secondary parameters* were defined, such as *textural energy, textual edginess, vector dispersion, relative extreme density,* etc. For stochastic texture representation *discrete Markov random field models* have been preferably used.

Along with *texture feature extraction, texture segmentation* is also a substantial subject of *texture analysis.* It consists in segmentation of the given image into a number of regions of specific *homogeneous* textures.

Raghu, Poongody, and Yegnanarayana (1994), [91], described a method of automated texture classification based on texture feature extraction using *self-organizing map* and *multilayer perceptron* [93]. The self-organizing map used maps the n-dimensional feature vectors of different sizes, orientations, and frequencies, generated by a set of *Gabor filters* [94–95], onto an m-dimensional output vector. The transformed feature vectors are, in the sequence, fed to a multi-layer perceptron network for classification purposes. In the ideal case, when the input data is noise-free and unequivocal, the self-organizing map itself can carry out the classification. In engineering practice, because the input data is not noise-free, the classification might be fuzzy, requiring a subsequent two-layer neural network as a *crisp classifier* [96].

Automated inspection of *textile fabrics,* based on a *Gaussian Markov Random Field* for modeling the image of a nondefective fabric texture, can help *detect* and *locate* various kinds of defects that might be present on fabric inspection sample [97]. The approach is applicable to the textures present in textile, jute, wood, tile, ceramic, and metallic surfaces.

The results in texture analysis of *woven fabrics,* achieved by Sardy, Ibrahim, and Yasuda (1993), [74], are remarkable. Using the *Neighboring Greylevel Dependence Matrix* and the *Greylevel Run Length Matrix* the identification of *plain, twill,* and *sateen waves,* and the detection of defects such as *reed mark, dirties, pick's inhomogenous* or *broken,* etc. was possible. The texture was considered from the statistical point of view, and described by statistical rules or characteristics underlying the distribution and the relation of gray levels. Also the extraction of textural features was carried out using statistical methods. For classification of defect categories, based on the extracted texture features, a three-layer back-propagation network was used. For each defect category the features were extracted several times and their average values taken as representative.

After being trained on real texture examples of woven fabrics, the network was capable to identify the plain, twill, and sateen with an accuracy of 84.4, 99.2, and 99.6 percent respectively. For reed mark, dirty, and pick's inhomogeneity, and breaks the identification accuracy of 85.4, 84.2, 83.4, and 98.4 percent respectively has been achieved.

Sardy and Ibrahim (1994), [98], also used a modular network for defects detection on textile products, based on textural feature extraction. In the network (Fig. 7), a Kohonen's self-organizing map (SOM) was cascaded with a 20-10-4 structured multi layer perceptron (MLP). After building the power spectrum of 128×128 pixel texture images, 20 textural features were selected and forwarded to the input of the self-organizing network for preliminary classification, based on a recursive learning algorithm

$$Wk + 1 = Wk + 0.7 \, (I - Wk)$$

where I is a 20-dimensional input vector of the network. The outputs of the network are forwarded to the cascaded multilayer perceptron for final classification. It was established by experiment that the cascaded networks give better classification results then any of them used separately.

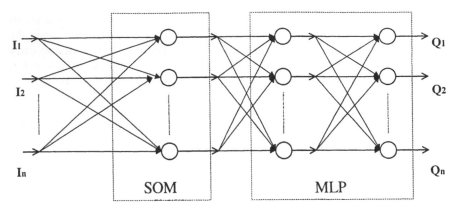

Figure 7   Cascaded neural network.

H. Automated Acoustic Inspection

Acoustic inspection refers to the analysis of sound, noise, and/or vibration, produced by a system under test and acquired by an *acoustic* or by a *vibration sensor*. In mechanical engineering, the inspection is mainly applicable to the devices with rotating components, such as electrical motors, combustion engines, pumps, gears, compressors, etc., producing signals that can be analyzed and used for trouble shooting, on-line monitoring, or quality control. In some industries, such as in ceramics, porcelain, glass, brick, and tile, the quality of products is analyzed by the sound reflected by the products. In this way, the structural inhomogenities, such as cracks or air bubbles, can be detected.

Originally, acoustic inspection in the industry was carried out by human labor, i.e. by human experts, trained to distinguish between the OK and NOT-OK status of the sound, noise, or vibration. The rate of correct diagnosis depends here on field experience, current internal disposition, and the tiredness of the inspecting expert. "Sound fatigue" was here one of the most crucial factors because a relatively high number of products are to be inspected *on-line*, in *real-time*, and usually under unfavorable environmental (noise-contaminated) conditions, requiring short work shifts of the inspecting experts. In all events, the quality assessments of the products made by the human inspector are subjective, nonquantifiable, and to a certain degree "vague." Therefore, there have been many good reasons to automate the inspection process. Nevertheless, because the human ear is much better qualified to accomplish the acoustic inspection than any signal analyzer using traditional mathematical approaches, much more had to be done in order to really apply the automated acoustic inspection in the industry. Some appropriate approaches, proposed for this purpose will be discussed in what follows.

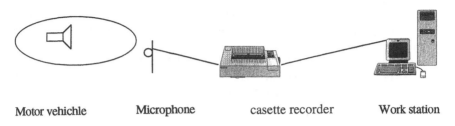

| Motor vehichle | Microphone | casette recorder | Work station |

Figure 8   Automated acoustic inspection.

I. Principles of Automated Acoustic Inspection

Figure 8 shows a general layout of a typical set up for automated acoustic inspection. A microphone or any kind of vibration sensor is used for acquisition of signals, generated by systems under inspection. The acquired signals have first to be smoothed or filtered to reduce the noise components present in the signals, and then to be digitized and stored for further processing needed for *pattern recognition* and *pattern classification*. Based on this, the inspection set up should identify whether the sample under test is OK or NOT_OK. In case of NOT_OK statement, it should further specify the *quality class degradation* to which the sample belongs. In this way the set up will finally:

- Increase the inspection and classification reliability
- Enable a higher inspection rate

When designing a computer-based setup for automated acoustic inspection, care should be taken that it:

- Needs a relatively simple training form (that is, attributes of simple training with an acceptable learning efficiency based on a relatively small number of training samples)
- Guarantees a high degree of adaptability to the new test experiment
- Requires only a simple procedure when operating in the field
- Has a relatively low prime cost

Currently, various setups that meet the majority of the criteria listed above are commercially available. Furthermore, the professional setups offer some additional features important for:

- Establishment and management of inspection data files
- Automated specific access to the databases

- Statistical analysis of inspection data formulation and on-line adaptation of quality criteria to meet the current product test and current status of the production line
- On-line training of automated device on new products
- Possibility of direct intervention in product quality control via process interfaces and related control software

J. Application Examples

Methods of signal analysis used in acoustic inspection predominantly rely on analysis of frequency spectrum, containing the signals' *characteristic features* to be extracted. The extracted features, usually the amplitude values of around 250 frequencies, selected as components of the signal *features vector,* are then forwarded to a *pattern classifier,* implementing a conventional pattern recognition algorithm, or to a neural network as a *cognitive pattern recognizer* [43]. In this way the automatic diagnosis of system status is possible. For instance, using neural networks for implementation of a feature extractor and pattern classifier it was possible to identify the class of internal errors of a combustion engine based on the explosion sound records of the engine under operation [99,100].

Neural networks have been successfully applied for acoustic control of pneumatic valves [101] and small motors [102]. For instance, using neural networks the problem of screening of small electric motors, where the motors inspection by aural cues is required for distinguishing between the *good* and the *bad* motors, was solved [103]. The problem is crucial because absolute motor screening is a prerequisite for a successful quality inspection and a reliable status diagnosis, otherwise achievable only by highly skilled inspection personnel. The inspection test requires a soundproof room, where the sound produced by the energized motors is picked up by microphones and evaluated by an expert, equipped with a sensitive headset. For proper training of the expert, a large number of recorded sound samples of *good* and *bad* motors are required as well as some expensive high-tech instruments, such as spectrum analyzer, sound recorder, signal processor, etc. All the above represents also a good provision for implementation of automated, neural network-based motor screening.

Neuro-classifiers have also been used in noise analysis of *planetary gears* [104]. The possibility of early discovery of gear wear, as indicated by a characteristic noise, is substantial for the gear steadfastness. Good classification results have been achieved using a back-propagation network as an *adaptive classifier,* driven by a discrete spectrum of 120 frequencies representing the entry feature vector of the network.

Chow, Mangum, and Yee (1991), [105], used a neural network for real-time condition monitoring of an *induction motor,* in order to support the early discovering of *incipient faults* and to prevent possible later motor failures. Preliminary simulation studies have shown that the neural networks as *incipient fault detectors* can help detect two most frequent incipient induction motor failures:

- *Stator winding fault,* mainly the deterioration of the stator winding isolation caused by excessive temperatures, vibrations, etc.
- *Rotor bearings wear* under constant load torque conditions, caused by contamined lubrication, misalignment, etc.

Based on simultaneous measurement of stator current and rotor speed and using the mathematical model of the motor, a neural network trained on a large number of input-output samples collected in the inspection field, the motor status classification as *good, fair,* or *bad* was possible.

REFERENCES

1. N Singh. *Systems Approach to Computer-Integrated Design and Manufacturing.* New York: Wiley, 1996.
2. Y Dotan, D Ben-Arieh. Modeling Flexible Manufacturing System: The Concurrent Logic Programming Approach, IEEE Trans. On Robotics and Autom., 7(1): 135-148, 1991.
3. R Anella, PR Kumar..Optimal Control of Production Rate in a Failure Prone Manufacturing System, IEEE Trans. of Autom. Control, vol. AC-31 (2): 116-126, 1986.
4. J Kimemia, SB Gerschwin. An Algorithm for the Computer Control of a Flexible Manufacturing System, IEEE Trans. On Autom. Control 28(12): 353-362, 1983.
5. SB Gerschwin. Hierarchical Flow Control: A Framework for Scheduling and Planning Discrete Events in Manufacturing Systems, Proc. IEEE 77(1): 195-208, 1989.
6. E Khmelnitsky, K Kogan, O Miamon. A Modified Iterative Method for Optimal Flow Control in Flexible Manufacturing Systems, Proc. 3rd IEEE Conf. on Control Application, Glasgow, U.K., 1994, 1769-1773.
7. AP Sage. *Optimum Systems Control, 2nd ed.,* Englewood Cliffs, New Jersey: Prentice-Hall, 1977.
8. FL Pereira, JB Sousa. Optimal Flow Control of Manufacturing Systems. Preprints of IFAC World Congress, Sydney, vol. 2, 1993, 107-110.
9. J Long. A Dynamic Flow Control Policy for Failure Prone Flexible Manufacturing Systems, Proc. 3rd IEEE Conf. on Control Appl., Glasgow, UK, 1994, 1499-1504.
10. YT Leung, M Kamata..Performance Analysis of Synchronous Production Lines, IEEE Trans. on Robotics and Automation 7(1): 1-8, 1991.
11. SE Ramaswamy, SB Joshi. Deadlock Avoidance in Automated Manufacturing Workstations - A Scheduling Approach, Proc. IEEE Intnl. Conf. on Robotics and Automation, San Diego, Calif., vol. 3,1994, 1992-1997.
12. YL Chen, TH Sun, and LC Fu. A Petri-Nets Based Hierarchical Structure for Dynamic Scheduler of an FMS: Rescheduling and Deadlock Avoidance, Proc. IEEE Intnl. Conf. on Robotics and Automation, San Diego, Calif., vol. 3, 1994, 1998-2004.
13. V Gopalakrishna, N Wiswanadham, KR Pattipati. Sensitivity Analysis of Failure-Prone Flexible Manufacturing Systems, Proc. IEEE Intnl. Conf. on Robotics and Automation, San Diego, Calif., vol. 1, 1994, 181-186.

14. G Petuelli, U Mueller..Optimization of the Performance of Manufacturing Systems by Integrated Tool Preparation Scheduling, Proc. 4th Intnl. Conf. on Control, Automation, Robotics, and Vision, ICARCV '96, Singapore, pp. 27-32, 1996.

15. PN Rao, NK Tewari, TK Kundra. *Computer Aided Manufacturing*. Tata McGraw-Hill, New Delhi, India, 1993.

16. DBarschdorf..Case Studies in Adaptive Fault Diagnosis Using Neural Networks, Proc. IMACS Annals of Computing and Appl. Math., Brussel, Sept. 3-7, 1990.

17. CH Dagli, ed. *Artificial Neural Networks for Intelligent Manufacturing*. New York: Chapman and Hall, 1994.

18. C-H Chen.. *Signal Processing Handbook*. New York: Marcel Dekker, Inc., 1988.

19. LE Atlas, GD Bernard, SB Narayanan. Application of Time-Frequency Analysis to Signals from Manufacturing and Machine Monitoring Sensors, Proc. IEEE 84(9): 1319-1329, 1996.

20. BP Bogert, MJR Healy, JW Tukey. The Quefrency Alanysis of Time Series for Echoes: Cepstrum, Pseudo-Autocovariance, Cross-Spectrum, and Saphe Cracking. Proc. Symp. on Time Series Analysis, Rosenblatt, M. ed., New York: Wiley, 209-243. 1963.

21. DG Childers, DP Skinner, RC Kemerait..The Cepstrum: A Guide to Processing, IEEE Proc. 65(10): 1428-1443, 1977.

22. RB Randall.Cepstrum Analysis and Gearbox Fault Diagnosis, Maintenance Management International, No. 3, 1982, 183-208.

23. J MacInture et al. Neural Networks Applications in Condition Monitoring, Proc. 9th Intnl. Conf. AIENG/94, Pennsylvania, pp. 37-48, 1994.

24. PD Hewitt, PJC Skitt, RC Witcomb. *A Self-organizing Feedforward Network Applied to Acoustic Data, New Developments in Neural Computing*. JG Taylor, CLT Mannion, eds, Bristoland, New York: Adam Hilger, 1989.

25. F Filippetti, G Franceschini, C Tassoni. Neural Networks Aided On-Line Diagnostics of Induction Motor Rotor Faults, Trans. IEEE on Industr.Appl. 31(4): 892-899, 1995.

26. PV Goode, M Chow. Neural Fuzzy System for Incipient Fault Detection in Induction Motor, In: Patric, 1996.

27. R Milne, 1990. Ametist: Rotating Machinery Condition Monitoring, Proc. Amer. Artif. Intell. AA90 Conf., Washington, 1990.

28. T Harris. 1993.An Introduction to Neural Networks, Proc. 6th Intl. Conf. on Joining of Materials, Helsingor, Denmark, 1993.

29. I Mayes. 1993.Use of Neural Networks for On-Line Vibration Monitoring, Proc. Sem. Appl.of Expert Syst. in the Power Generation Industry, Inst. Mech. Eng., London, 1993.

30. E Govekar, I Grabec, J Peklenik. Monitoring of a Drilling Process by a Neural Network, 21st CIRP Intl. Seminar on Manufacturing Systems, Stockholm, June 5-6, 1989.

31. L Monostori, J Nacsa. On the Application of Neural Nets in Real-Time Monitoring of Machine Process, 22nd CIRP Intnl. Seminar on Manufacturing Systems, Enscchede, Netherlands, June 11-12, 1990.

32. PD Wasserman, A Unal, S Haddad. *Neural Networks For On-Line Machine Condition Monitoring, Intelligent Eng. Syst. Through Artif. Neural Networks*. ASME Press, New York, 1991, 693-700.

33. M Javed, G Littlefair. Neural Networks Based Condition Monitoring Systems for Rotating Machinery, Proc. 5th Intnl. Congress on Cond. Monitoring and Diagnostic Engg. Management COMADEM, Bristol, England, 1993.

34. T Harris, J MacIntyre, P Smith. Neural Networks and Their Application toVibration Analysis, Proc. of the Structural Dynamics and Vibration Symposium, New Orleans, 1994.
35. SV Kamarthi. et al. *On-Line Tool Wear Monitoring Using a Kohonen's Feature Map, Intelligent Engineering Systems Through Artificial Neural Networks.* New York: ASME Press, 1991, 639-644.
36. S Elanayar, YC Shin. Tool Wear Estimation in Turning Operations Based on Radial Basis Functions, In: Intelligent Engineering Systems Through Artificial Neural Networks, CH Dadli, SRT Kumara, YC Shin, eds., New York: ASME Press, 1991, 685-692.
37. TI Liu, KS Anantharaman. On-Line Sensing of Drill Wear Using Neural Network Approach. In: Patric, 1996.
38. TI Liu, SM Wu..On-Line Detection of Drill Wear, ASME J. of Engg. for Industry, 112: 299-302, 1990.
39. RC Luo, MG Kay..Multisensor Integration and Fusion in Intelligent Systems, IEEE Trans. on Syst., Man, and Cyb.19(5): 901-931, 1989.
40. J Rajapakse, R Acharia. Multi Sensor Data Fusion with Hierarchical Neural Networks, IJCNN-90, San Diego, Calif., 1990, II-17–II-22.
41. M Eggers, T Khuon..Neural Network Data Fusion Concepts and Applications, IJCNN-90, San Diego, Calif., 1990, II-7–II-16.
42. TY Young, K-S Fu, eds. *Handbook of Pattern Recognition and Image Processing.* New York: Academic Press, 1986.
43. D Popovic, VP Bhatkar. *Methods and Tools for Applied Artificial Intelligence.* New York : Marcel Dekker, Inc., 1994.
44. RT Chin, CA Harlow. Automated Visual Inspection: A Survey, IEEE Trans. Pattern Anal. Mach. Intel., PAMI-4, No.-6, 1982, 557-573.
45. G Taguchi, EA Elsayed, TC Hsiang. *Quality Engineering in Production Systems.* New York: McGraw-Hill, 1989.
46. C Wang, DJ Cannon. Neural Network and Skeleton-Based Inspection of Surface Flaws, Proc. IEEE Intnl. Conf. on Robotics and Autom., San Diego, Calif., vol. 2, 1994, 1126-1131.
47. PLiatsis, et al..A Versatile Visual Inspection Tool for the Manufacturing Process, Proc. 3rd IEEE Conf. on Control Appl., Glasgow, U.K., 1994, 1505-1510.
48. JL Lamprecht. Implementing the ISO 9000 Series, New York: Marcel Dekker, 1993.
49. T Pyzdek, RW Berger. *Quality Engineering Handbook.* Tata McGraw-Hill, New Delhi, India, 1995.
50. RT Chin..Algorithms and Techniques for Automated Visual Inspection. In: *Handbook of Pattern Recognition and Image Processing*, TY Young, K-S Fu, eds., New York: Academic Press, ch. 24, 1986, 587-612.
51. DE Glover..Neural Nets in Automated Inspection, Synapse Connection The Digest of Neural Computing 2(6): 1-13, 17, 1988.
52. JF Jarvis..Visual Inspection Automation, Computer, May 1980b, 32-38.
53. AJ Baker, RA Brook. A Design Study of an Automatic System for On-line Detection and Classification of Surface Defects on Cold-Rolled Steel Strip, Optical Acta 25(12): 1187-1196, 1978.
54. JL Mundy.Visual Inspection of Metal Surfaces, Proc. 5th Intnl. Conf. on Pattern Recognition, Miami Beach, November 1980.

55. RC Grove, BE Rapp, CW Souder. A Real-Time Image Processing System for Automated Inspection of Drilled Holes, Proc. SPIE, Real-Time Signal Processing III, Vol. 241, San Diego, July 1980.

56. BG Batchelor, GAWilliams. Defect Detection on the Internal Surface of Hydraulics Cylinder for Motor Vehicles, Proc. SPIE Imaging Applications for Automatic Industrial Inspection and Assembly, Vol. 182, April 1979, 65-78.

57. S Sugiyama, N Takahashi. Optical Measuring Device for Interior Dimensions of Automobiles, Proc. SPIE, Adv. in Laser Eng. Appls., 1980, Vol. 247.

58. EL Graminski, RA Kirsh..Image Analysis in Paper Manufacturing, Proc. IEEE Conf. on Patt. Recogn. and Image Proc., 1977 , 137-143.

59. T Kasvand. Experiments on Automatic Extraction of Paper Pulp Fibers, Proc. 4th Intl. Joint Conf. on Pattern Recogn., Kyoto, November 1978, 958-959.

60. H Ikeda, T Inagaki, Y Nishimura.Ceramic Surface Inspection Using Laser Technique, Proc. OCO Conf. on Optical Methods in Sc. and Industrial Measurements, Tokyo, 1974, 487-492.

61. P Drews. Automated On-Line Quality Inspection in Ceramic Product Manufacturing Based on Parallel Computing System, Transputer Applications and Systems '95, IOS Press, 1995.

62. A Kuni. et al. Study on Automatic Inspection of Defects on Contact Parts, IFAC, Information Control Problems in Manufacturing Technology, 1977, 63-70.

63. Pau, L.F. 1980.Integrated Testing and Algorithms for Visual Inspection of Semiconductor IC's, Proc. 5th Int. Conf. Patt. Recogn., Miami Beach., Nov. 1980.

64. Clarke, L.T. 1978.The Use of Self-Scanned Arrays in the Glass Industry, Proc. SPIE, Industry. Appl. of Solid-State Image Scanners, Vol.145, Sira, London, pp.28-36.

65. Giebel, H. et al. 1982.A System for Automatic Inspection of Glass Bottles using texture analysis procedure, Proc. IEEE Sixth. Conf. Pattern Recogn., 1982, Vol. 2.

66. Ghosh, J. 1994.Vision Based Inspection, In: Artificial Neural Networks for Intelligent Manufacturing, Dagli, C.H., Ed., Chapman & Hall, London, 1994.

67. A Koenig, A. et al. Visual Inspection in Industrial Manufacturing, IEEE Micro, June 1995: 26-30.

68. Huang, S.H. and Zhang, H.C. Neural Networks in Manufacturing: A Survey In.: Patric, 1996.

69. H Don. et al. Metal Surface Inspection Using Image Processing Techniques, IEEE Trans. Syst. Man, Cybern., Vol. 14, Febr. 1984.

70. KW Khawaja et al.1994.Automated Assembly Inspection Using a Multiscale Algorithm Trained on Synthetic Images, Proc. IEEE Intnl. Conf. on Robotics and Automation, San Diego, Calif., vol. 4, pp. 3530-3536.

71. RN West..Automatic Inspection of Photographic Materials, Optical Acta 25(12): 1207-1214, 1978.

72. H Beck, D McDonald, D Brzakovic. A Self-Training Visual Inspection System with a Neural Network Classifier, International Joint Conference on Neural Networks, Vol. I, 1988, 307-311.

73. H Koshimizu. Fundamental Study on Automatic Fabric Inspection by Computer Image Processing, Proc. SPIE Imaging Applications for Automatic Industrial Inspection and Assembly, Vol. 182, 1979.

74. S Sardy, L Ibrahim, Y Yasuda. An Application of Vision System for the Identification and Defect detection on Woven Fabrics by Using Artificial Neural Network, Proc. 1993 Int. Joint Conf. on Neural Networks, 1993, 2141-2144.

75. R Conners et al..Identifying and Locating Surface Defects in Wood, IEEE Trans. Pattern Anal. and Machine Intell. 5(6), 1983.

76. L Lieberman. Model-Driven Vision for Industrial Automation. In: Advances in Digital Image Processing Theory, Applications, Implementation, P Stucki, ed., New York: Plenum Press, 1979, 235-246.

77. CC Yang et al..Active Visual Inspection Based on CAD Models, Proc. IEEE Intnl. Conf. on Robotics and Automnation, San Diego, Calif., vol. 2, 1994, 1120-1125.

78. RC Gonzales. *Digital Image Processing*. New York: Addison-Wesley Publ. Co. Inc., 1992.

79. JG Daugman. Complete Discrete 2-D Gabor Transform by Neural Networks for Image Analysis and Compression, *IEEE Trans. On Acoustic, Speech, and Signal* 36(7): 1169-1179, 1988.

80. PH Carter. Texture Discrimination Using Wavelets, SPIE Applications of Digital Image Processing XIV, Vol. 1567, 1991, 432-438.

81. D Hush. Classification with Neural Networks: A Performance Analysis, Proc. IEEE Intnl. Conf. on Syst. Engg., 1989, 277-280.

82. R Lippmann..A Critical Overview of Neural Network Classifiers, In: Neural Networks for Signal Processing: Proc. 1991 IEEE Workshop, BH Juang, SY Kung, CA Kamm, eds., IEEE Press 1991, 266-278.

83. AJ Maren, CT Harston, RM Pap. *Handbook of Neural Computing Applications*. New York: Academic Press, 1990.

84. BE Dom and V Brecher. Recent Advances in Automatic Inspection of Integrated Circuits for Pattern Defects, Machine Vision and Appl. 8, 5–19, 1995.

85. GT Stubbendieck, WJB Oldham. A Fractal Dimension Feature Extraction Technique for Detecting Flaws in Silicon wafers, IJCNN 1992, Baltimore.

86. S Maeda, T Hiroi, H Makihira, and H Kubota.Automated Visual Inspection of LSI Wafer Patterns Using a Derivative polarity Comparison Algorithm, Applications of Digital Image Processing XIV, San Diego, Calif., SPIE 1567:100-109, 1991.

87. JF Jarvis..A Method for Automating the Visual Inspection of Printed Wiring Boards, IEEE Trans. Patt. Anal. Mach. Intell. 2(1): 77-82,1980a..

88. HF Spence. Printed Circuit Board Diagnosis using Artificial Neural Networks and Circuit Magnetic Fields. In: Patric, 1996.

89. A Knoll. Automatic Fabric Inspection, Textile Inst. Industry, Jan. 1985.

90. B Bourgeois, C Walker..Texture Estimation with Neural Networks, IEEEConf. on Neural Networks for Oceanic Engg., 99-106, 1991.

91. PP Raghu, R Poongodi, Yegnanarayana.Texture Classification Using Combined Self-Organizing Map and Multilayer Perceptron, Intl. Conf. on Computer Systems in Education, IISc., 145-153, 1994.

92. RM Haralick. Statistical and Structural Approach to Texture, Proc. IEEE 65(5): 786-804, 1979.

93. G Carpenter, S Grossberg. *Pattern Recognition by Self-Organizing Neural Networks*. Cambridge, Massachusetts: The MIT Press, 1991.

94. B Kosko. *Neural Networks for Signal Processing*. New Jersey: Prentice-Hall International, 1992.

95. M Turner. Texture Discrimination by Gabor Functions, Biol. Cybern. 55: 71-82. 1986.

96. A Visa. A Texture Classifier Based on Neural Network Principles, Internat. Joint Conf. on Neural Networks, San Diego, , Vol. I, pp. 491-496, 1990.

97. FS Cohen, Z Fan, S Attali. Automated Inspection of Textile Fabrics Using Textural Models, *IEEE Trans. on Patt. Anal. And Mach. Intell.*13(8): 803-808. 1991.

98. S Sardy, S. and L Ibrachim, L. 1994. An Application of Neural Networks for Defect Detection on Textile Products Based on Texture Feature Extractions, The Intnl. Conf. on Signal Proc. Appl.& Technplogy, Boston, Oct. 24-26, 1995, pp. 1169-1172.
99. KA Marko, LA Feldkamp, GV Puskorius..Automotive Control System Diagnostics Using Trainable. Classifiers: Statistical Testing and Paradigm Selection, IJCNN, Vol. II, pp.472-477, 1990.
100. D Popovic, J Meier, H Wesemann..Acoustic Inspection of a Car Engine Using Neural Networks, Univ. of Bremen, Technical Report in German, 1996.
101. H Shimodaira, Y Sakaguchi, K Nakano..Application of Neural Computation to Sound Analysis for Valve Diagnosis, Proc. IEEE 86(1): 177-182, 1991.
102. M Kotani, Y Ueda, H Matsumoto, and T Kanagawa. .Hybrid Neural Networks for Acoustic Control, Proc, 1993 Intl. Conf. on Neural Networks, vol. I, 931-934, 1993.
103. SL Murphy, SI Sayegh. Application of Neural Networks to Acoustic Screening of Small Electric Motors, IEEEInternational Conference on Neural Networks, Vol. II., 472-477, 1992.
104. D Krall, E Martschew, J Witt..Use of Adaptive Classifiers in Analysis of Planetary Gears Noise, Intell. Softw. Technologien 3: 22-27, 1991.
105. M Chow, PM Mangum, SO Yee. A Neural Network Approach to Real-Time Condition Monitoring of Induction Motors, IEEE Trans. on Industry. Electronics 38(6): 448-453, 1991.

11
System Software

Stuart Bennett
The University of Sheffield, Sheffield, United Kingdom

I. BASIC CONCEPTS

Real-time systems have many common features as illustrated in Figure 1. This system controls the speed of a servo-motor and performs some logical control. It is also part of a larger system and is linked to this system by a communications network. There is provision for local communication with an operator. The speed measurement transducer provides an analog signal which has to be sampled and converted to a digital signal by an ADC (analog-to-digital-converter) and the motor controller requires an analog value which has to be converted from the internal digital representation by the DAC (digital-to-analog-converter). The logic control communicates via the digital input and digital output registers. Communication with the operator is via the input device and display device controllers and it is assumed that data is transferred between these devices in the form of eight bit (byte) character codes. The data transferred to and from the communications controller is in the form of messages which are made up of blocks of ASCII characters which are transferred character by character to the communications controller.

The various software modules shown (speed controller, logic control, system messages and operator communications) have to deal with different types of information and with interfaces which operate in different ways. For example, the input to the ADC is an analog signal which is continuous in time and continuous in value. When instructed to by the speed controller, the ADC samples the analog signal and converts it to a discrete (digital) number.

A. Data Types and Data Structures

The data that a real-time system has to handle can be summarized as follows:

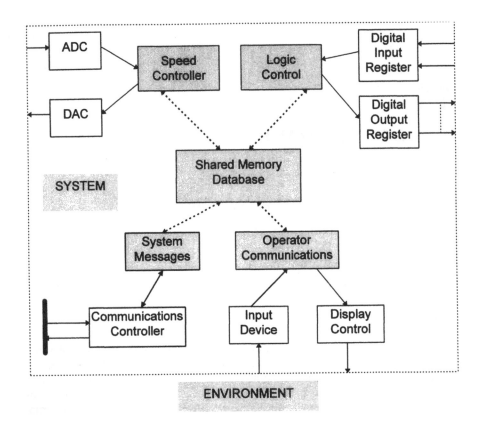

Figure 1 Example of a simple real-time control system.

1. Continuous value, continuous time
2. Discrete value, continuous time
3. Continuous value, discrete time
4. Discrete value, discrete time

A digital processor operates with only one of the above—discrete value, discrete time—although by using fast processing, real values and suitable interface devices, a reasonable approximation of the other data types can be obtained. However, it is important to realize that within the computer, processing is performed on discrete value, discrete time quantities and thus processing of all other types of data involves approximations.

The mapping of system data onto the typical data types supported by programming languages is illustrated below:

System Data Type	Language Data Type
continuous value	REAL
logic signal (single bit)	BITSET, BOOLEAN
logic register	BITSET, INTEGER
discrete value	INTEGER
character (individual)	CHAR
character string	STRING, ARRAY, FILE
communication blocks	RECORD

B. Communication and Synchronization [1]

Communication between the system and its environment can be initiated by:

* The occurrence of an event in the environment to which the system is expected to respond
* By some internal event in the system
* At predetermined intervals

Communication and synchronization may therefore be either event-based or clock-based (also referred to as cyclic or periodic).

Communication from the system to the environment is normally simple: in response to an internal event or at a predetermined time the system initiates the transfer process. If the environment process can accept the transfer then it proceeds; otherwise, either the system waits for a signal from the environment that it is now ready to accept the transfer, or it suspends the transfer and tries to restart it at some later time. Communication from the environment to the system is more difficult in that the system has to be made aware that a transfer needs to take place. Two basic techniques are used:

* Polling
* Interrupts

Polling

Polling is the name given to the technique that requires the system to check whether an environmental process is ready to make a data transfer. If it is ready, then the transfer takes place; otherwise the system continues with some other work. Polling is a simple technique to implement and is suitable for applications with the following characteristics:

* All events are equally probable and all have response times that are non-critical.

- The execution time of the code segment initiated in response to an event is of the same order as the interval between events.
- All inputs that might generate an event signal can be checked within the desired response time of any event.

Interrupts

An interrupt is a mechanism by which the flow of the program can be temporarily stopped to allow a special piece of software—an interrupt service routine, (also called an interrupt handler)—to run. When this routine has finished, the program which was temporarily suspended is resumed. Any type of event can be connected to an interrupt mechanism but interrupts are most commonly used for events which require a rapid response, for example:

- *Real-time clock:* The hardware provides a signal at regularly spaced intervals of time; the interrupt service routine counts the signals and hence is able to maintain a real-time clock.
- *Alarm inputs*: Various sensors can be used to provide a change in a logic level in the event of an alarm. Since alarms should be infrequent, but may need rapid response times, the use of an interrupt provides an effective and efficient solution.
- *Manual override*: Use of an interrupt can allow external control of a system to allow for maintenance and repair.
- *Hardware failure indication*: Failure of external hardware or of interface units can be signaled to the processor through the use of an interrupt.
- *Debugging aids*: Interrupts are frequently used to insert breakpoints or traces in the program during program testing.
- *Operating system*: Interrupts are used to force entry to the operating system before the end of a time slice.
- *Power failure warning*: It is simple to include in the computer system a circuit that detects the loss of power in the computer. If this circuit is connected to an interrupt that takes precedence over all other operations in the computer there can be sufficient time to carry out the necessary instructions to close the system down in an orderly fashion.

In a typical real-time system some external events will be handled by interrupts and others by polling.

Saving and Restoring Registers

Since an interrupt can occur at any point in a program, precautions have to be taken to prevent information which is being held temporarily in the CPU registers from being overwritten. All CPUs automatically save the contents of the program counter; this is vital. If the contents were not saved then a return to the point in the program at which the interrupt occurred could not be made. Some CPUs, however, do more and save all the registers. The methods commonly used are:

- Store the contents of the registers in a specified area of memory.
- Store the registers on the memory stack. This is a simple, widely used method which permits multilevel interrupts: the major disadvantage is the danger of stack overflow.
- Use of an auxiliary set of registers. Some processors provide two sets of the main registers and an interrupt routine can switch to the alternative set. If only two sets are provided multi-level interrupts cannot be handled. An alternative method is to use a designated area of memory as the working registers. Then an interrupt only requires a pointer to be changed to change the working register set.

The use of automatic storage of the working registers is an efficient method if all registers are to be used: it is inefficient if only one or two will be used by the interrupt routine. For this reason fully automatic saving is usually restricted to CPUs with only a few working registers in the CPU; systems with many working registers provide an option either to save or not to save. Unless response time is critical it is good engineering practice to save all registers: in this way there is no danger, in a subsequent modification to the interrupt service routine, of using a register which is not being saved, and failing to add it to the list of registers to be saved.

The machine status must be restored on exit from the interrupt routine; this is straightforward for all methods except that which uses the stack to save registers: in this case the registers are restored in the opposite order than that in which they were saved. Systems providing automatic saving also provide automatic restore on exit from the interrupt.

An example of the framework of an interrupt service routine is shown below.

```
INT1:   CALL SAVREG ; SAVREG is routine which saves working reg-
                isters
            ;
            ;code for interrupt handling is inserted here
            ;
        CALL RESREG ;RESREG is routine which restores working
                registers
        EI      ;enable interrupts
        RETI    ;return from interrupt routine
```

The above routine is suitable for a system in which interrupts are not allowed to be interrupted; hence the EI instruction which enables interrupts is not executed until immediately prior to the return from interrupt. The return from an interrupt routine has to be handled with care to prevent unwanted effects. For example the EI instruction does not re-enable interrupts until after the execution of the instruction which immediately follows it. Therefore, by using the EI/RETI combination a pending interrupt cannot take affect until after the return from the previous one has been completed. In this example it is assumed that the microprocessor automatically disables interrupts on acknowledgment of an interrupt and they remain disabled until an EI instruction is executed. Some CPUs operate by disabling interrupts for only one instruction following an interrupt acknowledge; it thus becomes the responsibility of the programmer to disable interrupts as the first instruction of the interrupt service routine if only a single level of interrupt is to be supported.

Interrupts Combined with Polling

It is possible to combine a number of events through an OR function to generate a single interrupt. The interrupt handler then polls each event input in turn to find the ones which are TRUE. This approach has the advantage over normal polling systems in that at least one of the inputs is guaranteed to be active. It is clearly, however, not a very satisfactory system if large numbers of devices have to be checked. The load can be reduced by first testing the devices that interrupt most frequently, but this may conflict with response time requirements in that a device that interrupts infrequently may require a rapid response time and hence should be checked first. If an equal response time, on average, for each device is required it will be necessary to rotate the order in which the devices are checked. Doing this can also prevent one device which interrupts frequently from locking out all others. The method does provide a flexible way of allocating priority to the various devices that can generate interrupts.

In most real-time systems a single interrupt level is unacceptable: the whole purpose of interrupts is to get a fast response and this would be prevented if a low priority interrupt could lock out a high priority one. It should be obvious that the ability to interrupt an interrupt service routine should be restricted to interrupts which are of higher priority than the routine executing. In order to do this there has to be some facility for masking out (or inhibiting) interrupts of lower priority.

Polling, with either busy-wait or periodic checks on device status, provides the simplest method of data transfer, in terms of the programming requirements and in the testing of programs. The use of interrupts results in software that is much less structured than a program with explicit transfers of control; there are potential transfers of control at every point in the program.

Interrupt-driven systems are much more difficult to test since many of the errors may be time-dependent. A simple rule is to check the interrupt part of the program if irregular errors are occurring. The generation of appropriate test routines for interrupt systems is difficult; for proper testing it is necessary to generate random interrupt patterns and to carry out detailed analysis of the results.

At high data-transfer rates the use of interrupts is inefficient because of the overheads involved in the interrupt service routine—saving and restoring the environment—hence polling is often used. An alternative for high rates of transfer is to substitute hardware for software control and use direct memory access techniques.

C. Timing Constraints [1–3]

Real-time systems are typically divided into two categories:

1. *Hard real-time*: These are systems that must satisfy the deadlines on each and every occasion.
2. *Soft real-time*: These are systems for which an occasional failure to meet a deadline does not compromise the correctness of the system. The general form of this type of constraint is that the computer must respond to the event within some specified maximum time.

An automatic bank teller provides an example of a system with a soft real-time constraint. A typical system is event driven in that it is started by the customer placing their card in the machine. The time constraint on the machine responding will be specified in terms of an average response time say, of 10 seconds, with the average being measured over a 24-hour period. (Note if the system has been carefully specified there will also be a maximum time, say 30 seconds, within which the system should respond). The actual response time will vary: if you have used such a system you will have learned that the response time obtained between 12 and 2 p.m. on a Friday is very different from that at 10 a.m. on a Sunday.

In practice the above categories are only guides; for example, in the speed control system an occasional missed sample would not seriously affect the performance, neither would a variation in sampling interval such that inequality $39.5 < T_s < 40.5$ ms with a mean of $T_s = 40$ ms was satisfied. The system would not be satisfactory, however, if the sampling interval T_s was in the range $1 < T_s < 1000$ ms with a mean of 10 ms over a 24-hour period. Similarly, the automatic bank teller would not be satisfactory if at a busy time customers had to wait ten minutes, even if it achieved a mean response measured over a 24-hour period of 20 seconds. A typical specification might be a mean response measured over a 24-hour period of 15 seconds, with 95% of requests being satisfied within 30 seconds and no response time greater than 60 seconds.

A hard real-time constraint obviously represents a much more severe constraint on the performance of the system than a soft real-time constraint and such systems present a difficult challenge both to hardware and to software designers. Most real-time systems contain a mixture of activities that can be classified as clock-based, event-based, and interactive with both hard and soft real-time constraints (they will also contain activities which are not real-time). A system designer will attempt to reduce the number of activities (tasks) that are subject to a hard real-time constraint.

We can formally define the constraints as follows:

Hard		Soft	
Periodic (Cyclic)	Aperiodic (Event)	Periodic (Cyclic)	Aperiodic (Event)
$t_c(i) = t_s \pm a$	$t_e(i) \le T_e$	$\frac{1}{n}\sum_{i=1}^{n} t_c(i) = t_s \pm a$ $n = T/t_s$	$\frac{1}{n}\sum_{i=1}^{n} t_e(i) \le T_a$ $n = T/t_s$

$t_c(i)$	the interval between the i and $i-1$ cycles
$t_e(i)$	the response time to the ith occurrence of event e
t_s	the desired periodic (cyclic) interval
T_e	the maximum permitted response time to event e
T_a	the average permitted response time to event e measured over some time interval T
n	the number of occurrences of event e within the time interval T, or the number of cyclic repetitions during the time interval T
a	a small timing tolerance

For some systems and tasks timing constraints may be combined in some form or other, or relaxed in some way. For example, soft aperiodic (event) time constraints are often combined with a hard real-time constraint such that $t_c(i) \le T_m$ where $T_m \gg T_a$. In some circumstances an occasional missed periodic hard real-time constraint might not be serious; for example, in a feedback control system missing the occasional sample will not result in a major disturbance of the controlled system. (Note: this is the case only if the manipulated variable is held constant at the previous value.)

D. Design Approaches

Examining Figure 2 we can see that the software has to perform the following functions:

- Speed control
- Logic control
- Operator display
- Operator input
- Communication of messages

Although not shown in Figure 2 the following will also have to be provided:

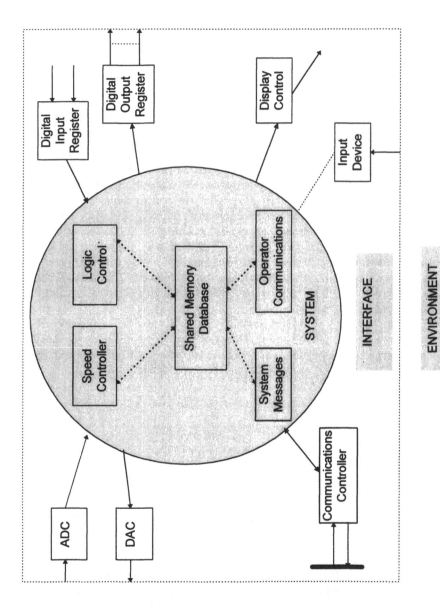

Figure 2

- System start up and shut down
- Clock function

Let us assume that the following time constraints apply:

- Speed control   hard constraint   run every 40 ms
- Logic control   hard constraint   respond within 100 ms
- Operator display   soft constraint   every 5s
- Operator input   soft constraint   respond within 1s
- System messages   soft constraint   respond within 200 ms

In practice the constraint on speed control may be relaxed a little say 40 ms $\pm$ 1 ms, with an average value over 1 minute say 40 ms $\pm$ 0.5 ms. In general the sampling time can be specified as $T_s \pm e_s$ with an average value, over time, T, of $T_s \pm e_a$. The requirement may also be relaxed to allow, for example, one sample in 100 to be missed. These constraints will form part of the test specification.

Let us also assume that the clock module must run every 20 ms in order not to miss a clock pulse. The operator display, as specified, has a hard constraint in that an update interval of 5 seconds, is given. Common sense suggests that this is unnecessary and an average time of 5 seconds should be adequate; however, a maximum time would also have to be specified, say 10 seconds. The startup module does not have to operate in real-time and hence can be considered as a standard interactive module.

There are obviously several different activities which can be divided into sub-problems. The sub-problems will have to share a certain amount of information and how this is done and how the next stages of the design proceed will depend upon the general approach to implementation. There are three possibilities:

- Single program
- Foreground-background system
- Multitasking

Single Program Approach

Using the standard programming approach the modules shown in Figure 2 are treated as procedures or subroutines of a single main program. The flow of such a program is illustrated in Figure 3. This structure is easy to program. However, it imposes the most severe of the real-time constraints, that is, the requirement that the clock module must run every 20 ms on all of the modules. For the system to work the clock module and any one of the other modules must complete their operations within 20 ms. If t_1, t_2 t_3, t_4, t_5 and t_6 are the *maximum* execution times for the modules clock, speed control, logic control, operator display, operator input and message communication respectively then a requirement for the system to work can be expressed as:

$$t_1 + \text{MAX} (t_2, t_3, t_4, t_5, t_6) < 20 \text{ ms}$$

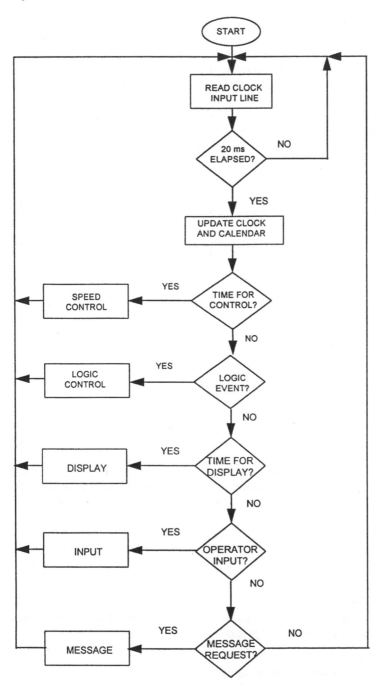

Figure 3 Single task approach to software design.

(Note: the values of t_1, t_2, t_3, t_4, t_5 and t_6 must include the time taken to carry out the tests required and t_1 must also include the time taken to read the clock input line.)

The single program approach is effective for simple, small systems and it leads to a clear and easily understandable design, with a minimum of both hardware and software. The interface with the environment is based on polling and such systems are usually easy to test. As the number of inputs, outputs and software segments increases the single program approach rapidly becomes unwieldy, inefficient and difficult to test. There is a tendency at the design stage to split modules, not because they are functionally different, but simply to enable them to complete within the required time interval.

Foreground-Background System

The load on the system can be reduced by using interrupts instead of polling. The environment can thus signal that it requires attention rather that the system having to check if an event needing attention has occurred. There are also obvious advantages—less module interaction, less tight real-time constraints—if the modules with hard real-time constraints can be separated from, and handled independently of, the modules with soft real-time constraints or no real-time constraints. The modules with hard real-time constraints, are run in the so-called *foreground* and the modules with soft constraints (or no constraints) are run in the *background*. The foreground modules, or *tasks* as they are usually termed, have a higher priority than the background tasks and a foreground task must be able to interrupt a background task.

A foreground-background approach can be implemented on any processor which supports interrupts. Also any single program operating system, for example MS-DOS, can be adapted to support foreground-background operation. However, care is required when implementing a foreground-background system on top of an existing operating system in that most operating systems already have within themselves some form of foreground-background operation and this may conflict with the application structure.

The terminology *foreground* and *background* can be confusing: much of the literature concerned with non–real-time software uses foreground to refer to the application software and background to refer to interrupt routines that are hidden from the user.

Using the foreground-background approach the structure shown in Figure 3 can be modified to that shown in Figure 4. There is now a very clear separation between the two parts of the system. A requirement for the foreground part to work is that

$$t_1 + t_2 < 20 \text{ ms}$$

where t_1 = maximum time for clock module and t_2 = maximum time for the control module. A requirement for the background part to work is that

1. Max $(t_4, t_5, t_6) < 10$ s
2. Display module runs on average every 5 s
3. Operator input responds in < 10 s

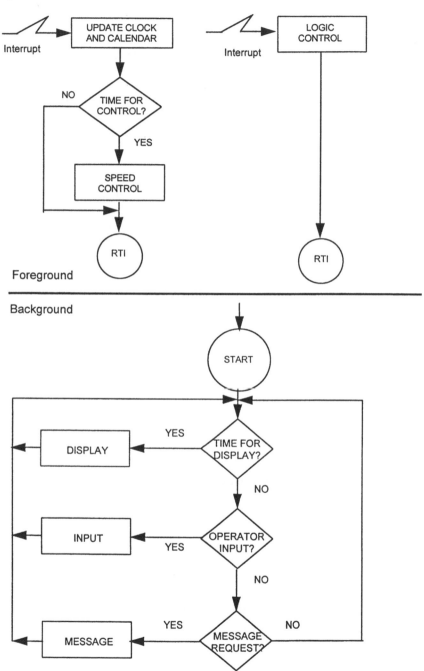

Figure 4 Foreground-background approach to software design.

Although the real-time constraints have been relaxed the measurements to be made in order to check the performance are more complicated than in the single program case and hence the evaluation of the performance of the system has been made more difficult.

Although the foreground-background approach separates the *control* structure of the foreground and background modules, the modules are still linked through the data structure. The linkage occurs because they share data variables; for example, the control tasks, the display task and the operator input task all require access to the controller parameters. In the single program (sometimes called single-tasking) there was no difficulty in controlling access to the shared variable since only one module (task) was active at any one time, whereas in the foreground-background system tasks may operate in parallel, i.e., one foreground module and one background module may be active at the same time. (Note: active does not mean *running* since if one CPU is being used only one task can be using it at any instant; however, both the foreground and background tasks may have the potential to run.)

In this particular example the variables can be shared between the control, display, and operator input modules without any difficulty since only one module writes to any given variable. The operator input module writes the controller parameters and set point variables, the clock module writes to the time variables and the control module writes to the plant data variables. However, the input from the operator must be buffered and only transferred to the shared storage when it has been verified as is illustrated below.

```
MODULE SpeedControl;
VAR
  p1, p2, p3 : REAL; (* Controller parameters declared as global variables *)
PROCEDURE GetParameters (VAR x, y, z : REAL);
BEGIN
  ...
  (* get new parameters from terminal and store in x, y, z *)
  ...
END GetParameters;
PROCEDURE OperatorInput;
VAR
  x, y, z : REAL;
BEGIN
  GetParameters (x, y, z);
  (* insert code to verify here *)
  p1: = x; (* transfer parameters to global variables *)
  p2 : = y;
  p3 : = z;
END OperatorInput;
BEGIN
  (* main program *)
END SpeedControl.
```

To understand the reasons for buffering, let us consider what would happen if a new value was stored directly in the shared data areas when it was entered. Suppose the controller was operating with p1 = 10, p2 = 5 and p3 = 6 and it was decided that the new values of the control parameters should be p1 = 20, p2 = 3 and p3 = 0.5. As soon as the new value of p1 is entered the controller begins to operate with p1 = 20, p2 = 5, p3 = 6, i.e., neither the old nor the new values. This may not matter if the operator enters the values quickly. But what happens if, after entering p1, the telephone rings or the operator is interrupted in some other way and consequently forgets to complete the entry? The plant could be left running with a completely incorrect (and possibly unstable) controller.

The method used in the example above is not strictly correct and safe since an interrupt could occur between transferring x to p1 and y to p2, in which case an incorrect controller would be used. For a simple feedback controller this would have little effect since it would be corrected on the next sample. It may be more serious if the change were to a sequence of operations for the logic controller. The potential for serious and possibly dangerous consequences is not great in small, simple systems (a good reason for keeping systems small and simple whenever possible): it is much greater in large systems.

The transfer of data between the foreground and background tasks, i.e., the statements:

```
p1:=x;
p2:=y;
p3:=z;
```

form what is known as a *critical section* of the program and should be an indivisible action. The simple way of ensuring this is to inhibit all interrupts during the transfer:

```
InhibitInterrupts;
p1:=x;
p2:=y;
p3:=z;
EnableInterrupts;
```

However, it is undesirable for several separate modules each to have access to the basic hardware of the machine and each to be able to change the status of the interrupts. From experience we know that modules concerned with the details of the computer hardware are difficult to design, code and test, and have a higher error rate than the average module. It is good practice to limit the number of such modules. Ideally transfers should take place at a time suitable for the controller module which implies that the operator module and the controller module should be synchronized or should rendezvous.

Multitasking Approach

The design and programming of large real-time systems is eased if the foreground-background partitioning can be extended into multiple partitions to permit many active tasks. This means that at the preliminary design stage each activity can be considered as a separate task (or process). The implications of this approach are that each task may be carried out in parallel and there is no assumption made at the preliminary design stage as to how many processors will be used in the system.

The implementation of a multitasking system requires the ability to:

- Create separate tasks.
- Schedule running of the tasks, usually on a priority basis.
- Share data between tasks.
- Synchronize tasks with each other and with external events.
- Prevent tasks corrupting each other.
- Control the starting and stopping of tasks.

The facilities to perform the above actions are typically provided by a Real-Time Operating System (RTOS) or a combination of RTOS and a real-time programming language. Multitasking, whether on a single processor or using multiple processors leads to issues concerned with resource sharing and synchronization.

E. Resource Sharing and Task Synchronization

For simplicity we will assume that we are using only one CPU and that the use of this CPU is time-shared between the tasks. We also assume that a number of so-called *primitive instructions* exist. These are instructions which are part of a programming language or the operating system and their implementation and correctness is guaranteed by the system. All that is of concern to the user is that an accurate description of the syntax and semantics is made available. In practice, with some understanding of the computer system, it should not be difficult to implement the primitive instructions. Underlying the implementation of primitive instructions will be an eventual reliance on the system hardware. For example, in a common memory system some form of arbiter will exist to provide for mutual exclusion in accessing an individual memory location.

14.1.5.1 Mutual exclusion

A multitasking operating system allows the sharing of resources between several concurrently active tasks. This does not imply that the resources can be used simultaneously. The use of some resources is restricted to only one task at a time. The restriction to "one task at a time" has to be made for resources such as input and output devices; otherwise there is a danger that input intended for one task could get corrupted by input for another task. Similarly problems can arise if two tasks share a data area and both tasks can write to the data area. This is illustrated by the example below.

Two software modules, bottle_in_count, and bottle_out_count are used to count pulses issued from detectors which observe bottles entering and leaving a processing area. The two modules run as independent tasks. The two tasks operate on the same variable bottle_count. Module bottle_in_count increments the variable and bottle_out_count decrements it. The modules are programmed in a high level language and the relevant program language statements are:

```
bottle_count := bottle_count + 1; (bottle_in_count)
bottle_count := bottle_count - 1; (bottle_out_count)
```

At assembler code level the high level instructions become:

```
{bottle_in_count}              {bottle_out_count}
LD A, (bottle_count)           LD A, (bottle_count)
ADD 1                          SUB 1
LD (bottle_count), A           LD (bottle_count), A
```

Now if variable bottle_count contains the value 10 and bottle_in_count is running and executes the statement LD A, (bottle_count) then the processor **A** register is loaded with the value 10. If the operating system now reschedules and bottle_out_count runs it will also pick up the value 10, subtract one from it and store 9 in bottle_count. When execution of bottle_in_count resumes its environment will be restored and the **A** register will contain the value 10, one will be added and the value 11 stored in bottle_count. Thus the final value of bottle_count after adding one to it and subtracting one from it will be 11 instead of the correct value 10.

In abstract terms mutual exclusion can be expressed in the form

```
BEGIN
  remainder 1
  pre-protocol (necessary overhead)
    critical section
  post-protocol (necessary overhead)
  remainder 2
END
```

Remainder 1 and remainder 2 represent sequential code that do not require access to a particular resource or to a common area of memory. The critical section is the part of the code that must be protected from interference from another task. Pre-protocols and post-protocols called before and after the critical sections are code segments that will ensure that the critical section is executed so as to exclude all other tasks. To benefit from concurrency both the critical section and the protocols must be much shorter than the remainders, so that the remainders represent a significant body of code that can be overlapped with other tasks. The protocols represent an overhead that has to be paid in order to obtain concurrency.

A commonly used approach based on the idea of a flag variable, which is illustrated below, does not provide a safe solution.

```
MODULE MutualExclusion1;
(* Mutual exclusion problem Condition Flag solution 1 *)
VAR
  deviceInUse: BOOLEAN;
PROCEDURE Task; (* task assumed to be running in parallel with other tasks *)
BEGIN
  (* remainder1 *)
  WHILE deviceInUse DO
    (* test and wait until available *)
  END (* while *)
  deviceInUse:= TRUE; (* claim resource *)
  (* use the resource—critical section *)
  deviceInUse := FALSE;
  (* remainder2 *)
END Task;
  (* main program *)
END MutualExclusion1.
```

In the above the flag variable deviceInUse is set to FALSE when the device is available and TRUE when it is in use. A task wishing to access the resource has to test this flag before using the resource. If the flag is FALSE then the resource is available and the task sets the flag TRUE and uses the resource. In the above solution there are two problems.

1. The WHILE statement forms a *busy wait* operation which relies on a pre-emptive interrupt to escape from the loop. If the task which has already claimed the resource cannot interrupt the *busy wait* then the task will continue to use the CPU and will exclude all other tasks.
2. The testing and setting of the flag are separate operations and hence the task could be suspended and replaced by another task between checking the flag and set the resource unavailable. A consequence could be that two tasks could both claim the same resource.

Semaphore

The most widely used form of primitive instruction for the purposes of mutual exclusion is the binary semaphore. A binary semaphore is a condition flag that records whether or not a resource is available. For a binary semaphore s, if s = 1 then the resource is available and the task may proceed, if s = 0 then the resource is unavailable and the task must wait. To avoid the processor wasting time while a task is waiting for a resource to become available there has to be a mechanism for suspending the running of a task when it is waiting and for recording that the task is waiting for a particular semaphore. A typical mechanism for doing this is to associate with each semaphore a queue (often referred to as a condition queue) of tasks that are waiting for a particular semaphore.

There are only three permissible operations on a semaphore: Initialize, Secure, and Release and the operating system must provide the following procedures.

Initialize (s:ABinarySemaphore, v:INTEGER): set semaphore s to value
of v (v=0 or 1)
Secure(s): if s=1 then set s:=0 and allow the task to proceed, otherwise
suspend the calling task
Release(s): if there is no task waiting for semaphore s then set s:=1 otherwise resume any task that is waiting on semaphore s.

The operations Secure(s) and Release(s) are system primitives which are carried out as indivisible operations and hence the testing and setting of the condition flag are performed effectively as one operation. As an example consider a task where wishes to access a printer.

```
(* Mutual exclusion problem—use of binary semaphore *)
VAR
  printerAccess: SEMAPHORE;
PROCEDURE Task;
BEGIN
  (* remainder1 *)
  Secure(printerAccess) (* if printer is not available task will be suspended at this
point *)
    (* printer available—critical section *)
    (* do output *)
  Release(printerAccess)
  (* remainder2 *)
END Task ;
```

Task Synchronization Without Data Transfer

A frequent requirement is the need to inform another task that an event has occurred, or to set a task to wait for an event to occur. No data needs to be exchanged by the tasks. A mechanism that enables this to be done is the so-called signal. A signal s is defined as a binary variable such that if s = 1 then a signal has been sent but has not yet been received. Associated with a signal is a queue and the permissible operations on a signal are:

Initialize (s:Signal: v:INTEGER)set s to the value of v (0 or 1).
Wait(s) if s = 1 then s:=0 else suspend the calling task and place it in the
condition queue s.
Send(s)if the condition queue s is empty then s:=1 else transfer the first
task in the condition queue to the ready queue.

Clearly a signal is similar to a semaphore: in fact the difference between the two is not in the way in which they are implemented but in the way in which they are used. A semaphore is used to secure and release a resource and as such the calls will both be made by one task: a signal is used to synchronize the activities of two tasks and one task will issue the send and the other task the wait. (Note: a signal is sometimes implemented such that if a task is not waiting it has no effect; that is, the receipt of a signal is not remembered.)

In practice two important additions to the basic signal mechanism are required: one is the facility to check to see if a task is waiting to send a signal, and the other is to be able to restrict the length of time which a task waits for a signal to occur. In a real-time application it is rarely correct for a task to be committed to wait indefinitely for an event to occur. The standard Wait(s) commits a task to an indefinite wait.

Data Transfer Without Synchronization

There are two approaches to the transfer or sharing of data between tasks: these are the pool and the channel.

- A pool is used to hold data common to several tasks, for example, tables of values or parameters, which tasks periodically consult or update. The write operation on a pool is destructive and the read operation is nondestructive. The shared variable bottle_count used in the mutual exclusion example (see above) is a pool. To avoid the problem of two or more tasks writing to a pool simultaneously, mutual exclusion on pools is required.
- A channel supports communication between producers and consumers of data. It can contain one or more items of information. Writing to a channel adds an item without changing items already in it. The read operation is destructive in that it removes an item from the channel. A channel can become empty and also, because in practice its capacity is finite, it can become full.

Channels provide a direct communication link between tasks, normally on a one-to-one basis. The communication is like a pipe down which successive collections of items of data—messages—can pass. Normally they are implemented so that they can contain several messages and so they act as a buffer between the tasks. One task is seen as the *producer* of information and the other as the *consumer*. Because of the buffer function of the channel the producer and consumer tasks can run asynchronously.

There are two basic implementation mechanisms for a channel:

- Queue (linked list)
- Circular buffer

The advantage of the queue is that the number of successive messages held in the channel is not fixed. The length of the queue can grow, the only limit being the amount of available memory. The disadvantage of the queue is that as the length of the queue increases, the access time, that is the time to add and remove items from the queue, increases. For this reason, and because it is not good practice to have undefined limits on functions in real-time systems, queues are rarely used.

The circular buffer uses a fixed amount of memory, the size being defined by the designer of the application. If the producer and consumer tasks run normally they would typically add and remove items from the buffer alternately. If, for some reason, one or the other is suspended for any length of time the buffer will either fill up or empty. The tasks using the buffer have to check, as appropriate, for buffer full and buffer empty conditions and suspend their operations until the empty or full condition changes.

As an example let us consider an alarm scanning task which for a period of time produces data at a rate much greater than that at which the logging task can print it out. A buffer is needed to store the data until the consuming task is ready to take it. We assume that the buffer is bounded, that is of finite size, and that the operation of storing an item of data in it is performed by the call

    Put(x)

and the item of the data is removed by the call

    Get(x)

Since the buffer is of finite size it is necessary to know when it is full and when it is empty. The following function calls are used:

    Full—which returns the value TRUE if the buffer is full
    Empty—which returns the value TRUE if the buffer is empty

Let us assume that the producer and consumer are formed by separate tasks that share a common buffer area.

```
(* Producer-consumer problem—solution 1 *)
VAR commonBuffer : buffer;
TASK Producer;
VAR x:data;
BEGIN
  LOOP
    Produce(x);
    WHILE Full DO
    Wait
    END (* while *);
    Put(x);
  END (* loop *);
END Producer;
TASK Consumer;
VAR x:data;
```

```
BEGIN
 LOOP
   WHILE Empty DO
   Wait
   END (* while *);
   Get(x);
   Consume(x);
 END (* loop *);
END Consumer;
```

The producer operates in an endless cycle producing some item x and wait-ing until the buffer is not full to place x in the buffer; the consumer also operates in an endless cycle waiting until the buffer is not empty and removing item x from the buffer. The above solution is not satisfactory for two reasons:

1. The Put(x) and Get(x) are both operating on the same buffer and for secu-rity of the data simultaneous access to the buffer cannot be allowed—the mutual exclusion problem.
2. Both the producer and the consumer use a *busy wait* in order to deal with the buffer full and buffer empty problem.

The first problem can be solved using the semaphore, with the operations secure and release. The second problem can be solved by using the signal mechanism described above. Because of the need to test for empty and full and to suspend the task if one or the other conditions appertains then, although transfer of a data item is not synchronized, there is task synchronization on the buffer full and buffer empty conditions. This is illustrated below:

```
(* Data transfer problem—solution 2 using semaphores and signals *)
VAR commonBuffer : Abuffer;
  bufferAccess : ABinarySemaphore;
  nonFull, nonEmpty : Signal;
TASK Producer;
VAR x:data;
BEGIN
 LOOP
   Produce(x);
   Secure(bufferAccess);
   IF Full THEN
     Release(bufferAccess); (* see note 1 below *)
     Wait(nonFull);
     Secure(bufferAccess);
   END (* if *);
   Put(x);
   Release(bufferAccess);
   IF Full THEN
     Wait(nonFull); (* see note 2 below *)
```

```
  END (* if *);
  Send(nonEmpty);
 END (* loop *);
END Producer;
TASK Consumer;
VAR x:data;
BEGIN
 LOOP
  Secure(bufferAccess);
  IF Empty THEN
   Release(bufferAccess); (*see note 1 below *)
   Wait(nonEmpty);
   Secure(bufferAccess);
  END (* if *);
  Get(x);
  Release(bufferAccess);
  IF Empty THEN
   Wait(nonEmpty); (* see note 2 below *)
  END (* if *);
  Send(nonFull);
  Consume(x);
 END (* loop *);
END Consumer;
```

Note 1: In the above the critical code is enclosed between secure and release operations but it is essential that the bufferAccess semaphore is released before executing the Wait(nonFull) or Wait(nonEmpty) primitives. If this is not done the system will deadlock. For example if the Producer executes Wait(nonFull) while holding the access rights to the buffer then the buffer can never become non-full since the only way it can is for the Consumer to remove an item of data, but the Consumer cannot gain access to it until it is released by the Producer.

Note 2: The additional Wait(nonFull) and Wait(nonEmpty) have to be included to avoid the following possibility. Assume that the buffer has space in it. Then as soon as the consumer task gets an item it will execute Send(nonFull); this sets the nonFull signal to 1. If, now the consumer task is preempted and the producer task puts data in the buffer until it is full then the Full condition becomes true and without the additional Wait(nonFull) at this point the Wait(nonFull) in the first IF statement would be executed and since nonFull = 1 then the producer task would not wait but would attempt to put one more item in the buffer. Alternative solutions can be devised, for example, by setting the Full = TRUE when the buffer reaches maximum size—1.

Both semaphores and signals can be generalized to allow a semaphore or a signal variable to have any non-negative integer value—in this form they are sometimes referred to as counting semaphores.

Synchronization with Data Transfer

There are two main forms of synchronization involving data transfer. The first involves the producer task placing a message in a designated known area, either in a channel or a pool, and signaling to the consumer that a message has been produced and is waiting to be collected, and then continuing to execute. The second is for a task to signal that it is ready to exchange data and to wait for the corresponding task to reach a point where the two tasks can exchange the data.

The first method can be implemented by a simple extension of the mechanism used in the example in the previous section to signal that a channel was empty or full. Instead of signaling these conditions, a signal is sent each time a message is placed in the channel. Either a generalized semaphore or signal that counts the number of sends and waits, or a counter has to be used. It can be further extended into a mailbox system in which the producer task deposits the data (message) together with an identifier specifying the recipient (or possible recipients). The mailbox system signals to the recipients that a message is waiting. With the mailbox system a task may also check to see if a message for it is in the mailbox.

The second method requires that the tasks meet or rendezvous, that is the two tasks can only exchange data when both of them reach a predetermined point in their execution. The rendezvous method forms the basis of task synchronization in the language Ada.

II. DESIGN TOOLS: AN OVERVIEW

A. Introduction

The production of robust, reliable, software of high quality for real-time computer control applications is a difficult task that requires the application of engineering methods. In recent years methods for developing and implementing real-time systems have proliferated. Pyle et. al. [5] lists some 35 such methods and Gomaa provides an overview of five major methods and a detailed study of ADARTS and CODARTS [6]. Gomaa classified the methods as follows:

- *Methods based on functional decomposition*: These methods largely derive from the original Yourdon approach and are exemplified by the Ward and Mellor [7], and Hatley and Pirbhai [8] methodologies. These methods have been widely used and are supported by a variety of CASE tools.
- *Methods based on concurrent task structuring*: Typical of these methods is DARTS the method developed by Gomaa and its derivatives CODARTS and ADARTS. MASCOT also falls into this category and at the implementation level the language occam.
- *Methods based on information hiding:* Examples of this approach which emphasize the production of modifiable, maintainable, and reusable code

are the Naval Research Laboratory method, HOOD, and a growing number of object oriented methods.

- *Methods based on modeling in the problem domain*: An early advocate of this approach was Jackson with the Jackson System Development Method. The basic idea is to produce a network of entities which directly relate to the problem domain and to map these onto software tasks. Some OOA and OOD methods fall into this category.

All the methods seek to address the issue of increasing the formalizing of the specification, design, and construction of such software.

B. Modeling

All of the methodologies address the problem in three distinct phases. The production of a *logical* or *abstract* model—the process of *specification*; the development of an *implementation* model for a *virtual machine* from the logical model—the process of *design*; and the construction of software for the virtual machine together with the implementation of the virtual machine on a physical system—the process of *implementation*. These phases, although differently named correspond to the phase of development generally recognized in software engineering texts. Their relationship to each other is shown in Figure 5.

- *Abstract model*: equivalent of a requirements specification, the result of the requirements capture and analysis phase.
- *Implementation model*: equivalent of the system design, product of the design stages—architectural design and detail design.
- *Implementation*: the process of mapping the implementation model onto the physical hardware, identifying the software modules, and coding them.

Although there is a logical progression from abstract model to implementation model to implemented software, and although three separate and distinct artifacts—abstract model, implementation model, and deliverable system—are produced, the phases overlap in time. The phases overlap because complex systems are best handled by a hierarchical approach: determination of the detail of the lower levels in the hierarchy of the logical model must be based on knowledge of higher level design decisions, and similarly the lower level design decisions must be based on the higher level implementation decisions. Another way of expressing this is to say that the higher level design decisions determine the requirements specification for the lower levels in the system.

The strength of this approach is that it supports the *layering* of the design and implementation, that is the system designers target the design for a *virtual machine* and the implementation of this virtual machine on a real machine be-

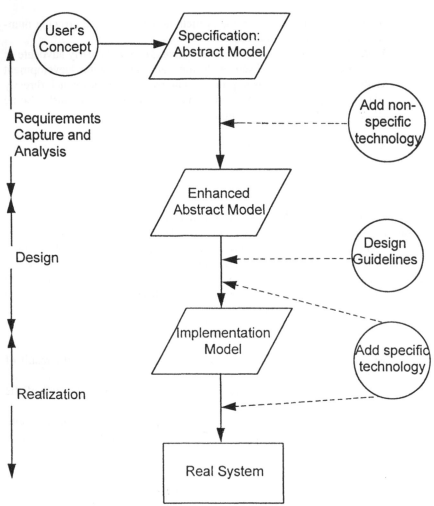

Figure 5 Software modeling.

comes a separate problem. In some methodologies the nature of the virtual machine is closely specified. For example, the MASCOT 3 methodology requires that a virtual machine be defined and specifies the characteristics of an appropriate machine (in MASCOT 2 the virtual machine characteristics were mandatory), and the HOOD system is based on an assumption that the virtual machine will be the Ada environment [9,10]. The object oriented methodologies assume that there is an underlying machine that will support the needs of the object oriented paradigm, in particular message passing and an inheritance structure. The Yourdon methodologies place the least restrictions on the characteristics of a virtual machine requiring only that there is support for multitasking.

The virtual machine approach is sometimes considered to be unsuitable for real-time systems in that it is said to impose layers of unnecessary code between the application software and the hardware. However, most methodologies provide means for directly accessing hardware features. For example, MASCOT 3 supports entities called *servers* and *access interfaces,* and Ada supports *representation specifications* all of which provide a means of programming using machine code and hence bypassing high level layers in the virtual machine.

C. Tool Support

All the methodologies require the support of CASE (computer aided software engineering) tools for their effective use. Without such tools the methods become too laborious for use on large systems and many of their benefits in terms of enforcing consistency are lost. The number and range of CASE tools available is growing rapidly and some simple ones are now available for PCs. Many of the tools support full or partial code generation.

The weaknesses of all the methodologies lie largely in two areas: one is the mapping of the application domain specification into the abstract model or software specification, and the other is in analyzing the implementation model prior to code generation. An approach known as the development framework [11], seeks to address this problem (see Figure 6). The system integrates commercial CASE

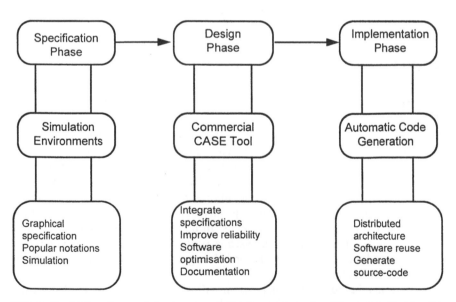

Figure 6 Major phases of the framework development system. (Figure provided by JM Bass.)

tools with simulation and automatic source-code generation tools and the environment supports the three principle developments phases: specification, design and implementation. The design phase supports the integration of continuous-time and discrete-state simulation models [12].

Simulation tools enable the designer to gain understanding of the functional behavior of the proposed system working within the application domain. Simulation of the plant with tentative (or candidate) control system designs using what-if scenarios allows exploration of the design space. Simulation models become more detailed as the design effort progresses. When all aspects of the required functional behavior are modeled the simulation can be seen as a specification of the control system.

The specification is automatically translated into a CASE notation using Framework tools. The CASE notation forms the basis of design optimization and analysis tools. Execution-time analysis using a building block approach is possible, since the CASE notation is a graphical representation of a more formal notation. Additional tools support clustering, replication and mapping of software architectures onto hardware. Automatic source-code generation is possible from the CASE notation. Application code selected from a reusable library of source-code templates is combined with process intercommunication code for each design. Process intercommunication harness software is required to support multiprocessor implementations. The environment supports the integration of continuous simulation models developed in SIMULINK (Trademark: The Mathworks) and discrete-state models developed in Statemate (Trademark: iLogix) see Figure 7. The CASE notation can be displayed using Software through Pictures (Trademark: IDE).

III. REAL-TIME OPERATING SYSTEMS

A. Introduction

Software design is simplified if details of the lower levels of implementation on a specific computer using a particular language can be hidden from the designer. An operating system for a given computer converts the hardware of the system into a virtual machine with characteristics defined by the operating system. Operating systems were developed, as their name implies, to assist the operator in running a batch processing computer, they subsequently developed to support both real-time systems and multi-access on-line systems. In addition to supporting and controlling the basic activities, operating systems provide various utility programs, for example, loaders, linkers, assembler and debuggers, as well as run-time support for high-level languages.

A general purpose operating system will provide some facilities that are not required in a particular application, and to be forced to include them adds unnecessarily to the system overheads. Usually during the installation of an operating sys-

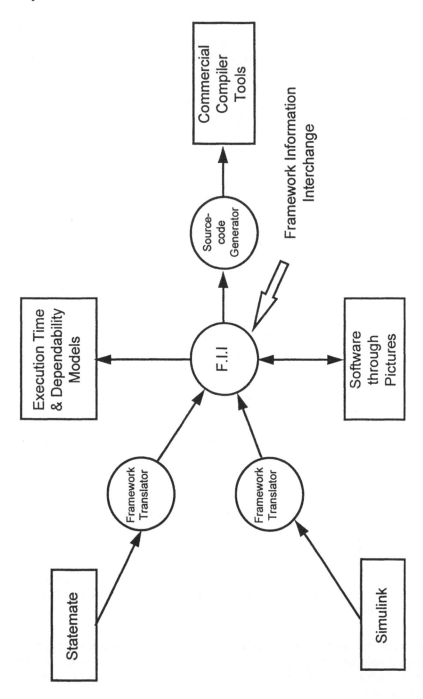

Figure 7 Tool integration for the framework development system. (From JM Bass.)

tem certain features can be selected or omitted; hence the operating system can be "tailored" to meet a specific application requirement.

With the development of the microprocessor came the idea of providing a minimum operating system, typically referred to as a monitor or executive program. This was taken further by Wirth [13,14] when he developed the languages Modula and Modula-2, which provide only a minimum kernel or nucleus; additional features can be added by the applications programmer writing in a high-level language. In this type of operating system the distinction between the operating system and the application software becomes blurred. The approach has many advantages for applications that involve small, embedded systems.

In summary a real-time multitasking operating system (RTOS) has to support the resource sharing and the timing requirements of the tasks and the functions can be divided as follows:

- *Task management*: the allocation of memory and processor time (scheduling) to tasks.
- *Memory management*: control of memory allocation.
- *Resource control*: control of all shared resources other than memory and CPU time.
- *Inter-task communication and synchronization*: provision of support mechanisms to provide safe communication between tasks and to enable tasks to synchronize their activities.

In addition to the above the system has to provide standard features such as support for disc files, basic input/output device drivers and utility programs.

The overall control of the system is provided by the *task management* module which is responsible for allocating the use of the CPU. This module is often referred to as the *monitor* or as the *executive control program* (or more simply the executive).

B. Task Management

The basic functions of the task management module or executive are:

1. To keep a record of the state of each task
2. To schedule an allocation of CPU time to each task
3. To perform the context switch, that is to save the status of the task that is currently using the CPU and restore the status of the task that is being allocated CPU time

In most real-time operating systems the executive dealing with task management functions is split into two parts: a scheduler that determines which task is to run next and which keeps a record of the state of the tasks, and a dispatcher that performs the context switch.

Task States

With one processor only one task can be running at any given time and hence the other tasks must be in some other state. The number of other states, the names given to the states, and the transition paths between the different states varies from operating system to operating system. A typical state diagram is given in Figure 8 and the various states are described below (names in parentheses are commonly used alternatives).

- RUNNING (ACTIVE) this is the task which has control of the CPU. It will normally be the highest priority of those ready to run.
- READY (RUNNABLE, ON) there may be several tasks in this state. The attributes of the task and the resources required to run the task must be available for the task to be placed in the READY state.
- SUSPENDED (WAITING, LOCKED OUT, DELAYED) the execution of tasks placed in this state has been suspended because the task requires some resource which is not available or because the task is waiting for some signal from the plant, for example, input from the analog-to-digital-converter, or the task is waiting for the elapse of time.
- EXISTENT (DORMANT, OFF) the operating system is aware of the existence of this task; however, the task has not been allocated a priority and has not been made runnable.
- NONEXISTENT (TERMINATED) the operating system has not as yet been made aware of the existence of this task, although it may be resident in the memory of the computer.

The status of the various tasks may be changed by actions within the operating system—a resource becoming available or unavailable—or by commands from the application tasks. A typical command is START (ID) which transfers a task from EXISTENT to READY state, where ID is the name by which the task is known to the operating system. It should be noted that the transition from READY to RUNNING can only be made at the behest of the dispatcher.

Task Descriptor

Information about the status of each task is held in a block of memory by the RTOS. This block is referred to by various names: *task descriptor* (TD), *process descriptor* (PD), *task control block* (TCB) or *task data block* (TDB). The information held in the TD will vary from system to system, but will typically consist of the following:

- Task identification (ID)
- Task priority (P)
- Current state of task

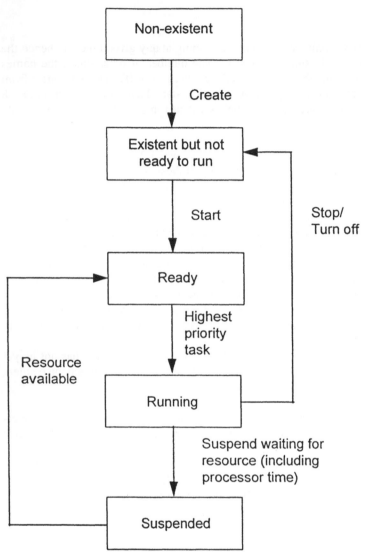

Figure 8 State diagram for task management.

- Area to store volatile environment (or a pointer to an area for storing the volatile environment)
- Pointer to next task in a list

The reason for including the last item in the list above is that the task descriptors are usually held in a linked list structure. The executive keeps a set of lists, one for each task state as shown in Figure 9. There is one active task (task ID = 10) and three tasks that are ready to run IDs = 20, 9, and 6. The entry held in the executive for the ready queue head points to task 20, which in turn points to task 9 and so on.

The advantage of the list structure is that the actual task descriptor can be located anywhere in the memory and hence the operating system is not restricted to a fixed number of tasks as was often the case in the older operating systems that used fixed length tables to hold task state information. With the list structure moving tasks between lists, re-ordering the lists, creating and deleting tasks can all be achieved simply by changing pointers. There is no need to copy or move the task descriptors themselves.

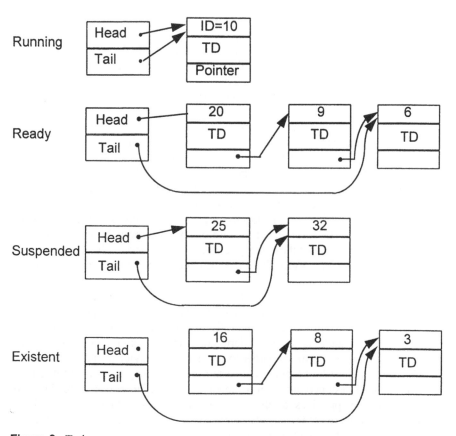

Figure 9 Task queues.

C. Task Scheduling

There are two main approaches to the allocation of time to a task when running
several tasks on a single CPU. One approach, referred to as cyclic scheduling,
allocates the CPU to a task in turn. The task uses the CPU for as long as it
wishes. When it no longer requires it the scheduler allocates it to the next task in
the list. This is a very simple strategy which is highly efficient in that it minimizes the time lost in switching between tasks. It is an effective strategy for
small embedded systems for which the execution times for each task run are
carefully calculated (often by counting the number of machine instruction cycles
for the task) and for which the software is carefully divided into appropriate task
segments. In general this approach is too restrictive since it requires that the task
units have similar execution times. It is also difficult to deal with random events
using this method.

The other main approach is to allow one task to preempt another, that is to
force a reallocation of the CPU from the currently running task to another task.
There are many preemptive strategies but all involve the possibility that a task will
be interrupted, hence the term preemptive, before it has completed a particular
invocation. A consequence of this is that the executive has to make provision to
save the volatile environment for each task, since at some later time it will be allocated CPU time and will want to continue from the exact point at which it was
interrupted. This process is called *context switching* and a mechanism for supporting it is described below.

The simplest form of pre-emptive scheduling is to use a time slicing approach (sometimes called a round robin method). Using this strategy each task is
allocated a fixed amount of CPU time—a specified number of *ticks* of the clock—
and at the end of this time it is stopped and the next task in the list is run. Thus
each task in turn is allocated an equal share of the CPU time. If a task completes
before the end of its time slice the next task in the list is run immediately.

The majority of existing RTOS use a priority scheduling mechanism. Tasks
are allocated a priority level and at the end of a predetermined time slice the task
with the highest priority of those ready to run is chosen and is given control of the
CPU. Note that this may mean that the task which is currently running continues
to run.

Task priorities may be fixed, a *static* priority system, or may be changed
during system execution, a *dynamic* priority system. Dynamic priority schemes
can increase the flexibility of the system, for example, they can be used to increase
the priority of particular tasks under alarm conditions. Changing priorities is, however, risky as it makes it much harder to predict the behavior of the system and to
test it. There is the risk of locking out certain tasks for long periods of time. If the
software is well designed and there is adequate computing power there should be
no need to change priorities; all the necessary constraints will be met. If it is badly
designed and/or there are inadequate computing resources then dynamic allocation
of priorities will not produce a viable, reliable system.

Whatever scheduling strategy is adopted the task management system has to deal with the handling of interrupts. These may be hardware interrupts caused by external events, or software interrupts generated by a running task. An interrupt forces a context switch. The running task is suspended and an interrupt handler is run. The interrupt handler should only contain a small amount of code and should execute very quickly. When the handler terminates, either the task that was interrupted is restored *or* the scheduler is entered and it determines which task should run. The RTOS designer has to decide which approach to adopt.

D. Priority Structures

In a real-time system the designer has to assign priorities to the tasks in the system. The priority will depend on how quickly a task will have to respond to a particular event. An event may be some activity of the process or may be the elapsing of a specified amount of time. Most RTOS provide facilities such that tasks can be separated into three broad levels of priority as follows:

1. *Interrupt level*: At this level are the service routines for the tasks and devices that require very fast response—measured in milliseconds. One of these tasks will be the real-time clock task and clock level dispatcher.
2. *Clock level*: at this level are the tasks that require repetitive processing, such as the sampling and control tasks, and tasks which require accurate timing. The lowest priority task at this level is the base level scheduler.
3. *Base level*: tasks at this level are of low priority and either have no deadlines to meet or are allowed a wide margin of error in their timing. Tasks at this level may be allocated priorities or may all run at a single priority level—that of the base level scheduler.

Interrupt Level

As we have already seen an interrupt forces a rescheduling of the work of the CPU and the system has no control over the timing of rescheduling. Because an interrupt generated re-scheduling is outside the control of the system it is necessary to keep the amount of processing to be done by the interrupt handling routine to a minimum. Usually the interrupt handling routine does sufficient processing to preserve the necessary information and to pass this information to a further handling routine which operates at a lower priority level, either clock level or base level. Interrupt handling routines have to provide a mechanism for task swapping, that is they have to save the volatile environment. On completion the routine will either simply restore the volatile environment and hence will return to the interrupted task, or it may exit to the scheduler.

Within the interrupt level of tasks there will be different priorities and there will have to be provision for preventing interrupts of lower priority interrupting higher priority interrupt tasks.

Clock Level

One interrupt level task will be the real-time clock handling routine that will be entered at some interval, usually determined by the required activation rate for the most frequently required task. Typical values are 1 to 200 ms. Each clock interrupt is known as a *tick* and represents the smallest time interval known to the system. The function of the clock interrupt handling routine is to update the time-of-day clock in the system and to transfer control to the dispatcher. The scheduler selects which task is to run at a particular clock tick. Clock level tasks divide into two categories:

- CYCLIC: tasks which require accurate synchronization with the outside world.
- DELAY: these tasks simply wish to have a fixed delay between successive repetitions or to delay their activities for a given period of time.

The CYCLIC tasks are ordered in a priority which reflects the accuracy of timing required for the task, those which require high accuracy being given the highest priority. Tasks of lower priority within the clock level will have some jitter since they will have to await completion of the higher level tasks.

DELAY tasks that wish to delay their activities for a fixed period of time, either to allow some external event to complete (for example, a relay may take 20 ms to close) or because they only need to run at certain intervals (for example, to update the operator display) usually run at the base level. When a task requests a delay its status is changed from runnable to suspended and remains suspended until the delay period has elapsed.

One method of implementing the delay function is to use a queue of task descriptors, say identified by the name *DELAYED*. This queue is an ordered list of task descriptors: the task at the front of the queue being that whose next running time is nearest to the current time. When a task delays itself it calls an executive task which calculates the time when the task is next due to run and inserts the task descriptor in the appropriate place in the queue.

A task running at the clock level checks the first task in the *DELAYED* queue to see if it is time for that task to run. If the task is due to run it is removed from the *DELAYED* queue and placed in the runnable queue. The task which checks the *DELAYED* queue may be either the dispatcher which is entered every time the real-time clock interrupts or it may be another clock level task which runs at less frequent intervals, say every 10 ticks, in which case it is then frequently part of the base level scheduler. Many real-time operating systems do not support the cycle operation and the user has to create an accurate repetitive timing for the task by using the delay function.

Base Level

The tasks at the base level are initiated on demand rather that at some predetermined time interval. The demand may be user input from a terminal, some process event or some particular requirement of the data being processed. The way in which the tasks at the base level are scheduled can vary. One simple way is to use time slicing on a round-robin basis. In this method each task in the runnable queue is selected in turn and allowed to run until either it suspends or the base level scheduler is again entered. For real-time work in which there is usually some element of priority this is not a particularly satisfactory solution. It would not be sensible to hold up a task, which had been delayed waiting for a relay to close but was now ready to run, in order to let the logging task run.

Priority Strategies

Most real-time systems use a priority strategy even for the base level tasks. This may be either a fixed level of priority or a variable level. The difficulty with a fixed level of priority is in determining the correct priorities for satisfactory operation. The ability to change priorities dynamically allows the system to adapt to particular circumstances. Dynamic allocation of priorities can be carried out using a high level scheduler or can be done on an *ad hoc* basis from within specific tasks. The high-level scheduler is an operating system task which is able to examine the use of the system resources. It may for example check how long tasks have been waiting and increase the priority of the tasks which have been waiting a long time. The difficulty with the high-level scheduler is that the algorithms used can become complicated and hence the overhead in running can become significant.

Alternatively priorities can be adjusted in response to particular events or under the control of the operator. For example, alarm tasks will usually have a high priority and during an alarm condition tasks such as the log of plant data may be delayed with the consequence that the output of the log lags behind real-time (note the data will be stored in buffer areas inside the computer). So that the log can catch up with real-time quickly it may be advisable to increase, temporarily, the priority of the printer output task.

Scheduler and Real-Time Clock Interrupt Handler

The real-time clock handler and the scheduler for the clock level tasks must be carefully designed as they run at frequent intervals. Particular attention has to be paid to the method of selecting the tasks to be run at each clock interval. If a check of all tasks were to be carried out then the overheads involved could become significant. The dispatcher/scheduler has two entry conditions:

1. The real-time clock interrupt and any interrupt that signals the completion of an input/output request
2. A task suspension due to a task delaying, completing, or requesting an input/output transfer

In response to the first condition the scheduler searches for work starting with the highest priority task and checking each task in priority order. Thus if tasks with a high repetition rate are given a high priority they will be treated as if they were clock level tasks, i.e., they will be run first during each system clock period. In response to the second condition a search for work is started at the task with the next lowest priority to the task that has just been running. There cannot be another higher priority task ready to run since a higher priority task becoming ready always preempts a lower priority running task.

E. Resource Management

In real-time systems tasks are designed to fulfill a common purpose and hence they need to communicate with each other. However, they may also be in competition for resources of the computer system and this competition must be regulated. The basic problems which arise—mutual exclusion, synchronization and data transfer—were considered earlier, as were some of the solutions based on using semaphores and signals.

Although the binary and general semaphores provide a simple and effective means of enforcing mutual access they have one weakness: in use they are scattered around the code. Each task that requires access to a particular resource has to know the details of the semaphore used to protect that resource and to use it. The onus on correct use is placed on the designer and implementer of each task. In a small system this causes few problems but in larger more complex systems where the tasks involved may be divided among several people, use of the semaphore becomes more difficult and the probability of introducing errors increases.

An alternative solution which associates the control of mutual exclusion with the resource rather than with the user task is the monitor. A monitor is a set of procedures that provide access to data or to a device. The procedures are encapsulated inside a module that has the special property that only one task at a time can be actively executing a monitor procedure. It can be thought of as providing a fence around critical data. The operations that can be performed on the data are moved inside the fence as well as the data itself. The user task thus communicates with the monitor rather than directly with the resource.

The advantage of a monitor over the use of semaphores or other mechanisms to enforce mutual exclusion is that the exclusion is implicit; the only action required by the programmer of the task requiring to use the resource is to invoke the entry to the monitor. If the monitor is correctly coded then an applications program cannot use a resource protected by a monitor incorrectly.

IV. REAL TIME PROGRAMMING

A. Introduction

Embedded real-time systems can be implemented using a wide variety of languages, from standard sequential languages running in a multitasking environment to special purpose real-time languages such as Ada, CORAL, or PEARL. For effective programming the language or the supporting operating system needs to provide additional features to those commonly provided for standard programming, namely the ability to:

- Create and manage tasks, including timing and synchronization
- Handle interrupts
- Directly access external hardware ports
- Transfer data between tasks
- Support mutual exclusion
- Handle exceptions, including trapping of application software errors that would otherwise lead to the halting of program execution

Few languages or language plus operating system combinations provide good support for all the requirements listed above and deciding on the best, or even a suitable language, for a given application is not a simple job. An ideal language would combine the following characteristics:

- Security
- Readability
- Flexibility
- Simplicity
- Portability
- Efficiency

Unfortunately many of the above characteristics are not compatible with each other. For example, assembly languages provide flexibility and efficiency but they do not provide security or portability and are not easy to read. Languages such as Ada and Modula-2 emphasize security and readability but are not simple. For most real-time applications the choice seems to be between using C (or C++) and obtaining flexibility and efficiency at the expense of security, or using Ada for major aerospace and defense related projects. Languages such as Modula-2 which seek to provide a good balance of characteristics find little favor. For many, simple small scale systems such as FORTH and the real-time BASICS can be good choices, particularly if small numbers are required.

Additional problems arise since many embedded systems run on simple processors with either none, or limited, software support and hence software development has to be carried out on a host machine and the resulting code downloaded to the target machine. The language implementation therefore must support partitioning of the loadable program into a code segment that can be loaded into ROM and a data segment that can be loaded into RAM.

B. Assemblers, Compilers, Linkers, and Loaders [15]

Assemblers and Compilers

Assemblers and compilers both *translate* statements in the programming language, referred to as a source-code statement, into some other format that eventually leads to a binary code that the processor can execute. The simplest form of assembler converts a mnemonic name to the appropriate binary code, for example, the statement LD A,10 to a bit pattern 01011010. The basic assembly statement pattern is

operation code	destination operand	source operand
LD	A	10

Most assemblers supplement the above by allowing the use of symbolic names for addresses, labels, and constants, with, in addition, means of allocating storage areas and support for modularization of the code into segments with control of scope and visibility of symbols. Some other assembler features are

- *Absolute addressing*: Assemblers of this determine *all* addresses at the time of assembly and the object-code produced can be directly loaded into memory and run. This type of assembler is suitable for small systems but poses difficulties if there is a need to divide the source-code into separate modules.
- *Relocatable code*: The object-code produced does not provide the final addresses to be used in the binary code, the calculation of addresses and the resolution of symbols is done at a later stage known as *linking* and possible further address modifications are done by the *loader*.
- *Macros*: Some assemblers provide support for macros. These are code units with a similar function to procedures in that they are used several times at different points in the program The difference is that in using a macro the code that the macro represents is inserted in line each time the macro command is inserted in the source-code. Thus the code for a procedure appears once in the object-code, but the code for a macro appears several times at different places in the object-code. The macro provides increased execution speed at the expense of additional memory.
- *Conditional assembly*: This allows specific parts of the source-code to be included or excluded by using directives. This facility is useful in being able to include additional code segments during testing and exclude them after testing without having to re-edit the source-code file. It is also useful if several different versions of the code are needed, say to suit different customers. A simple customer identifier can be changed and the source-code reassembled to produce the correct object-code for the customer.

Assemblers are low-level translators which typically convert one source-code statement to one object-code statement. A compiler is a much higher level translator which supports statements with greater semantic content such that one source-code statement results in many object-code statements. Compilers produce object-code in a wide range of formats: from machine code, assembly code, and in some cases another high-level language. If not in machine code then a further stage of translation is required. In addition to providing statements of greater semantic content than assembler statements, compilers also support a more extensive syntax (grammar), which helps to make the source-code more easily read than assembly code and also enables the compiler to do more extensive checking of the source-code for errors.

Segmentation Support

Support for subdividing a program into several segments are important for high quality professional work. Compilers and any assemblers provide mechanisms that enable parts of the source-code to be compiled or assembled separately. These mechanisms have to provide a means of storing and retaining entities (variables, constants, procedures, symbols) that are referenced in more than one segment. The three main approaches are:

1. *Full entity retention*: All entities from the previously translated segments are retained and are made available for use when the next segment is translated. This is a simple solution but is cumbersome in use. It is not recommended for large programs. One problem that arises is in maintaining distinct entities without a clash of names. This approach is now rarely used.
2. *Global entity retention*: Entities can be marked as global and such entities are retained and made available to subsequently translated segments. All other entities are local and are available only for the segment being translated. This idea is simple in concept but there are practical difficulties in implementation: for example, how are name conflicts for global entities handled? Does local reuse of a name hide the global entity? It is normal practice for one person in the design/coding team to control the allocation of names for global entities and to keep track of the names used.
3. *Public/private entity lists*: Some assemblers and compilers support the explicit declaration of scope of an entity. At the beginning of each segment two declarations are made, the first listing any entities defined in the current segment that to be available to other segments, and the second listing the names of entities that are not defined in the current segment and hence have to be imported.

Linkers and Loaders

To create a loadable object module the segments have to be combined and the program that does this is called a linker (see Figure 10). The linker has to take

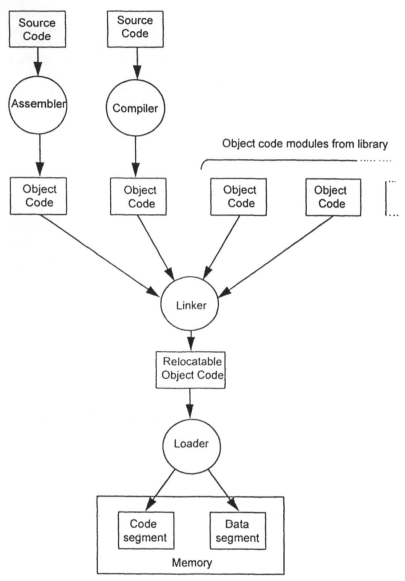

Figure 10 Processes involved in generating loadable code from source code.

segments of object-code, which may have been produced by different assemblers and compilers, and resolve all the segment linkages, calculate addresses for the links between segments and produce as output code which is loadable into memory. Linkers may have to search through code libraries several times since each segment added may require additional entities that are located in other object segments.

Linkers may produce as output object-code which is directly loadable, that is code which has been bound to specific memory locations, or it may preclude binary relocatable code. In the majority of cases binary relocatable code is produced since this allows the final decision on the actual location of the code in memory to be made by the loader. The loader program then calculates the absolute addresses for the code and inserts any modifications to the code needed to permit it to run at the specified memory location.

C. Debugging [16]

Software

Having obtained loadable code it is necessary to debug it. If we can assume that the hardware is working correctly then the software debug can be done by:

- Running in the target machine
- Running in simulation mode on a host machine
- Running on the target machine with debugging tools running on the host machine

If the target machine is the computer on which the software has been developed then it is likely to have, as part of the language support, extensive debugging tools, it is much more difficult if the target machine has limited facilities and say the code has been put into an EPROM. One approach is to add to the target machine a simple monitor program. This may be vendor supplied or developed for the application.

Monitor

The monitor will normally provide communications links with say a PC and will enable software running on the PC to

- Examine the contents of memory locations on the target machine
- Display the contents of the registers
- Change the contents of memory locations and registers
- Single step the program
- Execute the program (including running with breakpoints inserted)

Monitors are limited in what they can provide and a major difficulty in their use is that the absolute addresses of program entities are needed. Modern debugging systems provide the facilities to run the software in simulation mode and on the target machine with a link to the host machine. They provide the following features:

- *Symbolic level debugging*: The debugger maintains a list of entities and their absolute addresses. The user communicates using the names of the entities and information is displayed in terms of the names. This can be for either assembler level or high level source-code. Changing entity values is thus simple since it is done using source-code names.
- *Trace facilities*: A wide variety of trace facilities can be used, statement trace, procedure call trace, register trace, trace on a specific variable, etc.
- *Breakpoints*: A wide variety of breakpoints can be set including conditional breakpoints including, for example, an attempt to write to a specified range of addresses, or when the value of a particular variable is changed.

Software and Hardware

It is frequently the case that the cause of problems that arise cannot easily be assigned to the hardware or to the software. They may be caused by the interaction of the two. Logic state analyzers can be used to capture the status of signals on bus lines and this, combined with single step operation, can often provide detailed information that can assist in diagnosing the cause of an error. A more powerful tool is the in circuit emulator (ICE) which is inserted into the system in place of the actual microprocessor. Using an ICE enables the host computer to exercise much greater control over the operation of the software but the rest of the system is unaware that it is being controlled by an ICE and not the actual microprocessor.

V. SOFTWARE PROGRAMMING EXAMPLES

A. Basic PID algorithm

In this section the programming of a PID control algorithm is used to illustrate some of the practical implementation issues. The programming language used is Modula 2 and anyone familiar with a high level language should be able to follow the examples.

The basic PID algorithm can be expressed as

$$m(n) = K_p e(n) + K_i s(n) + K_d[e(n) - e(n-1)]$$
$$s(n) = s(n-1) + e(n)$$
where
$$K_i = K_p(T_s/T_i)$$
$$K_d = K_p(T_d/T_s)$$

and K_p = proportional gain; T_s = sampling interval; T_i = integral action time; and T_d derivative action time. It is simple to program this algorithm using a high level language as is shown below:

```
VAR  sn, mn, en, enOld, Kp, Ki, Kd : REAL;
BEGIN
  sn := sn + en;
  mn := Kp*en + Ki*sn + Kd*(en - enOld);
  enOld := en;
END.
```

If we assume a system as shown in Figure 11 then a simple program to implement PID control is as given below:

```
MODULE PID_1;
FROM MyLibrary IMPORT
  InputFromProcess,  (* Function which returns measured value, yn, of the system
    as a real value *)
  OutputToActuator; (* Procedure which sends manipulated variable, mn, to the
    actuator *)
FROM IO IMPORT
  KeyPressed ; (* Returns value TRUE if a key has been pressed *)
(* Procedures and functions listed above are assumed to be available in a library
    and are imported into the program *)
VAR  en, enOld, Kd, Ki, Kp, mn, sn, sp, yn : REAL;
PROCEDURE PIDControl;
BEGIN
  sn := sn + en;
  mn := Kp*en + Ki*sn + Kd*(en - enOld);
  enOld := en;
END PIDControl;
PROCEDURE Control;
  sn: = 0.0;
  enOld := sp - InputFromProcess();
  REPEAT
    yn := InputFromProcess():
    en := sp -yn;
    PIDController(en, mn);
    OutputToActuator(mn);
  UNTIL KeyPressed();
END Control;
BEGIN (* Main *)
(* Set initial conditions *)
  Kp := 1.0;
  Ki := 1.0;
  Kd := 1.0;
  sp := 1.0;
  Control;
END PID_1.
```

The above program ignores several practical problems that can be summarized as follows:

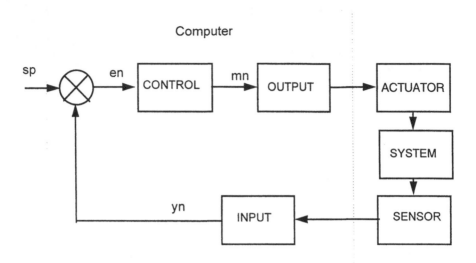

Figure 11 Control system for programming example.

- Timing: The values of Ki and Kd are based on the assumption that the system runs a particular sampling interval Ts. The above program samples yn and calculates an output value mn at intervals determined solely by the execution speed of the code.
- Scaling of the calculations: The sensor converts some physical quantity in the system to, say, a voltage which is then converted by the ADC to a number, for example, a temperature in the system with a range of 50°c to 200 °c may be converted to 0 to 10 volts which is then converted by the ADC to a number in the range 0 to 4095.
- Actuator limiting: The DAC and the actuator which it drives will have physical limits on the values that they can accept; hence precautions must be taken to avoid attempts to drive the system beyond normal limits.
- Integral action wind-up: If there is an error which the system cannot correct because the input to the system is already at its maximum value, for example, a temporary abnormal load, then the integral action term sn will continue to increase. Should the abnormal load continue for some time then the integral term can grow large and this results in a slow response from the system.
- Bumpless transfer between manual and automatic operation.

B. Timing of the Control Loop

The PID controller is normally designed to run at a specific sampling rate and hence the general form of the control program is

```
PROCEDURE Control;
(* declarations *)
BEGIN
  (* Set initial conditions *)
    LOOP
      (* timing synchronization *)
      (* read input data *)
      (* control calculation *)
      (* output control value to system *)
      (* check for exit condition *)
    END (* loop *)
  (* closing down housekeeping *)
END Control;
```

For accurate control and the avoidance of jitter on the timing of the controller output the following structure is often adopted.

```
PROCEDURE Control;
(* declarations *)
BEGIN
  (* Set initial conditions *)
    LOOP
      (* timing synchronization *)
      (* output mn(k-1) value to system *)
      (* read input data for sample k, i.e. yn(k)*)
      (* control calculation for mn(k) *)
      (* check for exit condition *)
    END (* loop *)
  (* closing down housekeeping *)
END (* Control *);
```

As is shown in Figure 12 this second form assumes that the execution times of the routines used to output the manipulated variable *mn* (the *OutputToActuator* routine) and to obtain the system output *yn* (the *InputFromProcess* routine) remain constant from sample to sample; thus the times at which these operations occur are closely synchronized with the timing signal. Any variation in the execution time of the control calculation will not affect the timing of these actions (providing the calculation is complete before the next sample time). It should be noted that with this arrangement the value mn output at sample interval k is computed using the value of *yn* obtained at the sample interval *k-1*. Hence there is a delay of one sample interval in the control computation. This is normally compensated for by including this delay in the design of the control algorithm or in the values chosen for the integral and derivative action times chosen for the PID controller.

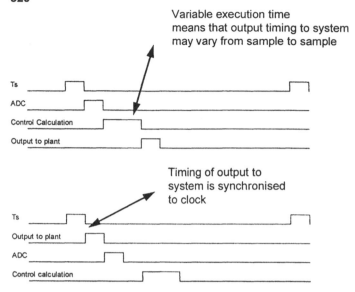

Figure 12   Operation timing showing potential for "jitter."

The synchronization can be obtained in several ways and these are outlined below.

Timing Using Software Loops

The simplest form of timing is to use software-based delay loops. These are created by forming a loop that runs for a given number of times before terminating. For example

```
PROCEDURE SoftDelay(count : INTEGER);
  VAR delayCount : INTEGER;
BEGIN
  delayCount := count;
  WHILE delayCount>0 DO
    delayCount := delayCount - 1;
  END (* while *);
END SoftDelay;
```

This method has the advantage of minimizing the amount of hardware required. However, it requires careful analysis of the program code to establish the execution times of the different paths' code. If the times for the different paths vary significantly then the paths may require balancing through the insertion of "ballast" code.

Timing Based on Polling

If an external timing device is available which sets a logic signal TRUE when it is time for the program to run then a simple way of controlling the program timing is for the program to use a routine which repeatedly samples this line. This is referred to as polling. A simple routine is shown below. The external timer would normally be based on a programmable timing device and this would have to be set to the correct interval when the system was initialized.

```
PROCEDURE Polling;
  VAR clockBit [address]: BOOLEAN; (* this assumes that the variable clockBit can
    be mapped to an address in the input memory space *)
BEGIN
  WHILE NOT(clockBit) DO
    NOP; (* null operation *)
  END (* while *);
  clockBit := FALSE;
END;
```

A major problem with the above methods is that the processor cannot be doing anything else while it is waiting for the next sample time at which the controller has to run. The following two approaches provide a means by which other operations can be carried out during the controller waiting time.

Timing Using a Hardware Interrupt

As with the polling method above, an external timing device can be used but instead of the program using a routine which samples the status of the timer, the output of the timer is connected to a hardware interrupt line. As illustrated in Figure 13, on receiving an interrupt, control is transferred from the running program to the PID controller that now runs. On completion it returns control to the point in the background program at which it was interrupted.

The details of implementing a program as an interrupt routine vary according to the computer system and the programming language being used. The following illustrates how such a system can be constructed to run on a DOS based PC, assuming that the PC timer calls interrupt 1CH at the DOS clock interval (approximately 18.2 Hz). This interrupt is set to do nothing but simply return from interrupt. By replacing the return from interrupt instruction by another instruction the user can force the execution of a user defined interrupt routine. Note: the code provided by the user must not result in a call to any DOS routine since DOS is not re-entrant.

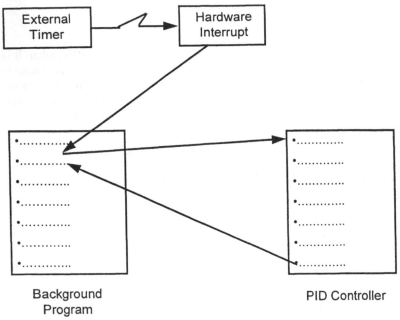

Figure 13 Example showing transfer of program control on an interrupt.

```
MODULE PID_2;
FROM MyLibrary IMPORT
  InputFromProcess, (* Function which returns measured value, yn, of the system
    as a real value *)
  OutputToActuator; (* Procedure which sends manipulated variable, mn, to the
    actuator *)
FROM IO IMPORT (* IO is a language support libarary of input and output proce-
    dures *)
  KeyPressed ; (* Returns value TRUE if a key has been pressed *)
FROM SYSTEM IMPORT (* SYSTEM is a language support library *)
NEWPROCESS, IOTRANSFER, TRANSFER; (* For an explanation of these see
    note below *)
CONST
  BIOSClockHook = 1CH;
  WorkSpaceSize = 5000;
VAR
  WorkSpace: ARRAY[1.. WorkSpaceSize] OF BYTE;
  ISR, Main: ADDRESS;
  ClockTicks, Ts: CARDINAL;
  TimeToStop, PrepareToStop: BOOLEAN;
  en, enOld, Kd, Ki, Kp, mn, sn, sp, yn: REAL;
```

```
PROCEDURE PIDControl;
BEGIN
  sn := sn + en;
  mn := Kp*en + Ki*sn + Kd*(en - enOld);
  enOld := en;
END PIDControl;
PROCEDURE Control;
  yn := InputFromProcess():
  en := sp -yn;
  PIDControl;
  OutputToActuator(mn);
END Control;
PROCEDURE InterruptServiceRoutine;
BEGIN
  LOOP
    IOTRANSFER(ISR, Main, BIOSClockHook);
    INC(ClockTicks)
    IF ClockTicks MOD Ts = 0 THEN
      Control;
    END (* IF*)
    IF ClockTicks > MaxCardinal THEN
      ClockTicks := 0;
    END (* IF *)
    IF PrepareToStop THEN EXIT
  END (* LOOP *)
  TimeToStop := TRUE;
  TRANSFER(ISR, Main);
END InterruptServiceRoutine;
BEGIN (* Main *)
(* Set initial conditions *)
NEWPROCESS(InterruptServiceRoutine, ADR(WorkSpace),
    SIZE(WorkSpaceSize), ISR);
  TRANSFER(Main, ISR);
  LOOP
    (* Do anything *)
    IF KeyPressed() THEN
      PrepareToStop:= TRUE
    END (* IF *)
    IF TimeToStop THEN
      EXIT
    END (* IF *)
  END (* Loop *)
END PID_2.
```

Note:
```
PROCEDURE NEWPROCESS(P: PROC; A: ADDRESS; S: CARDINAL; VAR P1:
    ADDRESS);
```

is a procedure provided by the Modula 2 language support libraries that converts a parameterless procedure into a coprocess. In the above program the procedure P is InterruptServiceRoutine, the parameter A is workspace for the process which has been declared as an arrray of size WorkSpaceSize of bytes; the Modula 2 implicit functions ADR and SIZE are used to provide the address of this space and its size. The parameter P1 returns the reference name by which the process formed by the procedure InterruptServiceRoutine is know to the system.

PROCEDURE TRANSFER(VAR P1, P2: ADDRESS)

The current process which is running is suspended and assigned to the process reference name P1 and then the process with the reference name P2 is run. The first call to TRANSFER in the program above is made in the main segment of the program immediately after the call NEWPROCESS. At this point the main body of the program is given the reference name MAIN and its execution is suspended at this point. Control is transferred to the process with the reference name ISR which is the procedure InterruptServiceRoutine.   At the exit from the InterruptServiceRoutine a call TRANSFER(ISR, Main) is made which stops the execution of InterruptServiceRoutine and starts main running at the point immediately following the call to TRANSFER.

PROCEDURE IOTRANSFER(VAR P1,P2 : ADDRESS; I: CARDINAL)

This procedure associates the current process (i.e the procedure in which is placed) with the interrupt number specified in the parameter I. It then suspends the current process giving it the reference name P1 and transfers control to the process with the reference name P2. When an interrupt associated with process P1 occurs the running process is suspended and is given the process reference name P2 and the process with the reference name P1 is run. In the above example, the initial transfer on setting up IOTRANSFER is made to the main program.

Timing Using a Real-Time Clock

There are many ways in which timing can be controlled using a real-time clock. In the following example it is assumed that in addition to the basic clock there is operating system support providing a function (getTime()) that returns the value of the current time in seconds (as a REAL number).The system also provides a procedure delayUntil(time) enabling a task to delay itself until a specified time.

```
MODULE PID_3;
FROM MyLibrary IMPORT
   InputFromProcess,
   OutputToActuator;
FROM IO IMPORT
   KeyPressed ;
FROM Timing IMPORT
      getTime: Time; (* Returns value of time in seconds *)
```

```
delayUntil; (* Suspends the task until the absolute value of time given in nextTime
    is reached *)
FROM Process IMPORT
  StartProcess; (* StartProcess is a procedure provided by the language support
    systems which combines the operations of declaring a procedure as a process
    and also transferring control to it so that it runs. Process provides a simple, pri-
    ority based scheduling system *)
CONST
  WorkSpaceSize = 5000;
VAR
  currentTime, nextTime, sampleInterval : Time; (* Time is assumed to be derived
    from type REAL and is expressed in seconds *)
  WorkSpace: ARRAY[1.. WorkSpaceSize] OF BYTE;
  en, enOld, Kd, Ki, Kp, mn, sn, sp, yn : REAL;
PROCEDURE PIDControl;
BEGIN
  sn := sn + en;
  mn := Kp*en + Ki*sn + Kd*(en - enOld);
  enOld := en;
END PIDControl;
PROCEDURE Control;
  currentTime:= getTime();
  nextTime := currentTime + sampleInterval;
  yn := InputFromProcess(); (* read input data *)
  enold := sp - yn; (* calculate error *)
  LOOP
    delayUntil (nextTime);
    nextTime := nextTime + sampleInterval;
    OutputToActuator(mn) (* output control value to system *)
    yn := InputFromProcess(); (* read input data *)
    en := sp - yn; (* calculate error *)
    PIDControl (); (* control calculation *)
  END (* loop *)
END Control;
BEGIN (* Main *)
  (* Set initial conditions *)
  StartProcess (Control, workspace, ControlPriority);
  LOOP (* this forms an idle process which runs while Control is delayed until next
    sample is due *)
  IF KeyPressed() THEN EXIT; (* check for exit condition *)
  END (* loop *)
(* closing down housekeeping *)
END PID_3.
```

C. Scaling and Bumpless Transfer

The meaning of the gain value K_p shown in the equation is dependent on the engineering units of the measurement and actuation devices as expressed in the quan-

tities $e(n)$ and $m(n)$. Industrial practice is to express the gain term in non-dimensional form by appropriate scaling of $e(n)$ and $m(n)$. This can be done by expressing $e(n)$ and $m(n)$ as fractional changes of the span of the measurement device and the actuator movement as appropriate. For example, if the measuring device has a range of 100°C to 150 °C then the span is 50 °C and hence a deviation (e(n)) of 10°C is represented by a fractional value of 0.2. The changes to the basic computations needed to express the gains in nondimensional form are illustrated in Figure 14, where ms = measurement instrument span and cs = control actuator span [17].

Figure 14 also shows how the offset to the value is incorporated into the system when the actuator is required to be set to some value other than zero when the error is zero.

Incorporating these changes into the Module PID_3 requires changes as follows:

1. Additional variables such that
   VARcs, en, enOld, Kd, Ki, Kp, M, mn, mna, ms, sn, sp : REAL;
2. Modification of the calculation of en
   en := (sp - yn)/ms;
3. Modification of the calculation of mn
   mn := mna*cs

D. Integral Action Windup

There are several methods for preventing or compensating for integral action windup, a simple method is to limit the value of the integral action term sn to some maximum value (and minimum value). With this scheme the computation $sn := sn + en$ is still carried out but if the result exceeds the maximum (minimum) value sn is reset to the maximum (minimum) value. This is illustrated below:

```
BEGIN (* PIDControl *)
  sn := sn + en;
  IF sn > smax THEN
     sn := smax
   ELSE IF sn < smin THEN
     sn := smin
  END; (* IF ELSE sn > smax *)
  mn := M + Kp*en + Ki*sn + Kd*(en - enOld);
  enOld := en;
END; (* end PIDControl *)
```

E. PID Controller Example

Putting together the various code fragments given above gives a simple controller module based on using a real-time clock as shown below:

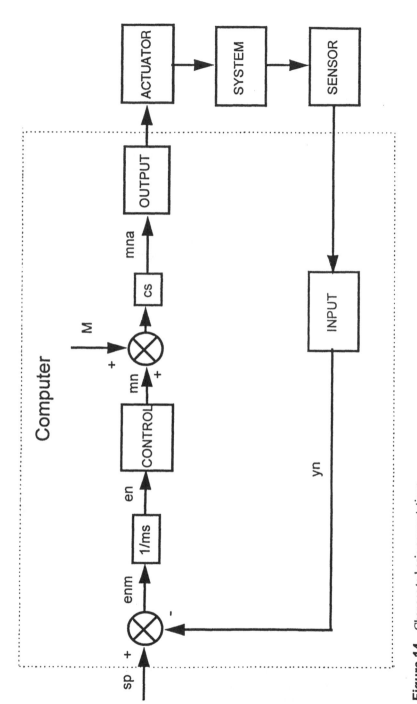

Figure 14 Changes to basic computations.

```
MODULE PID_4;
FROM MyLibrary IMPORT
  InputFromProcess,
  OutputToActuator;
FROM IO IMPORT
  KeyPressed ;
FROM Timing IMPORT
    getTime(): Time; (* Returns value of time in seconds *)
  delayUntil(nextTime); (* Suspends the task until the absolute value of time given
    in nextTime is reached *)
  CONST
  WorkSpaceSize = 5000;
VAR
  cs, en, enOld, Kd, Ki, Kp, M, mn, mna, ms, sn, sp, yn : REAL;
  currentTime, nextTime, sampleInterval : Time; (* Time is assumed to be derived
    from type REAL and is expressed in seconds *)
  WorkSpace: ARRAY[1.. WorkSpaceSize] OF BYTE;
PROCEDURE Initialize;
BEGIN
  Kp := 1.0;
  Ki := 1.0;
  Kd := 1.0;
  sp := 1.0;
  sn: = 0.0;
END Initialize;
PROCEDURE PIDControl;
BEGIN (* PIDControl *)
  sn := sn + en;
  IF sn > smax THEN
    sn := smax
    ELSE IF sn < smin THEN
    sn := smin
  END; (* IF ELSE sn > smax *)
  mn := M + Kp*en + Ki*sn + Kd*(en - enOld);
  enOld := en;
END PIDControl;
PROCEDURE Control;
    currentTime:= getTime();
    nextTime := currentTime + sampleInterval;
    yn := InputFromProcess(); (* read intial system output data *)
    enold := sp - yn; (* calculate initial setting of error *)
    LOOP
      delayUntil (nextTime);
      nextTime := nextTime + sampleInterval;
      OutputToActuator(mn) (* output control value to system *)
      yn := InputFromProcess(); (* read input data *)
      en := sp - yn; (* calculate error *)
      PIDControl; (* control calculation *)
    END (* loop *)
```

```
END Control;
BEGIN (* Main *)
  Initialize;  (* Set initial conditions *)
  StartProcess (Control, WorkSpace, ControlPriority); (*starts the Control process
    running *)
  LOOP (* this forms an idle process which runs while Control is delayed until next
    sample is due *)
  IF KeyPressed() THEN EXIT; (* check for exit condition *)
  END (* loop *)
(* closing down housekeeping *)
END PID_4.
```

REFERENCES

1. S Bennett. *Real-Time Computer Control: an Introduction. 2nd ed.* Englewood Cliffs, NJ: Prentice Hall, 1994.
2. PA Laplante. *Real-Time Systems Design and Analysis: An Engineer's Handbook.* New York: IEE Computer Society Press, 1992.
3. JA Stankovic, K Ramamritham. *Hard Real-Time Systems.* New York: IEEE Computer Society Press, 1988.
4. A Burns, A Wellings. *Real-Time Systems and their Programming Languages. 2nd ed.* Wokingham: Addison-Wesley, 1997.
5. IC Pyle, P Hruschka, M Lissandre, K Jackson. *Real-Time Systems: Investigating Industrial Practice.* Chichester: Wiley, 1993.
6. H Gomaa. *Software Design Methods for Concurrent and Real-Time Systems.* Reading, MA: Addison-Wesley, 1993.
7. PT Ward, SJ Mellor. *Structured Development for Real-Time Systems.* New York: Yourdon Press, 1986.
8. DJ Hatley, IA Pirbhai. *Strategies For Real-Time System Specification.* New York: Dorset House, 1988.
9. MASCOT. The Official Handbook of MASCOT, version 3.1, Computing Division, RSRE Malvern, (1987) available from Defence Research Information Centre. 65 Brown Street, Glasgow, G2 8EX.
10. PJ Robinson. *Hierarchical Object Oriented Design.* Englewood Cliffs, NJ: Prentice Hall, 1993.
11. JM Bass, AR Browne, MS Hajji, DG Marriott, PR Croll, PJ Fleming. Automating the Design of Real-Time Control Software. IEEE Parallel and Distributed Technology 2: 9-19, 1994.
12. MS Hajji, JM Bass, AR Browne, PJ Fleming. Design Tools for Hybrid Control Systems. International Workshop HART '97, Grenoble, France, Oded Maler (ed.), LNCS 1201, Springer March 1997, pp. 87-92.
13. N Wirth. *Programming in Modula-2.* Berlin: Springer-Verlag, 1986.
14. N Wirth. Towards a discipline of real-time programming. Comm. ACM, 22: 577-583, 1977.
15. K Jobes. *Software Tools and Techniques for Electronic Engineers.* London: McGraw-Hill, 1994.

16. JE Cooling. *Software Design for Real-Time Systems*. London: Chapman & Hall, 1991, re-issued London: International Thompson, 1995.
17. J Golten, A Verwer. *Control System Design and Simulation*. London: McGraw-Hill, 1991.

12
Mechatronic System Applications

Makoto Kajitani
University of Electro-Communications, Tokyo, Japan

I. BASIC CONCEPT IN MECHATRONIC DESIGN

A. Introduction

Our lives and the economic activities of society are closely related to the supply and demand of material, energy, and information. The history of humanity has experienced revolutions in the technology of material, energy, and information. The creation of agriculture was the first revolution involving materials. The Industrial Revolution, which was the energy revolution, took place in Europe in the 18th century. The present society, in which the information revolution is in progress, is now becoming an information society.

Needless to say, natural resources and energy are limited to the current amount available on earth. On the other hand, information can be produced by the mental activities of human beings. When people are satisfied with their material lifestyle, human beings seek mental pleasure, such that the importance of information increases relatively in comparison with material and energy. As long as mankind exists, information will be produced indefinitely for satisfying the mind. The value of material and energy depends on the amount of them, but that of information depends on its freshness. In other words, the former will be assessed on the basis of an integral view, the latter on a differential view.

Proceeding to the advanced information society described above, advanced technology to create energy-saving, resource-saving, and high intelligent systems has evolved. The new technology required by advanced information society is mechatronics created by integrating mechanics, electronics, and software.

The most striking features of mechatronic products are :

1. Precision mechanism
2. Software control by a microprocessor

Therefore, in addition to precision mechanical technology, information technology including a very wide range of engineering fields such as electronics, computers, software, control, etc. is required in order to design the mechatronic products.

B. Merging Information into Machines

Until the 1960s, mechanical products and electrical products used to be considered entirely different from industrial and technological standpoints. The dawn of the age of microcomputer opened in 1971 with the introduction of the Intel 4-bit general-purpose microprocessor. However, the energy crisis in October 1973 forced Japan to confront serious energy and material crises and to organize a drastic change of the industrial structure. This difficult problem was overcome by Mechatronics, a revolution in technology that merged information into machines.

Figure 1 shows the historical development of mechanical technology. Mechanization meant to save a lot of time and labor by machines that could give full power or energy. The progress of electronic technology especially semiconductor technology, made it possible for mechanical systems to integrate with electronic systems known as electro-mechanical integration. In addition, the emergence of microcomputers allowed machines to integrate with information by taking computer or information technology into mechanical technology thanks to very small inexpensive computers. Thus, mechanical technology has evolved into mechatronics owing to merging information technology into mechanical technology. It should be emphasized that the integration of information with machines is the essence of mechatronics.

How do the machine and information relate with each other? The information takes an active part on three stages as shown in Figure 2.

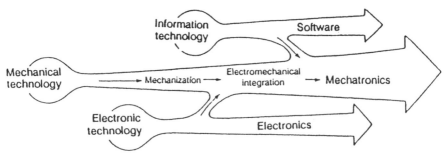

Figure 1   Passage from mechanical technology to mechatronics.

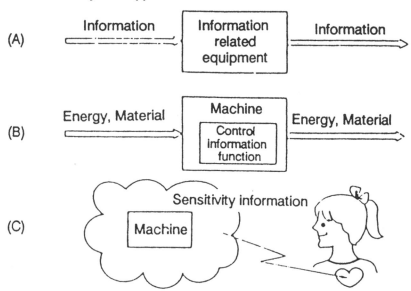

Figure 2 Relation between machine and information.

1. Development of machines to handle information (information related equipment). In recent years, the development and growth of so-called information related equipment such as computers, videocassette recorders, compact discs, laser printers, magnetic disk devices, etc., have been amazing. Moreover, on an as needed basis new various information equipment or systems have appeared one after another, for example mobile phone, digital cameras, digital video discs, intelligent toilets, etc.
2. Advancement of control function of machines. For example, the self-control function of automobiles was poor in the past. However, the introduction of car electronics in recent years has been remarkable. There are many devices and systems such as NC machine tools, robots, air-conditioners, etc. equipped with microcomputers in order to perform energy saving, resource saving, and intelligent functions.
3. High value addition by "sensitivity information."

The commercial value of the passenger car does not depend on only its original purpose, namely the transport function (traveling performance). It rather depends on its appeal to the human senses, for example, style, color, etc. Any machine sends out information to stimulate the five senses (sensitivity information). It is natural that the consumer wants to purchase a machine of beautiful design and good taste. Products which were originally designed to output sensitivity information such as musical instruments, toys, dolls, etc. have become more and more important in the advanced information society.

C. System Concept

Mechatronics itself does not aim at independent proper technology and engineering. It is synthesis technology using not only conventional mechanical technology but also existing engineering technology such as electronics, computer engineering, etc. freely for the required purposes. An example of the system concept for the mechatronic system is described below.

The engineering system consists of four internal functions as shown in Figure 3 from the standpoint of functional configuration. The main function is to perform the purpose of the system, that is, the function of converting the original material, energy, and information into the desired ones. The power function handles energy, lubrication, cooling, etc. required for the operation of the system. Energy sources must be supplied from the outside to the system.

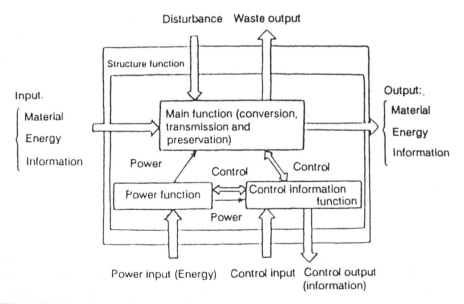

Figure 3 Functional configuration of mechatronic system.

The control/information function controls the entire system. It plays the functions of the brain and five senses of humans. The elements of the system must be arranged and assembled in space from the structural standpoint. In many cases, the structure has a great influence on the functions of the system. It should be noted that any disturbance from the outside may have a deleterious effect on the system. The waste output is often harmful to the external environment and other systems. The above mentioned system configuration can be applied to general engineering systems. Figure 4 summarizes the functions and the elements of the mechatronic system. The comparison of the five elements of a mechatronic system to those of human being brought the following correspondence. The computer corresponds to the brain, the sensors to the five senses, the actuators to the muscle, the mechanism to the skeleton, and the power source to the metabolism. The goal of the mechatronic system is the well-balanced and organic combination of each element shown in Figure 4.

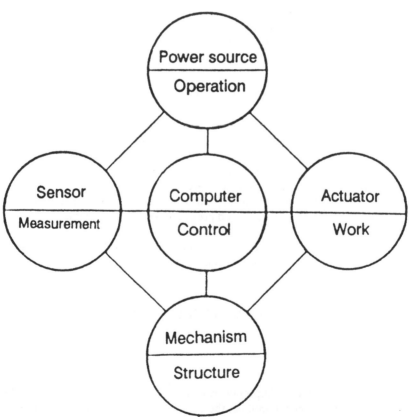

Figure 4 Five basic functions of mechatronic system.

D. Interface Concept

The mechatronic system is composed of various elements and devices such as mechanical elements, electronic devices, programming modules for computer, etc. It is very important to design for interfacing among them, because they have absolutely different physical properties.

In regard to the interface specifications, the following four items must be considered.

1. Mechanical specifications: The interface junction is required to connect two joints mechanically, so mechanical dimensions, shape, accuracy, etc. must be identical at the interface joints.
2. Physical specifications: Physical characteristics must be prescribed for energy, material, and information passing through the interface joints.
3. Information specifications: Logical and information requirements for the interface junction must be consistent with regulations such as specifications, standard, law, language, code, etc.
4. Environmental specifications: Environmental conditions in which interfaces must operate within permissible range of temperature, relative humidity, radioactivity, vibration , dust, etc.

Interface ability is classified into the following four stages according to its level:

1. Zero interface: There is no special function for matching.
2. Passive interface: No energy for matching is required but passive conversion, transmission and preservation are performed.
3. Active interface: Active functions are performed.
4. Intelligent interface: The microprocessors, etc. are mounted and the interface conditions can be modified programmable and adaptable.

Two basic concepts, namely, the system concept and the interface concept are required to merge different technologies under a single organizational design methodology.

II. MECHATRONIC DEVELOPMENT IN GEAR MEASURING TECHNOLOGY

A. Introduction

Gear, as you know, play an important role in mechatronic systems such as the transmission system of an automobile, the driving system of an aircraft, the servo mechanism of robot, the precise mechanism of information equipment and so on.

The requirements for gear accuracy have become increasingly. A guarantee for a high accuracy of gear is dependent on a high precision gear measuring system.

About 30 years ago, gear measuring machines were increasingly required to meet the demand for precision gears by Japanese mechanical industry, principally by the automobile industry. However, the conventional gear tester of those days was based on the mechanical generation method. Subsequently, the conventional method had not been able to satisfy a demand for high accurate measurement, automatic measurement, and quick data processing. The author et al. proposed the new method in order to meet the above requirements. Some systems based on the new idea were successfully produced on a commercial basis. This paragraph has provided a general outline of a rotary encoder developed for use in gear measuring system and has described two gear measuring systems developed by us, that is, the involute tooth profile measuring system with no involute generating mechanism and the automatic inspection system for tooth contact pattern2.

B. Development of Rotary Magnetic Scale [1,2]

Gear measuring systems require highly accurate angle measurements because an important role of gears lies in the function of transferring rotary motion. 30 years ago in Japan, a highly accurate rotary encoder was not easily affordable. So we decided to develop a rotary magnetic scale (or encoder) in order to put our new idea for a gear measuring method into practice.

A magnetic scale using magnetic recording techniques is a measuring system for measuring the longitudinal or angular displacement. The most important requirement for the production of magnetic scales is the ability to record accurately the magnetic pattern of uniform wavelength. The problem to be solved associated closely with both precision mechanical engineering and electronics, even though mechanical engineers and electronics engineers were regarded as different specialists at that time. That was exactly a mechatronic subject.

The traditional analysis and technique of the magnetic recording process cannot be used for magnetic scales, because the accuracy of the absolute position of the magnetization pattern has never been taken into consideration. As a result of theoretical and experimental study on basic magnetic recording characteristics, a special recording method, called pulse-type magnetic recording, was proposed for magnetic scales by the authors. The most important feature of this method lies in its capability to record independently of the velocity of the magnetic medium movement.

A recording method for magnetic scales is schematically shown in Figure 5. Positive and negative pulses are alternately passed through a recording head at equal distance d on the medium, so that the sinusoidal magnetization pattern of wavelength $\lambda = 2d$ is recorded. In order to record accurate sinusoidal magnetization with high amplitude, we have to find the optimum relation between a recording current and a recording distance d. Theoretical and experimental investigations could give the procedure for it.

Recording pulse current

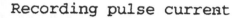

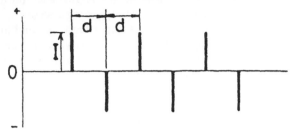

Magnetic pattern caused by
individual isolated pulses

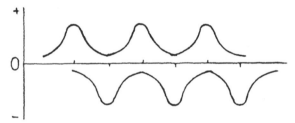

Resultant magnetic pattern

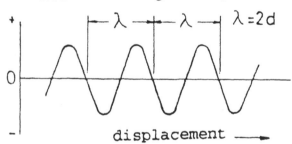

displacement ⟶

Figure 5 Pulse-type recording method for magnetic scales.

C. Tooth Profile Measuring System [3,4]

In 1968, the authors read a paper on a new involute tooth profile measuring system, which was the first success in excluding an involute generating mechanism in the world, at an academic annual meeting sponsored by Japan Society of Precision Engineers. 1968 was the epoch-making year in the history of gear measuring technique in Japan, because information processing based on digital electronic technique had been substituted for mechanical information generating. From that time on, this new technology was applied and developed for not only tooth profile measuring but also transmission error test and so on. Of course, a startling evolution in computer technology (especially microcomputer) fostered the development of gear measuring technique.

The measurement of tooth profile errors is one of the most important requirements for checking gears. Conventional involute measuring machines can be divided into two types, although the principle of operation is essentially the same, that is, a reference involute curve is mechanically generated. The one type requires a different base disc for each gear of different dimensions. In addition, the base disc must be accurately made with a high degree of finish and concentricity. The main disadvantage of this type lies in such tedious and costly preliminaries. In the other type, the required motion can be generated by a complex and high precision linkage mechanism that enables the base radius of the generated curve to be varied by adjustment of the linkage bar positions. It is necessary, therefore, to carry out periodic checks on its performance.

The principle based on the new measuring system lies in the fact that the length of the base circle tangent on an involute does vary in a linear manner with the rotating angle of the base circle, and that departures from linearity of the measured curve will represent involute profile errors. In practice, profile errors can be determined on a point-to-point basis by means of linear displacement measurements, coupled with incremental rotation of the gear about its axis. It should be noted that this method, not depending upon the mechanical generation of involute, is the most essential and straightforward metrology. Figure 6 shows the schematic diagram of the prototype new involute measuring system, and Figure 7 is a photograph of it. This system consists of three important stages as follows: electronic digital scales for linear and rotary measuring, a special-purpose computer and a recorder.

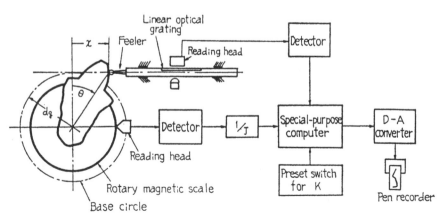

Figure 6 Schematic diagram of the digital involute measuring system.

Figure 7 Photograph of the prototype instrument. 1, Test gear; 2, feeler; 3, rotary magnetic scale; 4, reading head; 5, special-purpose computer; and 6, preset dial.

A feeler is fixed on a slide which can traverse freely in a parallel direction to a tangent to the base circle. During rotation of a gear, the flank of a tooth causes a feeler to traverse, so that a linear moire-fringe grating mounted on the slide will provide electrical pulses due to linear displacement. On the other hand, a rotary magnetic scale, fixed to the test gear, is used to provide a train of pulses, each of which defines a constant angular increment. The rotary magnetic scale had been developed by the authors as previously described. The pulse signals thus obtained are passed to the special-purpose digital computer, which instantaneously calculates departures from the theoretical relationship. The computing circuit (special-purpose computer) was newly developed by using digital integrated circuits that had gradually been available on the market. The most remarkable advantage of this system consisted in presetting on the panel switch with the base circle radii instead of preparing the base disc for a tested gear. It should be noted that this happened before the arrival of both the microcomputer and the personal computer.

As a result of this study, practical usage of this new system was ensured. The fruit of this labor was Osaka Seimitsu Kikai Co. succeeding in producing the first digital electronic involute checker in the world on a commercial basis in 1971. After that, a CNC (Computer-controlled Numerical Control) full automatic gear measuring machine was developed for the first time by the same company in 1976. This new measuring system, which evolved from the above mentioned electronic involute checker, could measure not only involute curve but also lead and pitch.

D. Automatic Visual Inspection of Tooth Contact Pattern of Gears [5]

The tooth bearing conditions in meshing gears has a close relation to their performance, such as service life, vibration, noise and so on. An evaluation of tooth bearing can be obtained by measuring tooth contact patterns generated on the contact surfaces. Contact pattern measurements are currently carried out on the basis of manual visual inspection. But it is remarkable that a skilled operator manually runs the testing machine and identifies whether gears are acceptable or rejected.

Figure 8 is a schematic diagram of the automatic systems for inspecting the tooth contact pattern. This measuring system consists of a TV camera, an image memory unit, microcomputer system, and gear meshing stand. The meshing gears, one of which is coated with a thin film of paint, are turned, so that contact marking is generated as a result of paint transferring from one tooth to its mating tooth. TV camera scans the tooth surfaces of the gears and converts reflected light into video signals. The computer takes the digitized values and extracts the required information from the image. As a result, the system enables inspection of several geometric quantities to be determined with sufficient accuracy to allow quality control.

Experimental tests were carried out on transmission gears for trucks. The graphic display during the inspections is shown in Figure 9. This means we can

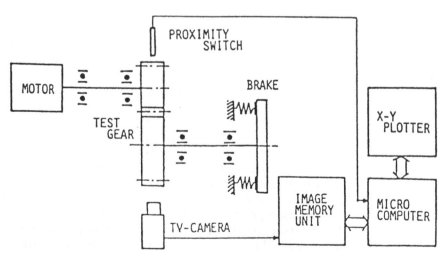

Figure 8   Schematic diagram of automatic inspection system for inspecting the tooth contact pattern of gears.

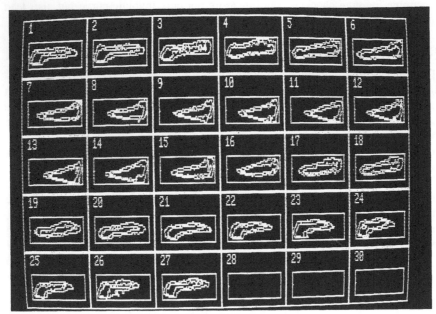

Figure 9 Example of monitor TV display during measurements.

monitor the measured contact pattern in real time. In addition, the system computes geometric features such as contact area ratio, contact pattern width, contact pattern height, centroid deviation of width of contact pattern and centroid deviation of height of contact pattern. After the derivation of these features, each value is classified into required grades.

The contact configuration of tooth contact surfaces in gearing is very complicated, but the following two situations can be considered: (a) two surfaces are in perfect contact, (b) two surfaces are not in contact with slight gap between them. Incidentally, tooth contact marking patterns provide information not only about Situation (a) but also about Situation (b), because the paint used does not have zero thickness. Therefore, the brightness of the tooth contact pattern depends on the thickness of transferred paint that indicates the gap between surfaces. On the other hand, the video output from the camera provides analog signals proportional to the brightness of the contact pattern. If binary conversion is performed, based on different threshold levels, a contact map of a tooth contact pattern can be produced as shown in Figure 10. This contour map pattern corresponds to the size of the gap between the contact surfaces, so we call this pattern a "tooth contact strength pattern."

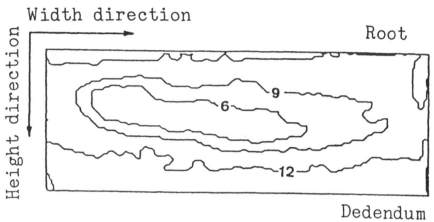

Figure 10   Contour map of tooth contact strength.

III. AUTOMATIC CALIBRATION SYSTEM FOR ANGULAR ENCODER (6)

A.   Introduction

The calibration of rotary encoders has usually been performed manually by comparison to a mechanical reference standard, such as a precise index table or a polygon mirror. It is very difficult to calibrate all graduations of the encoder by means of this conventional method and the calibration work requires a high level of skill and also takes a long time. In the process of developing a rotary magnetic scale described previously in Section II.B, an automatic calibration system with high accuracy and high efficiency for rotary encoders was empirically required.

Our system, together with a new self-checking method, consists of five subsystems: a precise rotary table with an air bearing, a direct-coupling dc servo motor that rotates the table at a constant speed, a rotary magnetic scale as a reference angular standard, time interval counters to detect slight angular differences by a time-conversion method and a microcomputer system that controls the operation of measurement and the data processing. Angular encoders under test are compared with the rotary magnetic scale, which is calibrated by a new self-checking method that we called the "multi-reading-head method."

B.   Principle of the Multi-Reading-Head Method

Generally the angle position error $\varepsilon(\theta)$ of angular encoders can be represented by a Fourier series such as Eq. (1)

$$\varepsilon(\theta) = \sum En\sin(n\theta + \alpha n) \tag{1}$$

where θ is the angular position (0 to 2π), n is a positive integer, En and αn are, respectively, the amplitude and phase angle of the nth Fourier component. By setting the two reading heads on the magnetic scale with an angular separation ϕ as shown in Figure 11, the angular difference $\delta(\theta,\phi)$, which is called the "relative-angle-position-error," between the output signal of the two heads can be expressed as follows:

$$\delta(\theta,\phi) = \varepsilon(\theta) - \varepsilon(\theta + \phi) \tag{2}$$

Substituting Eq. (1) into Eq. (2), gives Eq. (3). It can be seen that the Fourier components included in $\delta(\theta,\phi)$ vary depending upon the relative angle ϕ.

$$\delta(\theta,\phi) = 2\sum En\sin(-n\phi/2)\cos(n\theta + \alpha n + n\phi/2) \tag{3}$$

If we locate the reading head at the relative angle ϕk as shown in Eq. (4) and we define Tk(θ) as Eq. (5),

$$\phi k = \pi/2(k-1)(k = 1,2,3,\ldots,K) \tag{4}$$

$$Tk(\theta) = 1/2(k-1)\sum \delta(\theta + 2l\phi k, \phi k) \tag{5}$$

and by substituting Eq. (3) into Eq. (5), Tk(θ) becomes Eq. (6).

$$Tk(\theta) = \sum En\sin(n\theta + \alpha n) \qquad (m = 1,2,\ldots) \tag{6}$$

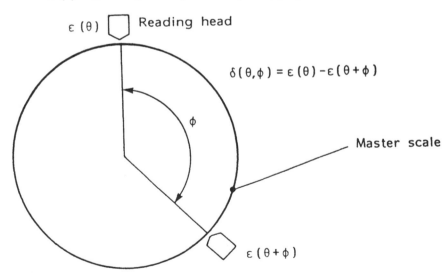

Figure 11   Relation between relative-angular-position-error $\delta(\theta,\phi)$ and relative angle ϕ of two reading heads.

From this it can be seen that $T_k(\theta)$ is the sum total of $n = (2m-1) \, 2(k\sim1)$ Fourier components of angular-positional-error $\varepsilon(\theta)$. The relationship between the relative angle ϕ_k, $T_k(\theta)$ and the number of Fourier components included in $T_k(\theta)$ are summarized in Table 1. If the $(k + I)$ reading heads are placed in positions of $\phi_k(k = 1,2,...,K)$ and $\delta(\theta,\phi)$ corresponding to each k can be measured at a time, $T_k(\theta)$ is obtained by Eq. (5). In accordance with the above principle, $\varepsilon(\theta)$ determined from Eq. (7) would be closely equal to the error $\varepsilon(\theta)$ under the conditions mentioned below. This method is called the "multi-reading-head method."

$$\varepsilon c(\theta) = \sum T_k(\theta) \tag{7}$$

All positive integers are divided into $n = (2m-I) \, 2(K\sim1)$ and $n' = m2K$. $\varepsilon(\theta)$ has n Fourier components of $\varepsilon(\theta)$, but does not have n' components. If K is selected so that n' components can be ignored in practical use, it can be assumed that $\varepsilon(\theta)= \varepsilon c(\theta)$.

C. System Configuration

Figure 12 shows a schematic diagram of the system and Figure 13 shows the actual system. As shown in Figure 14, the rotary magnetic scale, which is the master scale is mounted coaxially on a precision rotary air bearing in which its center movement is less than $0.1\mu m$. Also various angular encoders can be installed coaxially. The direct-coupling dc servo-motor controls the air bearing to rotate at a constant speed within a range of 0.03 to 30 rpm by the feedback of signals from the master scale. The fluctuation of velocity at 5 rpm is approximately 0.2%.

The rotary magnetic scale is superimposed on a brass drum 412.53 mm in diameter. The circumferential surface is coated with magnetic material and an N= 6,480 magnetic sine wave pattern is recorded on it: $1\mu m$ in length on the scale circumference corresponds to 1 in. in angle and 1 wavelength is $200\mu m$. Reading

Table 1 Equation for $T_k(\theta)$ and Fourier Components

k	ϕ_k	Equation $T_k(\theta)$	Fourier Components
1	π	$\dfrac{1}{2}\delta(\theta,\pi)$	$1,3,5,7,...$
2	$\pi/2$	$\dfrac{1}{4}[\delta(\theta,\pi/2)+\delta(\theta+\pi/2)]$	$2,6,10,14,...$
3	$\pi/4$	$\dfrac{1}{8}[\delta(\theta,\pi/4)+\delta(\theta+\pi/2,\pi/4)$ $+\delta(\theta+\pi,\pi/4)+\delta(\theta+3\pi/2,\pi/4)]$	$4,12,20,28,...$
k	$\pi/2^{(k-1)}$	$\dfrac{1}{2^k}\sum_{l=0}^{2^{(k-1)}-1}\delta\left(\theta+2l\dfrac{\pi}{2^{(k-1)}},\dfrac{\pi}{2^{(k-1)}}\right)$	$(2m-1)\,2^{(k-1)}$ $m = 1,2,3,...$

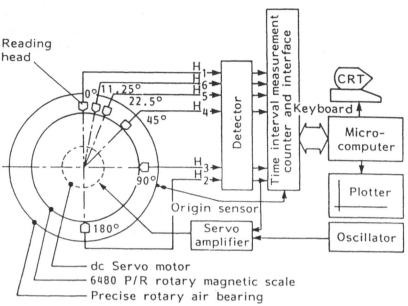

Figure 12 Schematic diagram of the calibration system.

Figure 13 General view of the calibration system.

heads can be placed at any optional positions around the scale. In the present system, six heads are placed at relative angles $\phi k(k = 1 - 5)$ as in Eq. (4) and as shown in Figure 12 for the multi-reading-head method. Therefore, the Fourier components of the master scale error which can be detected are 3,139 components, subtracting 32m components from the 3,240 components which the master scale may have. However, it has been confirmed that the size of 32m components can be ignored on the magnetic scale of the system.

For measurements of the relative-angular-positional-error δ in the multi-reading-head method and the relative measurement of the test scale referred to the master scale, a time-conversion method is used to detect slight angular differences automatically with high resolution. The basic principle of the time-conversion method is based on measuring the time intervals between pulses of the master scale and the test scale when they are rotating at a constant speed. Seven counter circuits are provided that can measure simultaneously the time intervals be-

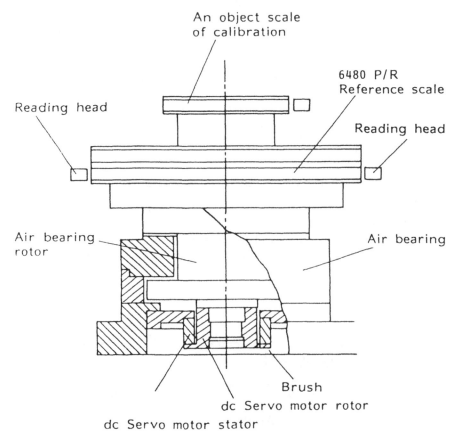

Figure 14 Precise rotary mechanism made up the calibration system.

tween the pulses from the reading head H1 and other heads (H ,...,H6 and so on) as shown in Figure 12. The counter counts the clock pulses of the crystal oscillator and the counted values are transmitted to the microcomputer. Since the clock used a frequency of 8 MHz, the angular resolution becomes 0.0135 in. at 5 rpm.

Measurement operation and data processing are controlled by the microcomputer. When calibrating the master scale by the multi-reading-head method, six sets of 6480 pieces of 16-bit data are acquired in about 12 s at 5 rpm. Then the relative-angular-positional-error δ is calculated by the time-conversion method. The calibrated values $\varepsilon(\theta)$ are determined by Eqs. (5) and (7). The data processing and plotting-out of the results takes about 5 min. When calibrating an angular encoder, the measurement of the relative-angular-positional-error between the master scale and the encoder are performed. Then the true error of the encoder is determined by correcting with the master scale error $\varepsilon(\theta)$.

Figure 15 shows the plot of the 6,480 calibration values $\varepsilon c(\theta)$ of the master scale, which were calibrated at the same time by the multi-reading-head method and the time-conversion method. It can be seen that the size of the angular-positional-error of the master scale is about 3 in. By spectrum analysis of the calibration values, it was found that the size of components higher than the 31st was 0.03 in.or less.

Figure 16 shows the calibration of a shaft type photoelectric encoder with 5,000 pulses per revolution (P/R). This is a result obtained by comparing the encoder with the master scale and correcting against the calibration values of the master scale. This encoder is constructed so that any odd Fourier components of the error are canceled by the two detectors, which are set at an interval of 180°. The effect can be seen in Figure 16.

IV. CONSTRUCTION ROBOT MARKING ON THE CEILING BOARD [7,8]

A. Introduction

Construction sites require the installation of fixtures such as air diffusers, lighting fixtures, speakers, etc. on the ceiling board. The marking of the installation

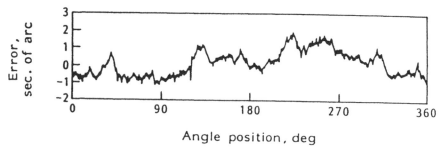

Figure 15 Master scale error measured by multi-reading-head method.

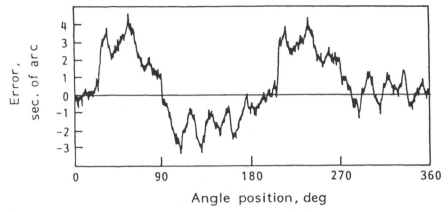

Figure 16 Example of the calibration of the photoelectric encoder.

position is currently done manually. The workers do not like such a marking job in large buildings because the operations require not only a level of sophistication but also working at dangerous heights. Therefore, the need is growing for robots to accomplish these tasks.

In order to accomplish such jobs, the robot must have the following two capabilities:

1. Self-positioning: it must be able to locate its own position automatically. This requires a technique that enables the robot to measure coordinates indicating its position.
2. Marking: it must be able to draw specific marks at designated positions on the ceiling board.

We have developed a prototype marking robot with an autonomous mobile system that could leave specific marks on ceiling boards in large buildings.

B. System Configuration and Marking Process

This robot performs basically three functions, that is, a self-positioning measurement, autonomous traveling and marking. The self-positioning measurement system can obtain precisely the coordinates of its own location. The basic principle behind this system is based on triangulation. The most important feature of this system is that it dose not require any exclusive arrangements for reference objects but uses existing pillars on the floor. On the floor of a large building, pillars stand in known positions at intervals of a few meters. They can serve as reference objects.

The mobile system transports other systems to the target point given in advance by a human being or CAD system. The robot moves by means of dead reckoning depending on rotational angles of two independent driving wheels. On its arrival at the target, the self-positioning measurement system confirm its own coordinates. If necessary, the robot moves slightly to adjust position. After reconfirming that it accurately stands on designated point, the lifting unit raises the drawing unit up to the ceiling. Specified figures, such as lines, circles, quadrangles, letters and symbols, are then drawn on the ceiling board by the drawing unit. This work is continued until all marking for a floor is completed. Figure 17 shows a prototype robot, which has a size of 800 x 754 x 2065 mm 3and a weight of 300 kg. It can work continuously maintaining a speed of 2 km/h for about 2 hours between battery recharges.

C. Self-Positioning Measurement System

Suppose that the measuring system can recognize the surrounding pillars and measure two angles, $\theta 21$ and $\theta 32$, whose center is coincident with the center of the measuring system, by detecting the edges of three pillars (one edge for each pillar) as shown in Figure 18. The coordinates of the measuring system center R(U,V) are given by the following equations according to the principle of triangulation:

$$U=U2 - AB/(B2 + C2) \tag{8}$$

$$V=V2 - AC/(B2 + C2) \tag{9}$$

$$\begin{aligned}
A&= \{(U1 - U2)(U3 - U2)+(V1 - V2)(V3 - V2)\} \ (\tan\theta 21 + \tan\theta 32) \\
&\quad +\{(U3 - U2)(V1 - V2) - (U1 - U2)(V3 - V2)\} \ (1 - \tan\theta 21 \ \tan\theta 32) \\
B&= (V1 - V3) \tan\theta 21 \ \tan\theta 32 - (U3 - U2) \tan\theta 21 - (U1 - U2)\tan\theta 32 \\
C&= (U3 - U1) \tan\theta 21\tan\theta 32 - (V3 - V2) \ \tan\theta 21 - (V1 - V2) \tan\theta 32 \\
\theta 21 &= \theta 2-\theta 1, \ \theta 32 = \theta 3 - \theta 2
\end{aligned} \tag{10}$$

where the coordinates of these edges, (U1,V1), (U2,V2), and (U3,V3) are known by the design specifications. The robot can detect two edges of each of the four pillars for a total of eight edges. Therefore, there are four combinations when it uses the right side of the pillars, and the same number for the left side, for eight measured positions. Then, the average position is taken as its final position.

Figure 19 shows the system organization of the self-positioning measurement system, which consists of a laser range finder and an angle measuring unit. The laser range finder employs a laser diode (wavelength: 670 nm, output: 10 mW) as the source. The reflected laser beam receiver consists of lenses, a bandpass filter, and a CCD linear sensor. The laser beam emitted from the laser diode is reflected back through the lenses and the band-pass filter to the linear CCD sensor (1,024 elements) where the reflected light forms an image. If the CCD sensor is placed off the optical axis of the emitted laser beam, the position of the image formed on the CCD sensor changes with the distance to the object. Therefore, the distance that the reflected laser beam has traveled is found by reading CCD sensor output and determining the position where the reflected laser beam is received or the address of the elements, as shown in Figure 20.

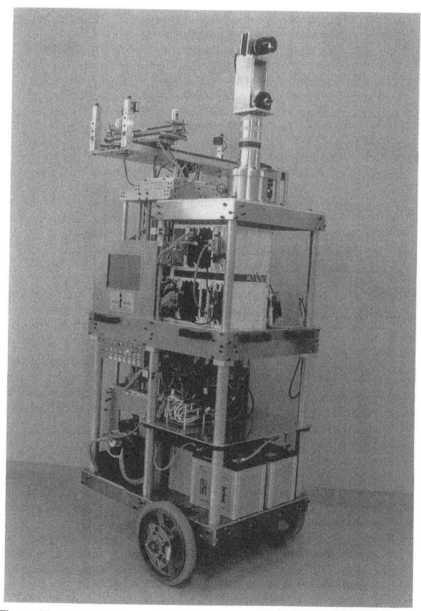

Figure 17 Overview of the marking robot.

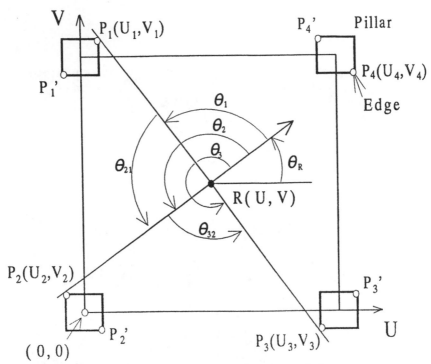

Figure 18 Principle of measurement method based on triangulation.

The laser diode and receiver are designed to rotate as a single unit around the center axis of the robot. Its angle of rotation is detected by the rotary encoder (angular resolution: 14.4 arc seconds), giving the direction in which laser beam is emitted, or the direction in which measurement is made. Outputs of the linear CCD sensor and the rotary encoder are entered into a computer.

The steps of the measurement process are as follows:

1. The laser range finder and the angle measuring unit are rotated as a single unit by the DC motor while the robot is idle.
2. Output data of both the CCD sensor and the rotary encoder are sampled at regular intervals.
3. The address of the CCD element, whose output is 50 percent of the maximum output of 1,024 CCD elements, A1 and A2, is found as shown in Figure 12.20. A1 is the position where output increases, and A2 is the position where it decreases.
4. Center address A0, calculated from the relation A0 = (A1+A2)/2, is taken as the position where the reflected laser beam is received.

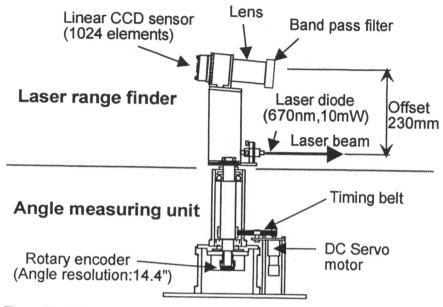

Figure 19 Self-positioning measurement system.

5. The distance that the reflected light has traveled is determined from a preset equation that relates the distance to the beam receiving position.
6. Steps 2–5 are repeated until the encoder completes one full rotation.
7. From distance data in all directions obtained by the laser range finder, the system identifies four pillars. Then, the place where distance pattern changes abruptly is taken as the edge of the pillar, so that the bearings of their edges are measured by the angle measuring unit.
8. From edges and their known coordinates, the robot determines its own position according to Eq. (1–3).

D. Marking System

The marking system, which is the other major component of this robot to put a mark on the ceiling board, uses lifting and drawing units in addition to mobile unit.

The mobile unit has two independent driving wheels and a caster wheel. It uses DC servo motors as actuators. The robot can work continuously at a speed of 1.5 km/h for about 2 hours on a single charge of batteries. The lifting unit has a 1,275 mm stroke to elevate the drawing unit and reaches to ceiling boards up to 3,070 mm height. Because most ceilings in office are about 2,700 mm, the robot can be used at most work construction sites. It has a telescopic three-stage mechanism composed mainly of DC servo motor, ball-screw,

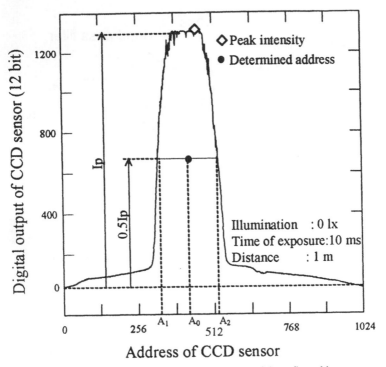

Figure 20   Procedure of determining the distance of the reflected laser.

and pulley as shown in Figure 21. The drawing unit is raised up at high speed until the distance between the ceiling and the system is within 150 mm to 200 mm. Then the photoelectric switch fit on the top of the drawing system turns on, and speed slows to avoid hitting the ceiling. As the drawing unit touches the ceiling, the touch switches are activated and elevation operation ends.

The drawing unit draws various and relatively large figures accurately on the ceiling board. To make the system compact for mobile installation, we developed X-Y plotter equipment whose x-axis is divided into two equal parts as shown in Figure 22. The maximum drawing range in the x-axis direction is 650 mm and 700 mm in the y-axis direction.

Experimental results for the prototype have proved that it is able to measure its own position within an error of a few millimeters in the case of four pillars that stand on the floor at intervals of 6m. The robot autonomously moved to the target point and drew the designated lines, circles, quadrangles, letters, and symbols such as a square on the ceiling board with an accuracy of 10 mm. The time required to make a mark, including measuring, moving, lifting, and drawing, was about 8min. Based on these experiments, we are confident that the marking robot will operate efficiently at actual building construction sites.

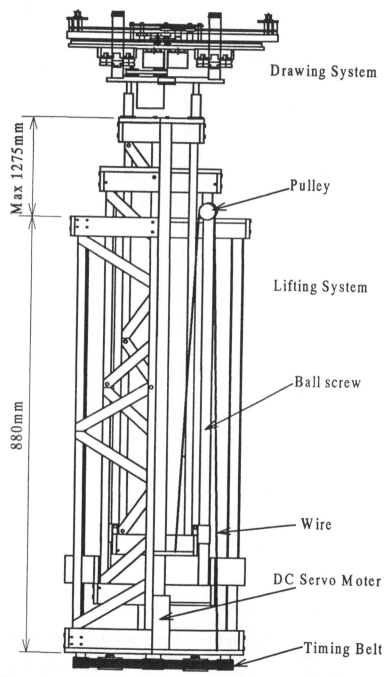

Drawing System

Pulley

Lifting System

Ball screw

Wire

DC Servo Moter

Timing Belt

Max 1275mm

880mm

Figure 21 Mechanism of marking system.

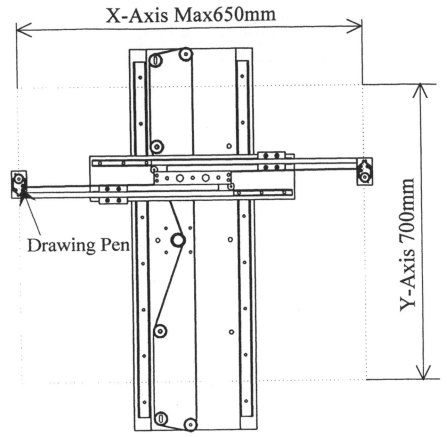

Figure 22   Drawing unit.

V. MUSICIAN ROBOTS [9,10]

A. Introduction

The research and development of Musician Robot (MUBOT) involves the following three objectives:

1. Case study of robotics. MUBOT is set as a case study of robotics. In solving problems posed from the case study it is looking forward to contributing to the advancement of robotics.
2. Application as an amusement robot. MUBOT provides musical performances or various entertainment. Even one who cannot play a musical instrument, can enjoy playing a musical instrument by means of MUBOT.

3. Application as the simulator of a musical performance. The use of MUBOT is effective in experiments of studies on musical engineering.

The MUBOT has been developed on the condition that a musical instrument has not been required to modify itself.

Figure 23 shows the relationship among the MUBOT, musical instrument, and performer. The performer is a human.

A musical performance requires not only a musical score but also a performer's idea for the music. The originality of the performer lies in how to cook and season the score, as does the expression of an artistic talent. The MUBOT acts directly on the instrument for providing a musical performance intended by the performer. The MUBOT never gives a performance of its own free will.

We have developed musician robots each of which is an expert in recorder, violin and cello. And they can also play in concert.

B. Recorder MUBOT

A recorder is performed by expiration and finger motion. Expiration of air requires a suitable flow according to each pitch of a sound. An electro-pneumatic converter controls the rate of air flow supplied from an air compressor. The relationship between the pitch of a sound and the flow has been previously found by experiment. Vibratos can be also applied by changing a flow with a certain fre-

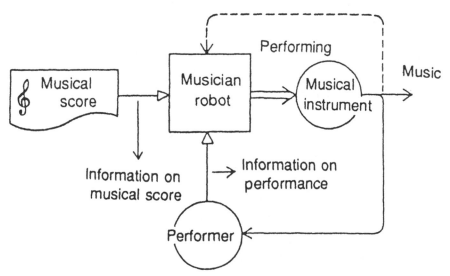

Figure 23 Relationship among musician robot, musical instrument, and performer.

quency. Tonguing is done by opening and closing the air valve to the mouth-piece. Each sound hole of the recorder is opened and closed by a rubber fingertip actuated by a pencil-shaped air cylinder. The rear hole (octave hole) is made to provide overall and half opening, and closing. Figure 24 shows the picture of the recorder MUBOT.

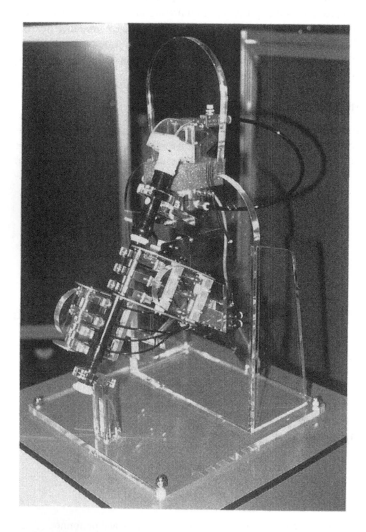

Figure 24 Recorder MUBOT.

C. Violin MUBOT

The playing system of the violin MUBOT consists of the following mechanisms:

1. Bow operating mechanism
2. Fingering mechanism

The bow operation requires selection of an active string, pressure of bow against a string, and reciprocation of a bow. A bow must make contact with any desired one from among four strings, and move rectilinearly with appropriate pressure against a string. For these three movements, individual mechanisms and actuators are provided. The hand holding the bow is only linearly reciprocated on a fixed track. Then, the body of a violin is rotated around its long axis such that a desired string appears on the operating track of the bow. The reciprocation of the bow is performed by the arm consisting of a pantograph mechanism. The holder of the bow has a mechanism to control the pressure of the bow.

To determine a musical interval, the mechanical fingers of the left hand are set at every half tone. They consist of the pencil-shaped air cylinders whose tip is covered with a small rubber. Figures 25 and 26 show the pictures of the violin MUBOT.

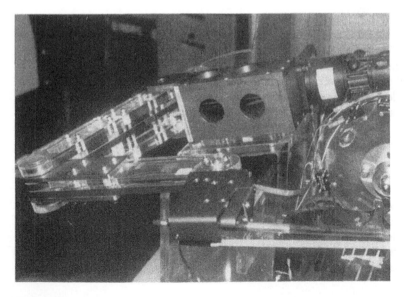

Figure 25 Violin MUBOT.

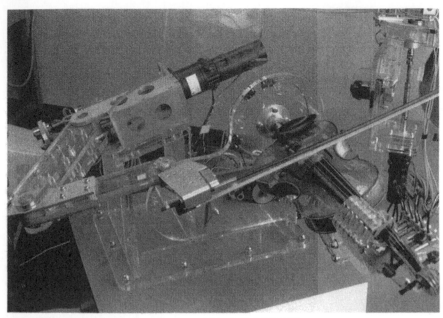

Figure 26 Bow operating mechanism of the violin MUBOT.

D. Cello MUBOT

The cello MUBOT has basically the same system as the violin MUBOT. However, a cello is heavier than a violin, such that the mechanism of the cello MUBOT
is concretely different than that of the violin MUBOT. Similar to the violin
MUBOT mentioned above, the bow operating mechanism requires three movements: selection of an active string, pressure of bow against a string and reciprocation of a bow. The reciprocation of the bow is a simple mechanism, as shown in
Figure 27. The whole mechanism moves on the arc-shaped rail to a position at
which a desired string can be touched as shown in Figure 28. In contrast to a violin, the body of a cello does not move. The pressure of the bow is applied by
pressing the wooden portion of the bow in the vicinity of the point of contact of
the bow and the string. The fingering mechanism consists of the pencil-shaped air
cylinders as in the case of violin MUBOT. They are arranged in a 6 x 6 array at
the interval of a half tone, totaling 24. The holder of the fingers is capable of
moving along the direction of the strings. The effect of vibrato is enabled by reciprocating the holder with the motor and eccentric cam. Figure 12.29 shows the
overall view of the cello MUBOT.

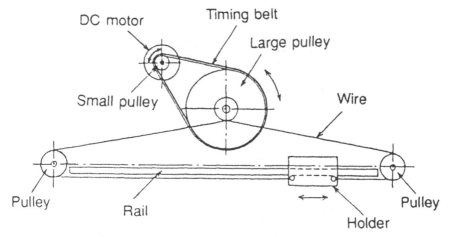

Figure 27 Mechanism of reciprocating a bow in the cello MUBOT.

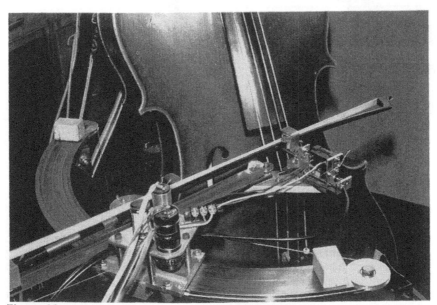

Figure 28 Bow operating mechanism of the cello MUBOT.

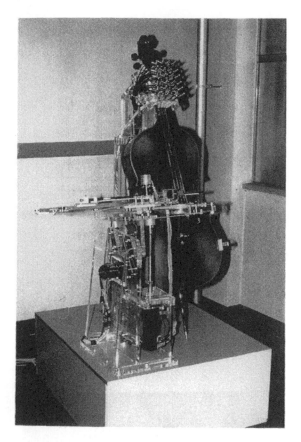

Figure 29   Cello MUBOT.

REFERENCES

1. M Kajitani, J Ishikawa, Y Sugizaki. Development of high accurate rotary magnetic scales and its applications. Proc Int Conf on Production Engineering, Tokyo, 1974.
2. M Kajitani, T Masuda. Recording method for high accurate rotary magnetic scales. Mechatronics 2(2):471–482, 1992.
3. M Kajitani, J Ishikawa, S Imai. A new involute tooth profile measuring system with special-purpose computer. Bull JSPE 4(2):47–48, 1970.

4. M Kajitani, J Ishikawa, S Imai. A new involute tooth profile measuring system with special-purpose computer. Bull T1T 106:157–168, 1971.
5. M Kajitani, N Shimizu, T Masuda, J Ishikawa. Automatic visual inspection of tooth contact pattern of gears. Bull JSPE 21(2):101–106, 1987.
6. T Masuda, M Kajitani. An automatic calibration system for angular encoders. Precision Engineering 11(2):95–100, 1989.
7. K Tanaka, M Kajitani, C Kanamori, H Ito, Y Abe, Y Tanaka. Development of marking robot working at building sites. Proceedings of the 12th International Symposium on Automation and Robotics in Construction, pp. 235–242, 1995.
8. Y Abe, M Kajitani, K Tanaka, C Kanamori, Y Tanaka. Development of a construction robot marking on the ceiling board. Proceedings of the 2nd International Conference on Mechatronics and Machine Vision in Practice, pp. 119–123, 1995.
9. M Kajitani. Development of musical robots. J Rob Mech 1(1):254–255, 1989.
10. M Kajitani. Simulation of musical performance. J Rob Mech 4(6):462–465, 1992.

13

An Operator's Model for Control and Optimization of Mechatronic Processes

George Vachtsevanos and Sungshin Kim[*]

Georgia Institute of Technology, Atlanta, Georgia

I. INTRODUCTION

Large scale industrial processes entail complex dynamics at start-up, shut-down, as well as under abnormal or emergency operating conditions. Parts of these dynamics are typically governed under closed loop control, particularly in normal steady state or quasi steady state conditions. However, when the processes enter into abnormal conditions because of a fault or disturbance, plant operators play an important role in managing the system's operating states in terms of bringing it back to a stable mode. Plant start up and shut-down are other examples of abnormal states where an operator's presence is required. Complexity of the plant dynamics under such conditions renders closed-loop or open-loop control infeasible. The operator is called upon to intervene in such cases and take charge of the plant's operation.

Procedures for improving the performance of control systems based on operator's experience have attained a considerable interest recently [14]. Sawaragi *et al.* [9] have attempted to develop an algorithm for analyzing an operator's control action sequence based on AI techniques. Nakayama *et al.* [7] and Hiraga *et al* [4] have used fuzzy neural networks to acquire control actions in a ship steering problem. Their approach considers only normal operating conditions and is focused on ship steering direction commands.

---

[*]*Current affiliation*: Pusan National University, Pusan, Korea.

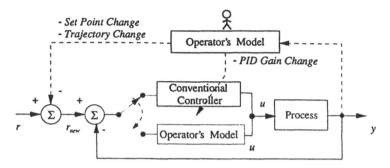

Figure 1 The operator's model as an element in the control hierarchy.

This contribution suggests the development of an operator's model based on the polynomial fuzzy neural network (PFNN) architecture. The learning algorithm for the operator's model, as shown in Figure 1, is run in parallel with an expert process operator and their responses are used to train the operator's model. Empirical data, heuristic information, and operator's knowledge are employed in the development of the model. Such a model imitates the operator's actions under abnormal conditions and it can also be employed as an operator's assistant or for training purposes.

As shown in Figure 1, the operator's model may be exploited at two stages of the control architecture: as part of the direct (inner) control loop substituting for or adjusting gains of the controller or in a supervisory configuration (outer loop). The latter's main function is to provide new setpoints or trajectories to inner loop controllers so that robustly stable performance is maintained in the event of large disturbance. In such circumstances, the available control authority is not capable of accommodating disturbance fronts and a setpoint change may assist to maintain the closed-loop system within the prescribed stability envelope. Either one or both functions may be desirable and practical for an operator's model depending upon accessibility requirements and process characteristics.

The development, implementation, and utility of an operator's model involve several fundamental steps: (1) knowledge acquisition, (2) knowledge representation, (3) model implementation, and (4) model utilization in the control hierarchy. In this chapter, the first two tasks are briefly reviewed in the next section while attention is focused primarily on the implementation algorithm and the control routines.

II. KNOWLEDGE ACQUISITION AND REPRESENTATION

In a typical industrial enterprise, knowledge about a specific process' operations resides at three levels:

1. The skilled operator's level—here, the operator builds through experience mental behavioral models of effect-cause relationships and employs them

in a reactive control mode to take fast corrective action in the event of an emergency.

2. The process engineer level—at this level of the hierarchy, the human "operator" possesses some fundamental knowledge of the process dynamics. He (or she) is familiar with the subsystem interactions and the primary objective is to maintain overall process operations at acceptable levels. The process engineer takes into consideration a broad category of interacting system variables. He can reason and explain his actions and can project effects and determine cause→control action relationships over longer time horizons.

3. The design/control engineers level—his final level of the hierarchy involves long-term optimization and control functions; usually, the expertise resides far from the operating plant and can only be utilized for fine-tuning and optimizing the control algorithm.

Knowledge acquisition may be based on event reports, observations of actual control room performance, polling of operators and other experts, operator's manuals and standard procedures published by the equipment manufacturers, etc. [1]. Operators base their decisions on their knowledge of key system properties at different levels of abstraction depending on their perception of the system's immediate control requirements. The operator attempts to maintain a normal operating state, to change operating states when called upon or to synchronize and coordinate states when a system reconfiguration strategy is dictated by prevailing conditions. Under a severe disturbance condition, the operator's main concern is to devise a suitable control strategy for the system's safe recovery. Knowledge acquisition usually relies upon similarity and frequency criteria, that is, the control action most likely to be selected by the operator is determined by the resemblance of the current local situation to situations framed in memory and the frequency distribution of these situations in past circumstances.

Knowledge acquired using the procedural steps described above can be categorized into three types: *system*, *contextual*, and *task*. Based on this knowledge set, human problem solving takes place at three levels: recognition and classification, planning, and execution. Thus, when an abnormal situation develops, the first task is to detect that the problem exists and to categorize it (recognition and classification). An approach or plan to solving the problem must then be developed (planning), and finally the plan must be implemented (execution). This provides an appropriate means for describing and evaluating human performance and behavior in controlling and monitoring a plant. Such a performance and behavior, if captured successfully, pave the way for an operator's model to be directly incorporated into the control loop of a plant.

In the artificial intelligence arena, several techniques have been suggested for knowledge representation. Among them are schemata, scripts, frames, and production rules. They provide easily accessible solution frameworks experienced in past circumstances and transcribed as mental models or simple crisp or fuzzy look-up tables. They operate in a local environment by matching descriptive elements of process behavior to situational control actions. In this contribution, the preferred methodology for knowledge representation is fuzzy production rules. The latter are of the form: IF a current process indicator is *Positive Small* and a past process indicator is *Zero*, THEN the suggested control action is *Negative Small*.

III. KNOWLEDGE IMPLEMENTATION

A. Neuro-Fuzzy Architecture for Operator's Model

The fuzzy inference engine, based on the polynomial fuzzy neural network (PFNN) architecture, as shown in Figure 2, is combined with a fuzzy system [12] through a polynomial neural network (PNN). A PNN is a feed-forward network that computes a polynomial function of a set of parallel inputs in order to generate an output. A PNN [11] is employed in the defuzzification scheme to improve output performance and select rules. Ivakhnenko [5] describes an approximation method that is determined by successively fitting polynomials and employing a mean squared error at each layer to determine the best fit at that layer.

The PFNN consists of a set of if-then rules with appropriate membership functions whose parameters are optimized via a Genetic Algorithm (GA) [2]. For n inputs, x_i, with m_i membership functions, $i = 1, \ldots, n$, and one output, $\hat{y}$, the total number of rules, r, is defined as: $r = \Pi_{i=1}^{n} m_i$. A typical rule with fuzzy if-then structure is expressed by:

$$Rule_{(i)} = \text{IF } x_1 \text{ is } A_1^{i_1} \text{ and } \cdots \text{ and } x_n \text{ is } A_n^{i_n} \text{ THEN } y_i \text{ is } p_i, \tag{1}$$

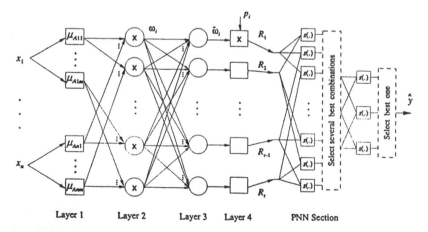

Figure 2 Polynomial fuzzy neural network (PFNN) system with n inputs and one output.

where x_j, $j = 1, \ldots, n$, are nonfuzzy input variables, $A_k^{i_k}$, $1 < i_k < m_k$, are fuzzy variables related to the k-th input with i_k-th membership function, and p_i, $i = 1, \ldots, r$, are constants in the consequent. The membership functions for A_i^j, $\mu_{A_i^j}(x_i)$, for example, can be of the Gaussian type. The outputs of layer 4, R_i, are inferred as follows:

$$R_i = \frac{\Pi_{k=1}^n \mu_{A_k^j}(x_k) \cdot p_i}{\Sigma_{j=1}^r \Pi_{k=1}^n \mu_{A_k^j}(x_k)} \tag{2}$$

In the PNN section, a least-squares fit of the training data for the function to be learned is determined for all pairs of the input variables, R_i, using polynomials up to the second degree. For example, the output of the bottom left module is a function of the inputs R_{r-1} and R_r:

$$s(R_{r-1}, R_r; a_i) = a_0 + a_1 R_{r-1} + a_2 R_r + a_3 R_{r-1} R_r + a_4 R_{r-1}^2 + a_5 R_r^2. \tag{3}$$

A decision rule is used after the first level that chooses only those polynomials that provide the best fit to the training data. The outputs of this reduced set of polynomials are then used as inputs to a second layer of the model. Here again all pairs of polynomials up to the second degree are tested, and only those that provide the best fit are kept as inputs to the next layer. Therefore, the PFNN structure can increase the performance of the defuzzification scheme and reduce the number of rules since all of the R_is are not employed to produce the final output. The performance of the PFNN modeling method is discussed in [13].

The proposed PFNN fuzzy inference system is a suitable structure capable of not only identifying the operator's rules from empirical data and capturing the operator's knowledge but also is a viable construct to optimize membership functions for improved performance. The identified fuzzy rules can be incorporated into the control structure of the overall complex system.

B. Hybrid Genetic Optimization

As suggested in the previous section, the parameters of the PFNN are optimized via a genetic algorithm. Genetic algorithms are a class of general purpose search strategies that strike a near optimal balance between exploration and exploitation of a parameter space [2]. The major disadvantage of a genetic algorithm is the excessive number of runs of the design code required for convergence.

The search space in a GA is discretized by its resolution. In the binary coding method, the bit length L_i and the corresponding resolution R_i are related by

$$R_i = \frac{UB_i - LB_i}{2^{L_i} - 1}, \tag{4}$$

where UB_i and LB_i are the upper and lower bounds of the parameter x_i, respectively. As a result, the parameter set can be transformed into a binary string (chromosome) with length $l = \Sigma_{i=1}^{n} L_i$, where n is the number of parameters to be optimized and L_i is different for each parameter. By adding k bits to the parameters, the resolution is improved by approximately 2^k times. On the other hand, the search space is dramatically increased to $(2^k)^n$ times.

To overcome this problem, a hybrid genetic optimization method that combines a GA with Nelder and Mead's simplex method is introduced. First, a GA is used to search the coarse search space that is predefined by (4) and to find the basin of attraction of the global solution. Second, the result of the GA is passed to a simplex method as an initial condition. In an n-input, one-output system, its genotype, that is the complete sets of chromosomes, is illustrated in Figure 3.

IV. APPLICATION EXAMPLES

A. pH Process

System Description and Dynamics

A continuously stirred tank reactor (CSTR) for pH titration is shown in Figure 4. Even this simple reactor configuration entails electromechanical monitoring and actuation devices and, in some sense, may be classified as belonging to the general mechatronics area. The dynamic model for the stirred tank as derived by McAvoy [6] is given by:

$$\left.\begin{array}{l} \dfrac{Vd\xi}{dt} = F_1 C_1 - (F_1 + F_2)\xi \\[2mm] \dfrac{Vd\zeta}{dt} = F_2 C_2 - (F_1 + F_2)\zeta \\[2mm] [H^+]^3 + (K_a + \zeta)[H^+]^2 + \\[2mm] \{K_a(\zeta - \xi) - K_w\}[H^+] - K_a K_w = 0 \\[2mm] \mathrm{pH} = -\log_{10}[H^+] \end{array}\right\} \qquad (5)$$

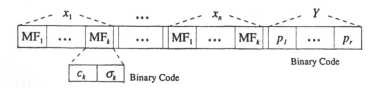

Figure 3 Coding scheme for a fuzzy system.

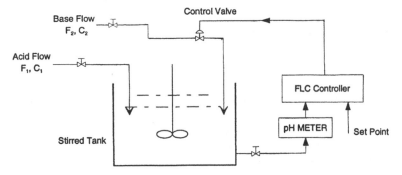

Figure 4 A continuously stirred tank reactor for pH titration control.

The physical meaning of each variable and its associated initial values are listed in Table 1.

The essential part of the process is a 1-liter stirred tank having two inlets which neutralize each other and an overflow outlet. One of the inlets, F_1, contains acetic acid of concentration C_1 and the other, F_2, sodium hydroxide of concentration C_2. The acid flow rate, F_1, is held constant in the nondisturbance case. The flow rate of hydroxide, F_2, is used to control the pH value of the solution. The acid flow rate is changed in the disturbance case with a variance of 2.

The development of a control structure for this process incorporating the operator's rules and leading to an expert system requires analysis of activities and methods used by operators and engineers in the plant. Observation and analysis of operator's actions and activities under normal and abnormal conditions for the control of such a system leads to the development of an operator's rule base.

In the following section the operator's models for the inner loop and the outer loop are described and simulation results are presented. The performance of the operator's model in the inner loop is compared with a PID controller for a step

Table 1 The Physical Parameters Used in the Study of the pH Problem

Term	Meaning	Initial setting
V	Volume of tank	1,000 cc
F_1	Flow rate of acid	81 l/min
F_2	Flow rate of base	519 l/min
C_1	Concentration of acid	0.320 moles/l
C_2	Concentration of base	0.05005 miles/l
K_a	Acid equilibrium constant	1.8×10^{-5}
K_w	Water equilibrium constant	1.0×10^{-14}

change in the input reference. Under a large disturbance, the effective performance of the operator's model in the outer loop is demonstrated in the comparison with the inner loop controllers.

Simulation Results

   Operator's Model in the Inner Loop.   The operator's model has two inputs, $e(t)$ and $\Delta e(t)$, where $e(t) = r - y$ and $\Delta e(t) = e(t) - e(t - 1)$, and one output as shown in Figure 1. The two input variables have 5 Gaussian membership functions each, and their universes of discourse are normalized. The output is the flow rate of hydroxide, F_2, with 5 membership functions.

   The performance of a tuned PID controller is depicted in Figure 5. In this case, the input reference is changed from pH 7 to pH 9. It can be compared the performance of the operator's model controller in the inner loop, as shown in Figure 6.

   In the disturbance case, as shown in Figure 7(c), the performance of the inner loop PID controller and operator's model controller are shown in Figure 7(a) and Figure 7(b), respectively.

   Operator's Model in the Outer Loop   The outer loop operator's model has the same two inputs as the inner loop operator's model. The two input variables have 3 Gaussian membership functions each, and their universes of discourse are normalized. The output is the change in the reference, r, to the new reference, r_{new}, with 5 membership functions. The set of rules for the outer loop model are as follows:

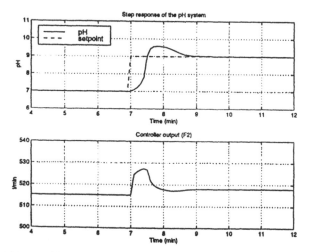

Figure 5   Step response of the pH system controlled by a PID inner loop.

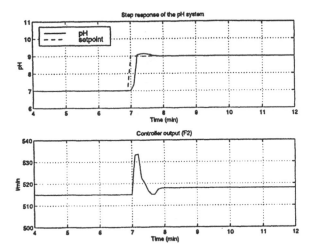

Figure 6 Step response of the pH system controlled by a fuzzy operator's model in the inner loop.

If $\Delta error$ is NB and *error* is NB then Δu is ZE
If $\Delta error$ is NB and *error* is ZE then Δu is NS
If $\Delta error$ is NB and *error* is PB then Δu is NB
If $\Delta error$ is ZE and *error* is NB then Δu is PS
If $\Delta error$ is ZE and *error* is ZE then Δu is ZE
If $\Delta error$ is ZE and *error* is PB then Δu is NS
If $\Delta error$ is PB and *error* is NB then Δu is PB
If $\Delta error$ is PB and *error* is ZE then Δu is PS
If $\Delta error$ is PB and *error* is PB then Δu is ZE

where NB = negative big, NS = negative small, ZE = zero, PS = positive small, PB = positive big, and Δu equals $\Delta setpoint,$ that is, the change in the setpoint.

Certain situations (large disturbances, mode transitioning, reconfiguration, etc.) require that the operator's model be placed in the outer loop, as shown in Figure 1. The corrective actions taken by the operator involve changes in the setpoint reference input. This type of operator's action not only reduces the present overshoot but also improves the settling time and results in robustly stable operation.

The outer loop operator's model assists in improving the *overshoot* and *settling time* of the response. The rule-based operator's model changes the reference setpoint to r_{new} resulting in a better response, as shown in Figure 8.

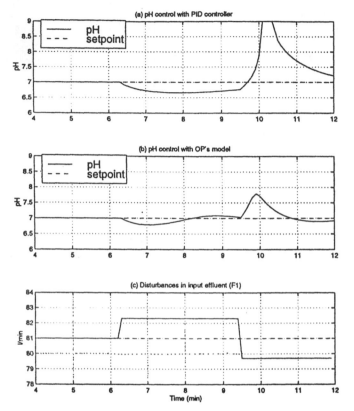

Figure 7 pH control performance for large disturbances in F1 with an active inner loop: (a) PID controller, (b) fuzzy operator's model, (c) disturbance in F1.

B. PID Gain Tuner Based on Fuzzy Rules

A proportional, integral, derivative (PID) controller works with three gain values: proportional gain, integral gain, and derivative gain. Appropriate selection of these gains is very important for good closed-loop performance of the system.

The PID gain tuner is based on a fuzzy operator's model as shown in Figure 1. Twenty fuzzy if-then rules are developed for computing the incremental values of the proportional, integral, and derivative gains as functions of damping, frequency, and steady-state error. The PID gain tuner determines the increments to the values of the three gains represented by

$$\Delta K = \begin{bmatrix} \Delta k_p \\ \Delta k_i \\ \Delta k_d \end{bmatrix} = \begin{bmatrix} F_{fuzzy}(damping, frequency) \\ F_{fuzzy}(sserror, frequency) \\ F_{fuzzy}(damping) \end{bmatrix} \tag{6}$$

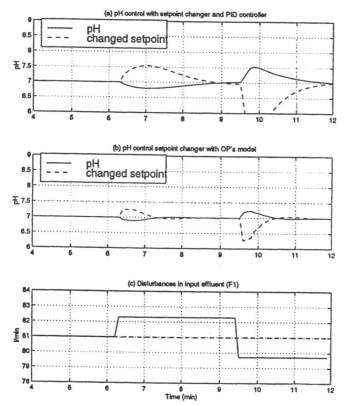

Figure 8 pH control performance for large disturbances in F1 with external fuzzy setpoint changes and an active inner loop: (a) PID controller, (b) fuzzy operator's model, (c) disturbance in F1.

The input variables, damping, frequency, and steady-state error, have seven, three, and three Gaussian membership functions, respectively. The output variables Δk_p, Δk_i, and Δk_d have seven Gaussian membership functions each. The universes of discourse for these variables are normalized. The final PID gains are modified as: $k = k + \Delta k$, based on the following set of rules:

> If (*damping* is mf1) then (Δk_p is mf7)
> If (*damping* is mf2) then (Δk_p is mf6)
> If (*damping* is mf3) then (Δk_p is mf5)
> If (*damping* is mf4) then (Δk_p is mf4)
> If (*damping* is mf5) or (*frequency* is mf1) then (Δk_p is mf3)

If (*damping* is mf6) or (*frequency* is mf2) then (Δk_p is mf2)
If (*damping* is mf7) or (*frequency* is mf3) then (Δk_p is mf1)
If (*sserror* is mf1) then (Δk_i is mf4)
If (*sserror* is mf2) then (Δk_i is mf5)
If (*sserror* is mf3) then (Δk_i is mf6)
If (*frequency* is mf1) then (Δk_i is mf1)
If (*frequency* is mf2) then (Δk_i is mf2)
If (*frequency* is mf3) then (Δk_i is mf3)
If (*damping* is mf1) then (Δk_d is mf1)
If (*damping* is mf2) then (Δk_d is mf2)
If (*damping* is mf3) then (Δk_d is mf3)
If (*damping* is mf4) then (Δk_d is mf4)
If (*damping* is mf5) then (Δk_d is mf5)
If (*damping* is mf6) then (Δk_d is mf6)
If (*damping* is mf7) then (Δk_d is mf7)

Autotuning of the PID gains based on the operator's model is tested for the following third order plant [10]:

$$G(s) = \frac{400}{s(s^2 + 30s + 200)} \tag{7}$$

First, the Ziegler-Nichols method is applied to determine the PID gains of the controller. The PID gains are calculated as

$$k_p = 0.6k_m, \quad k_d = \frac{k_p \pi}{4\omega_m}, \quad k_i = \frac{k_p \omega_m}{\pi} \tag{8}$$

where k_m is the gain at which the proportional system oscillates, and ω_m is the oscillation frequency. The *rlocus* and *rlocfind* MATLAB commands can be used to find the crossover gain of $k_m = 15$ and crossover frequency of $\omega_m = 14$ rad/sec. Then the Ziegler—Nichols equations in (8) are used to find the gain parameters: $k_p = 9$, $k_d = 0.5$, $k_i = 40$. The resulting system response is shown in Figure 9.

Second, a fuzzy tuner based on the operator's model is applied to autotune the PID gains. The progression of the PID gains is shown in Table 2. The final PID gains are shown in the last row of Table 2 and the corresponding response is depicted in Figure 10. The stepwise improvement in the system response is illustrated in Figure 11.

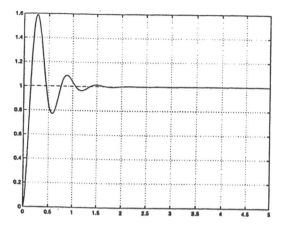

Figure 9 Step response using Ziegler-Nichols method.

Table 2 System Response Parameters and Corresponding PID Gains of the Fuzzy Autotuner

Count	k_p	k_i	k_d	Damping	ss Error	Frequency
1	10.000	0.000	0.000	0.700	−0.004	1.613
2	9.022	0.001	0.032	0.611	0.001	1.563
3	7.278	0.002	0.062	0.527	0.000	1.515
4	7.278	0.003	0.089	0.450	0.000	1.471
5	6.513	0.004	0.111	0.379	0.000	1.429
6	5.777	0.005	0.129	0.312	0.000	1.351
7	5.073	0.006	0.144	0.247	0.000	1.282
8	4.420	0.006	0.157	0.185	0.000	1.191
9	3.826	0.007	0.166	0.130	0.000	1.111
10	3.291	0.008	0.173	0.081	0.000	1.000
11	2.803	0.008	0.176	0.039	−0.001	0.862
12	2.832	0.009	0.178	0.010	−0.001	0.667
13	2.313	0.009	0.178	0.007	−0.001	0.625
14	2.289	0.009	0.179	0.006	−0.001	0.602
15	2.276	0.009	0.179	0.005	−0.001	0.595
16	2.265	0.009	0.179	0.005	−0.001	0.558
17	2.256	0.008	0.179	0.005	−0.001	0.575
18	2.248	0.008	0.180	0.004	−0.001	0.575
19	2.241	0.008	0.180	0.004	−0.001	0.568
20	2.235	0.008	0.180	0.004	−0.001	0.562

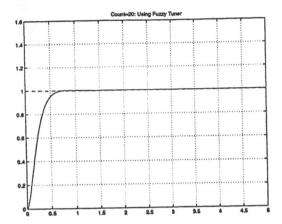

Figure 10   Step response using fuzzy tuner.

B. Gear Ratio Control in a Continuously Variable Transmission

System Description and Model Based on Experimental Data

The continuously variable transmission (CVT) is characterized by a precise constitutive relationship governing the mechanism of slippage. A schematic diagram of a CVT is shown in Figure 12. The physical system of the CVT consists of a

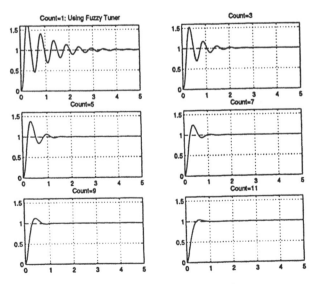

Figure 11   Transition of a fuzzy tuner at each step.

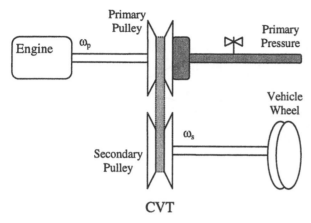

Figure 12 Schematic diagram of CVT module.

metal v-shaped belt between two split plate pulleys, which are hydraulically actuated. The hydraulic system, specifically referring to the hydraulic actuator pulley plates of the CVT, is highly nonlinear and requires a good model for purposes of ratio control [8]. The system consists of a primary pulley indirectly linked to the engine and a secondary pulley ultimately leading to the rear end axle and wheels.

Compared with a conventional automatic transmission, which shifts among up to five gear ratios, a CVT uses the entire range of ratios between low and high gears. The CVT does not shift. It achieves better fuel economy than a conventional automatic transmission by constantly changing ratios to keep the engine running in its most efficient rpm range based on driver demands.

A model for the CVT gear ratio profile is derived on the basis of a Dynamic Polynomial Neural Network (DPNN) [3,11]; a schematic diagram of the DPNN model depicted in Figure 13. The output y is the gear ratio and u is the primary pressure. The model consists of time delay terms of y and u. Figure 14 shows a measured gear ratio response and the model gear ratio response.

Simulation Results

Operator's Model in the Inner Loop The operator's model has two inputs, $e(t)$ and $\Delta e(t)$, where $e(t) = r - y$ and $\Delta e(t) = e(t) - e(t - 1)$, and one output as shown in Figure 1. The output is the primary pressure to the primary pulley. The membership functions of the inputs and output are the same as in Section IV.A.

The performance of a tuned PID controller is depicted in Figure 15(a). In this case, the desired gear ratio profile is smoothly ramping up and down. It may be compared with the performance of the operator's model controller in the inner loop, as shown in Figure 15(b).

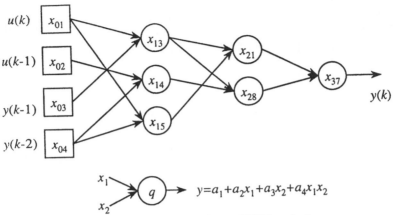

Figure 13 Model structure of CVT based upon DPNN method.

With a disturbance term introduced between 12sec and 15sec in the desired gear ratio profile, the performance of the inner loop PID controller and the operator's model controller are shown in Figures 16(a) and Figure 16(b), respectively.

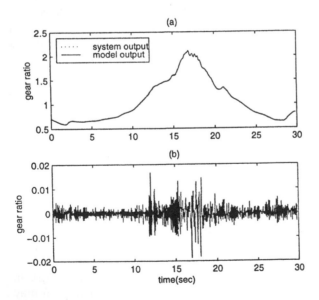

Figure 14 (a) Model ratio response and measured ratio response to the same input sequence, (b) error.

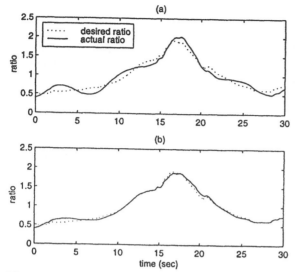

Figure 15 (a) Output of the gear ratio controlled by a PID inner loop, (b) output of the fear ratio controlled by a fuzzy operator's model in the inner loop.

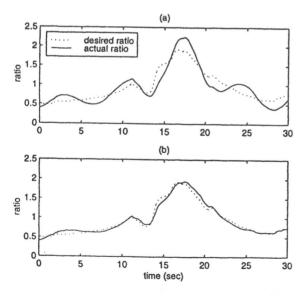

Figure 16 Gear ratio control performance for a disturbance between 12sec and 15sec with an active inner loop: (a) PID controller, (b) fuzzy operator's model.

Operator's Model in the Outer Loop   The outer loop operator's model has the same two inputs as the inner loop operator's model. The output of the outer loop is the change in the reference, r, to the new reference, r_{new}. In this example, r is the desired gear ratio profile. The membership functions of the inputs and output and the set of rules for the outer loop model are the same as in Section IV.A.

The corrective actions for certain situations (large disturbances, etc.) taken by the outer loop involve changes in the setpoint reference input. The outer loop operator's model assists in improving the *overshoot* and *settling time* of the response as mentioned in Section IV.A. In the disturbance case, the gear ratio control output of Figure 17(a) shows a better response than the results in Figure 16. The rule-based operator's model changes the gear ratio profile to the new one shown in Figure 17(a).

V. CONCLUSIONS

The operator's model is used in two ways: replacing the conventional controller (inner loop), and taking a more supervisory role (outer loop). The operator's model performs better than a PID controller when employed in the inner loop. Simulation results indicate better response times and a larger stability envelope for the operator's model controller. In case of large disturbances, where a conventional con-

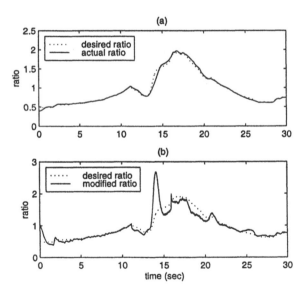

Figure 17   Gear ratio control performance for a disturbance between 12sec and 15sec with external fuzzy setpoint changes and an active inner loop: (a) fuzzy operator's model, (b) modified input gear ratio profile.

troller cannot perform satisfactorily, the outer loop operator's model controller brings the system successfully back to normal conditions. A similar model may be developed for fault detection and diagnosis purposes in industrial plants. The diagnostic methodology used by operators and engineers consists of observing symptoms, formulating hypotheses and attempting to prove them in order to explain a particular malfunction. This strategy can be captured via a similar modeling approach.

The modeling approach introduced in this contribution is generic and applicable to a variety of industrial, manufacturing and other dynamic large-scale processes where the operator is called upon to exercise corrective actions based upon his (her) experience. In the mechatronics arena, complex motion control or automated machining and assembly tasks typically incorporate an operator in the loop. In the automotive and aerospace industries, such "intelligent" methodologies may assist to capture the expert's knowledge and to assure "better" system performance under emergency situations.

REFERENCES

1. HJ Chris. *Knowledge based systems methods*. Prentice Hall, 1995.
2. D Goldberg. *Genetic Algorithms in Search, Optimization, and Machine Learning*. Addison-Wesley, 1989.
3. S Farlow, ed. *Self Organizing Methods in Modeling: GMDH-Type Algorithm*. New York: Marcel Dekker, Inc., 1984.
4. I Hiraga, T Furuhashi, Y Uchikawa. Acquisition of operator's rules for collision avoidance using fuzzy neural networks. IEEE Trans. on Fuzzy Systems 3(3): 280-287, Aug. 1995.
5. G Ivakhnenko. Polynomial theory of complex systems. IEEE Trans. Syst. Man and Cybern. vol. SMC-12: 364-378, 1971.
6. TJ McAvoy. The optimal and Ziegler-Nichols control. Ind. Eng. Chem. Process Des. Develop. 11(1): 71-78, 1972.
7. S Nakayama, S Horikawa, T Furuhashi, Y Uchikawa. Knowledge acquisition of strategy and tactics using fuzzy neural networks. Intl. Joint Conf. on Neural Networks. Baltimore, Maryland: 751-756, June 1992.
8. K Sato, R Sakakiyama, H Nakamura. Development of electronically controlled CVT system equipped with CVTip. Proc. of Int. Conf. Continuously Variable Power Transmissions: 53-58, Sept., 1996.
9. T Sawaragi, K Ogawa, O Katai, S Iwai. Skill acquisition for intelligent fuzzy controller by analyzing operator's control action sequence. SICE'89: 1369-1372, 1989.
10. B Shahian and M Hassul, Control System Design using Matlab. New York: Prentice Hall, 1993, 171-173.
11. S Shrier, RL Barron, and LO Gilstrap. Polynomial and neural networks: analogies and engineering applications. IEEE First Int. Conf. Neural Networks. vol. 2: 431-439, 1987.
12. M Sugeno and T Yasukawa. A fuzzy-logic-based approach to qualitative modeling. IEEE Trans. Fuzzy Systems. 1(1): 7-31, 1993.

13. Sungshin Kim and G Vachtsevanos. A polynomial fuzzy neural network for identification and control. North American Fuzzy Information Proc. Society: 5-9, June 19-22, 1996.

14. S Yamamooto, R Kimura and H Miyahara. Automatic knowledge acquisition system for blast furnace distribution operation. Proceedings of the 12th Triennial World Congress of the Internal Federation of Automatic Control. Sydney Australia: 179-184, 1993.

14
Ethics in Product Design

Donald R. Searing and Michael J. Rabins
Texas A & M University, College Station, Texas

I. WHY ARE ETHICS IMPORTANT IN ENGINEERING?

As engineers, we hold vast power in our hands. We have the opportunity to provide products and services that can affect the lives of vast numbers of people. These products and services can affect the public in beneficial or harmful ways. Early on, the organizing bodies of our engineering professional societies realized this fact. They created lists of rules to which all engineers would be held accountable. These rules were established to ensure that the public was not placed in danger by the engineering profession and its practitioners. These sets of rules, for the guidance of engineers, are known as the Codes of Ethics for Engineers.

All professions (medicine, law, and engineering) have codes of ethics. It is a requirement that arises from the idea of a social contract between professionals and society. According to this social contract, professionals are "the guardians of a public trust" (Harris et al. [1], pp. 30-31). Professionals have skills and knowledge that are deemed essential to the functioning of society. Thus, these implicit social contracts set down the terms of the arrangement between professionals and society. For engineers, such a contract is also laid out in the various professional codes of ethics. Society allows professionals to be self-regulated, autonomous, and affords them significant social status. In turn, the professionals agree to police their own ranks and to serve society to their utmost capacity. This service to society includes "promoting the well-being of the general public" (Harris et al. [1], pp. 30-31) and "ensuring that professionals are devoted to the public welfare even when it conflicts with self-interest" (Harris et al. [1], pp30-31).

The reason that service to society is of such paramount interest should be obvious. As professionals, we hold people's lives in our hands. Mistakes made by professionals are sometimes measured in the number of lives lost. The doctor who makes a mistake loses his patient, the lawyer who errs can send his client to prison, the engineer who makes an error in judgment can cost hundreds of people their lives. There is a long list of famous engineering errors; the collapse of the Grand Teton dam, the disintegration of the Tacoma Narrows Bridge, the Kansas City Hyatt Regency walkway collapse, the loss of the Space Shuttle Challenger, and the list goes on. As engineers, we owe it to society, the users of our products and services, to practice our craft with the greatest care. The Codes of Ethics for Engineers helps us achieve that goal by ensuring that all professionals work with certain tenets in mind. Society can be assured that individuals who are doing engineering work have achieved a certain level of proficiency in their field and that these professionals are beholden to protect the safety and welfare of the public.

Being a professional has its benefits, but it can also cause some problems. The autonomy allowed to professionals is often at odds with the realities of life as an employee or contractor. Oftentimes, the obligations placed upon one as an engineer will be in direct conflict with the obligations required by an employer or by oneself. In this case it is important to have tools available which can help you determine the appropriate course of action. Fortunately for engineers, the skills needed to wade through these ethical problems are very similar to the skills used in the design process. The problem solving and design techniques taught to engineers are especially applicable to problems in the realm of ethics.

This chapter will seek to introduce the reader to the realm of ethical problem solving, emphasizing the similarities between it and the design process most engineers are familiar with. It will be shown that ethical considerations can both drive and constrain engineering design. The chapter will conclude with a look at some areas in which design and ethics play equally important roles in engineering: specifically, risk analysis and environmental concerns. It will conclude with an examination of a case directly related to the field of mechatronics.

II. A BRIEF INTRODUCTION TO AN ETHICAL PROBLEM SOLVING METHODOLOGY

A. A Case Study—The Challenger Incident

The following is a summary of events as outlined during the Rogers Commission investigation into the Challenger accident [2].

On the night of 27 January, 1986, Roger Boisjoly, an engineer at Morton Thiokol, faced a crisis (Harris et al.[1] pp. 1-2). He was sitting in the audience of a teleconference between his superiors and NASA management. The topic of discussion at this teleconference was the readiness to launch of the space shuttle. Boisjoly had just presented his findings on o-ring resiliency at low temperatures. He had noticed an ominous trend in his data. Booster rocket o-rings that had been subjected to the coldest launch on record (53°F) showed a significant amount of leakage of hot exhaust gases. Tomorrow's launch temperature had been forecast to be 26°F, the coldest launch yet. He had argued that at that temperature the o-rings would not be capable of sealing their joint and a major catastrophe could result.

Robert Lund, Boisjoly's superior, agreed that it was not safe for the shuttle to fly the next day and recommended against the launch going ahead. NASA officials questioned the completeness of the o-ring leakage report and the resulting no-launch recommendation. The Morton Thiokol management requested a suspension of the teleconference so that they could reassess their recommendation. It was NASA policy not to fly without the approval of their contractors, such as Morton Thiokol, and Morton Thiokol management would not give their approval without the agreement of their engineers.

Jerald Mason, senior vice-president at Morton Thiokol, knew that NASA needed a successful launch. The space shuttle program was way behind schedule and significantly over budget. In addition to those concerns, the President of the United States was also planning on referring to Christa McAuliffe, the first teacher in space, in his State of the Union Address that was to take place within the next week. Morton Thiokol needed a successful launch as well. Their billion-dollar booster rocket contract was soon due for renewal and a recommendation against launching would not help Thiokol obtain the contract. Mason also knew that the evidence on which the engineers were basing their conclusions was weak. There were only a few data points on the chart, and the lowest one was at 53°F. The engineers were relying on the apparent correlation between temperature and the resiliency of the o-rings.

Management knew they had to come up with a decision before the teleconference resumed. Jerald Mason turned to Robert Lund and said, "Take off your engineering hat, and put on your management hat." Management came to their decision quickly after that. When the teleconference resumed, NASA officials were informed that Morton Thiokol had reversed their decision and that now the company would give a launch recommendation.

Roger Boisjoly retreated to his office and refused to talk to anyone. The next day, he joined his coworkers in watching a telecast of the launch. He was watching when the space shuttle Challenger exploded 73 seconds into its flight. It was later determined that the o-rings, which had been the subject of debate at their teleconference, were directly responsible for the loss of the vehicle.

The case outlined above is one in which an engineer has been drawn into an ethical dilemma. Should Roger Boisjoly have done more, or was he right in being loyal to his employer? How do you resolve situations in which several of your obligations come into conflict? Situations like the one above can be managed with a consistent ethical problem solving methodology. This should sound familiar, since engineers utilize a similar problem solving methodology in the design process. Both ethical and design problems deal with the same sort of complex multi-constraint, multi-objective situations. Thus, each stage of the ethical problem solving methodology that will be outlined below will be introduced along with its design methodology counterpart.

B. Framing the Problem

The first step in any problem solving process is to define the problem in terms of concepts relevant to the methodology that will be used to solve the problem. In the design field, the problem definition is created through a cyclical repetition and refinement process. The problem is first defined in terms of needs (very broad statements of goals and constraints that will influence the design). These needs are then refined into requirements (specific statements of design goals and constraints). The requirements are then processed into specifications. These specifications outline the problem in a very quantifiable way that the design process can use to measure progress and completion. The process of developing the specifications is very important, since the specifications will significantly influence the design.

The ethical problem solving methodology begins in a similar fashion. The definition of the problem is paramount. The first step in defining an ethics problem involves outlining the facts that are relevant to the case. This step is important for two reasons. First, it is important to have as many of the facts in the case as possible to provide you with a stable base for your decision. Second, it is important to list the facts because this process could uncover facts that are not known. Listing the unknown, relevant facts, or factual issues (Harris et al. [1], p. 103) is important because these are often critical pieces of information that can help you resolve your conflict. For example, if Roger Boisjoly had known whether or not he would be fired if he had argued more vehemently, then his decision on whether to back down would have been much easier. Sometimes this information is readily available to the individual after significant research or investigation. Oftentimes, though, the resolution of these factual issues is difficult. Many facts are, for some reason, unknown. Either they are impossible to know (as in the knowledge of the consequences of Boisjoly's arguments) or the information is in dispute (e.g., does the information show that the o-rings will fail at 29°F?). In these cases, individuals dealing with ethical problems must do what every good designer does: make assumptions. You must make a reasonable guess at the information even when it is not available. Only by resolving these issues can you proceed with the analysis.

The second step in the analysis of the problem domain is that of finding and resolving what are known as conceptual issues (Harris et al. [1], p. 107). A conceptual issue is a definition or concept that is in dispute or unknown. This may be a difficult concept for engineers to accept since in engineering when a specification says "acceleration not to exceed 3gs" it is generally not necessary to define what is meant by "acceleration" or "exceed." But, in ethics where words are the very essence of the argument, it is necessary to ensure that the words are interpreted as they are meant. Thus, identifying and defining terms that may be misunderstood is an important step in the process. As with factual issues, it may be necessary for you to assume a reasonable definition if one is not available. An example of a conceptual issue would be, "What is meant by 'loyalty to your employer' in the sample case above?" This concept is crucial in framing the problem. If 'loyalty to your employer' is defined as looking out for the best interests of the corporation which employs you, then Roger Boisjoly would be proper in defying his superiors if their actions were demonstrably harmful to the corporation. If the term is defined as being a loyal employee who accepts the chain of command in the interests of the company, then the results would be different. It would be impossible to defend any action, such as Boisjoly's hypothetical defiance, that violated the chain of command. Thus, it is of great importance to define the terms of the problem.

The last step in defining the problem is that of identifying and resolving the moral issues (Harris et al. [1], p. 97) of the case. A moral issue is a question about the morality of some action. For example, asking, "Is allowing your product to kill a person a morally justifiable action?" is a moral issue. It is at issue because while most people may answer this question in a certain way, others will have a different answer. Most may say it is never morally permissible to kill a person. Historically, there have been engineers who chose to let their design kill people because it was too expensive to fix the design (Harris [1], pp. 184-185). The various interpretations of the morality of the situation can lead to conflict. Thus, it is important to state the moral ground that will be utilized in the analysis, so that there can be no misunderstanding when the analysis is reviewed. The statement of the moral ground is done by resolving the moral issues that arise in this process of problem definition. The resolution of the moral issues usually takes on the form of a statement of moral permissibility or impermissibility (a moral rule). For example, the moral issue stated above could be resolved with a statement along the lines of, "The killing of a person is never morally permissible, regardless of the cost." This resolution actively conveys the permissibility of the action in question as well as stating the conditions in which the rule does or does not hold.

So, in an ethical analysis, as in design problems, it is important to have a highly detailed problem definition. Oftentimes, the problem being examined will be resolved in one of the stages of problem definition. This is often the case because many situations that seem to be ethical problems are often actually disagreements over the resolution of factual, conceptual, and moral issues. But, there are cases where the problem definition phase just serves to focus the analyzer on the moral dilemma that resides at the heart of the situation. When this is the case, more problem-solving tools are needed to come to a resolution of the problem.

C. Problem Solving Methods

In the design field there are many methods of proceeding from the specifications (detailed problem statement) to the final design. One might use one of the various functional analysis methodologies to achieve a design solution. These methods provide an algorithm for processing the information from the problem definition stage and for providing a framework to achieve a functional design solution. The ethical problem solving methodology has a similar algorithm. In fact it has two algorithms, since there are two primary types of ethical problems. The first type of problem has been called a "Line Drawing Problem" in the literature (Harris et al. [1], p. 133). This is a problem where a line of ethical permissibility must be drawn somewhere in a range of situations. The other type of ethical problem encountered is known as a "Conflict Problem" (Harris et al. [1], p. 135). In this type of problem, there are two or more competing obligations that must be satisfied in order to achieve a satisfactory resolution of the problem. These two types of problems cover the majority of ethical problems that will be encountered.

Line Drawing Problems

The Line Drawing Problem can best be demonstrated by considering several situations involving gifts from vendors (Harris et al. [1], p. 133). Most industrial vendors or their salespeople will give a person who uses their services a gift to show their appreciation. It is common practice in industry (How many of you are using pens or pads of paper which were given to you by a vendor in whose product you showed interest?). To demonstrate the situation in which the Line Drawing Algorithm is used, several situations have to be outlined.

Situation 1: A vendor representative gives you a pen with his contact information on it after a preliminary meeting.

Situation 2: A vendor representative gives you an expensive wristwatch after a preliminary meeting.

Situation 3: A vendor representative gives you a free trip to some tropical destination after you decide to recommend the use of the vendor's product.

Situation 4: Before you decide which vendor to buy from, one vendor representative promises to give you a cash reward if you specify that vendor's products.

Each one of these situations contains the same main moral problem: a vendor representative is giving a gift to a decision-maker who may decide to use that vendor's product. The difference in each of these situations is slight. It may be a difference in the timing of the gift or a difference in the value of the gift. In general, most people see no moral problem in accepting a pen or pad of paper from a vendor representative (Situation 1). Those same people do have a moral problem with accepting a bribe (Situation 4). What is the difference? Both activities involve a gift from the vendor to a potential customer. The difference lies in the value of the gift. So, since Situation 1 is permissible and Situation 4 is impermissible, there must be a line drawn somewhere that separates the morally permissible actions from the morally impermissible actions. Drawing that line is often a challenge, though a thorough corporate policy on the receipt of gifts will help. It is easy to separate two situations that lie so far apart on the line of permissibility. It is more difficult to draw a line between Situation 1 and Situation 2. The algorithm that will be presented will aid in drawing this line.

The Line Drawing Algorithm consists of several steps. The first step is to construct a list of the features of the case. The features are generalized facts and factual issues from the case. For example, in the gift scenarios presented above some features that could be used are "size of the gift" and "timing of the gift." The list of features should cover all of the relevant aspects of the case. These features should be entered into a chart (Table 1) using the first column.

Table 1 Line Drawing Analysis Chart

Feature	Negative paradigm	50/50	Positive paradigm
Size of the gift	Very large		No value
Timing of the gift	Before decision		After decision

In the second and the fourth column of the chart are what are known as paradigm situations. The paradigm situations represent the limits of the permissibility scale. The negative paradigm is a situation that is incontestably morally impermissible. The positive paradigm is a situation that is incontestably morally permissible (Harris et al. [1], p. 127). The paradigm situations are created by finding the positive and negative limits of each feature. For example, in the chart above, the feature, "size of the gift" would have a negative paradigm value of "very large." Everyone would agree that if a vendor representative offered a decision-maker a one million-dollar "gift" that this would obviously affect the decision process and thus would be morally impermissible. The positive paradigm value of the feature would be "no value" since everyone would agree that a gift of no value would be morally permissible. The likelihood of a gift with no value affecting someone's decision making process would be very low. Each feature is examined individually and their limits are determined. The paradigm situations are thus developed. Each situation is actually a collection of the limiting cases of each of the features. Thus, the negative paradigm situation would be the case where the size of the gift was very large and the gift was given before the decision had been made. The positive paradigm situation would be one in which the gift had no value and it was given after the decision had been made.

After the positive and negative paradigms have been developed, it is time to begin the most important stage of the process. It is now possible to locate the current situation being faced within the context developed by these paradigms. The current situation is evaluated feature by feature and is given a value based on the distance it lies from each of the paradigms. For example, if, in the case above, the gift were a moderately expensive wristwatch, then the feature "size of gift" would be evaluated to have a value of somewhere near the middle of the range on the positive side. This is due to a fact that a watch is neither a gift of "no value" nor is it a gift of "immense value." Its value falls in the middle of the range. I have elected to value it as slightly to the positive side of the midline in this feature. If it is assumed that the gift had been given before the decision was made, then there is a clear-cut valuation on the second feature, "timing of the gift."

Thus, the chart from Table 1 would now look like the one in Table 2.

The Xs represent the relative position of the actual case in the context defined by the paradigms. The resolution of this problem can be easily found through use of the last step of the methodology.

Table 2 Line Drawing Analysis Chart with Values from the Case

Feature	Negative paradigm	50/50		Positive paradigm
Size of the gift	Very large		X	No value
Timing of the gift	Before decision	X		After decision

The last step of the Line Drawing Analysis is essentially an easy one. The permissibility of the actual situation is determined by the balance of the Xs. If more Xs are closer to the negative paradigm than they are to the positive paradigm, then the action being considered leans toward being morally impermissible. While the case may not be the absolutely worst case in terms of moral permissibility, it is sufficiently close to warrant extreme caution. If more Xs are closer to the positive paradigm than the negative paradigm, then the action leans toward being morally permissible. The degree of closeness as well as the relative importance of that feature measured against the others are also important factors to consider. For example, in Table 2, there is one X that is barely on the positive side of the midpoint line. There is also one X which resides next to the negative paradigm. Thus, when considering this case, the extremely negative X in the second feature would outweigh the slightly positive X. The conclusion that would be drawn from the values given in Table 2 would be that the gift of the wristwatch would lean towards being morally impermissible.

There are several ways to make this test more accurately portray a given situation. The first method of improving this test is to add more features to the analysis. The accuracy of the method increases with the number of features that are used. A second method of improving the accuracy is the involvement of weighting factors. In most situations, it is the case that a few of the features are far more important than the other features. These features and their values should be given a greater weight in the determination of the resolution.

Conflict Problems

The second type of ethical problem encountered is one in which there are conflicting obligations that must be met to resolve the problem. In the case mentioned earlier, Roger Boisjoly faced a situation in which he had conflicting obligations. He had an obligation to protect the public, which in this case are the astronauts riding in the shuttle who were uninformed of the o-ring problem. He also had an obligation to be a loyal employee. Both of these obligations are outlined in the Engineering Code of Ethics that Boisjoly, as a professional engineer, is required to uphold. The Engineering Codes of Ethics contain many such obligations that can conflict with each other. This type of problem is quite a common occurrence for most engineers. The primary conflict usually concerns the obligations to protect the public and to be a loyal employee. These problems arise because businesses are not held to as strict an ethical standard as engineers are.

Conflict problems are familiar to engineers. All design work is done within the confines of three conflicting goals: features, costs, and schedule. What makes engineering such an interesting field is the balancing of these interests. In an ideal world, a product must simultaneously have all of the features desired by the market, a cost which makes it reasonable to manufacture, and a schedule for the design process which will bring the product to market before a competitor's product. Fortunately for engineers, this is not an ideal world. The customer always wants more features than can be provided, the product always costs more than the market will bear, and the schedule is always too long. Faced with this reality, engineers must either use their creativity to create a solution which can meet all of the constraints or use their judgment and engineering skills to determine which constraint is paramount and meet that to the fullest while letting the other constraints remain to some extent unfulfilled. These are also the exact processes used to solve conflict problems of an ethical nature.

The first step in the solution of an ethical conflict problem is to determine the obligations that must be met. Usually, the obligations faced by an engineer are outlined in the Engineering Codes of Ethics. Obligations to uphold the public safety, to be a loyal employee, to be honest, not to practice outside the engineer's area of competence, and not to disclose confidential information of clients are all listed in the Engineering Codes of Ethics (Harris et al. [1], p. 389). There are also other obligations that an engineer may face which do not stem from the codes of ethics. These obligations include the obligation to protect their own interests and to protect their families' interests. It is easy to see how some of these obligations can come into conflict with others. Each problem has a unique set of competing obligations which all need to be met to develop a satisfactory solution to the ethical problem being faced.

The second step in the resolution of an ethical conflict problem is to brainstorm solutions to the problem. There are three types of solutions that can come from such a brainstorming session. The first type of solution is an "easy choice" (Harris at al. [1], p. 135). In this case, it is apparent that one of the conflicting obligations greatly outweighs all of the other obligations, and the solution revolves around satisfying the one paramount obligation. A case where "easy choice" solutions are present is usually not considered to be a dilemma. The second type of solution that can occur is known as a "hard choice" (Harris at al. [1], p. 135). A "hard choice" solution is one in which, despite the engineer's best efforts, it is impossible to meet all of the obligations faced in the problem. In this solution, the engineer must sacrifice one or more of their obligations in order to satisfy the others. This is the least desirable type of solution for obvious reasons, but sometimes is the only option an engineer can have. The third type of solution is known as the "creative middle-way" solution (Harris at al. [1], p. 135). The "creative middle-way" solution is one in which the engineer creates a course of action that manages to meet all of the competing obligations. These types of solutions are exactly what engineers are trained to find in their technical fields of expertise.

Engineers are trained to take competing requirements and find a solution that resides within the remaining design space. The process used in design is the same one used to find "creative middle-way" solutions to ethical problems. The goal at this stage of the solution process is to compile a list of possible solutions to a problem. For example, if we were to analyze Roger Boisjoly's situation, we could develop several other courses of action. He could have appealed to someone higher in the chain of command than those present in the preflight briefing. He also could have called someone higher in the chain of command at NASA. He could have spoken to the astronauts' spouses, informed them of the possible dangers, and had them assist in stopping the launch. Boisjoly could have threatened to quit if they launched the space shuttle against his recommendation. Some of these actions may sound a bit far-fetched, but they are the results of actual brainstorming sessions (Texas A&M University [3]) and should be included at this stage.

The third step in the process of solving an ethical conflict problem is to enumerate which obligations are met by which solutions. Each solution should be analyzed to determine if it meets each obligation. For example, the solution that consisted of Roger Boisjoly finding someone at Morton Thiokol senior to those in the room would meet all of the obligations faced by Boisjoly. His obligation to protect the health and safety of the public (which consists of the astronauts) would be met if he pressed this solution until he found a sympathetic ear and stopped the launch. His obligation to be a loyal employee would be met (depending on the definition to the conceptual issue of "loyalty") because he has his employer's best interest in mind: Morton Thiokol would be tremendously affected if a shuttle were to be lost. Also, by working inside the company and not going public with this information his employer's interests are protected. Thus, it can be said that this course of action meets both obligations faced by Roger Boisjoly and can be considered a good "creative middle-way" solution. An example of a solution which is not a "creative middle-way" solution is one in which Roger Boisjoly had spoken to the media and reported that Morton Thiokol was endangering the lives of the astronauts. If the launch was stopped due to media attention, then Roger Boisjoly's obligation to protect the public would have been met. But, his obligation to be a loyal employee would be violated because he betrayed his employer's trust in him. Thus, only one obligation was satisfied while the other was sacrificed and the solution would be of the "hard choice" type.

The last step in the process of finding a solution is to select a solution which meets all of the obligations and which is feasible to perform. The solution which should be chosen is the one which best meets all of the obligations and which is the most likely to actually succeed. For example, several of the solutions listed above for the problem Roger Boisjoly faced are less feasible than the others. The option that includes calling the astronauts' families is much more difficult to perform than the option that includes the threat to quit. Feasibility is important because the action listed must be able to be performed to have any effect on the situation. Perfect solutions that cannot be implemented are useless in this process.

There may not be a solution that meets all of the obligations. If that is the case, the engineer has but one recourse and that is to either make a "hard choice" by choosing one of the suboptimal solutions or the engineer can spend more time searching for a solution which meets all of the obligations. Oftentimes, the "hard choice" is the only way out of the solution and must be selected. Unfortunately, someone, usually the engineer, always gets hurt in a "hard choice" situation. That is why the authors advocate only making "hard choice" decisions as an absolute last choice.

The methodology for resolving conflict problems provides another useful result. Not only does the method point to the best solution, it also supplies a list of solutions ranked by their feasibility and number of obligations met. This ranked list can be used to formulate contingency-based courses of action. The best solution can be attempted and, if it fails, then the next solution on the list can be tried. This allows one to not only develop the best solution, but it enables one to plan a comprehensive course of action encompassing all solutions. For example, if Roger Boisjoly is unable to convince someone higher up in his corporation of the dangers of a failure of the o-rings, then he can try the next course of action, which would probably be trying to get in contact with someone at NASA who outranks the people he has previously dealt with. If he is unsuccessful with NASA, then he should try to contact the astronauts' families and enlist their aid in his fight. If that is unsuccessful, then Boisjoly should continue applying his list of solutions until he reaches his last solution, which in this case would be to blow the whistle and alert the media. Using the ranked list as a guide for his actions, Boisjoly would have exhausted all of his other options before finally having to fall back onto the action of whistleblowing, which is always the option of last resort.

The two ethical problem-solving approaches outlined above will not give you the solution to the ethical problems that will arise, but, much like the design process, they will show the way to the solution. These problem-solving methodologies will not provide the moral basis needed to solve ethical problems. That moral basis must be provided by the user of the methodologies. The fundamental assumption that is being used here is that all engineers will seek to uphold the Engineering Codes of Ethics and will use the tenets included in these codes as the core of their moral basis for action. In the next section, the link between these problem-solving methods and the design process will be shown.

III. PUTTING THE PIECES TOGETHER: DESIGN AND ETHICS

The process of engineering is the process of creating solutions that can simultaneously achieve the goals of the problem and remain within certain constraints. These goals and constraints are referred to as "design drivers." The three most common "design drivers" are the cost of the solution, the schedule of the project and the features of the solution. These drivers can act as either goals or constraints, depending on the situation. For example, in the design of a personal computer system, the product's features are usually the goals, while scheduling and costs are the constraints. The design process for the computer focuses on providing features (the right amount of memory, the right amount of disk space, the proper sound and video cards, whether or not a modem should be included) within the bounds of the cost (must generate profits for the manufacturer at the current market prices) and the schedule (must be on the market before the holidays). An example of a case where cost would become a goal would be if in the previous example, it had been stated in the project description that the computer to be designed would be the most affordable on the market.

Ethical considerations can similarly be viewed as "design drivers," either constraints or goals. Ethical behavior is either something to be sought or something used to limit the design space. The technical features referred to in the previous paragraph are analogous to the moral issues at stake in an ethical case. These two topics will be examined in detail in the next subsection.

A. Ethics as a Design Constraint

In its role as a "design driver," ethics can sometimes act as a source for multiple constraints on the design process. In fact, engineers have been using their ethics in this manner for years. Out of their concern for the safety of the public, engineers have long used factors of safety to ensure the safety of their designs. This over-design by design limits the solution space considerably. If it is known that a bridge must be designed to withstand six times the maximum service load, then the selection of materials and design techniques is highly limited. This may seem to be only indirectly linked to ethics, but the basic reason that factors of safety exist is that engineers have an ethical code which holds paramount the heath, safety, and welfare of the public (Harris et al. [1], p. 389). This code of ethics would seem to drive engineers to be as conservative as possible in designs which could harm many people if there were to be a failure (e.g. bridges, stadiums and tall buildings). Thus, while the ethical considerations are not directly related to the design process , they do spawn constraints that are applicable to the design.

Another way that ethics can be used as a source for design constraints is in cases involving environmental impacts. Everything designed by engineers will have an environmental impact of some sort. The engineers' personal ethics involving the environment will directly impact the design solutions. The world has been becoming more environmentally conscious in the recent past and this trend is not about to end. Thus, as engineers, it is important to observe this trend and design products accordingly. Environmental concerns can create quite a few constraints on the design. For example, automobile manufacturers are under increasing pressure to reduce the environmental effects of their automobiles. This limitation has ramifications that affect most of the systems involved in an automobile. The exhaust system and engine must be optimized to reduce engine emissions and noise. The air conditioning system must use refrigerants which do not use the chloroflourocarbons that may be harming the ozone layer. These types of constraints tend to be very costly, but many forward-thinking manufacturers are developing designs that take many of these environmental factors into consideration (see Section III.E, Areas of Interest: The Environment, for more information on this type of application).

Ethical considerations can create constraints for any design. The constraints due to safety and environmentalism stem from ethical concerns and thus are highly influenced by the moral beliefs and ethical reasoning of the designers. Engineers and corporations with different ethical outlooks will develop differing constraints and thus different solutions. Efforts to standardize the approach of engineering in general to these areas of concern are slowly gaining momentum. The codes already address the issue of safety and decree that it be the paramount concern. Several engineering societies have added amendments to their codes to ensure that their constituents have a professional obligation to protect the environment (e.g. the Institute of Electronic and Electrical Engineers and the American Society of Civil Engineers) (Harris et al. [1], p. 394). This is a step in the right direction, since it can serve to place a lower limit on the level of emphasis on the environment in engineering practice.

B. Ethics as a Design Goal

Ethical considerations can also drive the design process by being goals to be sought. There are two ways in which ethics can drive design. The first is similar to the previous discussion of design constraints becoming goals. The second involves the conflict problem resolution methodology and the development of creative middle-way solutions. The first can be simply stated as the inclusion of ethical considerations as goals in the design process. For example, it may be stated in the problem statement that the solution will be environmentally friendly and safety-oriented. Thus, this makes the ethical considerations an important objective that must be achieved during the design process.

The second, in which ethics can drive design, is by formulating the problem for the design process. In the conflict problem resolution methodology outlined above, one step requires the engineer to develop solutions to the ethical conflict the engineer faces. In this way, the design process can be used to solve an ethical dilemma. The design process is applicable to more than just technical engineering problems. It is actually quite a powerful decision-making technique in its own right. Thus, when an ethical dilemma is faced, such as the one faced by Roger Boisjoly in the case of the space shuttle booster o-rings, engineers should be able to use their design skills to develop courses of action for resolving their dilemma. A case which illustrates this idea rather dramatically is one that is used in the Engineering Ethics class at Texas A&M University [2]. This case concerns the routing of high-tension power lines from a power plant in the country to a substation in the city. The power lines would be run through a rather impoverished area of the city. The citizens in this area complained that they would be affected by the magnetic fields emanating from the lines. The class was asked to develop an alternative solution to the routing problem that would meet all of the obligations listed in the case. The solution to this case required the students to start a design process. They had to assess the need (Why did the city need the power lines?), research the problem domain (Are the magnetic fields harmful?, What other alternatives are there to running high-tension power lines?) and then they had to synthesize a feasible solution. Many of the students developed interesting solutions. One student developed a solution which rerouted the power lines through land that was sparsely populated. This solution was more expensive than the original solution because the path of the power lines was longer. But, the student reasoned that this cost could be offset by turning the power line path and buffer zone into a park for which the city would help pay. Other students determined that the solution to this problem may lie in building a smaller power plant in the city to help out during times when the city needed more power. Thus, more power could be brought to the city without running it through power lines. They reasoned that if the city was growing, as was stated in the case, that in a few years, the city would need more power than the lines could provide. A power plant closer to the city would be the most cost-effective solution. This type of thinking was exactly what we were looking for in our students. By assessing the ethical problem first, the engineers saw opportunities that would not have necessarily occurred to them with just a design problem. The ethical analysis performed on the situation allowed them to challenge the premises and ask questions such as "Why are we building these power lines?" or "How could we make a rerouting of the lines less expensive?". These questions encouraged the students to think in a more holistic sense. They focused on the details of the problem before leaping into the details of the possible solutions.

Using an ethical problem to spawn design also encourages engineers to approach design tasks in context. Since no design happens in a vacuum, there are always external interfaces which must be met. In these types of situations, these external interfaces are defined in terms of the considerations examined in the ethical analysis. The ethical analysis, in this sort of situation, can be seen as a very thorough needs analysis for the design problem. This thoroughness and top-down approach assists engineers in developing creative solutions that solve the technical problems and are easily implemented in the environment surrounding the design. This expansion of the considerations examined in the design process helps engineers develop better solutions, which means better products and fewer problems.

C. Area of Interest: Ethics and Design

There are several areas where ethics and design play complementary roles. Two of the most important areas are risk analysis and environmentally-conscious design. Both of these areas require knowledge of design and ethics. Risk analysis is a convergence of design and safety issues. Environmentally-conscious design is a convergence of design and environmental ethics. Both risk analysis and environmentally-conscious design are fields of increasing importance. The more that is learned about systems, the more it is understood that any change made to a system has effects which resonate throughout the system. Society and the environment are some of the most complex and chaotic systems, so changes made to them need to be made with caution and as full an understanding of the possible effects as is possible. Thus, these two areas of interest are of great importance to engineers who want to minimize the adverse effects of their designs on the world.

D. Area of Interest: Risk Analysis

Every endeavor that humans undertake contains risks of some sort, whether it be risks to the financing of the endeavor, or risks of injury and death to the participants. Something can always go wrong. This can be especially disastrous when considering engineering designs. Thus, it is of the utmost importance to evaluate all of the risks associated with any design. The process of determining the risks associated with any design is known as risk analysis. The first step in any risk analysis is the identification of the all of the risks faced by a system. This step is usually accomplished with either a fault-tree analysis or a failure modes and effects analysis (FMEA) (Harris et al. [1], pp. 230-235). These two analysis methods provide a systematic approach for identifying all possible failure modes of a design, though it is impossible to account for all possible failures. For example, the National Aeronautics and Space Administration (NASA) has binders several feet thick that detail the failure modes for the space shuttle system. Each of these failure modes is given a criticality level which measures the severity of the failure with respect to the mission and the ability of the craft to return to earth [Texas A&M University [4]). During the design process, this list of possible failure modes assisted the engineers in reducing the risks to a level they considered acceptable. Determining which risks are acceptable is the next step in the process.

The second step in risk analysis is to determine the level of acceptable risk. This is more difficult than it would seem, since the level of acceptable risk depends on the public being considered and the type of risk being applied. For example, it is widely known that statistically it is safer to fly in a commercial airliner than it is to drive in an automobile. But, many people still prefer to drive instead of fly. Apparently, these people assume the death that would occur in a plane crash is far more horrific than the death encountered in an automobile crash. Thus, a risk analysis which would rely only on the statistical results would not be an acceptable risk assessment for the majority of the public. Thus, the risk analysis must take into account the attitudes of the affected public as well as the statistical risk data.

There is another factor to consider in a risk analysis and that is the fundamental type of risk being assumed. There are two approaches to measuring risk that can be used. The first approach to measuring risk comes from a utilitarian philosophy. Utilitarianism focuses on maximizing the overall benefit of any given action. Thus, an ethical action is one that increases the overall benefit to society. Utilitarianism is attractive to engineers because it can be used to quantitatively weigh all of the benefits and costs incurred by an action and thus turns a decision into an equation. The utilitarian approach does tend to neglect considerations such as justice and individual rights. When you attempt to optimize benefits to society overall, there is no room for concerns about the individual. A utilitarian approach to risk would focus on the probability and balance of costs and benefits to society.

The second approach is known as the respect for persons philosophy (Harris et al. [1], p. 165). This approach focuses on protecting the rights of individuals rather than optimizing overall benefit. The problem with the respect for persons approach is that there is no quantifiable method to help with the processing of respect for persons concerns. But, it forms the perfect balance for the utilitarian approach. The best method by which to approach risk is to first use a utilitarian approach, measuring the costs and benefits. Once the risks are understood from an overall standpoint, then respect for persons concerns should be evaluated. In this way, not only are the overall benefits maximized, but all parties whose rights may be infringed upon are accounted for in the process. The respect for persons approach to risk would be concerned with the risk of rights violations to any individual affected by a design.

The approaches to risk differ because the basis for assessing those risks is different. For example, in the power line case previously mentioned, the utilitarian standpoint would advocate the original route of the power lines. There would be a great benefit to the population of the city now that they would have a consistent power supply. The costs to the relatively few people who would possibly be affected by the magnetic fields would be small compared to the amount of benefit the power lines would deliver. Thus, the risks to society in general are minimal while there is a great deal of benefit. The respect for persons approach would hold paramount the rights of the affected individuals. Thus, since the people in the neighborhood would be facing possible dire consequences (disease and possibly death), the respect for persons approach would declare this situation to be unacceptable. In this manner, respect for persons concerns often temper actions that would be considered acceptable by utilitarian analysis.

A good risk analysis should contain all of the processes listed above. Some sort of formal approach should be used to quantify and enumerate the risks incurred by a design. The risks should be evaluated for their acceptability based on utilitarian concerns, respect for persons concerns, and the affected audience's attitudes towards the risks they may incur. These risks should be considered during the design process and the risk assessment process should be used in the design review process to point out omissions or unacceptable risks. The design can then be modified to reduce the risks that were considered unacceptable. In this way, ethics and design work concurrently to create safer products as well as a better understanding of the unintended effects of a given design.

E. Area of Interest: The Environment

Since it is impossible to design something with zero environmental impact, it is important for those involved in the design process to understand the issues that a concern for the environment brings to the process. There are several approaches to the environment that must be understood to really accomplish environmentally-friendly design. Currently, most engineering codes of ethics approach the environment from an anthropocentric point of view. In this approach, components of the environment are given value because they are viewed as important to humankind. Thus, those who advocate establishing a national park because they enjoy camping and want to protect a beautiful site for camping are arguing from an anthropocentric point of view. The converse to this approach is a nonanthropocentric approach. In this approach, the environment is believed to have intrinsic value. A nonanthropocentric approach to the national park scenario would consist of the argument that certain parts of nature should be set aside for its own sake. The IEEE ethics code is the first to code to include a nonanthropocentric-based clause that advocates the protection of the environment. This is bound to change as several other professional societies are considering similar clauses (e.g. the American Society of Mechanical Engineers) (Texas A&M University [3]). Thus, it may soon become a professional moral obligation to hold the safety of the environment to be as paramount as the safety of the public.

There are three main components to consider in the environmental effects of a design. They correspond to the three tenets espoused in most environmental literature: reduce, reuse, and recycle. Each of these tenets have been applied to designs to minimize their environmental impact. In the following subsections, each tenct will be discussed and a relevant example or examples of its application in design will be given.

Reduce

In engineering, the concept espoused in the reduction tenet can be applied in several ways. The production of a product incurs resource use not only in the item itself, but in the manufacturing process as well. Thus, it is important to address reduction issues as they concern both use in the product and use in the manufacturing of a product.

Reduction in the use of resources in the actual product is a natural engineering task. In a design, materials selection is always a critical task. The cost of materials must be weighed against their performance. So integrating a concern for reduction of use of natural resources into the current materials selection process should be simple. As more engineers embrace the idea of using more renewable resource materials, these materials should become cheaper and easier to purchase.

Mercedes-Benz has instituted just such a renewable resource materials selection policy for their 1997 E-class sedans. Their door linings will now be made exclusively of two plant materials; flax and sisal (Mercedes Benz [5]). These plants are renewable natural resources, and thus, they reduce the dependence of

Mercedes Benz on non-renewable resources such as petroleum-based plastics. If more renewable resource materials find their way into engineering designs and products, then the consumption rate of natural resources can be reduced significantly.

The second type of reduction concerns the reduction of resources used in production of the products. This concern can be expanded to include the reduction of waste products generated by the production processes. The concerns surrounding the issue of manufacturability are already part and parcel for most engineering design teams. It does no one any good to have a perfect design that cannot be manufactured. Thus, the concern for reduction of resource use and minimization of waste generation can be rather easily integrated into the current process.

The 3M Corporation has seen tremendous success with their pollution reduction program known as the 3P program (Pollution Prevention Pays). This program focuses on the redesign of their current manufacturing processes to reduce their pollution below the current legal limits. Their initiative has reaped them great rewards in cost savings and marketable technology for pollution reduction (Texas A&M University [3]).

BMW, the automobile manufacturer, is currently phasing in the use of water-based paint in the manufacture of their cars (BMW [6]). The new paints reduce the emission of volatile organic compounds during the construction of the automobiles. They also discontinued used of chlorinated fluorocarbons (CFCs) in the fabrication of the foam plastics used in their automobiles (BMW [6]). Thus, as stated above, their engineers decided to use materials in their production process that minimized the adverse effects on the environment.

Reuse

Design for reuse is another way to reduce the effect of a product on the environment. It is more difficult for the engineer to accomplish. A product must be capable of being reused in the same role without significant refurbishing or modification to fall into this classification. A design that can be reused is one which reduces the future need for natural resources. A good example of reusability is the canvas shopping bag. They are made by several manufacturers to allow groceries to be carried home from the store without the use of paper or plastic bags. Thus, the reusable bag replaces a vast number of disposable bags which in turn decreases the demand on the paper and plastic industries. Thus, less trees are cut down and less plastic is made, thus conserving those resources.

A good example of a product designed for reuse is the toner cartridges for laser printers. Companies such as Hewlett Packard request that users return their empty toner cartridges to be reconditioned and refilled. These companies offer rebates as incentives to have the cartridges returned. Without these programs, these toner cartridges would be filling up landfills around the world. These programs return the cartridges to service, thus reducing waste and reducing added resource consumption. It is a good idea for the companies as well, since they do not have to spend additional financial resources making and holding the toner cartridges.

Recycle

Design for recycling is the last resort in environmentally-friendly design. If a resource must be used and cannot be reused, then the last resort before internment in a landfill is recycling. Recycling involves the conversion of a product back into its raw materials. Many products can be recycled back into raw materials for a similar function: plastic jugs, bottles, aluminum cans, etc. Some recycling converts products into raw material for other products: paper recycling and conversions such as plastic jugs into jacket insulation. These are relatively simple products to recycle though. It is much more difficult to recycle something more complex, like an automobile.

Recycling something on the scale of an automobile requires quite a bit of planning during the initial design of the product. First, the recyclability of the components of the system must be considered. Some components can only be recycled into other uses (e.g. vulcanized tires cannot be recycled into tires, only into asphalt additives or playground cushioning material). Other components can be recycled into similar components (e.g. aluminum parts can be recycled into new aluminum parts). Some components corrode during use, thus making them useless for recycling (BMW switched to noncorrodable running gear to enable its recycling) (BMW [6]). Second, the ability to separate the components is a very important consideration when designing for recyclability. Components of differing materials that are permanently attached (e.g. welded or glued) are difficult to recycle. To facilitate recycling, materials should be attached using a method which is reversible (e.g. rivets or bolts). Third, a plan for the implementation of the recycling must be developed. Recycling does not happen by itself. Many companies who are designing for recyclability are also implementing plans to perform the recycling of their products. Mercedes Benz, a company who has instituted several environmentally-friendly policies including designing for recyclability, has just completed construction of a vehicle disassembly plant which will disassemble their automobiles for recycling and reuse (Mercedes Benz [7]).

F. Life-Cycle Engineering

A term, life-cycle engineering, has been coined to describe the process of designing a product and understanding the issues surrounding its uses and disposal. In life-cycle engineering, the engineer examines the processes the product will go through during and after its useful life. Issues such as emissions during the product's lifetime, possible wear patterns of the product, and the use of replacement parts and maintenance time are all considered while initially designing the product. Since most engineers who use this process are environmentally aware, care is usually taken to ensure that the product can be disposed of in an environmentally-friendly way (reused or recycled).

It is important for engineers to understand the issues involved in designing for the environment. Consumers are demanding products that are more environmentally-friendly and society is demanding that designers and builders practice their craft with as much respect for the environment as possible. Thus, it is up to engineers to develop new methods of approaching problems that concern the environment. Considering the life-cycle of the product is the most powerful method available for developing products in an environmentally-friendly manner.

IV. AN ETHICAL CASE IN MECHATRONICS—THERAC-25

The case of the Therac-25 accidents is a good demonstration of an ethical problem that can arise in the field of mechatronics. Nancy Levinson and Clark Turner state that the Therac-25 accidents are "the most serious computer-related accidents to date" [8]. The case centers on a medical linear accelerator and its control system (Leveson and Turner [8]). These accelerators are used to destroy tumors and leave the surrounding tissue relatively unharmed (Leveson and Turner [8]). The Therac-25 was designed to rely on its control software for tasks that had been controlled by mechanical means in previous models. The designers eliminated the mechanical interlocks and other safety devices and decided instead to implement them in software (Leveson and Turner [8]). As in most designs, the Therac-25 software design relied heavily on a previous medical accelerator's software system. The control software was based on software that was in use on the Therac-20 machine (Leveson and Turner [8]). As it turns out, there were errors in the Therac-20 software, but they had never manifested themselves, due to the mechanical safety devices present in its design. The Therac-25 did not have these mechanical safety features and thus the errors were able to occur during its operation (Leveson and Turner [8]). The most serious error was discovered after several patients exhibited severe radiation burns after their treatment with the Therac-25 (Leveson and Turner [8]). The control software and hardware had faults in the sensors and sensor measurements that indicated the position of the turntable which held the apertures for the beam (Leveson and Turner [8]). Since the Therac-25 was known as a "dual mode" accelerator this meant it could generate either a stream of electrons or a stream of X-rays depending on which aperture was placed in front of the main beam (Leveson and Turner [8]). In order to generate the X-rays, the accelerator would have to generate a significantly greater amount of energy. This energy would be focused on an X-ray target which would convert the energy into X-rays. If this target was not in place, then the high-energy beam would hit the patient exposing them to lethal amounts of radiation (Levenson and Turner [8]). Thus, the error in the software which allowed the turntable misalignment would cause serious injury to the patient being treated.

The software also had an intermittent error which sometimes allowed a full dose of radiation to be given to the patient while indicating to the operator that the dose had not been given (Leveson and Turner [8], pp. 18-41). So, in most of the accidents, the patients were not only being exposed to excessive radiation due to the misalignment of the turntable, but the operator would dose them repeatedly due to a display error.

There are several features of this case which make it an interesting ethical study. First, the software engineers who designed the control system made several assumptions which turned out to be false. They assumed that the software of the previous version of the accelerator was free of errors and that it would function properly without the previous design's mechanical interlocks. Second, the software engineers assumed that since they reused routines from the previous software, testing of the new system was mostly unnecessary (Leveson and Turner [8]). Third, the company that manufactured the Therac-25 denied that there was a problem even after there had been several injuries to patients undergoing treatment (Leveson and Turner [8]).

It is the authors' contention that the engineers involved did not realize the severity of the effects of a malfunction in their software. The lack of testing of the software indicates a general misunderstanding of the risks the design would place on its users. It is assumed that the engineers involved did not intend to injure the users of their design, but it is apparent that they did not approach this design with enough understanding of the seriousness of their task. If these engineers had followed some type of ethical analysis during the initial phases of their design, they may have been made more aware of the risks and may have been prompted to question some of the assumptions which had been made. Therein lies the power of an ethical analysis. It provides another layer of review during the design process. This review is especially important because it consists of an analysis of the effects of the design on the public and society. This level of analysis has often been overlooked in the past, and if it continues to be overlooked, accidents such as those surrounding the Therac-25 will continue to happen. As the world comes to rely on computers and software in more and more life-threatening situations, these accidents could begin to happen with increasing frequency. It is up to current and future engineers to ensure that errors such as those which caused the deaths of the Theras-25 patients do not happen.

V. CONCLUSION

Engineers are problem solvers. The design process is the tool used to approach and solve the technical problems that come along. Solving other types of problems, like ethical problems, should come naturally to all engineers. It takes a thorough problem description, some creative thinking to develop solutions, and judgment to select the best solution. Thus, the basic techniques learned in engineering can assist you in solving ethical problems.

Ethical problems arise frequently for engineers since their products have great influence in the world. An engineering solution can change the fabric of society, it can change the face of the planet, and it can perform immense amounts of good. But, engineering failures are often public catastrophes in which many people are injured or killed. Thus, it is important that we, as engineers, perform our work diligently and with the utmost care. The way to ensure that as few fatalities happen as possible is to adhere to the engineering codes of ethics and to practice our craft responsibly. Since the fundamental process of our craft is design, it is important that we design with ethics as our guide. We must be aware that every design has consequences, be they social, political or environmental. We must be able to deal with these consequences by either avoiding or minimizing them. For this we must look to ethics, where there are established lines of thought and procedures for dealing with these type of problems. In ethics, we find the tools we need to shoulder the heavy load placed upon us as engineers.

REFERENCES

1. CE Harris, M Pritchard, M Rabins. *Engineering Ethics: Concepts and Cases.* New York: Wadsworth Publishing Company, 1995.
2. Rogers Commission, "Report to the President by the Presidential Commission on the Space Shuttle Challenger Accident," Washington DC,1986.
3. Engineering Ethics Class at Texas A&M University (Catalog #: ENGR 482), Spring 1997.
4. Interview with Dr. Aaron Cohen, Texas A&M University, Fall 1996.
5. "E-class saloon." www.mercedes.com.
6. "BACKISSUE: BMW at the 'IAA'." www.bmw.com.
7. "News." www.mercedes.com.
8. N Leveson, C Turner. An Investigation of the Therac-25 Accidents. IEEE Computer 26(7): 18-41, July 1993.

Index

ISBN 0-8247-0226-3

EAN

9 780824 702267

90000>